Newnes Engineering Mathematics Pocket Book

Newnes Engineering Mathematics Pocket Book

Third edition

John Bird B.Sc(Hons), C.Eng, C.Math, MIEE, FIMA, FIIE, FCollP.

Newnes

OXFORD AUCKLAND BOSTON
JOHANNESBURG MELBOURNE NEW DELHI

Newnes
An imprint of Butterworth-Heinemann
Linacre House, Jordan Hill, Oxford OX2 8DP
225 Wildwood Avenue, Woburn, MA 01801-2041
A division of Reed Educational and Professional Publishing Ltd

First published as the *Newnes Mathematics for Engineers Pocket Book* 1983
Reprinted 1988, 1990 (twice), 1991, 1992, 1993
Second edition 1997
Third edition as the *Newnes Engineering Mathematics Pocket Book* 2001

British Library Cataloguing in Publication Data
A catalogue record for this book is available from the British Library

Library of Congress Cataloguing in Publication Data
A catalogue record for this book is available from the Library of Congress

ISBN 0 7506 4992 5

Typeset by Laser Words, Madras, India
Printed in Great Britain

Contents

Preface ix

Number and algebra 1
1. Basic arithmetic 1
2. Revision of fractions, decimals and percentages 4
3. Indices and standard form 10
4. Errors, calculations and evaluation of formulae 14
5. Algebra 16
6. Simple equations 25
7. Simultaneous equations 29
8. Transposition of formulae 32
9. Quadratic equations 35
10. Inequalities 40
11. Logarithms 46
12. Exponential functions 49
13. Hyperbolic functions 55
14. Partial fractions 61
15. Number sequences 64
16. The binomial series 67
17. Maclaurin's series 71
18. Solving equations by iterative methods 74
19. Computer numbering systems 80

Mensuration 86
20. Areas of plane figures 86
21. The circle and its properties 91
22. Volumes of common solids 95
23. Irregular areas and volumes and mean values 102

Geometry and trigonometry 109
24. Geometry and triangles 109
25. Introduction to trigonometry 115
26. Cartesian and polar co-ordinates 122
27. Triangles and some practical applications 125
28. Trigonometric waveforms 129
29. Trigonometric identities and equations 141
30. The relationship between trigonometric and hyperbolic functions 145
31. Compound angles 148

Graphs 155
32. Straight line graphs 155
33. Reduction of non-linear laws to linear form 160
34. Graphs with logarithmic scales 166
35. Graphical solution of equations 170

36. Polar curves 178
37. Functions and their curves 185

Vectors 199
38. Vectors 199
39. Combination of waveforms 207
40. Scalar and vector products 211

Complex numbers 219
41. Complex numbers 219
42. De Moivre's theorem 226

Matrices and determinants 231
43. The theory of matrices and determinants 231
44. The solution of simultaneous equations by matrices and determinants 235

Boolean algebra and logic circuits 244
45. Boolean algebra 244
46. Logic circuits and gates 255

Differential calculus 264
47. Introduction to differentiation 264
48. Methods of differentiation 271
49. Some applications of differentiation 276
50. Differentiation of parametric equations 283
51. Differentiation of implicit functions 286
52. Logarithmic differentiation 288
53. Differentiation of inverse trigonometric and hyperbolic functions 290
54. Partial differentiation 294
55. Total differential, rates of change and small changes 297
56. Maxima, minima and saddle points of functions of two variables 299

Integral calculus 305
57. Introduction to integration 305
58. Integration using algebraic substitutions 308
59. Integration using trigonometric and hyperbolic substitutions 310
60. Integration using partial fractions 314
61. The $t = \tan\frac{\theta}{2}$ substitution 316
62. Integration by parts 318
63. Reduction formulae 320
64. Numerical integration 326
65. Areas under and between curves 330
66. Mean and root mean square values 336
67. Volumes of solids of revolution 338
68. Centroids of simple shapes 340
69. Second moments of area of regular sections 346

Differential equations 353
70. Solution of first order differential equations by separation of variables 353
71. Homogeneous first order differential equations 357
72. Linear first order differential equations 358
73. Second order differential equations of the form $a\frac{d^2y}{dx^2}+b\frac{dy}{dx}+cy=0$ 360
74. Second order differential equations of the form $a\frac{d^2y}{dx^2}+b\frac{dy}{dx}+cy=f(x)$ 363
75. Numerical methods for first order differential equations 367

Statistics and probability 373
76. Presentation of statistical data 373
77. Measures of central tendency and dispersion 380
78. Probability 386
79. The binomial and Poisson distributions 389
80. The normal distribution 392
81. Linear correlation 398
82. Linear regression 400
83. Sampling and estimation theories 403

Laplace transforms 414
84. Introduction to Laplace transforms 414
85. Properties of Laplace transforms 416
86. Inverse Laplace transforms 419
87. The solution of differential equations using Laplace transforms 421
88. The solution of simultaneous differential equations using Laplace transforms 424

Fourier series 427
89. Fourier series for periodic functions of period 2π 427
90. Fourier series for a non-periodic function over range 2π 431
91. Even and odd functions and half-range Fourier series 433
92. Fourier series over any range 438
93. A numerical method of harmonic analysis 441

Index 448

Preface

Newnes Engineering Mathematics Pocket Book is intended to provide students, technicians, scientists and engineers with a readily available reference to the essential engineering mathematics formulae, definitions and general information needed during their studies and/or work situation — a handy book to have on the bookshelf to delve into as the need arises.

The text is divided, for convenience of reference, into fourteen main sections embracing number and algebra, mensuration, geometry and trigonometry, graphs, vectors, complex numbers, matrices and determinants, Boolean algebra and logic circuits, differential and integral calculus, differential equations, statistics and probability, Laplace transforms and Fourier series. Within the main sections are arranged 93 chapters and over 800 numerical examples and 400 diagrams are included to aid understanding.

The text assumes little previous knowledge and is suitable for a wide range of courses of study. It will be particularly useful for students studying mathematics within NVQ's and GNVQ's, National and Higher National technician certificates and diplomas, GCSE and A levels and for Engineering Degree courses.

John Bird
University of Portsmouth

Number and Algebra

1 Basic Arithmetic

Arithmetic operations

Whole numbers are called **integers**. +3, +5, +72 are called positive integers; −13, −6, −51 are called negative integers. Between positive and negative integers is the number 0 which is neither positive nor negative.
The four basic arithmetic operators are: add (+), subtract (−), multiply (x) and divide (÷)
For addition and subtraction, when **unlike signs** are together in a calculation, the overall sign is **negative.** Thus, adding minus 4 to 3 is $3 + -4$ and becomes $3 - 4 = -1$. **Like signs** together give an overall **positive sign.** Thus subtracting minus 4 from 3 is $3 - -4$ and becomes $3 + 4 = 7$
For multiplication and division, when the numbers have **unlike signs**, the answer is **negative**, but when the numbers have **like signs** the answer is **positive**. Thus $3 \times -4 = -12$, whereas $-3 \times -4 = +12$. Similarly

$$\frac{4}{-3} = -\frac{4}{3} \quad \text{and} \quad \frac{-4}{-3} = +\frac{4}{3}$$

For example, to add 27, −74, 81 and −19:

This example is written as $27 - 74 + 81 - 19$

Adding the positive integers:	27
	81
Sum of positive integers is:	108
Adding the negative integers:	74
	19
Sum of negative integers is:	93

Taking the sum of the negative integers from the sum of the positive integers gives:

$$\begin{array}{r} 108 \\ -93 \\ \hline 15 \end{array} \qquad \text{Thus} \quad \mathbf{27 - 74 + 81 - 19 = 15}$$

In another example, to subtract 89 from 123:
This is written mathematically as $123 - 89$

$$\begin{array}{r} 123 \\ -89 \\ \hline 34 \end{array} \qquad \text{Thus} \quad \mathbf{123 - 89 = 34}$$

In another example, to multiply 74 by 13:

This is written as 74×13

$$\begin{array}{rl} 74 & \\ 13 & \\ \hline 222 & \leftarrow 74 \times 3 \\ 740 & \leftarrow 74 \times 10 \\ \hline \text{Adding:} \quad 962 & \\ \hline \end{array}$$

Thus **74 × 13 = 962**

In another example, to divide 1043 by 7:

When dividing by the numbers 1 to 12, it is usual to use a method called **short division**.

$$\begin{array}{r} 1\ 4\ 9 \\ 7\overline{)10^{3}4^{6}3} \end{array}$$

Step 1. 7 into 10 goes 1, remainder 3. Put 1 above the 0 of 1043 and carry the 3 remainder to the next digit on the right, making it 34

Step 2. 7 into 34 goes 4, remainder 6. Put 4 above the 4 of 1043 and carry the 6 remainder to the next digit on the right, making it 63

Step 3. 7 into 63 goes 9, remainder 0. Put 9 above the 3 of 1043. Thus **1043 ÷ 7 = 149**

In another example, to divide 378 by 14:

When dividing by numbers which are larger than 12, it is usual to use a method called **long division**.

$$\begin{array}{lrr} & & 27 \\ & & 14\overline{)378} \\ (2) & 2 \times 14 \rightarrow & \underline{28} \\ & & 98 \\ (4) & 7 \times 14 \rightarrow & \underline{98} \\ & & \underline{\cdot\cdot} \end{array}$$

(1) 14 into 37 goes twice. Put 2 above the 7 of 378

(3) Subtract. Bring down the 8. 14 into 98 goes 7 times. Put 7 above the 8 of 378.

(5) Subtract.

Thus **378 ÷ 14 = 27**

Highest common factors and lowest common multiples

When two or more numbers are multiplied together, the individual numbers are called **factors.** Thus a factor is a number that divides into another number exactly. The **highest common factor (HCF)** is the largest number that divides into two or more numbers exactly.

A **multiple** is a number that contains another number an exact number of times. The smallest number that is exactly divisible by each of two or more numbers is called the **lowest common multiple (LCM)**.

For example, to determine the HCF of the numbers 12, 30 and 42:

Each number is expressed in terms of its lowest factors. This is achieved by repeatedly dividing by the prime numbers 2, 3, 5, 7, 11, 13... (where possible) in turn. Thus

$$12 = \boxed{2} \times 2 \times 3$$
$$30 = \boxed{2} \times 3 \times 5$$
$$42 = \boxed{2} \times 3 \times 7$$

The factors that are common to each of the numbers are 2 in column 1 and 3 in column 3, shown by the broken lines. Hence the **HCF is 2 × 3, i.e. 6**. That is, 6 is the largest number which will divide into 12, 30 and 42.

In another example, to determine the LCM of the numbers 12, 42 and 90: The LCM is obtained by finding the lowest factors of each of the numbers, and then selecting the largest group of any of the factors present.

Thus

$$12 = \boxed{2 \times 2} \times 3$$
$$42 = 2 \times 3 \times \boxed{7}$$
$$90 = 2 \times \boxed{3 \times 3} \times 5$$

The largest group of any of the factors present are shown by the broken lines and are 2 × 2 in 12, 3 × 3 in 90, 5 in 90 and 7 in 42

Hence **the LCM is 2 × 2 × 3 × 3 × 5 × 7 = 1260**, and is the smallest number which 12, 42 and 90 will all divide into exactly.

Order of precedence and brackets

When a particular arithmetic operation is to be performed first, the numbers and the operator(s) are placed in brackets. Thus 3 times the result of 6 minus 2 is written as $3 \times (6 - 2)$ or $3(6 - 2)$.

In arithmetic operations, the order in which operations are performed are:

(i) to determine the values of operations contained in brackets;
(ii) multiplication and division (the word 'of' also means multiply); and
(iii) addition and subtraction.

This **order of precedence** can be remembered by the word **BODMAS**, standing for **B**rackets, **O**f, **D**ivision, **M**ultiplication, **A**ddition and **S**ubtraction, taken in that order.

The basic laws governing the use of brackets and operators are shown by the following examples:

(i) $2 + 3 = 3 + 2$, i.e. the order of numbers when adding does not matter;
(ii) $2 \times 3 = 3 \times 2$, i.e. the order of numbers when multiplying does not matter;
(iii) $2 + (3 + 4) = (2 + 3) + 4$, i.e. the use of brackets when adding does not affect the result;
(iv) $2 \times (3 \times 4) = (2 \times 3) \times 4$, i.e. the use of brackets when multiplying does not affect the result;
(v) $2 \times (3 + 4) = 2(3 + 4) = 2 \times 3 + 2 \times 4$, i.e. a number placed outside of a bracket indicates that the whole contents of the bracket must be multiplied by that number;

(vi) $(2+3)(4+5) = (5)(9) = 45$, i.e. adjacent brackets indicate multiplication;

(vii) $2[3+(4\times 5)] = 2[3+20] = 2\times 23 = 46$, i.e. when an expression contains inner and outer brackets, the inner brackets are removed first.

For example, to find the value of $6+4\div(5-3)$:
The order of precedence of operations is remembered by the word BODMAS.

Thus

$$\begin{aligned} 6+4\div(5-3) &= 6+4\div 2 && \text{(Brackets)} \\ &= 6+2 && \text{(Division)} \\ &= \mathbf{8} && \text{(Addition)} \end{aligned}$$

In another example, to determine the value of $13-2\times 3+14\div(2+5)$:

$$\begin{aligned} 13-2\times 3+14\div(2+5) &= 13-2\times 3+14\div 7 && \text{(B)} \\ &= 13-2\times 3+2 && \text{(D)} \\ &= 13-6+2 && \text{(M)} \\ &= 15-6 && \text{(A)} \\ &= \mathbf{9} && \text{(S)} \end{aligned}$$

---

2 Revision of Fractions, Decimals and Percentages

Fractions

When 2 is divided by 3, it may be written as $\frac{2}{3}$ or 2/3. $\frac{2}{3}$ is called a **fraction**. The number above the line, i.e. 2, is called the **numerator** and the number below the line, i.e. 3, is called the **denominator**.

When the value of the numerator is less than the value of the denominator, the fraction is called a **proper fraction**; thus $\frac{2}{3}$ is a proper fraction. When the value of the numerator is greater than the denominator, the fraction is called an **improper fraction**. Thus $\frac{7}{3}$ is an improper fraction and can also be expressed as a **mixed number**, that is, an integer and a proper fraction. The improper fraction $\frac{7}{3}$ is equal to the mixed number $2\frac{1}{3}$
When a fraction is simplified by dividing the numerator and denominator by the same number, the process is called **cancelling**. Cancelling by 0 is not permissible.
For example, to simplify $\frac{1}{3}+\frac{2}{7}$:
The lowest common multiple (i.e. LCM) of the two denominators is 3×7, i.e. 21
Expressing each fraction so that their denominators are 21, gives:

$$\frac{1}{3}+\frac{2}{7} = \left(\frac{1}{3}\times\frac{7}{7}\right)+\left(\frac{2}{7}\times\frac{3}{3}\right) = \frac{7}{21}+\frac{6}{21} = \frac{7+6}{21} = \mathbf{\frac{13}{21}}$$

Alternatively:

$$\frac{1}{3}+\frac{2}{7} = \frac{\overset{\text{Step (2)}\downarrow}{(7\times 1)}+\overset{\text{Step (3)}\downarrow}{(3\times 2)}}{\underset{\uparrow\text{Step (1)}}{21}}$$

Step 1: the LCM of the two denominators;
Step 2: for the fraction $\frac{1}{3}$, 3 into 21 goes 7 times, 7 × the numerator is 7 × 1;
Step 3: for the fraction $\frac{2}{7}$, 7 into 21 goes 3 times, 3 × the numerator is 3 × 2

Thus $\frac{1}{3}+\frac{2}{7}=\frac{7+6}{21}=\frac{13}{21}$ as obtained previously.

In another example, to find the value of $3\frac{2}{3}-2\frac{1}{6}$:

One method is to split the mixed numbers into integers and their fractional parts. Then

$$3\tfrac{2}{3}-2\tfrac{1}{6}=\left(3+\tfrac{2}{3}\right)-\left(2+\tfrac{1}{6}\right)=3+\tfrac{2}{3}-2-\tfrac{1}{6}$$

$$=1+\tfrac{4}{6}-\tfrac{1}{6}=1\tfrac{3}{6}=\mathbf{1\tfrac{1}{2}}$$

Another method is to express the mixed numbers as improper fractions.

Since $3=\frac{9}{3}$, then $3\frac{2}{3}=\frac{9}{3}+\frac{2}{3}=\frac{11}{3}$

Similarly, $2\frac{1}{6}=\frac{12}{6}+\frac{1}{6}=\frac{13}{6}$

Thus $3\frac{2}{3}-2\frac{1}{6}=\frac{11}{3}-\frac{13}{6}=\frac{22}{6}-\frac{13}{6}=\frac{9}{6}=\mathbf{1\frac{1}{2}}$ as obtained previously.

In another example, to find the value of $\frac{3}{7}\times\frac{14}{15}$:

Dividing numerator and denominator by 3 gives:

$$\frac{{}^{1}\cancel{3}}{7}\times\frac{14}{\cancel{15}_{5}}=\frac{1}{7}\times\frac{14}{5}=\frac{1\times 14}{7\times 5}$$

Dividing numerator and denominator by 7 gives:

$$\frac{1\times{}^{2}\cancel{14}}{{}_{1}\cancel{7}\times 5}=\frac{1\times 2}{1\times 5}=\mathbf{\frac{2}{5}}$$

This process of dividing both the numerator and denominator of a fraction by the same factor(s) is called **cancelling**.

In another example, to simplify $\frac{3}{7}\div\frac{12}{21}$:

$$\frac{3}{7}\div\frac{12}{21}=\frac{\frac{3}{7}}{\frac{12}{21}}$$

Multiplying both numerator and denominator by the reciprocal of the denominator gives:

$$\frac{\frac{3}{7}}{\frac{12}{21}} = \frac{\frac{{}^{1}\cancel{3}}{{}_{1}\cancel{7}} \times \frac{\cancel{21}^{3}}{\cancel{12}_{4}}}{\frac{{}_{1}\cancel{12}}{{}_{1}\cancel{21}} \times \frac{\cancel{21}^{1}}{\cancel{12}_{1}}} = \frac{\frac{3}{4}}{1} = \mathbf{\frac{3}{4}}$$

This method can be remembered by the rule: invert the second fraction and change the operation from division to multiplication. Thus:

$$\frac{3}{7} \div \frac{12}{21} = \frac{{}^{1}\cancel{3}}{{}_{1}\cancel{7}} \times \frac{\cancel{21}^{3}}{\cancel{12}_{4}} = \mathbf{\frac{3}{4}} \text{ as obtained previously.}$$

Ratio and proportion

The **ratio** of one quantity to another is a fraction, and is the number of times one quantity is contained in another quantity **of the same kind**.
If one quantity is **directly proportional** to another, then as one quantity doubles, the other quantity also doubles. When a quantity is **inversely proportional** to another, then as one quantity doubles, the other quantity is halved.
For example, a piece of timber 273 cm long is cut into three pieces in the ratio of 3 to 7 to 11. To determine the lengths of the three pieces:
The total number of parts is $3 + 7 + 11$, that is, 21. Hence 21 parts correspond to 273 cm.

$$1 \text{ part corresponds to } \frac{273}{21} = 13 \text{ cm}$$

$$3 \text{ parts correspond to } 3 \times 13 = 39 \text{ cm}$$

$$7 \text{ parts correspond to } 7 \times 13 = 91 \text{ cm}$$

$$11 \text{ parts correspond to } 11 \times 13 = 143 \text{ cm}$$

i.e. **the lengths of the three pieces are 39 cm, 91 cm and 143 cm** (Check: $39 + 91 + 143 = 273$)

In another example, a gear wheel having 80 teeth is in mesh with a 25-tooth gear. The gear ratio is determined as follows:

$$\text{Gear ratio} = 80:25 = \frac{80}{25} = \frac{16}{5} = 3.2$$

i.e. gear ratio = **16 : 5** or **3.2 : 1**

In another example, an alloy is made up of metals A and B in the ratio 2.5 : 1 by mass. The amount of A which has to be added to 6 kg of B to make the alloy is determined as follows:

$$\text{Ratio } A:B::2.5:1 \text{ i.e. } \frac{A}{B} = \frac{2.5}{1} = 2.5$$

When $B = 6$ kg, $\frac{A}{6} = 2.5$ from which, $\mathbf{A = 6 \times 2.5 = 15\ kg}$

In another example, 3 people can complete a task in 4 hours. To determine how long it will take 5 people to complete the same task, assuming the rate of work remains constant:
The more the number of people, the more quickly the task is done, hence inverse proportion exists.
3 people complete the task in 4 hours,
1 person takes three times as long, i.e. $4 \times 3 = 12$ hours,
5 people can do it in one fifth of the time that one person takes, that is $\frac{12}{5}$ hours or **2 hours 24 minutes**

Decimals

The decimal system of numbers is based on the **digits** 0 to 9. A number such as 53.17 is called a **decimal fraction**, a decimal point separating the integer part, i.e. 53, from the fractional part, i.e. 0.17

A number which can be expressed exactly as a decimal fraction is called a **terminating decimal** and those which cannot be expressed exactly as a decimal fraction are called **non-terminating decimals**. Thus, $\frac{3}{2} = 1.5$ is a terminating decimal, but $\frac{4}{3} = 1.33333\ldots$ is a non-terminating decimal. $1.33333\ldots$ can be written as $1.\dot{3}$, called 'one point-three recurring'.

The answer to a non-terminating decimal may be expressed in two ways, depending on the accuracy required:

(i) correct to a number of **significant figures**, that is, figures which signify something, and
(ii) correct to a number of **decimal places**, that is, the number of figures after the decimal point.

The last digit in the answer is unaltered if the next digit on the right is in the group of numbers 0, 1, 2, 3 or 4, but is increased by 1 if the next digit on the right is in the group of numbers 5, 6, 7, 8 or 9. Thus the non-terminating decimal 7.6183... becomes 7.62, correct to 3 significant figures, since the next digit on the right is 8, which is in the group of numbers 5, 6, 7, 8 or 9. Also 7.6183... becomes 7.618, correct to 3 decimal places, since the next digit on the right is 3, which is in the group of numbers 0, 1, 2, 3 or 4.
For example, to evaluate $42.7 + 3.04 + 8.7 + 0.06$:
The numbers are written so that the decimal points are under each other. Each column is added, starting from the right.

$$\begin{array}{r} 42.7 \\ 3.04 \\ 8.7 \\ \underline{0.06} \\ \underline{54.50} \end{array}$$

Thus $\mathbf{42.7 + 3.04 + 8.7 + 0.06 = 54.50}$
In another example, to determine the value of 74.3×3.8:

When multiplying decimal fractions: (i) the numbers are multiplied as if they are integers, and (ii) the position of the decimal point in the answer is such that there are as many digits to the right of it as the sum of the digits to the right of the decimal points of the two numbers being multiplied together. Thus

(i)
$$\begin{array}{r} 743 \\ \underline{38} \\ 5944 \\ \underline{22\,290} \\ \underline{28\,234} \end{array}$$

(ii) As there are $(1 + 1) = 2$ digits to the right of the decimal points of the two numbers being multiplied together, $(74.\underline{3} \times 3.\underline{8})$, then

$$\mathbf{74.3 \times 3.8 = 282.34}$$

In another example, to evaluate $37.81 \div 1.7$, correct to (i) 4 significant figures and (ii) 4 decimal places:

$$37.81 \div 1.7 = \frac{37.81}{1.7}$$

The denominator is changed into an integer by multiplying by 10. The numerator is also multiplied by 10 to keep the fraction the same. Thus

$$37.81 \div 1.7 = \frac{37.81 \times 10}{1.7 \times 10} = \frac{378.1}{17}$$

The long division is similar to the long division of integers and the first four steps are as shown:

$$\begin{array}{r} 22.24117.. \\ 17\overline{)378.100000} \\ \underline{34} \\ 38 \\ \underline{34} \\ 41 \\ \underline{34} \\ 70 \\ \underline{68} \\ 20 \end{array}$$

(i) **37.81 ÷ 1.7 = 22.24, correct to 4 significant figures**, and
(ii) **37.81 ÷ 1.7 = 22.2412, correct to 4 decimal places**.

In another example, to convert 0.4375 to a proper fraction:

0.4375 can be written as $\frac{0.4375 \times 10\,000}{10\,000}$ without changing its value,

i.e. $0.4375 = \frac{4375}{10\,000}$

By cancelling $\frac{4375}{10\,000} = \frac{875}{2000} = \frac{175}{400} = \frac{35}{80} = \frac{7}{16}$

i.e. $\mathbf{0.4375 = \frac{7}{16}}$

In another example, to express $\frac{9}{16}$ as a decimal fraction:

To convert a proper fraction to a decimal fraction, the numerator is divided by the denominator. Division by 16 can be done by the long division method, or, more simply, by dividing by 2 and then 8:

$$2\overline{)9.00}^{\;4.50} \qquad 8\overline{)4.5000}^{\;0.5625} \qquad \text{Thus, } \mathbf{\frac{9}{16} = 0.5625}$$

Percentages

Percentages are used to give a common standard and are fractions having the number 100 as their denominators. For example, 25 per cent means $\frac{25}{100}$ i.e. $\frac{1}{4}$ and is written 25%.

For example, to express 0.0125 as a percentage:
A decimal fraction is converted to a percentage by multiplying by 100. Thus, 0.0125 corresponds to $0.0125 \times 100\%$, i.e. **1.25%**

In another example, to express $\frac{5}{16}$ as a percentage:
To convert fractions to percentages, they are (i) converted to decimal fractions and (ii) multiplied by 100
By division, $\frac{5}{16} = 0.3125$, hence $\frac{5}{16}$ corresponds to $0.3125 \times 100\%$, i.e. **31.25%**

In another example, it takes 50 minutes to machine a certain part. Using a new type of tool, the time can be reduced by 15%. The new time taken is determined as follows:

$$15\% \text{ of } 50 \text{ minutes} = \frac{15}{100} \times 50 = \frac{750}{100} = 7.5 \text{ minutes}$$

hence the **new time taken** is 50 – 7.5 = **42.5 minutes**
Alternatively, if the time is reduced by 15%, then it now takes 85% of the original time, i.e. 85% of 50 = $\frac{85}{100} \times 50 = \frac{4250}{100}$ = **42.5 minutes**, as above.

In another example, a German silver alloy consists of 60% copper, 25% zinc and 15% nickel. The masses of the copper, zinc and nickel in a 3.74 kilogram block of the alloy is determined as follows:
By direct proportion:

$$100\% \text{ corresponds to } 3.74 \text{ kg}$$

$$1\% \text{ corresponds to } \frac{3.74}{100} = 0.0374 \text{ kg}$$

$$60\% \text{ corresponds to } 60 \times 0.0374 = 2.244 \text{ kg}$$

$$25\% \text{ corresponds to } 25 \times 0.0374 = 0.935 \text{ kg}$$

$$15\% \text{ corresponds to } 15 \times 0.0374 = 0.561 \text{ kg}$$

Thus, the masses of the copper, zinc and nickel are **2.244 kg, 0.935 kg and 0.561 kg**, respectively.
(Check: $2.244 + 0.935 + 0.561 = 3.74$).

---

3 Indices and Standard Form

Indices

The lowest factors of 2000 are $2 \times 2 \times 2 \times 2 \times 5 \times 5 \times 5$. These factors are written as $2^4 \times 5^3$, where 2 and 5 are called **bases** and the numbers 4 and 3 are called **indices**.
When an index is an integer it is called a **power**. Thus, 2^4 is called 'two to the power of four', and has a base of 2 and an index of 4. Similarly, 5^3 is called 'five to the power of 3' and has a base of 5 and an index of 3.
Special names may be used when the indices are 2 and 3, these being called 'squared' and 'cubed', respectively. Thus 7^2 is called 'seven squared' and 9^3 is called 'nine cubed'. When no index is shown, the power is 1, i.e. 2 means 2^1.

Reciprocal

The **reciprocal** of a number is when the index is -1 and its value is given by 1 divided by the base. Thus the reciprocal of 2 is 2^{-1} and its value is $\frac{1}{2}$ or 0.5. Similarly, the reciprocal of 5 is 5^{-1} which means $\frac{1}{5}$ or 0.2

Square root

The **square root** of a number is when the index is $\frac{1}{2}$, and the square root of 2 is written as $2^{\frac{1}{2}}$ or $\sqrt{2}$. The value of a square root is the value of the base which when multiplied by itself gives the number. Since $3 \times 3 = 9$, then $\sqrt{9} = 3$. However, $(-3) \times (-3) = 9$, so $\sqrt{9} = -3$. There are always two answers when finding the square root of a number and this is shown by putting both a $+$ and a $-$ sign in front of the answer to a square root problem. Thus $\sqrt{9} = \pm 3$ and $4^{\frac{1}{2}} = \sqrt{4} = \pm 2$, and so on.

Laws of indices

When simplifying calculations involving indices, certain basic rules or laws can be applied, called the **laws of indices**. These are given below.

(i) When multiplying two or more numbers having the same base, the indices are added. Thus $3^2 \times 3^4 = 3^{2+4} = 3^6$
(ii) When a number is divided by a number having the same base, the indices are subtracted. Thus $\frac{3^5}{3^2} = 3^{5-2} = 3^3$
(iii) When a number which is raised to a power is raised to a further power, the indices are multiplied. Thus $(3^5)^2 = 3^{5 \times 2} = 3^{10}$
(iv) When a number has an index of 0, its value is 1. Thus $3^0 = 1$

(v) A number raised to a negative power is the reciprocal of that number raised to a positive power. Thus $3^{-4} = \frac{1}{3^4}$. Similarly, $\frac{1}{2^{-3}} = 2^3$.

(vi) When a number is raised to a fractional power the denominator of the fraction is the root of the number and the numerator is the power.

Thus $8^{\frac{2}{3}} = \sqrt[3]{8^2} = (2)^2 = 4$

and $25^{\frac{1}{2}} = \sqrt[2]{25^1} = \sqrt{25^1} = \pm 5$ (Note that $\sqrt{\ } \equiv \sqrt[2]{\ }$)

For example, to evaluate (a) $5^2 \times 5^3$ (b) $3^2 \times 3^4 \times 3$:

From law (i):

(a) $5^2 \times 5^3 = 5^{(2+3)} = 5^5 = 5 \times 5 \times 5 \times 5 \times 5 = \mathbf{3125}$
(b) $3^2 \times 3^4 \times 3 = 3^{(2+4+1)} = 3^7 = 3 \times 3 \times \cdots$ to 7 terms $= \mathbf{2187}$

In another example, to find the value of (a) $\frac{7^5}{7^3}$ and (b) $\frac{5^7}{5^4}$:

From law (ii):

(a) $\dfrac{7^5}{7^3} = 7^{(5-3)} = 7^2 = \mathbf{49}$ (b) $\dfrac{5^7}{5^4} = 5^{(7-4)} = 5^3 = \mathbf{125}$

In another example, to simplify: (a) $(2^3)^4$ (b) $(3^2)^5$, expressing the answers in index form:
From law (iii):

(a) $(2^3)^4 = 2^{3\times4} = \mathbf{2^{12}}$ (b) $(3^2)^5 = 3^{2\times5} = \mathbf{3^{10}}$

In another example, to evaluate $\dfrac{(10^2)^3}{10^4 \times 10^2}$:

From the laws of indices:

$$\frac{(10^2)^3}{10^4 \times 10^2} = \frac{10^{(2\times3)}}{10^{(4+2)}} = \frac{10^6}{10^6} = 10^{6-6} = 10^0 = \mathbf{1}$$

In another example, to evaluate (a) $4^{1/2}$ (b) $16^{3/4}$ (c) $27^{2/3}$ (d) $9^{-1/2}$:

(a) $4^{1/2} = \sqrt{4} = \pm\mathbf{2}$
(b) $16^{3/4} = \sqrt[4]{16^3} = (2)^3 = \mathbf{8}$

(Note that it does not matter whether the 4th root of 16 is found first or whether 16 cubed is found first—the same answer will result.)

(c) $27^{2/3} = \sqrt[3]{27^2} = (3)^2 = \mathbf{9}$
(d) $9^{-1/2} = \dfrac{1}{9^{1/2}} = \dfrac{1}{\sqrt{9}} = \dfrac{1}{\pm 3} = \pm\mathbf{\dfrac{1}{3}}$

In another example, to evaluate $\dfrac{3^3 \times 5^7}{5^3 \times 3^4}$:

The laws of indices only apply to terms **having the same base**. Grouping terms having the same base, and then applying the laws of indices to each of the groups independently gives:

$$\frac{3^3 \times 5^7}{5^3 \times 3^4} = \frac{3^3}{3^4} \times \frac{5^7}{5^3} = 3^{(3-4)} \times 5^{(7-3)}$$

$$= 3^{-1} \times 5^4 = \frac{5^4}{3^1} = \frac{625}{3} = \mathbf{208\frac{1}{3}}$$

In another example, to evaluate: $\dfrac{4^{1.5} \times 8^{1/3}}{2^2 \times 32^{-2/5}}$:

$$4^{1.5} = 4^{3/2} = \sqrt{4^3} = 2^3 = 8, \quad 8^{1/3} = \sqrt[3]{8} = 2,$$

$$2^2 = 4, \quad 32^{-2/5} = \frac{1}{32^{2/5}} = \frac{1}{\sqrt[5]{32^2}} = \frac{1}{2^2} = \frac{1}{4}$$

Hence $\dfrac{4^{1.5} \times 8^{1/3}}{2^2 \times 32^{-2/5}} = \dfrac{8 \times 2}{4 \times \frac{1}{4}} = \dfrac{16}{1} = \mathbf{16}$

Alternatively, $\dfrac{4^{1.5} \times 8^{1/3}}{2^2 \times 32^{-2/5}} = \dfrac{[(2)^2]^{3/2} \times (2^3)^{1/3}}{2^2 \times (2^5)^{-2/5}}$

$$= \frac{2^3 \times 2^1}{2^2 \times 2^{-2}} = 2^{3+1-2-(-2)} = 2^4 = \mathbf{16}$$

Standard form

A number written with one digit to the left of the decimal point and multiplied by 10 raised to some power is said to be written in **standard form**. Thus: 5837 is written as 5.837×10^3 in standard form, and 0.0415 is written as 4.15×10^{-2} in standard form.

When a number is written in standard form, the first factor is called the **mantissa** and the second factor is called the **exponent**. Thus the number 5.8×10^3 has a mantissa of 5.8 and an exponent of 10^3.

(i) Numbers having the same exponent can be added or subtracted in standard form by adding or subtracting the mantissae and keeping the exponent the same. Thus:

$$2.3 \times 10^4 + 3.7 \times 10^4 = (2.3 + 3.7) \times 10^4 = 6.0 \times 10^4, \text{ and}$$

$$5.9 \times 10^{-2} - 4.6 \times 10^{-2} = (5.9 - 4.6) \times 10^{-2} = 1.3 \times 10^{-2}$$

When the numbers have different exponents, one way of adding or subtracting the numbers is to express one of the numbers in non-standard form, so that both numbers have the same exponent. Thus:

$$2.3 \times 10^4 + 3.7 \times 10^3 = 2.3 \times 10^4 + 0.37 \times 10^4$$

$$= (2.3 + 0.37) \times 10^4 = 2.67 \times 10^4$$

Alternatively, $2.3 \times 10^4 + 3.7 \times 10^3 = 23\,000 + 3700$

$$= 26\,700 = 2.67 \times 10^4$$

(ii) The laws of indices are used when multiplying or dividing numbers given in standard form. **For example,**

$$(2.5 \times 10^3) \times (5 \times 10^2) = (2.5 \times 5) \times (10^{3+2})$$

$$= 12.5 \times 10^5 \text{ or } 1.25 \times 10^6$$

In another example, $\dfrac{6 \times 10^4}{1.5 \times 10^2} = \dfrac{6}{1.5} \times (10^{4-2}) = 4 \times 10^2$

In another example, to express in standard form: (a) 38.71 (b) 0.0124
For a number to be in standard form, it is expressed with only one digit to the left of the decimal point. Thus:

(a) 38.71 must be divided by 10 to achieve one digit to the left of the decimal point and it must also be multiplied by 10 to maintain the equality, i.e.

$$38.71 = \frac{38.71}{10} \times 10 = \mathbf{3.871 \times 10} \text{ in standard form}$$

(b) $0.0124 = 0.0124 \times \dfrac{100}{100}$

$$= \frac{1.24}{100} = \mathbf{1.24 \times 10^{-2}} \text{ in standard form}$$

In another example, to express in standard form, correct to 3 significant figures:

(a) $19\dfrac{2}{3}$ (b) $741\dfrac{9}{16}$

(a) $19\dfrac{2}{3} = 19.\dot{6} = \mathbf{1.97 \times 10}$ in standard form, correct to 3 significant figures

(b) $741\dfrac{9}{16} = 741.5625 = \mathbf{7.42 \times 10^2}$ in standard form, correct to 3 significant figures

In another example, to find the value of (a) $7.9 \times 10^{-2} - 5.4 \times 10^{-2}$ and (b) $9.293 \times 10^2 + 1.3 \times 10^3$ expressing the answers in standard form:

(a) $7.9 \times 10^{-2} - 5.4 \times 10^{-2} = (7.9 - 5.4) \times 10^{-2} = \mathbf{2.5 \times 10^{-2}}$

(b) Since only numbers having the same exponents can be added by straight addition of the mantissae, the numbers are converted to this form before adding.

Thus: $9.293 \times 10^2 + 1.3 \times 10^3 = 9.293 \times 10^2 + 13 \times 10^2$

$$= (9.293 + 13) \times 10^2$$

$$= 22.293 \times 10^2$$

$$= \mathbf{2.2293 \times 10^3} \text{ in standard form}$$

Alternatively, the numbers can be expressed as decimal fractions, giving:

$$9.293 \times 10^2 + 1.3 \times 10^3 = 929.3 + 1300 = 2229.3$$

$$= \mathbf{2.2293 \times 10^3} \text{ in standard form}$$

as obtained previously. This method is often the 'safest' way of doing this type of example.

---

4 Errors, Calculations and Evaluation of Formulae

Errors and approximations

In all problems in which the measurement of distance, time, mass or other quantities occurs, an exact answer cannot be given; only an answer that is correct to a stated degree of accuracy can be given. To take account of this an **error due to measurement** is said to exist.
To take account of measurement errors it is usual to limit answers so that the result given is **not more than one significant figure greater than the least accurate number given in the data**.
Rounding-off errors can exist with decimal fractions. For example, to state that $\pi = 3.142$ is not strictly correct, but '$\pi = 3.142$ correct to 4 significant figures' is a true statement. (Actually, $\pi = 3.14159265\ldots$).
It is possible, through an incorrect procedure, to obtain the wrong answer to a calculation. This type of error is known as **a blunder**.
An **order of magnitude error** is said to exist if incorrect positioning of the decimal point occurs after a calculation has been completed.
Blunders and order of magnitude errors can be reduced by determining **approximate values of calculations**. Answers that do not seem feasible must be checked and the calculation repeated as necessary.
An engineer will often need to make a quick mental approximation for a calculation. **For example,** $\dfrac{49.1 \times 18.4 \times 122.1}{61.2 \times 38.1}$ may be approximated to $\dfrac{50 \times 20 \times 120}{60 \times 40}$ and then, by cancelling, $\dfrac{50 \times \cancel{20}^{1} \times \cancel{120}^{\cancel{2}^{1}}}{{}_{1}\cancel{60} \times \cancel{40}_{\cancel{2}_{1}}} = 50$. An accurate answer somewhere between 45 and 55 could therefore be expected. certainly an answer around 500 or 5 would not be expected. Actually, by calculator $\dfrac{49.1 \times 18.4 \times 122.1}{61.2 \times 38.1} = 47.31$, correct to 4 significant figures.

Use of calculator

The most modern aid to calculations is the pocket-sized electronic calculator. With one of these, calculations can be quickly and accurately performed, correct to about 9 significant figures. The scientific type of calculator has made the use of tables and logarithms largely redundant.
To help you to become competent at using your calculator check that you agree with the answers to the following examples:

$21.93 \times 0.012981 = 0.2846733\ldots = \mathbf{0.2847}$, correct to 4 significant figures

$\frac{1}{0.0275} = 36.3636363\ldots = \mathbf{36.364}$, correct to 3 decimal places

$46.27^2 - 31.79^2 = 1130.3088 = \mathbf{1.130 \times 10^3}$, correct to 4 significant figures

$\sqrt{5.462} = 2.3370922\ldots = \mathbf{2.337}$, correct to 4 significant figures

$\sqrt{0.007328} = 0.08560373 = \mathbf{0.086}$, correct to 3 decimal places

$4.72^3 = 105.15404\ldots = \mathbf{105.2}$, correct to 4 significant figures

$\sqrt[3]{47.291} = 3.61625876\ldots = \mathbf{3.62}$, correct to 3 significant figures

---

Conversion tables and charts

It is often necessary to make calculations from various conversion tables and charts. Examples include currency exchange rates, imperial to metric unit conversions, train or bus timetables, production schedules and so on.
For example, some approximate imperial to metric conversions are shown in Table 4.1.

Table 4.1

length	1 inch = 2.54 cm
	1 mile = 1.61 km
weight	2.2 lb = 1 kg
	(1 lb = 16 oz)
capacity	1.76 pints = 1 litre
	(8 pints = 1 gallon)

9.5 inches = 9.5 × 2.54 cm = 24.13 cm

and 24.13 cm = 24.13 × 10 mm = **241.3 mm**

50 m.p.h. = 50 × 1.61 km/h = **80.5 km/h**

$300\text{ km} = \frac{300}{1.61}\text{ miles} = \mathbf{186.3\ miles}$

$30\text{ lb} = \frac{30}{2.2}\text{ kg} = \mathbf{13.64\ kg}$

42 kg = 42 × 2.2 lb = 92.4 lb

0.4 lb = 0.4 × 16 oz = 6.4 oz = 6 oz, correct to the nearest ounce

Thus 42 kg = **92 lb 6 oz**, correct to the nearest ounce

15 gallons = 15 × 8 pints = 120 pints

$120\text{ pints} = \frac{120}{1.76}\text{ litres} = \mathbf{68.18\ litres}$

$$40 \text{ litres} = 40 \times 1.76 \text{ pints} = 70.4 \text{ pints}$$

$$70.4 \text{ pints} = \frac{70.4}{8} \text{ gallons} = \mathbf{8.8\ gallons}$$

Evaluation of formulae

The statement $v = u + at$ is said to be a **formula** for v in terms of u, a and t. v, u, a and t are called **symbols**.
The single term on the left-hand side of the equation, v, is called the **subject of the formulae**.
Provided values are given for all the symbols in a formula except one, the remaining symbol can be made the subject of the formula and may be evaluated by using a calculator.
For example, velocity v is given by $v = u + at$. To find v, correct to 3 significant figures if $u = 9.86$ m/s, $a = 4.25$ m/s^2 and $t = 6.84$ s:

$$v = u + at = 9.86 + (4.25)(6.84)$$
$$= 9.86 + 29.07 = 38.93$$

Hence velocity v = 38.9 m/s, correct to 3 significant figures
In another example, the volume V cm^3 of a right circular cone is given by $V = \frac{1}{3}\pi r^2 h$. To find the volume, correct to 4 significant figures, given that $r = 4.321$ cm and $h = 18.35$ cm:

$$V = \tfrac{1}{3}\pi r^2 h = \tfrac{1}{3}\pi(4.321)^2(18.35) = \tfrac{1}{3}\pi(18.671041)(18.35)$$

Hence volume, V = 358.8 cm^3, correct to 4 significant figures

In another example, force F Newton's is given by the formula $F = \dfrac{Gm_1m_2}{d^2}$, where m_1 and m_2 are masses, d their distance apart and G is a constant. To find the value of the force given that $G = 6.67 \times 10^{-11}$, $m_1 = 7.36$, $m_2 = 15.5$ and $d = 22.6$, expressing the answer in standard form, correct to 3 significant figures:

$$F = \frac{Gm_1m_2}{d^2} = \frac{(6.67 \times 10^{-11})(7.36)(15.5)}{(22.6)^2}$$
$$= \frac{(6.67)(7.36)(15.5)}{(10^{11})(510.76)} = \frac{1.490}{10^{11}}$$

Hence force F = 1.49 $\times 10^{-11}$ Newtons, correct to 3 significant figures

---

5 Algebra

Basic operations

Algebra is that part of mathematics in which the relations and properties of numbers are investigated by means of general symbols. For example, the area of a rectangle is found by multiplying the length by the breadth; this is expressed algebraically as $A = l \times b$, where A represents the area, l the length and b the breadth.

The basic laws introduced in arithmetic are generalised in algebra. Let a, b, c and d represent any four numbers. Then:

(i) $a+(b+c)=(a+b)+c$
(ii) $a(bc)=(ab)c$
(iii) $a+b=b+a$
(iv) $ab=ba$
(v) $a(b+c)=ab+ac$
(vi) $\dfrac{a+b}{c}=\dfrac{a}{c}+\dfrac{b}{c}$
(vii) $(a+b)(c+d)=ac+ad+bc+bd$

For example, to find the value of $4p^2qr^3$, given that $p=2$, $q=\frac{1}{2}$ and $r=1\frac{1}{2}$:
Replacing p, q and r with their numerical values gives:

$$4p^2qr^3 = 4(2)^2\left(\tfrac{1}{2}\right)\left(\tfrac{3}{2}\right)^3$$
$$= 4\times 2\times 2\times \tfrac{1}{2}\times\tfrac{3}{2}\times\tfrac{3}{2}\times\tfrac{3}{2} = \mathbf{27}$$

In another example, to find the sum of $5a-2b$, $2a+c$, $4b-5d$ and $b-a+3d-4c$:
The algebraic expressions may be tabulated as shown below, forming columns for the $a's$, $b's$, $c's$ and $d's$. Thus:

$$\begin{array}{rrrr}
+5a & -2b & & \\
+2a & & +c & \\
 & +4b & & -5d \\
-a & +b & -4c & +3d \\
\hline
6a & +3b & -3c & -2d
\end{array}$$

Adding gives: $6a+3b-3c-2d$

In another example, to multiply $2a+3b$ by $a+b$:
Each term in the first expression is multiplied by a, then each term in the first expression is multiplied by b, and the two results are added. The usual layout is shown below.

$$\begin{array}{lrrr}
 & 2a & +3b & \\
 & a & +b & \\
\hline
\text{Multiplying by } a\rightarrow & 2a^2 & +3ab & \\
\text{Multiplying by } b\rightarrow & & +2ab & +3b^2 \\
\hline
\text{Adding gives:} & \mathbf{2a^2} & \mathbf{+5ab} & \mathbf{+3b^2}
\end{array}$$

In another example, to simplify $2p\div 8pq$:

$2p\div 8pq$ means $\dfrac{2p}{8pq}$. This can be reduced by cancelling, as in arithmetic.

Thus: $\dfrac{2p}{8pq}=\dfrac{\cancel{2}^1\times\cancel{p}^1}{{}_4\cancel{8}\times{}_1\cancel{p}\times q}=\mathbf{\dfrac{1}{4q}}$

Laws of indices

The laws of indices are:

(i) $a^m \times a^n = a^{m+n}$ (ii) $\dfrac{a^m}{a^n} = a^{m-n}$ (iii) $(a^m)^n = a^{mn}$

(iv) $a^{\frac{m}{n}} = \sqrt[n]{a^m}$ (v) $a^{-n} = \dfrac{1}{a^n}$ (vi) $a^0 = 1$

For example, to simplify $a^3b^2c \times ab^3c^5$:

Grouping like terms gives: $a^3 \times a \times b^2 \times b^3 \times c \times c^5$

Using the first law of indices gives: $a^{3+1} \times b^{2+3} \times c^{1+5}$

i.e. $a^4 \times b^5 \times c^6 = \boldsymbol{a^4b^5c^6}$

In another example, to simplify $\dfrac{a^3b^2c^4}{abc^{-2}}$ and evaluate when $a = 3$, $b = \frac{1}{8}$ and $c = 2$:

Using the second law of indices, $\dfrac{a^3}{a} = a^{3-1} = a^2$, $\dfrac{b^2}{b} = b^{2-1} = b$ and $\dfrac{c^4}{c^{-2}} = c^{4--2} = c^6$

Thus $\dfrac{a^3b^2c^4}{abc^{-2}} = \boldsymbol{a^2bc^6}$

When $a = 3$, $b = \frac{1}{8}$ and $c = 2$, $a^2bc^6 = (3)^2\left(\frac{1}{8}\right)(2)^6$

$$= (9)\left(\tfrac{1}{8}\right)(64) = \mathbf{72}$$

In another example, to simplify $\dfrac{x^2y^3 + xy^2}{xy}$:

Algebraic expressions of the form $\dfrac{a+b}{c}$ can be split into $\dfrac{a}{c} + \dfrac{b}{c}$. Thus

$$\frac{x^2y^3 + xy^2}{xy} = \frac{x^2y^3}{xy} + \frac{xy^2}{xy} = x^{2-1}y^{3-1} + x^{1-1}y^{2-1} = \boldsymbol{xy^2 + y}$$

(since $x^0 = 1$, from the sixth law of indices)

In another example, to simplify $\dfrac{(mn^2)^3}{(m^{1/2}n^{1/4})^4}$:

The brackets indicate that each letter in the bracket must be raised to the power outside.

Using the third law of indices gives:

$$\frac{(mn^2)^3}{(m^{1/2}n^{1/4})^4} = \frac{m^{1\times3}n^{2\times3}}{m^{(1/2)\times4}n^{(1/4)\times4}} = \frac{m^3n^6}{m^2n^1}$$

Using the second law of indices gives: $\dfrac{m^3n^6}{m^2n^1} = m^{3-2}n^{6-1} = \boldsymbol{mn^5}$

Brackets and factorisation

When two or more terms in an algebraic expression contain a common factor, then this factor can be shown outside of a **bracket**. For example

$$ab + ac = a(b + c)$$

which is simply the reverse of law (v) of algebra on page 17, and

$$6px + 2py - 4pz = 2p(3x + y - 2z)$$

This process is called **factorisation**.
In another example, to remove the brackets and simplify the expression:

$$2a - [3\{2(4a - b) - 5(a + 2b)\} + 4a]$$

Removing the innermost brackets gives: $2a - [3\{8a - 2b - 5a - 10b\} + 4a]$

Collecting together similar terms gives: $2a - [3\{3a - 12b\} + 4a]$

Removing the 'curly' brackets gives: $2a - [9a - 36b + 4a]$

Collecting together similar terms gives: $2a - [13a - 36b]$

Removing the outer brackets gives: $2a - 13a + 36b$
i.e. $\mathbf{-11a + 36b}$ or $\mathbf{36b - 11a}$ (see law (iii), page 17)
In another example, to factorise (a) $xy - 3xz$ (b) $4a^2 + 16ab^3$
(c) $3a^2b - 6ab^2 + 15ab$:
For each part of this example, the HCF of the terms will become one of the factors.
Thus: (a) $xy - 3xz = \mathbf{x(y - 3z)}$
(b) $4a^2 + 16ab^3 = \mathbf{4a(a + 4b^3)}$
(c) $3a^2b - 6ab^2 + 15ab = \mathbf{3ab(a - 2b + 5)}$

In another example, to factorise $ax - ay + bx - by$:
The first two terms have a common factor of a and the last two terms a common factor of b. Thus: $ax - ay + bx - by = a(x - y) + b(x - y)$
The two newly formed terms have a common factor of $(x - y)$. Thus:

$$a(x - y) + b(x - y) = \mathbf{(x - y)(a + b)}$$

Fundamental laws and precedence

The **laws of precedence** which apply to arithmetic also apply to algebraic expressions. The order is Brackets, Of, Division, Multiplication, Addition and Subtraction (i.e. **BODMAS).**
For example, to simplify $2a + 5a \times 3a - a$:
Multiplication is performed before addition and subtraction thus:

$$2a + 5a \times 3a - a = 2a + 15a^2 - a = a + 15a^2 = \mathbf{a(1 + 15a)}$$

In another example, to simplify $a \div 5a + 2a - 3a$:
The order of precedence is division, then addition and subtraction. Hence

$$a \div 5a + 2a - 3a = \frac{a}{5a} + 2a - 3a = \frac{1}{5} + 2a - 3a = \mathbf{\frac{1}{5} - a}$$

In another example, to simplify $3c + 2c \times 4c + c \div 5c - 8c$:
The order of precedence is division, multiplication, addition and subtraction.

Hence: $3c + 2c \times 4c + c \div 5c - 8c = 3c + 2c \times 4c + \dfrac{c}{5c} - 8c$

$$= 3c + 8c^2 + \frac{1}{5} - 8c$$

$$= \mathbf{8c^2 - 5c + \frac{1}{5}}$$

$$\text{or } \mathbf{c(8c - 5) + \frac{1}{5}}$$

Direct and inverse proportionality

An expression such as $y = 3x$ contains two variables. For every value of x there is a corresponding value of y. The variable x is called the **independent variable** and y is called the **dependent variable**.
When an increase or decrease in an independent variable leads to an increase or decrease of the same proportion in the dependent variable this is termed **direct proportion**. If $y = 3x$ then y is directly proportional to x, which may be written as $y \propto x$ or $y = kx$, where k is called the **coefficient of proportionality** (in this case, k being equal to 3).
When an increase in an independent variable leads to a decrease of the same proportion in the dependent variable (or vice versa) this is termed **inverse proportion**. If y is inversely proportional to x then $y \propto (1/x)$ or $y = k/x$. Alternatively, $k = xy$, that is, for inverse proportionality the product of the variables is constant.
Examples of laws involving direct and inverse proportional in science include:

(i) **Hooke's law**, which states that within the elastic limit of a material, the strain ε produced is directly proportional to the stress, σ, producing it, i.e. $\varepsilon \propto \sigma$ or $\varepsilon = k\sigma$
(ii) **Charles's law**, which states that for a given mass of gas at constant pressure the volume V is directly proportional to its thermodynamic temperature T, i.e. $V \propto T$ or $V = kT$
(iii) **Ohm's law**, which states that the current I flowing through a fixed resistor is directly proportional to the applied voltage V, i.e. $I \propto V$ or $I = kV$
(iv) **Boyle's law**, which states that for a gas at constant temperature, the volume V of a fixed mass of gas is inversely proportional to its absolute pressure p, i.e. $p \propto (1/V)$ or $p = k/V$, i.e. $pV = k$

Polynomial division

A **polynomial** is an expression of the form $f(x) = a + bx + cx^2 + dx^3 + \cdots$ and **polynomial division** is sometimes required when resolving into partial fractions—(see chapter 14, page 61).

For example, to divide $2x^2 + x - 3$ by $x - 1$:
$2x^2 + x - 3$ is called the **dividend** and $x - 1$ the **divisor**. The usual layout is shown below with the dividend and divisor both arranged in descending powers of the symbols.

$$\begin{array}{r|l} & 2x + 3 \\ \hline x - 1 & 2x^2 + x - 3 \\ & \underline{2x^2 - 2x} \\ & \quad\quad 3x - 3 \\ & \quad\quad \underline{3x - 3} \\ & \quad\quad \underline{\cdot \quad \cdot} \end{array}$$

Dividing the first term of the dividend by the first term of the divisor, i.e. $2x^2/x$ gives $2x$, which is put above the first term of the dividend as shown. The divisor is then multiplied by $2x$, i.e. $2x(x - 1) = 2x^2 - 2x$, which is placed under the dividend as shown. Subtracting gives $3x - 3$. The process is then repeated, i.e. the first term of the divisor, x, is divided into $3x$, giving $+3$, which is placed above the dividend as shown. Then $3(x - 1) = 3x - 3$ which is placed under the $3x - 3$. The remainder, on subtraction, is zero, which completes the process.
Thus $\mathbf{(2x^2 + x - 3) \div (x - 1) = (2x + 3)}$
[A check can be made on this answer by multiplying $(2x + 3)$ by $(x - 1)$ which equals $2x^2 + x - 3$]

In another example, to divide $(x^2 + 3x - 2)$ by $(x - 2)$:

$$\begin{array}{r|l} & x \;\; + 5 \\ \hline x - 2 & x^2 + 3x - 2 \\ & \underline{x^2 - 2x} \\ & \quad\quad 5x - 2 \\ & \quad\quad \underline{5x - 10} \\ & \quad\quad \underline{\quad\quad 8} \end{array}$$

Hence $\dfrac{x^2 + 3x - 2}{x - 2} = \mathbf{x + 5 + \dfrac{8}{x - 2}}$

The factor theorem

There is a simple relationship between the factors of a quadratic expression and the roots of the equation obtained by equating the expression to zero.
For example, consider the quadratic equation $x^2 + 2x - 8 = 0$. To solve this we may factorise the quadratic expression $x^2 + 2x - 8$ giving $(x - 2)(x + 4)$.
Hence $(x - 2)(x + 4) = 0$
Then, if the product of two numbers is zero, one or both of those numbers must equal zero. Therefore, either

$(x - 2) = 0$, from which, $x = 2$

or $(x + 4) = 0$, from which, $x = -4$

It is clear then that a factor of $(x - 2)$ indicates a root of $+2$, while a factor of $(x + 4)$ indicates a root of -4.

In general, we can therefore say that:

a factor of $(x - a)$ corresponds to a root of $x = a$

In practice, we always deduce the roots of a simple quadratic equation from the factors of the quadratic expression, as in the above example. However, we could reverse this process. If, by trial and error, we could determine that $x = 2$ is a root of the equation $x^2 + 2x - 8 = 0$ we could deduce at once that $(x - 2)$ is a factor of the expression $x^2 + 2x - 8$. We wouldn't normally solve quadratic equations this way — but suppose we have to factorise a cubic expression (i.e. one in which the highest power of the variable is 3). A cubic equation might have three simple linear factors and the difficulty of discovering all these factors by trial and error would be considerable. It is to deal with this kind of case that we use the **factor theorem**. This is just a generalised version of what we established above for the quadratic expression.
The factor theorem provides a method of factorising any polynomial, $f(x)$, which has simple factors.
A statement of the **factor theorem** says:

> **'if $x = a$ is a root of the equation $f(x) = 0$, then $(x - a)$ is a factor of $f(x)$'**

For example, to factorise $x^3 - 7x - 6$ and use it to solve the cubic equation $x^3 - 7x - 6 = 0$:

Let $f(x) = x^3 - 7x - 6$

If $x = 1$, then $f(1) = 1^3 - 7(1) - 6 = -12$

If $x = 2$, then $f(2) = 2^3 - 7(2) - 6 = -12$

If $x = 3$, then $f(3) = 3^3 - 7(3) - 6 = 0$

If $f(3) = 0$, then $(x - 3)$ is a factor — from the factor theorem.

We have a choice now. We can divide $x^3 - 7x - 6$ by $(x - 3)$ or we could continue our 'trial and error' by substituting further values for x in the given expression — and hope to arrive at $f(x) = 0$.
Let us do both ways. Firstly, dividing out gives:

$$\begin{array}{r|l} & x^2 + 3x + 2 \\ \hline x - 3 & x^3 + 0 - 7x - 6 \\ & \underline{x^3 - 3x^2} \\ & \quad 3x^2 - 7x - 6 \\ & \quad \underline{3x^2 - 9x} \\ & \qquad\quad 2x - 6 \\ & \qquad\quad \underline{2x - 6} \\ & \qquad\quad \underline{\;\cdot \quad \cdot\;} \end{array}$$

Hence $\dfrac{x^3 - 7x - 6}{x - 3} = x^2 + 3x + 2$

i.e. $x^3 - 7x - 6 = (x - 3)(x^2 + 3x + 2)$

$x^2 + 3x + 2$ factorises 'on sight' as $(x + 1)(x + 2)$

Therefore $\boldsymbol{x^3 - 7x - 6 = (x-3)(x+1)(x+2)}$

A second method is to continue to substitute values of x into $f(x)$.
Our expression for $f(3)$ was $3^3 - 7(3) - 6$. We can see that if we continue with positive values of x the first term will predominate such that $f(x)$ will not be zero.
Therefore let us try some negative values for x. Therefore
$f(-1) = (-1)^3 - 7(-1) - 6 = 0$; hence $(x+1)$ is a factor (as shown above).
Also $f(-2) = (-2)^3 - 7(-2) - 6 = 0$; hence $(x+2)$ is a factor (also as shown above).
To solve $x^3 - 7x - 6 = 0$, we substitute the factors, i.e.

$$(x-3)(x+1)(x+2) = 0$$

from which, $\boldsymbol{x = 3}$, $\boldsymbol{x = -1}$ and $\boldsymbol{x = -2}$
Note that the values of x, i.e. 3, -1 and -2, are all factors of the constant term, i.e. the 6. This can give us a clue as to what values of x we should consider.

The remainder theorem

Dividing a general quadratic expression $(ax^2 + bx + c)$ by $(x - p)$, where p is any whole number, by long division gives:

$$\begin{array}{r|l}
 & ax + (b + ap) \\
\hline
x - p & ax^2 + bx \qquad + c \\
 & \underline{ax^2 - apx} \\
 & \qquad (b + ap)x + c \\
 & \qquad \underline{(b + ap)x - (b + ap)p} \\
 & \qquad \qquad \underline{c + (b + ap)p}
\end{array}$$

The remainder, $c + (b + ap)p = c + bp + ap^2$ or $ap^2 + bp + c$
This is, in fact, what the **remainder theorem** states, i.e.

> **'if $(ax^2 + bx + c)$ is divided by $(x - p)$, the remainder will be $ap^2 + bp + c$'**

If, in the dividend $(ax^2 + bx + c)$, we substitute p for x we get the remainder $ap^2 + bp + c$

For example, when $(3x^2 - 4x + 5)$ is divided by $(x - 2)$ the remainder is $ap^2 + bp + c$, (where $a = 3$, $b = -4$, $c = 5$ and $p = 2$), i.e. the remainder is

$$3(2)^2 + (-4)(2) + 5 = 12 - 8 + 5 = \mathbf{9}$$

We can check this by dividing $(3x^2 - 4x + 5)$ by $(x - 2)$ by long division:

$$\begin{array}{r|l}
 & 3x + 2 \\
\hline
x - 2 & 3x^2 - 4x + 5 \\
 & \underline{3x^2 - 6x} \\
 & \qquad 2x + 5 \\
 & \qquad \underline{2x - 4} \\
 & \qquad \quad \underline{\;\;9}
\end{array}$$

Similarly, when $(x^2 + 3x - 2)$ is divided by $(x - 1)$, the remainder is $1(1)^2 + 3(1) - 2 = \mathbf{2}$

It is not particularly useful, on its own, to know the remainder of an algebraic division. However, if the remainder should be zero then $(x - p)$ is a factor. This is very useful therefore when factorising expressions.

The **remainder theorem** may also be stated for a **cubic equation** as:

> **'if $(ax^3 + bx^2 + cx + d)$ is divided by $(x - p)$, the remainder will be $ap^3 + bp^2 + cp + d$'**

As before, the remainder may be obtained by substituting p for x in the dividend.

For example, when $(3x^3 + 2x^2 - x + 4)$ is divided by $(x - 1)$, the remainder is $ap^3 + bp^2 + cp + d$ (where $a = 3$, $b = 2$, $c = -1$, $d = 4$ and $p = 1$), i.e. the remainder is $3(1)^3 + 2(1)^2 + (-1)(1) + 4 = 3 + 2 - 1 + 4 = \mathbf{8}$

Similarly, when $(x^3 - 7x - 6)$ is divided by $(x - 3)$, the remainder is $1(3)^3 + 0(3)^2 - 7(3) - 6 = 0$, which means that $(x - 3)$ is a factor of $(x^3 - 7x - 6)$.

Continued fractions

Any fraction may be expressed in the form shown below for the fraction $\frac{26}{55}$:

$$\frac{26}{55} = \frac{1}{\frac{55}{26}} = \frac{1}{2 + \frac{3}{26}} = \frac{1}{2 + \frac{1}{\frac{26}{3}}} = \frac{1}{2 + \frac{1}{8 + \frac{2}{3}}} = \frac{1}{2 + \frac{1}{8 + \frac{1}{\frac{3}{2}}}}$$

$$= \frac{1}{2 + \frac{1}{8 + \frac{1}{1 + \frac{1}{2}}}}$$

The latter factor can be expressed as:

$$\cfrac{1}{A + \cfrac{\alpha}{B + \cfrac{\beta}{C + \cfrac{\gamma}{D + \delta}}}}$$

Comparisons show that A, B, C and D are 2, 8, 1 and 2 respectively. A fraction written in the general form is called a **continued fraction** and the integers A, B, C and D are called the **quotients** of the continued fraction. The quotients may be used to obtain closer and closer approximations, being called **convergents**.

A tabular method may be used to determine the convergents of a fraction:

	1	2	3	4	5
a		2	8	1	2
$b\begin{cases} bp \\ bq \end{cases}$	$\frac{0}{1}$	$\frac{1}{2}$	$\frac{8}{17}$	$\frac{9}{19}$	$\frac{26}{55}$

The quotients 2, 8, 1 and 2 are written in cells $a2$, $a3$, $a4$ and $a5$ with cell a1 being left empty.
The fraction $\frac{0}{1}$ is always written in cell $b1$.
The reciprocal of the quotient in cell $a2$ is always written in cell $b2$, i.e. $\frac{1}{2}$ in this case.
The fraction in cell b3 is given by $\dfrac{(a3 \times b2p) + b1p}{(a3 \times b2q) + b1q}$,

i.e. $\dfrac{(8 \times 1) + 0}{(8 \times 2) + 1} = \dfrac{8}{17}$

The fraction in cell b4 is given by $\dfrac{(a4 \times b3p) + b2p}{(a4 \times b3q) + b2q}$,

i.e. $\dfrac{(1 \times 8) + 1}{(1 \times 17) + 2} = \dfrac{9}{19}$, and so on.

Hence the convergents of $\dfrac{26}{55}$ are $\dfrac{1}{2}, \dfrac{7}{17}, \dfrac{9}{19}$ and $\dfrac{26}{55}$, each value approximating closer and closer to $\dfrac{26}{55}$.
These approximations to fractions are used to obtain practical ratios for gearwheels or for a dividing head (used to give a required angular displacement).

6 Simple Equations

Expressions, equations and identities

$(3x - 5)$ is an example of an **algebraic expression**, whereas $3x - 5 = 1$ is an example of an **equation** (i.e. it contains an 'equals' sign)
An equation is simply a statement that two quantities are equal. For example,

$$1 \text{ m} = 1000 \text{ mm} \quad \text{or} \quad F = \tfrac{9}{5}C + 32 \quad \text{or} \quad y = mx + c$$

An **identity** is a relationship that is true for all values of the unknown, whereas an equation is only true for particular values of the unknown. For example, $3x - 5 = 1$ is an equation, since it is only true when $x = 2$, whereas $3x \equiv 8x - 5x$ is an identity since it is true for all values of x. (Note '$\equiv$' means 'is identical to').

Simple linear equations (or equations of the first degree) are those in which an unknown quantity is raised only to the power 1.
To '**solve an equation**' means 'to find the value of the unknown'.
Any arithmetic operation may be applied to an equation **as long as the equality of the equation is maintained**.
For example, to solve the equation $4x = 20$:
Dividing each side of the equation by 4 gives: $\frac{4x}{4} = \frac{20}{4}$
(Note that the same operation has been applied to both the left-hand side (LHS) and the right-hand side (RHS) of the equation so the equality has been maintained).
Cancelling gives: $\boldsymbol{x = 5}$, which is the solution to the equation.

In another example, to solve $\frac{2x}{5} = 6$:
The LHS is a fraction and this can be removed by multiplying both sides of the equation by 5. Hence $5\left(\frac{2x}{5}\right) = 5(6)$
Cancelling gives: $2x = 30$
Dividing both sides of the equation by 2 gives: $\frac{2x}{2} = \frac{30}{2}$ i.e. $\boldsymbol{x = 15}$
In another example, to solve $a - 5 = 8$:
Adding 5 to both sides of the equation gives:

$$a - 5 + 5 = 8 + 5$$

i.e. $\boldsymbol{a = 13}$

The result of the above procedure is to move the '-5' from the LHS of the original equation, across the equals sign, to the RHS, but the sign is changed to $+$.

In another example, to solve $6x + 1 = 2x + 9$:
In such equations the terms containing x are grouped on one side of the equation and the remaining terms grouped on the other side of the equation. Changing from one side of an equation to the other must be accompanied by a change of sign. Thus since $6x + 1 = 2x + 9$

then $6x - 2x = 9 - 1$

$$4x = 8$$

$$\frac{4x}{4} = \frac{8}{4}$$

i.e. $\boldsymbol{x = 2}$

In another example, to solve $4(2r - 3) - 2(r - 4) = 3(r - 3) - 1$:

Removing brackets gives: $8r - 12 - 2r + 8 = 3r - 9 - 1$

Rearranging gives: $8r - 2r - 3r = -9 - 1 + 12 - 8$

i.e. $3r = -6$

and $\boldsymbol{r} = \frac{-6}{3} = \boldsymbol{-2}$

Note that when there is only one fraction on each side of an equation, 'cross-multiplication' can be applied. **For example**, if $\frac{3}{x} = \frac{4}{5}$ then $(3)(5) = 4x$, from which, $\boldsymbol{x = \frac{15}{4}}$ **or** $\boldsymbol{3\frac{3}{4}}$

In another example, to solve $\sqrt{x} = 2$:

[$\sqrt{x} = 2$ is not a 'simple equation' since the power of x is $\frac{1}{2}$ i.e. $\sqrt{x} = x^{\frac{1}{2}}$; however, it is included here since it occurs often in practise].
Wherever square root signs are involved with the unknown quantity, both sides of the equation must be squared. Hence

$$(\sqrt{x})^2 = (2)^2$$

i.e. $\boldsymbol{x = 4}$

In another example, to solve $\left(\frac{\sqrt{b}+3}{\sqrt{b}}\right) = 2$:

To remove the fraction each term is multiplied by $\sqrt{b}$. Hence

$$\sqrt{b}\left(\frac{\sqrt{b}+3}{\sqrt{b}}\right) = \sqrt{b}(2)$$

Cancelling gives: $\sqrt{b} + 3 = 2\sqrt{b}$

Rearranging gives: $3 = 2\sqrt{b} - \sqrt{b} = \sqrt{b}$

Squaring both sides gives: $\mathbf{9 = b}$

In another example, to solve $x^2 = 25$:
This problem involves a square term and thus is not a simple equation (it is, in fact, a quadratic equation). However the solution of such an equation is often required and is therefore included here for completeness.
Whenever a square of the unknown is involved, the square root of both sides of the equation is taken. Hence

$$\sqrt{x^2} = \sqrt{25}$$

i.e. $x = 5$

However, $x = -5$ is also a solution of the equation because
$(-5) \times (-5) = +25$
Therefore, whenever the square root of a number is required there are always two answers, one positive, the other negative.
The solution of $x^2 = 25$ is thus written as: $\boldsymbol{x = \pm 5}$

Practical problems involving simple equations

For example, a copper wire has a length l of 1.5 km, a resistance R of 5 Ω and a resistivity of 17.2×10^{-6} Ωmm. To find the cross-sectional area, a, of the wire, given that $R = \rho l/a$:

Since $R = \rho l/a$ then $5\ \Omega = \dfrac{(17.2 \times 10^{-6}\ \Omega\ \text{mm})(1500 \times 10^3\ \text{mm})}{a}$

From the units given, a is measured in mm^2

Thus $\quad 5a = 17.2 \times 10^{-6} \times 1500 \times 10^3$

and $\quad a = \dfrac{17.2 \times 10^{-6} \times 1500 \times 10^3}{5}$

$$= \frac{17.2 \times 1500 \times 10^3}{10^6 \times 5} = \frac{17.2 \times 15}{10 \times 5} = 5.16$$

Hence the cross-sectional area of the wire is 5.16 mm²

In another example, the temperature coefficient of resistance α may be calculated from the formula $R_t = R_0(1 + \alpha t)$. To find α given $R_t = 0.928$, $R_0 = 0.8$ and $t = 40$:

Since $R_t = R_0(1 + \alpha t)$ then $\quad 0.928 = 0.8[1 + \alpha(40)]$

$$0.928 = 0.8 + (0.8)(\alpha(40))$$

$$0.928 - 0.8 = 32\alpha$$

$$0.128 = 32\alpha$$

Hence $\quad \boldsymbol{\alpha = \dfrac{0.128}{32} = 0.004}$

In another example, the distance s metres travelled in time t seconds is given by the formula $s = ut + \frac{1}{2}\text{a}t^2$, where u is the initial velocity in m/s and a is the acceleration in m/s^2. To find the acceleration of the body if it travels 168 m in 6 s, with an initial velocity of 10 m/s:

$s = ut + \frac{1}{2}at^2$, and $s = 168$, $u = 10$ and $t = 6$

Hence $\quad 168 = (10)(6) + \frac{1}{2}a(6)^2$

$$168 = 60 + 18a$$

$$168 - 60 = 18a$$

$$108 = 18a$$

$$a = \frac{108}{18} = 6$$

Hence the acceleration of the body is 6 m/s²

In another example, the extension x m of an aluminium tie bar of length l m and cross-sectional area A m^2 when carrying a load of F newtons is given by the modulus of elasticity $E = Fl/Ax$. To find the extension of the tie bar (in mm) if $E = 70 \times 10^9\ \text{N/m}^2$, $F = 20 \times 10^6$ N, $A = 0.1\ \text{m}^2$ and $l = 1.4$ m:
$E = Fl/Ax$, hence

$$70 \times 10^9\ \frac{\text{N}}{\text{m}^2} = \frac{(20 \times 10^6\ \text{N})(1.4\ \text{m})}{(0.1\ \text{m}^2)(x)}$$

(the unit of x is thus metres)

$$70 \times 10^9 \times 0.1 \times x = 20 \times 10^6 \times 1.4$$

$$x = \frac{20 \times 10^6 \times 1.4}{70 \times 10^9 \times 0.1}$$

Cancelling gives: $$x = \frac{2 \times 1.4}{7 \times 100} \text{ m}$$

$$= \frac{2 \times 1.4}{7 \times 100} \times 1000 \text{ mm}$$

Hence the extension of the tie bar, $x = 4$ mm

---

7 Simultaneous Equations

Introduction to simultaneous equations

Only one equation is necessary when finding the value of a **single unknown quantity** (as with simple equations in chapter 6). However, when an equation contains **two unknown quantities** it has an infinite number of solutions. When two equations are available connecting the same two unknown values then a unique solution is possible. Similarly, for three unknown quantities it is necessary to have three equations in order to solve for a particular value of each of the unknown quantities, and so on.
Equations that have to be solved together to find the unique values of the unknown quantities, which are true for each of the equations, are called **simultaneous equations**.
Two methods of solving simultaneous equations in two unknowns analytically are: (a) by **substitution**, and (b) by **elimination**.
(A graphical solution of simultaneous equations is shown in Chapter 35 and matrices and determinants are used in Chapter 44).

For example, to solve the following equations for x and y, (a) by substitution, and (b) by elimination:

$$x + 2y = -1 \qquad (1)$$

$$4x - 3y = 18 \qquad (2)$$

(a) **By substitution**
From equation (1): $x = -1 - 2y$
Substituting this expression for x into equation (2) gives:

$$4(-1 - 2y) - 3y = 18$$

This is now a simple equation in y.
Removing the bracket gives:

$$-4 - 8y - 3y = 18$$

$$-11y = 18 + 4 = 22$$

$$y = \frac{22}{-11} = -2$$

Substituting $y = -2$ into equation (1) gives:

$$x + 2(-2) = -1$$

$$x - 4 = -1$$

$$x = -1 + 4 = 3$$

Thus $x = 3$ and $y = -2$ is the solution to the simultaneous equations

(b) **By elimination**

$$x + 2y = -1 \quad (1)$$

$$4x - 3y = 18 \quad (2)$$

If equation (1) is multiplied throughout by 4 the coefficient of x will be the same as in equation (2), giving:

$$4x + 8y = -4 \quad (3)$$

Subtracting equation (3) from equation (2) gives:

$$\begin{array}{rcl} 4x - 3y = 18 & & (2) \\ \underline{4x + 8y = -4} & & (3) \\ \underline{0 - 11y = 22} & & \end{array}$$

Hence $y = \dfrac{22}{-11} = -2$

(Note, in the above subtraction, $18 - -4 = 18 + 4 = 22$).
Substituting $y = -2$ into either equation (1) or equation (2) will give $x = 3$ as in method (a). The solution $\boldsymbol{x = 3}$, $\boldsymbol{y = -2}$ is the only pair of values that satisfies both of the original equations.

In another example, to solve

$$7x - 2y = 26 \quad (1)$$

$$6x + 5y = 29 \quad (2)$$

When equation (1) is multiplied by 5 and equation (2) by 2 the coefficients of y in each equation are numerically the same, i.e. 10, but are of opposite sign.

5 × equation (1) gives: $35x - 10y = 130$ (3)
2 × equation (2) gives: $\underline{12x + 10y = 58}$ (4)
Adding equation (3) and (4) gives: $\underline{47x + 0 = 188}$

Hence $x = \dfrac{188}{47} = 4$

[Note that when the signs of common coefficients are **different** the two equations are **added**, and when the signs of common coefficients are the **same** the two equations are **subtracted**]
Substituting $x = 4$ in equation (1) gives:

$$7(4) - 2y = 26$$

$$28 - 2y = 26$$

$$28 - 26 = 2y$$

$$2 = 2y$$

Hence $y = 1$

Thus the solution is $x = 4$, $y = 1$, since these values maintain the equality when substituted in both equations.

Practical problems involving simultaneous equations

There are a number of situations in engineering and science where the solution of simultaneous equations is required.

For example, the law connecting friction F and load L for an experiment is of the form $F = aL + b$, where a and b are constants. When $F = 5.6$, $L = 8.0$ and when $F = 4.4$, $L = 2.0$. To find the values of a and b and the value of F when $L = 6.5$:

Substituting $F = 5.6, L = 8.0$ into $F = aL + b$ gives:

$$5.6 = 8.0a + b \qquad (1)$$

Substituting $F = 4.4, L = 2.0$ into $F = aL + b$ gives:

$$4.4 = 2.0a + b \qquad (2)$$

Subtracting equation (2) from equation (1) gives:

$$1.2 = 6.0a$$

$$\boldsymbol{a} = \frac{1.2}{6.0} = \mathbf{\frac{1}{5}}$$

Substituting $a = \frac{1}{5}$ into equation (1) gives:

$$5.6 = 8.0\left(\frac{1}{5}\right) + b$$

$$5.6 = 1.6 + b$$

$$5.6 - 1.6 = b$$

i.e. $$\boldsymbol{b = 4}$$

Hence $a = \frac{1}{5}$ and $b = 4$

When, say, $\mathbf{L = 6.5}$, $F = aL + b = \frac{1}{5}(6.5) + 4 = 1.3 + 4$, i.e. $\mathbf{F = 5.30}$

In another example, the resistance $R\ \Omega$ of a length of wire at t°C is given by $R = R_0(1 + \alpha t)$, where R_0 is the resistance at 0°C and α is the temperature coefficient of resistance in /°C. To find the values of α and R_0 if $R = 30\ \Omega$ at 50°C and $R = 35\ \Omega$ at 100°C:

Substituting $R = 30, t = 50$ into $R = R_0(1 + \alpha t)$ gives:

$$30 = R_0(1 + 50\alpha) \qquad (1)$$

Substituting $R = 35, t = 100$ into $R = R_0(1 + \alpha t)$ gives:

$$35 = R_0(1 + 100\alpha) \qquad (2)$$

Although these equations may be solved by the conventional substitution method, an easier way is to eliminate R_0 by division. Thus, dividing equation (1) by equation (2) gives:

$$\frac{30}{35} = \frac{R_0(1 + 50\alpha)}{R_0(1 + 100\alpha)} = \frac{1 + 50\alpha}{1 + 100\alpha}$$

'Cross-multiplying' gives:

$$30(1 + 100\alpha) = 35(1 + 50\alpha)$$

$$30 + 3000\alpha = 35 + 1750\alpha$$

$$3000\alpha - 1750\alpha = 35 - 30$$

$$1250\alpha = 5$$

i.e. $\boldsymbol{\alpha} = \dfrac{5}{1250} = \dfrac{1}{250}$ or **0.004**

Substituting $\alpha = \dfrac{1}{250}$ into equation (1) gives:

$$30 = R_0\left\{1 + (50)\left(\frac{1}{250}\right)\right\}$$

$$30 = R_0(1.2)$$

$$\mathbf{R_0} = \frac{30}{1.2} = \mathbf{25}$$

Thus the solution is $\boldsymbol{\alpha}$ = 0.004/°C and $\mathbf{R_0}$ = 25 Ω

---

8 Transposition of Formulae

When a symbol other than the subject is required to be calculated it is usual to rearrange the formula to make a new subject. This rearranging process is called **transposing the formula** or **transposition**.
The rules used for transposition of formulae are the same as those used for the solution of simple equations (see Chapter 6)—basically, **that the equality of an equation must be maintained**.

For example, to transpose $p = q + r + s$ to make r the subject:
The aim is to obtain r on its own on the left-hand side (LHS) of the equation. Changing the equation around so that r is on the LHS gives:

$$q + r + s = p \qquad (1)$$

Subtracting $(q + s)$ from both sides of the equation gives:

$$q + r + s - (q + s) = p - (q + s)$$

Thus $q + r + s - q - s = p - q - s$

i.e. $\boldsymbol{r = p - q - s}$ (2)

It is shown with simple equations, that a quantity can be moved from one side of an equation to the other with an appropriate change of sign. Thus equation (2) follows immediately from equation (1) above.

In another example, to transpose $v = f\lambda$ to make λ the subject:

Rearranging gives: $f\lambda = v$

Dividing both sides by f gives: $\frac{f\lambda}{f} = \frac{v}{f}$, i.e. $\boldsymbol{\lambda = \frac{v}{f}}$

In another example, to rearrange $I = \frac{V}{R}$ for V:

Rearranging gives: $\frac{V}{R} = I$

Multiplying both sides by R gives:

$$R\left(\frac{V}{R}\right) = R(I)$$

Hence $\boldsymbol{V = IR}$

In another example, to rearrange the formula $R = \frac{\rho l}{a}$ to make a the subject:

Rearranging gives: $\frac{\rho l}{a} = R$

Multiplying both sides by a gives:

$$a\left(\frac{\rho l}{a}\right) = a(R) \quad \text{i.e.} \quad \rho l = aR$$

Rearranging gives: $aR = \rho l$

Dividing both sides by R gives:

$$\frac{aR}{R} = \frac{\rho l}{R}$$

i.e. $\boldsymbol{a = \frac{\rho l}{R}}$

In another example, the final length, l_2 of a piece of wire heated through θ°C is given by the formula $l_2 = l_1(1 + \alpha\theta)$. Making the coefficient of expansion, α, the subject:

Rearranging gives: $l_1(1 + \alpha\theta) = l_2$

Removing the bracket gives: $l_1 + l_1\alpha\theta = l_2$

Rearranging gives: $l_1\alpha\theta = l_2 - l_1$

Dividing both sides by $l_{1\theta}$ gives: $\frac{l_1\alpha\theta}{l_1\theta} = \frac{l_2 - l_1}{l_1\theta}$

i.e. $\boldsymbol{\alpha = \frac{l_2 - l_1}{l_1\theta}}$

In another example, a formula for kinetic energy is $k = \frac{1}{2}mv^2$. To transpose the formula to make v the subject:

Rearranging gives: $\frac{1}{2}mv^2 = k$

Whenever the prospective new subject is a squared term, that term is isolated on the LHS, and then the square root of both sides of the equation is taken.

Multiplying both sides by 2 gives: $mv^2 = 2k$

Dividing both sides by m gives: $$\frac{mv^2}{m} = \frac{2k}{m}$$

i.e. $$v^2 = \frac{2k}{m}$$

Taking the square root of both sides gives:

$$\sqrt{v^2} = \sqrt{\left(\frac{2k}{m}\right)}$$

i.e. $$\boldsymbol{v = \sqrt{\left(\frac{2k}{m}\right)}}$$

In another example, the impedance of an a.c. circuit is given by $Z = \sqrt{R^2 + X^2}$. To make the reactance, X, the subject:

Rearranging gives: $$\sqrt{R^2 + X^2} = Z$$

Squaring both sides gives: $$R^2 + X^2 = Z^2$$

Rearranging gives: $$X^2 = Z^2 - R^2$$

Taking the square root of both sides gives: $\boldsymbol{X = \sqrt{Z^2 - R^2}}$

In another example, to transpose the formula $p = \dfrac{a^2x + a^2y}{r}$ to make a the subject:

Rearranging gives: $$\frac{a^2x^2 + a^2y}{r} = p$$

Multiplying both sides by r gives: $$a^2x + a^2y = rp$$

Factorising the LHS gives: $$a^2(x + y) = rp$$

Dividing both sides by $(x + y)$ gives: $$\frac{a^2(x + y)}{(x + y)} = \frac{rp}{(x + y)}$$

i.e. $$a^2 = \frac{rp}{(x + y)}$$

Taking the square root of both sides gives: $$\boldsymbol{a = \sqrt{\left(\frac{rp}{x + y}\right)}}$$

In another example, expressing p in terms of D, d and f given that $\dfrac{D}{d} = \sqrt{\left(\dfrac{f + p}{f - p}\right)}$:

Rearranging gives: $$\sqrt{\left(\frac{f + p}{f - p}\right)} = \frac{D}{d}$$

Squaring both sides gives: $$\left(\frac{f + p}{f - p}\right) = \frac{D^2}{d^2}$$

Cross-multiplying, i.e. multiplying each term by $d^2(f - p)$, gives:

$$d^2(f + p) = D^2(f - p)$$

Removing brackets gives: $\quad d^2 f + d^2 p = D^2 f - D^2 p$

Rearranging, to obtain terms in p on the LHS gives:

$$d^2 p + D^2 p = D^2 f - d^2 f$$

Factorising gives: $\quad p(d^2 + D^2) = f(D^2 - d^2)$

Dividing both sides by $(d^2 + D^2)$ gives:

$$\boldsymbol{p = \frac{f(D^2 - d^2)}{(d^2 + D^2)}}$$

---

9 Quadratic Equations

Introduction to quadratic equations

As stated in chapter 6, an **equation** is a statement that two quantities are equal and to '**solve an equation**' means 'to find the value of the unknown'. The value of the unknown is called the **root** of the equation.
A **quadratic equation** is one in which the highest power of the unknown quantity is 2. For example, $x^2 - 3x + 1 = 0$ is a quadratic equation.
There are four methods of **solving quadratic equations**.

These are: (i) by factorisation (where possible)
(ii) by 'completing the square'
(iii) by using the 'quadratic formula'
or (iv) graphically (see Chapter 35)

Solution of quadratic equations by factorisation

Multiplying out $(2x + 1)(x - 3)$ gives $2x^2 - 6x + x - 3$, i.e. $2x^2 - 5x - 3$. The reverse process of moving from $2x^2 - 5x - 3$ to $(2x + 1)(x - 3)$ is called **factorising**.
If the quadratic expression can be factorised this provides the simplest method of solving a quadratic equation.

For example, if $2x^2 - 5x - 3 = 0$, then, by factorising:
$(2x + 1)(x - 3) = 0$
Hence either $\quad (2x + 1) = 0 \quad$ i.e. $x = -\frac{1}{2}$

or $\quad (x - 3) = 0 \quad$ i.e. $x = 3$

The technique of factorising is often one of 'trial and error'

In another example, to solve the equations $x^2 + 2x - 8 = 0$ by factorisation:

The factors of x^2 are x and x. These are placed in brackets thus: $(x\quad)(x\quad)$
The factors of -8 are $+8$ and -1, or -8 and $+1$, or $+4$ and -2, or -4 and $+2$.
The only combination to give a middle term of $+2x$ is $+4$ and -2, i.e.

$$x^2 + 2x - 8 = (x + 4)(x - 2)$$

(Note that the product of the two inner terms added to the product of the two outer terms must equal the middle term, $+2x$ in this case.)
The quadratic equation $x^2 + 2x - 8 = 0$ thus becomes $(x + 4)(x - 2) = 0$.
Since the only way that this can be true is for either the first or the second, or both factors to be zero, then either $(x + 4) = 0$ i.e. $x = -4$
or $(x - 2) = 0$ i.e. $x = 2$

Hence the roots of $x^2 + 2x - 8 = 0$ are $x = -4$ and 2

In another example, to determine the roots of $x^2 - 6x + 9 = 0$ by factorisation:
Since $x^2 - 6x + 9 = 0$ then $(x - 3)(x - 3) = 0$, i.e. $(x - 3)^2 = 0$ (the left-hand side is known as **a perfect square**). Hence $\mathbf{x = 3}$ is the only root of the equation $x^2 - 6x + 9 = 0$.

In another example, to determine the roots of $4x^2 - 25 = 0$ by factorisation:
$4x^2 - 25 = 0$ (the left-hand side is **the difference of two squares**, $(2x)^2$ and $(5)^2$).

Thus $\quad (2x + 5)(2x - 5) = 0$

Hence either $\quad (2x + 5) = 0 \quad$ i.e. $\quad \mathbf{x = -\frac{5}{2}}$

or $\quad (2x - 5) = 0 \quad$ i.e. $\quad \mathbf{x = \frac{5}{2}}$

In another example, the roots of a quadratic equation are $\frac{1}{3}$ and -2. To determine the equation in x:
If the roots of a quadratic equation are α and β then $(x - \alpha)(x - \beta) = 0$.
Hence if $\alpha = \frac{1}{3}$ and $\beta = -2$, then

$$\left(x - \frac{1}{3}\right)(x - (-2)) = 0$$

$$\left(x - \frac{1}{3}\right)(x + 2) = 0$$

$$x^2 - \frac{1}{3}x + 2x - \frac{2}{3} = 0$$

$$x^2 + \frac{5}{3}x - \frac{2}{3} = 0$$

Hence $\quad \mathbf{3x^2 + 5x - 2 = 0}$

Solution of quadratic equations by 'completing the square'

An expression such as x^2 or $(x + 2)^2$ or $(x - 3)^2$ is called a perfect square.

If $x^2 = 3$ then $x = \pm\sqrt{3}$

If $(x + 2)^2 = 5$ then $x + 2 = \pm\sqrt{5}$ and $x = -2 \pm \sqrt{5}$

If $(x - 3)^2 = 8$ then $x - 3 = \pm\sqrt{8}$ and $x = 3 \pm \sqrt{8}$

Hence if a quadratic equation can be rearranged so that one side of the equation is a perfect square and the other side of the equation is a number, then the solution of the equation is readily obtained by taking the square roots of each side as in the above examples. The process of rearranging one side of a quadratic equation into a perfect square before solving is called **'completing the square'**.

$$(x+a)^2 = x^2 + 2ax + a^2$$

Thus in order to make the quadratic expression $x^2 + 2ax$ into a perfect square it is necessary to add (half the coefficient of $x)^2$ i.e. $\left(\frac{2a}{2}\right)^2$ or a^2

For example, $x^2 + 3x$ becomes a perfect square by adding $\left(\frac{3}{2}\right)^2$, i.e.

$$x^2 + 3x + \left(\tfrac{3}{2}\right)^2 = \left(x + \tfrac{3}{2}\right)^2$$

In another example, to solve $2x^2 + 5x = 3$ by 'completing the square':
The procedure is as follows:

1. Rearrange the equation so that all terms are on the same side of the equals sign (and the coefficient of the x^2 term is positive). Hence $2x^2 + 5x - 3 = 0$
2. Make the coefficient of the x^2 term unity. In this case this is achieved by dividing throughout by 2.

   Hence $\quad \dfrac{2x^2}{2} + \dfrac{5x}{2} - \dfrac{3}{2} = 0$

   i.e. $\quad x^2 + \dfrac{5}{2}x - \dfrac{3}{2} = 0$

3. Rearrange the equations so that the x^2 and x terms are on one side of the equals sign and the constant is on the other side. Hence $x^2 + \frac{5}{2}x = \frac{3}{2}$
4. Add to both sides of the equation (half the coefficient of $x)^2$. In this case the coefficient of x is $\frac{5}{2}$. Half the coefficient squared is therefore $\left(\frac{5}{4}\right)^2$. Thus

   $$x^2 + \tfrac{5}{2}x + \left(\tfrac{5}{4}\right)^2 = \tfrac{3}{2} + \left(\tfrac{5}{4}\right)^2$$

   The LHS is now a perfect square, i.e.

   $$\left(x + \tfrac{5}{4}\right)^2 = \tfrac{3}{2} + \left(\tfrac{5}{4}\right)^2$$

5. Evaluate the RHS. Thus $\left(x + \dfrac{5}{4}\right)^2 = \dfrac{3}{2} + \dfrac{25}{16} = \dfrac{24+25}{16} = \dfrac{49}{16}$
6. Taking the square root of both sides of the equation (remembering that the square root of a number gives a $\pm$ answer). Thus

   $$\sqrt{\left(x + \frac{5}{4}\right)^2} = \sqrt{\left(\frac{49}{16}\right)}$$

   i.e. $\quad x + \frac{5}{4} = \pm\frac{7}{4}$

7. Solve the simple equation. Thus $x = -\frac{5}{4} \pm \frac{7}{4}$

   i.e. $x = -\frac{5}{4} + \frac{7}{4} = \frac{2}{4} = \frac{1}{2}$

   and $x = -\frac{5}{4} - \frac{7}{4} = -\frac{12}{4} = -3$

   Hence $\boldsymbol{x = \frac{1}{2}}$ **or $\boldsymbol{-3}$** are the roots of the equation $2x^2 + 5x = 3$.

Solution of quadratic equations by formula

If $ax^2 + bx + c = 0$ then $\boxed{x = \dfrac{-b \pm \sqrt{b^2 - 4ac}}{2a}}$

This is known as the **quadratic formula**

For example, to solve $3x^2 - 11x - 4 = 0$ by using the quadratic formula: Comparing $3x^2 - 11x - 4 = 0$ with $ax^2 + bx + c = 0$ gives $a = 3$, $b = -11$ and $c = -4$.

$$\text{Hence,} \quad x = \frac{-(-11) \pm \sqrt{(-11)^2 - 4(3)(-4)}}{2(3)}$$

$$= \frac{+11 \pm \sqrt{121 + 48}}{6} = \frac{11 \pm \sqrt{169}}{6}$$

$$= \frac{11 \pm 13}{6} = \frac{11 + 13}{6} \text{ or } \frac{11 - 13}{6}$$

$$\textbf{Hence,} \quad x = \frac{24}{6} = \mathbf{4} \text{ or } \frac{-2}{6} = -\frac{\mathbf{1}}{\mathbf{3}}$$

Practical problems involving quadratic equations

There are many **practical problems** where a quadratic equation has first to be obtained, from given information, before it is solved.

For example, the height s metres of a mass projected vertically upwards at time t seconds is $s = ut - \frac{1}{2}gt^2$. To determine how long the mass will take after being projected to reach a height of 16 m (a) on the ascent and (b) on the descent, when $u = 30$ m/s and $g = 9.81$ m/s^2:
When height $s = 16$ m, $16 = 30t - \frac{1}{2}(9.81)t^2$
i.e. $4.905t^2 - 30t + 16 = 0$
Using the quadratic formula:

$$t = \frac{-(-30) \pm \sqrt{(-30)^2 - 4(4.905)(16)}}{2(4.905)}$$

$$= \frac{30 \pm \sqrt{586.1}}{9.81} = \frac{30 \pm 24.21}{9.81} = 5.53 \text{ or } 0.59$$

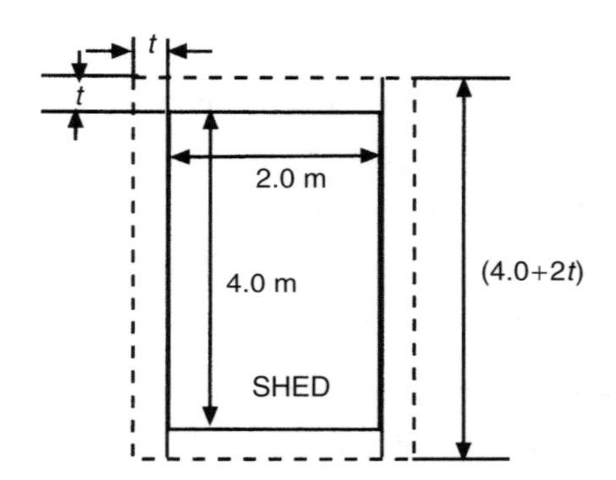

Figure 9.1

Hence the mass will reach a height of 16 m after 0.59 s on the ascent and after 5.53 s on the descent.

In another example, a shed is 4.0 m long and 2.0 m wide. A concrete path of constant width is laid all the way around the shed and the area of the path is 9.50 m^2. To calculate its width, to the nearest centimetre:
Figure 9.1 shows a plan view of the shed with its surrounding path of width t metres. Area of path $= 2(2.0 \times t) + 2t(4.0 + 2t)$

i.e. $\quad 9.50 = 4.0t + 8.0t + 4t^2$

or $\quad 4t^2 + 12.0t - 9.50 = 0$

Hence $\quad t = \dfrac{-(12.0) \pm \sqrt{(12.0)^2 - 4(4)(-9.50)}}{2(4)}$

$$= \frac{-12.0 \pm \sqrt{296.0}}{8} = \frac{-12.0 \pm 17.20465}{8}$$

Hence $t = 0.6506$ m or -3.65058 m
Neglecting the negative result which is meaningless, the width of the path, $\boldsymbol{t = 0.651}$ **m** or **65 cm**, correct to the nearest centimetre.

The solution of linear and quadratic equations simultaneously

Sometimes a linear equation and a quadratic equation need to be solved simultaneously.

For example, to determine the values of x and y which simultaneously satisfy the equations:

$$y = 5x - 4 - 2x^2 \quad \text{and} \quad y = 6x - 7$$

For a simultaneous solution the values of y must be equal, hence the RHS of each equation is equated. Thus

$$5x - 4 - 2x^2 = 6x - 7$$

Rearranging gives: $5x - 4 - 2x^2 - 6x + 7 = 0$

i.e. $-x + 3 - 2x^2 = 0$

or $2x^2 + x - 3 = 0$

Factorising gives: $(2x + 3)(x - 1) = 0$

i.e. $x = -\frac{3}{2}$ or $x = 1$

In the equation $y = 6x - 7$,

when $x = -\frac{3}{2}$, $y = 6\left(-\frac{3}{2}\right) - 7 = -16$

and when $x = 1$, $y = 6 - 7 = -1$

[Checking the result in $y = 5x - 4 - 2x^2$:

$$\text{when } x = -\frac{3}{2}, \; y = 5\left(-\frac{3}{2}\right) - 4 - 2\left(-\frac{3}{2}\right)^2$$

$$= -\frac{15}{2} - 4 - \frac{9}{2} = -16$$

as above; and when $x = 1$, $y = 5 - 4 - 2 = -1$ as above]

Hence the simultaneous solutions occur when $\boldsymbol{x = -\frac{3}{2}, y = -16}$ and when $\boldsymbol{x = 1, y = -1}$

---

10 Inequalities

Introduction to inequalities

An **inequality** is any expression involving one of the symbols $<$, $>$, $\leq$ or $\geq$

$p < q$ means p is less than q

$p > q$ means p is greater than q

$p \leq q$ means p is less than or equal to q

$p \geq q$ means p is greater than or equal to q

Some simple rules

(i) When a quantity is **added or subtracted** to both sides of an inequality, the inequality still remains. For example,

if $p < 3$ then $p + 2 < 3 + 2$ (adding 2 to both sides)

and $p - 2 < 3 - 2$ (subtracting 2 from both sides)

(ii) When **multiplying or dividing** both sides of an inequality by a **positive** quantity, say 5, the inequality **remains the same**. For example,

$$\text{if } p > 4 \text{ then } 5p > 20 \quad \text{and} \quad \frac{p}{5} > \frac{4}{5}$$

(iii) When **multiplying or dividing** both sides of an inequality by a **negative** quantity, say -3, **the inequality is reversed**. For example,

$$\text{if } p > 1 \text{ then } -3p < -3 \quad \text{and} \quad \frac{p}{-3} < \frac{1}{-3}$$

(Note $>$ has changed to $<$)

To **solve an inequality** means finding all the values of the variable for which the inequality is true.

Simple inequalities

For example, to solve the following inequalities:
(a) $3 + x > 7$ (b) $z - 2 \geq 5$
(a) Subtracting 3 from both sides of the inequality $3 + x > 7$ gives:

$$3 + x - 3 > 7 - 3 \text{ i.e. } \boldsymbol{x > 4}$$

Hence all values of x greater than 4 satisfy the inequality
(b) Adding 2 to both sides of the inequality $z - 2 \geq 5$ gives:

$$z - 2 + 2 \geq 5 + 2 \text{ i.e. } \boldsymbol{z \geq 7}$$

Hence all values of z equal to or greater than 7 satisfy the inequality

In another example, to solve the inequality $4x + 1 > x + 5$:
Subtracting 1 from both sides of the inequality $4x + 1 > x + 5$ gives:

$$4x > x + 4$$

Subtracting x from both sides of the inequality $4x > x + 4$ gives:

$$3x > 4$$

Dividing both sides of the inequality $3x > 4$ by 3 gives:

$$\boldsymbol{x > \frac{4}{3}}$$

Hence all values of x greater than $\frac{4}{3}$ satisfy the inequality $4x + 1 > x + 5$

Inequalities involving a modulus

The **modulus** of a number is the size of the number, regardless of sign; it is denoted by vertical lines enclosing the number.
For example, $|4| = 4$ and $|-4| = 4$ (the modulus of a number is never negative).

The inequality $|t| < 1$ means that all numbers whose actual size, regardless of sign, is less than 1, i.e. any value between -1 and $+1$.
Thus $|t| < 1$ **means $-1 < t < 1$**
Similarly, $|x| > 3$ means all numbers whose actual size, regardless of sign, is greater than 3, i.e. any value greater than 3 and any value less than -3.
Thus $|x|$ **> 3 means $x > 3$ and $x < -3$**

For example, to solve the following inequality $|3x + 1| < 4$
Since $|3x + 1| < 4$ then $-4 < 3x + 1 < 4$

$$-4 < 3x + 1 \text{ becomes } -5 < 3x \text{ and } -\tfrac{5}{3} < x$$

$$3x + 1 < 4 \text{ becomes } 3x < 3 \text{ and } x < 1$$

Hence these two results together become $-\frac{5}{3} < x < 1$ and mean that the inequality $|3x + 1| < 4$ is satisfied for any value of x greater than $-\frac{5}{3}$ but less than 1.

Inequalities involving quotients

If $\frac{p}{q} > 0$ then $\frac{p}{q}$ must be a positive value.
For $\frac{p}{q}$ to be positive, **either** p is positive **and** q is positive
or p is negative **and** q is negative
i.e. $\frac{+}{+} = +$ and $\frac{-}{-} = +$
If $\frac{p}{q} < 0$ then $\frac{p}{q}$ must be a negative value.
For $\frac{p}{q}$ to be negative, **either** p is positive **and** q is negative
or p is negative **and** q is positive
i.e. $\frac{+}{-} = -$ and $\frac{-}{+} = -$

For example, to solve the inequality $\dfrac{t+1}{3t-6} > 0$:

Since $\dfrac{t+1}{3t-6} > 0$ then $\dfrac{t+1}{3t-6}$ must be positive.

For $\dfrac{t+1}{3t-6}$ to be positive, **either** (i) $t + 1 > 0$ **and** $3t - 6 > 0$

or (ii) $t + 1 < 0$ **and** $3t - 6 < 0$

(i) If $t + 1 > 0$ then $t > -1$
and if $3t - 6 > 0$ then $3t > 6$ and $t > 2$
Both of the inequalities $t > -1$ and $t > 2$ are only true when $t > 2$, i.e. the fraction $\dfrac{t+1}{3t-6}$ is positive when $\boldsymbol{t > 2}$

(ii) If $t + 1 < 0$ then $t < -1$
and if $3t - 6 < 0$ then $3t < 6$ and $t < 2$

Both of the inequalities $t < -1$ and t < 2 are only true when $t < -1$, i.e. the fraction $\frac{t+1}{3t-6}$ is positive when $\boldsymbol{t < -1}$

Summarising, $\frac{t+1}{3t-6} > 0$ when $\boldsymbol{t > 2}$ or $\boldsymbol{t < -1}$

Inequalities involving square functions

The following two general rules apply when inequalities involve square functions:

(i) **if $x^2 > k$ then $x > \sqrt{k}$ or $x < -\sqrt{k}$** (1)

(ii) **if $x^2 < k$ then $-\sqrt{k} < x < \sqrt{k}$** (2)

For example, to solve the inequality $t^2 > 9$:
Since $t^2 > 9$ then $t^2 - 9 > 0$
i.e. $(t+3)(t-3) > 0$ by factorising
For $(t+3)(t-3)$ to be positive,

either (i) $(t+3) > 0$ **and** $(t-3) > 0$

or (ii) $(t+3) < 0$ **and** $(t-3) < 0$

(i) If $(t+3) > 0$ then $t > -3$
and if $(t-3) > 0$ then $t > 3$
Both of these are true only when $\boldsymbol{t > 3}$

(ii) If $(t+3) < 0$ then $t < -3$
and if $(t-3) < 0$ then $t < 3$
Both of these are true only when $\boldsymbol{t < -3}$
Summarising, $t^2 > 9$ when $\boldsymbol{t > 3}$ **or** $\boldsymbol{t < -3}$ which demonstrates rule (1) above

In another example, to solve the inequality $t^2 < 9$:
Since $t^2 < 9$ then $t^2 - 9 < 0$
i.e. $(t+3)(t-3) < 0$ by factorising
For $(t+3)(t-3)$ to be negative,

either (i) $(t+3) > 0$ **and** $(t-3) < 0$

or (ii) $(t+3) < 0$ **and** $(t-3) > 0$

(i) If $(t+3) > 0$ then $t > -3$
and if $(t-3) < 0$ then $t < 3$
Hence (i) is satisfied when $t > -3$ and $t < 3$ which may be written a $\boldsymbol{-3 < t < 3}$

(ii) If $(t+3) < 0$ then $t < -3$
and if $(t-3) > 0$ then $t > 3$
It is not possible to satisfy both $t < -3$ and $t > 3$, thus no values of t satisfies (ii)

Summarising, $t^2 < 9$ when $\boldsymbol{-3 < t < 3}$ which means that all values of t between -3 and $+3$ will satisfy the inequality, satisfying rule (2) above.

Quadratic inequalities

Inequalities involving quadratic expressions are solved using either **factorisation** or '**completing the square**'.

For example, $x^2 - 2x - 3$ is factorised as $(x + 1)(x - 3)$

and $6x^2 + 7x - 5$ is factorised as $(2x - 1)(3x + 5)$

If a quadratic expression does not factorise, then the technique of 'completing the square' is used. In general, the procedure for $x^2 + bx + c$ is:

$$x^2 + bx + c \equiv \left(x + \frac{b}{2}\right)^2 + c - \left(\frac{b}{2}\right)^2$$

For example, $x^2 + 4x - 7$ does not factorise; completing the square gives:

$$x^2 + 4x - 7 \equiv (x + 2)^2 - 7 - 2^2 \equiv (x + 2)^2 - 11$$

Similarly, $x^2 - 6x - 5 \equiv (x - 3)^2 - 5 - 3^2 \equiv (x - 3)^2 - 14$

For example, to solve the inequality $x^2 + 2x - 3 > 0$:
Since $x^2 + 2x - 3 > 0$ then $(x - 1)(x + 3) > 0$ by factorising
For the product $(x - 1)(x + 3)$ to be positive,

either (i)$(x - 1) > 0$ and $(x + 3) > 0$

or (ii)$(x - 1) < 0$ and $(x + 3) < 0$

(i) Since $(x - 1) > 0$ then $x > 1$
and since $(x + 3) > 0$ then $x > -3$
Both of these inequalities are satisfied only when $\boldsymbol{x > 1}$
(ii) Since $(x - 1) < 0$ then $x < 1$
and since $(x + 3) < 0$ then $x < -3$
Both of these inequalities are satisfied only when $\boldsymbol{x < -3}$

Summarising, $x^2 + 2x - 3 > 0$ is satisfied when either $\boldsymbol{x > 1}$ or $\boldsymbol{x < -3}$

In another example, to solve the inequality $y^2 - 8y - 10 \geq 0$:
$y^2 - 8y - 10 \equiv (y - 4)^2 - 10 - 4^2 \equiv (y - 4)^2 - 26$
$y^2 - 8y - 10$ does not factorise; completing the square gives:

The inequality thus becomes: $(y - 4)^2 - 26 \geq 0$

or $(y - 4)^2 \geq 26$

From equation 1, $(y - 4) \geq \sqrt{26}$

or $(y - 4) \leq -\sqrt{26}$

from which, $\boldsymbol{y \geq 4 + \sqrt{26}}$

or $\boldsymbol{y \leq 4 - \sqrt{26}}$

Hence $y^2 - 8y - 10 \geq 0$ is satisfied when $\boldsymbol{y \geq 9.10}$ **or** $\boldsymbol{y \leq -1.10}$ correct to 2 decimal places.

Regions

A **region** is a set of points on a graph that satisfies an inequality.
For example, in Figure 10.1(a), the shaded region is defined as $x > 3$ where the straight line $x = 3$ is shown as a broken line, and in Figure 10.1(b), the shaded region is defined as $x \geq 3$, where the straight line $x = 3$ is shown as a solid line. The region $x \geq 3$ includes all the points on the line $x = 3$ and to the right of it.
Similarly, in Figure 10.2, the shaded region is defined as $y \leq -2$
In Figure 10.3, the line $x + y = 4$ is shown as a broken line (note, if $x + y = 4$, then $y = -x + 4$, which is a straight line of gradient -1 and y-axis intercept 4); the shaded region is defined as $x + y < 4$

For example, to show on Cartesian axes the following regions: (a) $y > 3x - 2$ (b) $x + 2y < 8$

(a) Figure 10.4 shows the straight line $y = 3x - 2$ and the shaded region defines the inequality $y > 3x - 2$
(b) Figure 10.5 shows the straight line $x + 2y = 8$ (i.e. $2y = -x + 8$ or $y = -\frac{1}{2}x + 4$ which is a straight line of gradient $-\frac{1}{2}$ and y-axis intercept 4) as a broken line and the shaded region defines the inequality $x + 2y < 8$ (As a check, take any point, say, $x = 1$, $y = 1$; then $x + 2y = 1 + 2 = 3$ which is less than 8. The shaded area indicates all the points where $x + 2y < 8$).

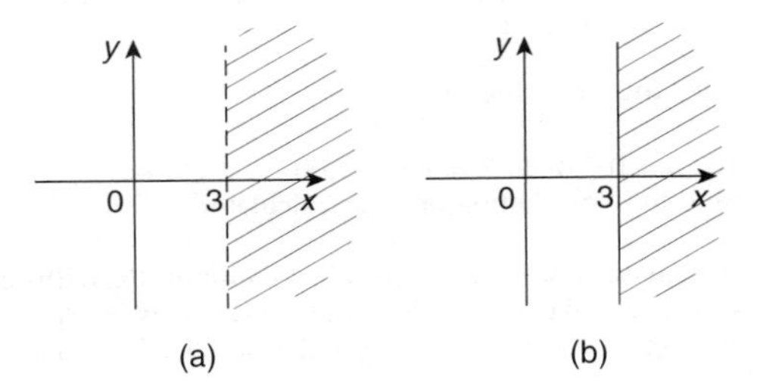

Figure 10.1

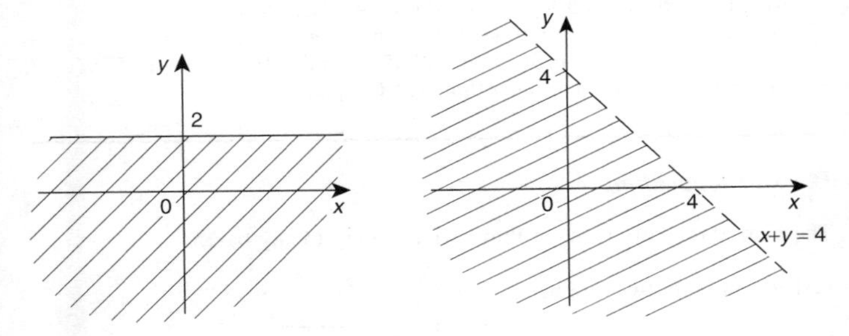

Figure 10.2

Figure 10.3

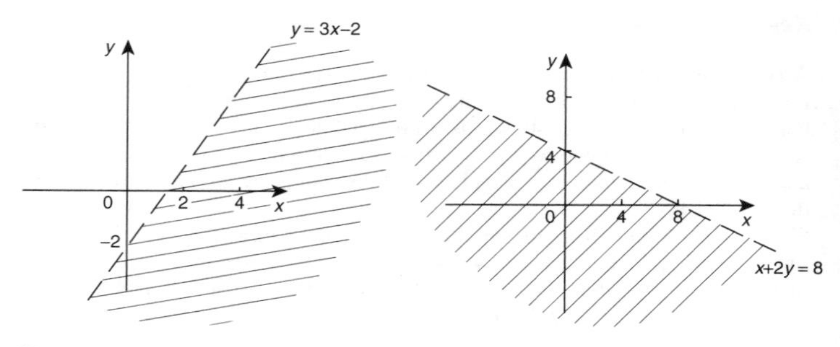

Figure 10.4 **Figure 10.5**

11 Logarithms

Introduction to logarithms

With the use of calculators firmly established, logarithmic tables are now rarely used for calculation. However, the theory of logarithms is important, for there are several scientific and engineering laws that involve the rules of logarithms.

If a number y can be written in the form a^x, then the index x is called the 'logarithm of y to the base of a',

i.e. $\boxed{\textbf{if } y = a^x \textbf{ then } x = \log_a y}$

Thus, since $1000 = 10^3$, then $3 = \log_{10} 1000$
Check this using the 'log' button on your calculator.

(a) Logarithms having a base of 10 are called **common logarithms** and $\log_{10}$ is usually abbreviated to lg. The following values may be checked by using a calculator: $\lg 17.9 = 1.2528\ldots$, $\lg 462.7 = 2.6652\ldots$ and $\lg 0.0173 = -1.7619\ldots$

(b) Logarithms having a base of e (where 'e' is a mathematical constant approximately equal to 2.7183) are called **hyperbolic**, **Napierian** or **natural logarithms**, and $\log_e$ is usually abbreviated to ln.
The following values may be checked by using a calculator:
$\ln 3.15 = 1.1474\ldots$, $\ln 362.7 = 5.8935\ldots$ and $\ln 0.156 = -1.8578\ldots$
For more on Napierian logarithms see Chapter 12.

Laws of logarithms

There are three laws of logarithms, which apply to any base:

(i) To multiply two numbers:

$$\boxed{\log(\mathrm{A} \times \mathrm{B}) = \log \mathrm{A} + \log \mathrm{B}}$$

(ii) To divide two numbers:

$$\boxed{\log\left(\frac{A}{B}\right) = \log A - \log B}$$

(iii) To raise a number to a power:

$$\boxed{\lg A^n = n \log A}$$

For example, to evaluate (a) $\log_3 9$ (b) $\log_{10} 10$ (c) $\log_{16} 8$:

(a) Let $x = \log_3 9$ then $3^x = 9$ from the definition of a logarithm, i.e. $3^x = 3^2$, from which $x = 2$
Hence $\mathbf{\log_3 9 = 2}$

(b) Let $x = \log_{10} 10$ then $10^x = 10$ from the definition of a logarithm, i.e. $10^x = 10^1$, from which $x = 1$
Hence $\mathbf{\log_{10} 10 = 1}$ (which may be checked by a calculator)

(c) Let $x = \log_{16} 8$ then $16^x = 8$, from the definition of a logarithm, i.e. $(2^4)^x = 2^3$, i.e. $2^{4x} = 2^3$ from the laws of indices, from which, $4x = 3$ and $x = \frac{3}{4}$
Hence $\mathbf{\log_{16} 8 = \frac{3}{4}}$

In another example, to evaluate (a) lg 0.001 (b) $\ln e$ (c) $\log_3 \dfrac{1}{81}$:

(a) Let $x = \lg 0.001 = \log_{10} 0.001$ then $10^x = 0.001$, i.e. $10^x = 10^{-3}$, from which $x = -3$
Hence $\mathbf{\lg 0.001 = -3}$ (which may be checked by a calculator)

(b) Let $x = \ln e = \log_e e$ then $e^x = e$, i.e. $e^x = e^1$ from which $x = 1$.
Hence $\mathbf{\ln e = 1}$ (which may be checked by a calculator)

(a) Let $x = \log_3 \dfrac{1}{81}$ then $3^x = \dfrac{1}{81} = \dfrac{1}{3^4} = 3^{-4}$, from which $x = -4$

Hence $\mathbf{\log_3 \dfrac{1}{81} = -4}$

In another example, to solve the equations:
(a) $\lg x = 3$ (b) $\log_2 x = 3$ (c) $\log_5 x = -2$:

(a) If $\lg x = 3$ then $\log_{10} x = 3$ and $x = 10^3$, i.e. $x = \mathbf{1000}$

(b) If $\log_2 x = 3$ then $x = \mathbf{2^3 = 8}$

(c) If $\log_5 x = -2$ then $x = 5^{-2} = \dfrac{1}{5^2} = \mathbf{\dfrac{1}{25}}$

In another example, to solve the equation:
$\log(x - 1) + \log(x + 1) = 2\log(x + 2)$:

$$\log(x-1) + \log(x+1) = \log(x-1)(x+1) \quad \text{from the first law of logarithms}$$

$$= \log(x^2 - 1)$$

$$2\log(x+2) = \log(x+2)^2 = \log(x^2+4x+4)$$

Hence if $\log(x^2-1) = \log(x^2+4x+4)$

then $x^2 - 1 = x^2 + 4x + 4$

i.e. $-1 = 4x + 4$

i.e. $-5 = 4x$

i.e. $\mathbf{x = -\frac{5}{4}}$ or $\mathbf{-1\frac{1}{4}}$

Indicial equations

The laws of logarithms may be used to solve certain equations involving powers — called **indicial equations**.

For example, to solve, say, $3^x = 27$, logarithms to a base of 10 are taken of both sides, i.e. $\log_{10} 3^x = \log_{10} 27$

and $x\log_{10} 3 = \log_{10} 27$ by the third law of logarithms

Rearranging gives $x = \dfrac{\log_{10} 27}{\log_{10} 3} = \dfrac{1.43136\ldots}{0.4771\ldots} = 3$ which may be readily checked.

$\left(\text{Note, } \dfrac{\log 27}{\log 3} \text{ is } \textbf{not} \text{ equal to } \log\dfrac{27}{3}\right)$.

In another example, to solve the equation $2^{x+1} = 3^{2x-5}$ correct to 2 decimal places:

Taking logarithms to base 10 of both sides gives:

$$\log_{10} 2^{x+1} = \log_{10} 3^{2x-5}$$

i.e. $(x+1)\log_{10} 2 = (2x-5)\log_{10} 3$

$$x\log_{10} 2 + \log_{10} 2 = 2x\log_{10} 3 - 5\log_{10} 3$$

$$x(0.3010) + (0.3010) = 2x(0.4771) - 5(0.4771)$$

i.e. $0.3010x + 0.3010 = 0.9542x - 2.3855$

Hence $2.3855 + 0.3010 = 0.9542x - 0.3010x$

$$2.6865 = 0.6532x$$

from which $\boldsymbol{x} = \dfrac{2.6865}{0.6532} = \mathbf{4.11}$, correct to 2 decimal places.

Graphs of logarithmic functions

A graph of $y = \log_{10} x$ is shown in Figure 11.1 and a graph of $y = \log_e x$ is shown in Figure 11.2. Both are seen to be of similar shape; in fact, the same general shape occurs for a logarithm to any base.

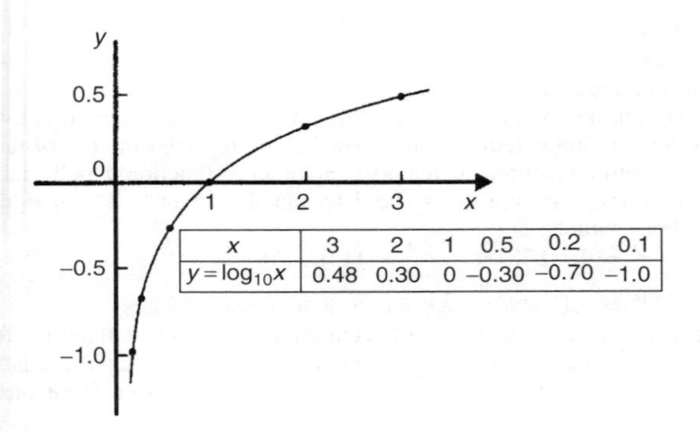

x	3	2	1	0.5	0.2	0.1
$y = \log_{10} x$	0.48	0.30	0	−0.30	−0.70	−1.0

Figure 11.1

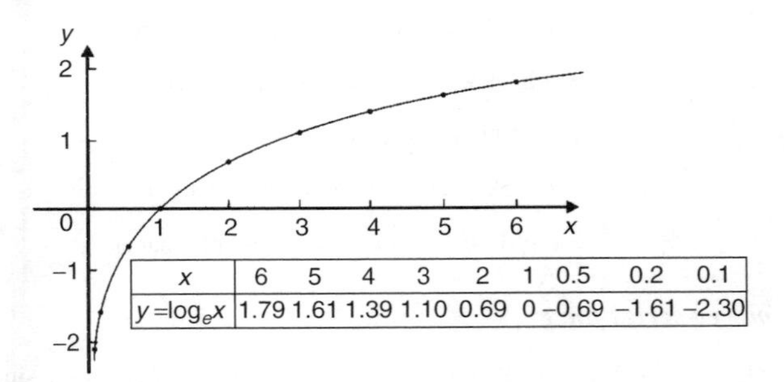

x	6	5	4	3	2	1	0.5	0.2	0.1
$y = \log_e x$	1.79	1.61	1.39	1.10	0.69	0	−0.69	−1.61	−2.30

Figure 11.2

In general, with a logarithm to any base a, it is noted that:

(i) $\log_a 1 = 0$
(ii) $\log_a a = 1$
(iii) $\log_a 0 \to -\infty$

12 Exponential Functions

The exponential function

An exponential function is one which contains e^x, e being a constant called the exponent and having an approximate value of 2.7183. The exponent arises from the natural laws of growth and decay and is used as a base for natural or Napierian logarithms.

Evaluating exponential functions

The value of e^x may be determined by using:

(a) a calculator, or
(b) the power series for e^x, or
(c) tables of exponential functions.

The most common method of evaluating an exponential function is by using a scientific notation **calculator**, this now having replaced the use of tables.

Most scientific notation calculators contain an e^x function which enables all practical values of e^x and e^{-x} to be determined, correct to 8 or 9 significant figures. For example,

$$e^1 = 2.7182818, \quad e^{2.4} = 11.023176$$

and $e^{-1.618} = 0.19829489$, correct to 8 significant figures

In practical situations the degree of accuracy given by a calculator is often far greater than is appropriate. The accepted convention is that the final result is stated to one significant figure greater than the least significant measured value.

Use your calculator to check the following values:

$$e^{0.12} = 1.1275, \text{ correct to 5 significant figures}$$

$$e^{-1.47} = 0.22993, \text{ correct to 5 decimal places}$$

$$e^{-0.431} = 0.6499, \text{ correct to 4 decimal places}$$

$$e^{9.32} = 11\,159, \text{ correct to 5 significant figures}$$

$$e^{-2.785} = 0.0617291, \text{ correct to 7 decimal places}$$

The power series for e^x

The value of e^x can be calculated to any required degree of accuracy since it is defined in terms of the following **power series**:

$$e^x = 1 + x + \frac{x^2}{2!} + \frac{x^3}{3!} + \frac{x^4}{4!} + \ldots \qquad (1)$$

(where $3! = 3 \times 2 \times 1$ and is called 'factorial 3')

The series is valid for all values of x.

The series is said to **converge**, i.e. if all the terms are added, an actual value for e^x (where x is a real number) is obtained. The more terms that are taken, the closer will be the value of e^x to its actual value. The value of the exponent e, correct to say 4 decimal places, may be determined by substituting $x = 1$ in the power series of equation (1). Thus

$$e^1 = 1 + 1 + \frac{(1)^2}{2!} + \frac{(1)^3}{3!} + \frac{(1)^4}{4!} + \frac{(1)^5}{5!} + \frac{(1)^6}{6!} + \frac{(1)^7}{7!} + \frac{(1)^8}{8!} + \cdots$$

$$= 1 + 1 + 0.5 + 0.16667 + 0.04167 + 0.00833 + 0.00139 + 0.00020 + 0.00002 + \cdots$$

$$= 2.71828$$

i.e. $\quad e = 2.7183$ correct to 4 decimal places

The value of $e^{0.05}$, correct to say 8 significant figures, is found by substituting $x = 0.05$ in the power series for e^x. Thus

$$e^{0.05} = 1 + 0.05 + \frac{(0.05)^2}{2!} + \frac{(0.05)^3}{3!} + \frac{(0.05)^4}{4!} + \frac{(0.05)^5}{5!} + \cdots$$
$$= 1 + 0.05 + 0.00125 + 0.000020833 + 0.000000260$$
$$+ 0.000000003$$

and by adding,

$$e^{0.05} = 1.0512711, \text{ correct to 8 significant figures}$$

In this example, successive terms in the series grow smaller very rapidly and it is relatively easy to determine the value of $e^{0.05}$ to a high degree of accuracy. However, when x is nearer to unity or larger than unity, a very large number of terms are required for an accurate result.

If in the series of equation (1), x is replaced by $-x$, then

$$e^{-x} = 1 + (-x) + \frac{(-x)^2}{2!} + \frac{(-x)^3}{3!} + \cdots$$

i.e. $$e^{-x} = 1 - x + \frac{x^2}{2!} - \frac{x^3}{3!} + \cdots$$

In a similar manner the power series for e^x may be used to evaluate any exponential function of the form ae^{kx}, where a and k are constants.

In the series of equation (1), let x be replaced by kx. Then

$$ae^{kx} = a\left\{1 + (kx) + \frac{(kx)^2}{2!} + \frac{(kx)^3}{3!} + \ldots\right\}$$

Thus $$5e^{2x} = 5\left\{1 + (2x) + \frac{(2x)^2}{2!} + \frac{(2x)^3}{3!} + \ldots\right\}$$

$$= 5\left\{1 + 2x + \frac{4x^2}{2} + \frac{8x^3}{6} + \ldots\right\}$$

i.e. $$5e^{2x} = 5\left\{1 + 2x + 2x^2 + \frac{4}{3}x^3 + \cdots\right\}$$

Graphs of exponential functions

Values of e^x and e^{-x} obtained from a calculator, correct to 2 decimal places, over a range $x = -3$ to $x = 3$, are shown in the following table.

x	−3.0	−2.5	−2.0	−1.5	−1.0	−0.5	0	0.5	1.0	1.5	2.0	2.5	3.0
e^x	0.05	0.08	0.14	0.22	0.37	0.61	1.00	1.65	2.72	4.48	7.39	12.18	20.09
e^{-x}	20.09	12.18	7.39	4.48	2.72	1.65	1.00	0.61	0.37	0.22	0.14	0.08	0.05

Figure 12.1 shows graphs of $y = e^x$ and $y = e^{-x}$.

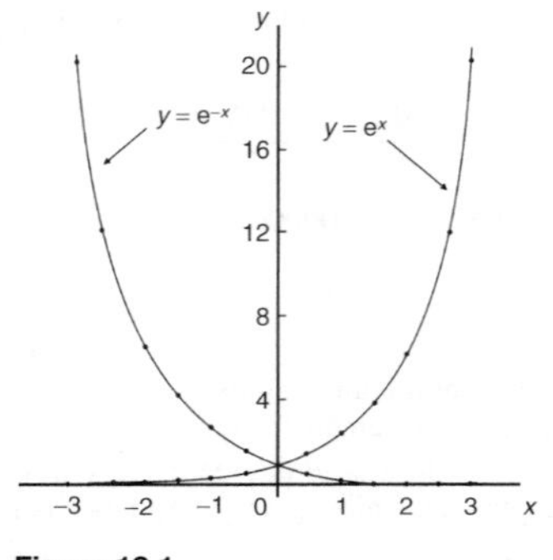

Figure 12.1

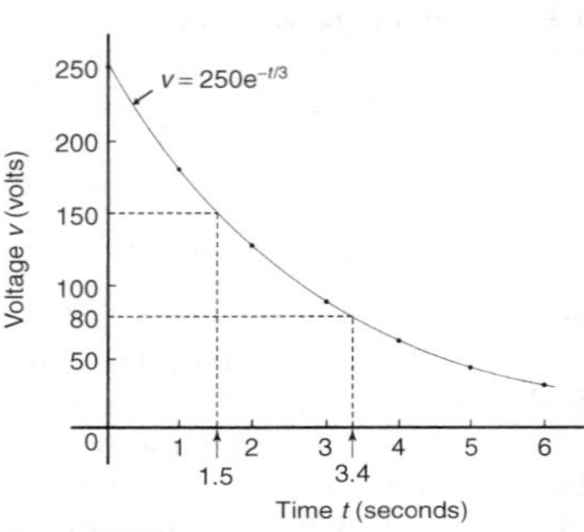

Figure 12.2

For example, the decay of voltage, v volts, across a capacitor at time t seconds is given by $v = 250e^{-t/3}$. To draw a graph showing the natural decay curve over the first 6 seconds:

A table of values is drawn up as shown below.

t	0	1	2	3	4	5	6
$e^{-t/3}$	1.00	0.7165	0.5134	0.3679	0.2636	0.1889	0.1353
$v = 250e^{-t/3}$	250.0	179.1	128.4	91.97	65.90	47.22	33.83

The natural decay curve of $v = 250e^{-t/3}$ is shown in Figure 12.2.
From the graph, when, say, time $t = 3.4$ s, voltage $\boldsymbol{v}$ **= 80 volts** and when, say, voltage $\boldsymbol{v}$ **= 150** volts, time $\boldsymbol{t}$ **= 1.5 seconds**.

Napierian logarithms

Logarithms having a base of e are called **hyperbolic, Napierian** or **natural logarithms** and the Napierian logarithm of x is written as $\log_e x$, or more commonly, $\ln x$.

Evaluating Napierian logarithms

The value of a Napierian logarithm may be determined by using:

(a) a calculator, or
(b) a relationship between common and Napierian logarithms, or
(c) Napierian logarithm tables

The most common method of evaluating a Napierian logarithm is by a scientific notation **calculator**, this now having replaced the use of four-figure tables, and also the relationship between common and Napierian logarithms,

$$\log_e y = 2.3026 \log_{10} y$$

Most scientific notation calculators contain a '$\ln x$' function which displays the value of the Napierian logarithm of a number when the appropriate key is pressed. Using a calculator,

$$\ln 4.692 = 1.5458589\ldots$$
$$= 1.5459, \text{ correct to 4 decimal places}$$

and $$\ln 35.78 = 3.57738907\ldots$$
$$= 3.5774, \text{ correct to 4 decimal places}$$

Use your calculator to check the following values:

$\ln 1.732 = 0.54928$, correct to 5 significant figures

$\ln 1 = 0$

$\ln 1750 = 7.4674$, correct to 4 decimal places

$\ln 0.00032 = -8.04719$, correct to 6 significant figures

$\ln e^3 = 3$

$\ln e^1 = 1$

From the last two examples we can conclude that

$$\mathbf{\log_e e^x = x}$$

This is useful when solving equations involving exponential functions.
For example, to solve $e^{3x} = 8$, take Napierian logarithms of both sides, which gives

$$\ln e^{3x} = \ln 8$$

i.e. $$3x = \ln 8$$

from which $$x = \tfrac{1}{3}\ln 8 = \mathbf{0.6931}, \text{ correct to 4 decimal places}$$

Laws of growth and decay

The laws of exponential growth and decay are of the form $y = Ae^{-kx}$ and $y = A(1 - e^{-kx})$, where A and k are constants. When plotted, the form of each of these equations is as shown in Figure 12.3. The laws occur frequently in engineering and science and examples of quantities related by a natural law include:

(i) Linear expansion $l = l_0 e^{\alpha\theta}$
(ii) Change in electrical resistance with temperature $R_\theta = R_0 e^{\alpha\theta}$
(iii) Tension in belts $T_1 = T_0 e^{\mu\theta}$
(iv) Newton's law of cooling $\theta = \theta_0 e^{-kt}$
(v) Biological growth $y = y_0 e^{kt}$
(vi) Discharge of a capacitor $q = Qe^{-t/CR}$
(vii) Atmospheric pressure $p = p_0 e^{-h/c}$
(viii) Radioactive decay $N = N_0 e^{-\lambda t}$

(ix) Decay of current in an inductive circuit $i = Ie^{-Rt/L}$
(x) Growth of current in a capacitive circuit $i = I(1 - e^{-t/CR})$

For example, the current i amperes flowing in a capacitor at time t seconds is given by $i = 8.0(1 - e^{-t/CR})$, where the circuit resistance R is 25×10^3 ohms and capacitance C is 16×10^{-6} farads. To determine (a) the current i after 0.5 seconds and (b) the time, to the nearest millisecond, for the current to reach 6.0 A:

(a) Current $\mathrm{i} = 8.0(1 - e^{-t/CR}) = 8.0[1 - e^{-0.5/(16\times10^{-6})(25\times10^3)}]$

$$= 8.0(1 - e^{-1.25}) = 8.0(1 - 0.2865047\cdots)$$

$$= 8.0(0.7134952\cdots) = \mathbf{5.71\ amperes}$$

(b) Transposing $i = 8.0(1 - e^{-t/CR})$ gives: $\dfrac{i}{8.0} = 1 - e^{-t/CR}$

from which, $e^{-t/CR} = 1 - \dfrac{i}{8.0} = \dfrac{8.0 - i}{8.0}$

Taking the reciprocal of both sides gives: $e^{t/CR} = \dfrac{8.0}{8.0 - i}$

Taking Napierian logarithms of both sides gives:

$$\frac{t}{CR} = \ln\left(\frac{8.0}{8.0 - i}\right)$$

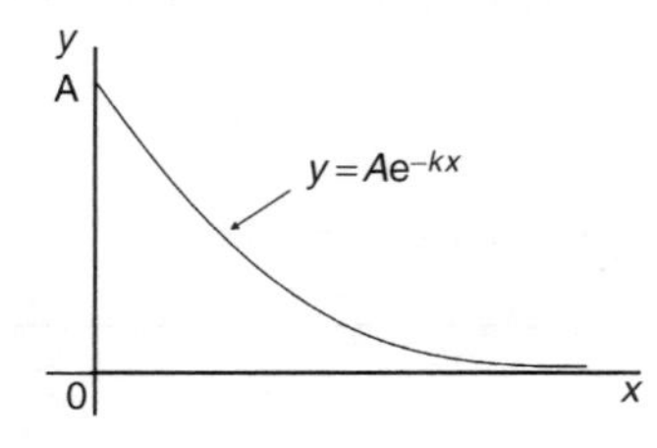

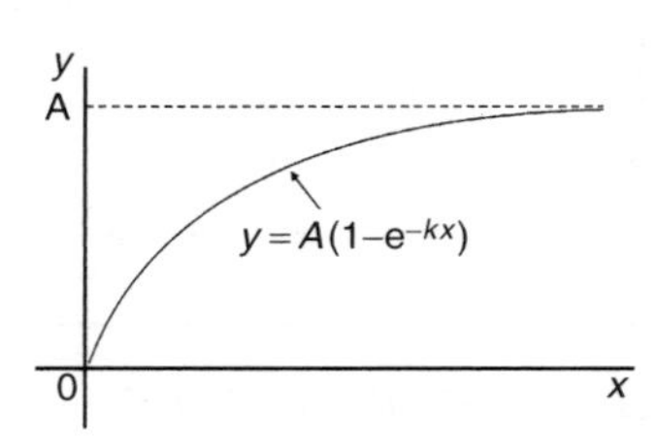

Figure 12.3

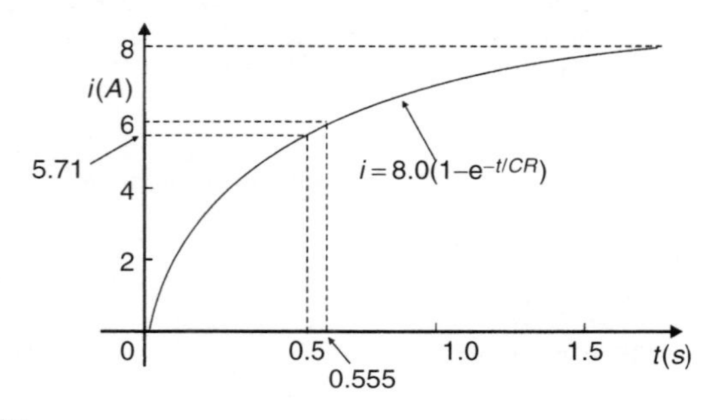

Figure 12.4

Hence $t = CR\ln\left(\frac{8.0}{8.0 - i}\right) = (16 \times 10^{-6})(25 \times 10^3)\ln\left(\frac{8.0}{8.0 - 6.0}\right)$

when $i = 6.0$ amperes,

i.e. $\quad \mathbf{t} = \frac{400}{10^3}\ln\left(\frac{8.0}{2.0}\right) = 0.4\ln 4.0 = 0.4(1.3862943\cdots)$

$= 0.5545\ s = \mathbf{555\ ms}$, to the nearest millisecond

A graph of current against time is shown in Figure 12.4.

---

13 Hyperbolic Functions

Introduction to hyperbolic functions

Functions which are associated with the geometry of the conic section called a hyperbola are called **hyperbolic functions**. Applications of hyperbolic functions include transmission line theory and catenary problems.

By definition:

(i) Hyperbolic sine of x, $\mathbf{\sinh x} = \dfrac{e^x - e^{-x}}{2}$ (1)

'$\sinh x$' is often abbreviated to 'sh x' and is pronounced as 'shine x'

(ii) Hyperbolic cosine of x, $\mathbf{\cosh x} = \dfrac{e^x + e^{-x}}{2}$ (2)

'$\cosh x$' is often abbreviated to 'ch x' and is pronounced as 'kosh x'

(iii) Hyperbolic tangent of x, $\mathbf{\tanh x} = \dfrac{\sinh x}{\cosh x} = \dfrac{e^x - e^{-x}}{e^x + e^{-x}}$ (3)

'$\tanh x$' is often abbreviated to 'th x' and is pronounced as 'than x'

(iv) Hyperbolic cosecant of x, $\mathbf{\text{cosech}\, x} = \dfrac{1}{\sinh x} = \dfrac{2}{e^x - e^{-x}}$ (4)

'cosech x' is pronounced as 'coshec x'

(v) Hyperbolic secant of x, $\mathbf{\text{sech}\, x} = \dfrac{1}{\cosh x} = \dfrac{2}{e^x + e^{-x}}$ (5)

'sech x' is pronounced as 'shec x'

(vi) Hyperbolic cotangent of x, $\mathbf{\coth x} = \dfrac{1}{\tanh x} = \dfrac{e^x + e^{-x}}{e^x - e^{-x}}$ (6)

'$\coth x$' is pronounced as 'koth x'

Some properties of hyperbolic functions

Replacing x by 0 in equation (1) gives:

$$\sinh 0 = \frac{e^0 - e^{-0}}{2} = \frac{1-1}{2} = 0$$

Replacing x by 0 in equation (2) gives:

$$\cosh 0 = \frac{e^0 + e^{-0}}{2} = \frac{1+1}{2} = 1$$

If a function of x, $f(-x) = -f(x)$, then $f(x)$ is called an **odd function** of x. Replacing x by $-x$ in equation (1) gives:

$$\sinh(-x) = \frac{e^{-x} - e^{-(-x)}}{2} = \frac{e^{-x} - e^x}{2} = -\left(\frac{e^x - e^{-x}}{2}\right) = -\sinh x$$

Replacing x by $-x$ in equation (3) gives:

$$\tanh(-x) = \frac{e^{-x} - e^{-(-x)}}{e^{-x} + e^{-(-x)}} = \frac{e^{-x} - e^x}{e^{-x} + e^x} = -\left(\frac{e^x - e^{-x}}{e^x + e^{-x}}\right) = -\tanh x$$

Hence **sinh x and tanh x are both odd functions**, as also are $\operatorname{cosech} x \left(= \frac{1}{\sinh x}\right)$ and $\coth x \left(= \frac{1}{\tanh x}\right)$

If a function of x, $f(-x) = f(x)$, then $f(x)$ is called an **even function** of x. Replacing x by $-x$ in equation (2) gives:

$$\cosh(-x) = \frac{e^{-x} + e^{-(-x)}}{2} = \frac{e^{-x} + e^x}{2} = \cosh x$$

Hence **cosh x is an even function**, as also is $\operatorname{sech} x \left(= \frac{1}{\cosh x}\right)$

Hyperbolic functions may be evaluated easiest using a calculator. Many scientific notation calculators actually possess sinh and cosh functions; however, if a calculator does not contain these functions, then the definitions given above may be used.

For example, to evaluate sinh 5.4, correct to 4 significant figures:

$$\sinh 5.4 = \tfrac{1}{2}(e^{5.4} - e^{-5.4}) = \tfrac{1}{2}(221.406416\ldots - 0.00451658\ldots)$$

$$= \tfrac{1}{2}(221.401899\ldots) = \mathbf{110.7}, \text{ correct to 4 significant figures}$$

In another example, to evaluate sech 0.86, correct to 4 significant figures:

$$\operatorname{sech} 0.86 = \frac{1}{\cosh 0.86} = \frac{1}{\frac{1}{2}(e^{0.86} + e^{-0.86})}$$

$$= \frac{2}{(2.36316069\ldots + 0.42316208\ldots)}$$

$$= \frac{2}{2.78632277\ldots} = \mathbf{0.7178}$$

Graphs of hyperbolic functions

A graph of $\boldsymbol{y = \sinh x}$ is shown in Figure 13.1. Since the graph is symmetrical about the origin, $\sinh x$ is an **odd function**.
A graph of $\boldsymbol{y = \cosh x}$ is shown in Figure 13.2. Since the graph is symmetrical about the y-axis, $\cosh x$ is an **even function**. The shape of $y = \cosh x$ is that of a heavy rope or chain hanging freely under gravity and is called a **catenary**. Examples include transmission lines, a telegraph wire or a fisherman's line, and is used in the design of roofs and arches. Graphs of $y = \tanh x$, $y = \operatorname{cosech} x$, $y = \operatorname{sech} x$ and $y = \coth x$ are shown in Figures 13.3 and 13.4.

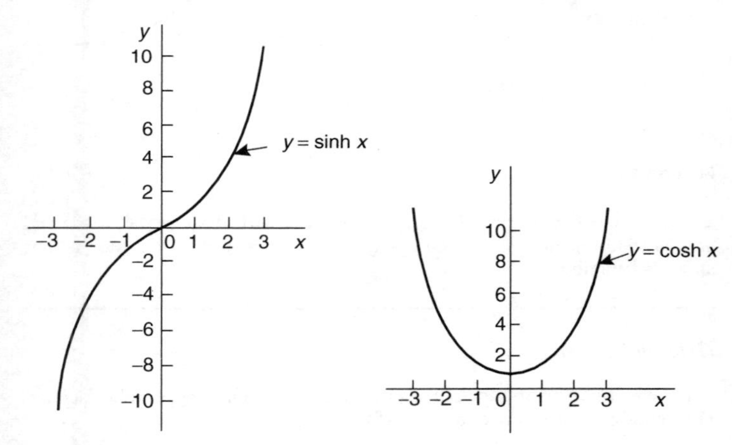

Figure 13.1

Figure 13.2

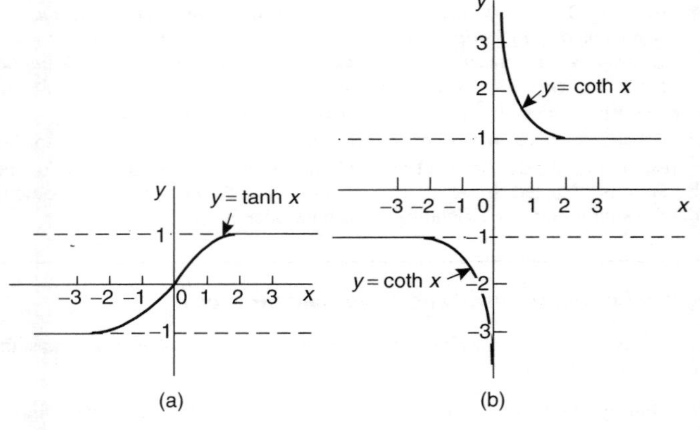

Figure 13.3

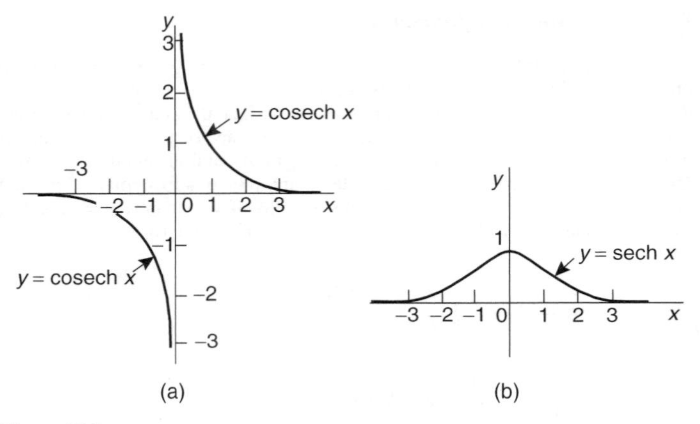

Figure 13.4

$y = \tanh$, $y = \coth x$ and $y = \operatorname{cosech} x$ are symmetrical about the origin and are thus odd functions. $y = \operatorname{sech} x$ is symmetrical about the y-axis and is thus an **even function**.

Hyperbolic identities

For every trigonometric identity there is a corresponding hyperbolic identity. **Hyperbolic identities** may be proved by either

(i) replacing $\operatorname{sh} x$ by $\dfrac{e^x - e^{-x}}{2}$ and $\operatorname{ch} x$ by $\dfrac{e^x + e^{-x}}{2}$, or

(ii) by using **Osborne's rule**, which states: '*the six trigonometric ratios used in trigonometrical identities relating general angles may be replaced by their corresponding hyperbolic functions, but the sign of any direct or implied product of two sines must be changed*'.

For example, since $\cos^2 x + \sin^2 x = 1$ then, by Osborne's rule, $\operatorname{ch}^2 x - \operatorname{sh}^2 x = 1$, i.e. the trigonometric functions have been changed to their corresponding hyperbolic functions and since $\sin^2 x$ is a product of two sines the sign is changed from $+$ to $-$. Table 13.1 shows some trigonometric identities and their corresponding hyperbolic identities.

Solving equations involving hyperbolic functions

Equations of the form $\boldsymbol{a} \operatorname{\mathbf{ch}} \boldsymbol{x} + \boldsymbol{b} \operatorname{\mathbf{sh}} \boldsymbol{x} = \boldsymbol{c}$, where a, b and c are constants may be solved either by:

(a) plotting graphs of $y = a \operatorname{ch} x + b \operatorname{sh} x$ and $y = c$ and noting the points of intersection, or more accurately,

Table 13.1

Trigonometric identity	*Corresponding hyperbolic identity*
$\cos^2 x + \sin^2 x = 1$	$\text{ch}^2 x - \text{sh}^2 x = 1$
$1 + \tan^2 x = \sec^2 x$	$1 - \text{th}^2 x = \text{sech}^2 x$
$\cot^2 x + 1 = \text{cosec}^2 x$	$\text{coth}^2 x - 1 = \text{cosech}^2 x$
Compound angle formulae	
$\sin(A \pm B) = \sin A \cos B \pm \cos A \sin B$	$\text{sh}(A \pm B) = \text{sh} A \,\text{ch} B \pm \text{ch} A \,\text{sh} B$
$\cos(A \pm B) = \cos A \cos B \mp \sin A \sin B$	$\text{ch}(A \pm B) = \text{ch} A \,\text{ch} B \pm \text{sh} A \,\text{sh} B$
$\tan(A \pm B) = \dfrac{\tan A \pm \tan B}{1 \mp \tan A \tan B}$	$\text{th}(A \pm B) = \dfrac{\text{th} A \pm \text{th} B}{1 \pm \text{th} A \,\text{th} B}$
Double angles	
$\sin 2x = 2 \sin x \cos x$	$\text{sh}\, 2x = 2 \,\text{sh}\, x \,\text{ch}\, x$
$\cos 2x = \cos^2 x - \sin^2 x$	$\text{ch}\, 2x = \text{ch}^2 x + \text{sh}^2 x$
$= 2\cos^2 x - 1$	$= 2\,\text{ch}^2 x - 1$
$= 1 - 2\sin^2 x$	$= 1 + 2\,\text{sh}^2 x$
$\tan 2x = \dfrac{2 \tan x}{1 - \tan^2 x}$	$\text{th}\, 2x = \dfrac{2\,\text{th}\, x}{1 + \text{th}^2 x}$

(b) by adopting the following procedure:

(i) Change $\text{sh}\, x$ to $\left(\dfrac{e^x - e^{-x}}{2}\right)$ and $\text{ch}\, x$ to $\left(\dfrac{e^x + e^{-x}}{2}\right)$

(ii) Rearrange the equation into the form $pe^x + qe^{-x} + r = 0$, where p, q and r are constants.

(iii) Multiply each term by e^x, which produces an equation of the form

$$p(e^x)^2 + re^x + q = 0 \text{ (since } (e^{-x})(e^x) = e^0 = 1)$$

(iv) Solve the quadratic equation $p(e^x)^2 + re^x + q = 0$ for e^x by factorising or by using the quadratic formula.

(v) Given $e^x = a$ constant (obtained by solving the equation in (iv)), take Napierian logarithms of both sides to give $x = \ln$ (constant)

For example, to solve the equation $2.6\,\text{ch}\, x + 5.1\,\text{sh}\, x = 8.73$, correct to 4 decimal places:

Following the above procedure:

(i) $2.6\,\text{ch}\, x + 5.1\,\text{sh}\, x = 8.73$

$$\text{i.e. } 2.6\left(\frac{e^x + e^{-x}}{2}\right) + 5.1\left(\frac{e^x - e^{-x}}{2}\right) = 8.73$$

(ii) $1.3e^x + 1.3e^{-x} + 2.55e^x - 2.55e^{-x} = 8.73$

$$\text{i.e. } 3.85e^x - 1.25e^{-x} - 8.73 = 0$$

(iii) $3.85(e^x)^2 - 8.73e^x - 1.25 = 0$

(iv) $e^x = \dfrac{-(-8.73) \pm \sqrt{[(-8.73)^2 - 4(3.85)(-1.25)]}}{2(3.85)}$

$$= \frac{8.73 \pm \sqrt{95.463}}{7.70} = \frac{8.73 \pm 9.7705}{7.70}$$

Hence $e^x = 2.4027$ or $e^x = -0.1351$

(v) $x = \ln 2.4027$ or $x = \ln(-0.1351)$ which has no real solution.

Hence $\boldsymbol{x = 0.8766}$, correct to 4 decimal places.

Series expansions for cosh x *and* sinh x

By definition, $e^x = 1 + x + \dfrac{x^2}{2!} + \dfrac{x^3}{3!} + \dfrac{x^4}{4!} + \dfrac{x^5}{5!} + \ldots$ from chapter 12

Replacing x by $-x$ gives:

$$e^{-x} = 1 - x + \frac{x^2}{2!} - \frac{x^3}{3!} + \frac{x^4}{4!} - \frac{x^5}{5!} + \ldots$$

$$\cosh x = \frac{1}{2}(e^x + e^{-x}) = \frac{1}{2}\left[\left(1 + x + \frac{x^2}{2!} + \frac{x^3}{3!} + \frac{x^4}{4!} + \frac{x^5}{5!} + \ldots\right) + \left(1 - x + \frac{x^2}{2!} - \frac{x^3}{3!} + \frac{x^4}{4!} - \frac{x^5}{5!} + \ldots\right)\right]$$

$$= \frac{1}{2}\left[\left(2 + \frac{2x^2}{2!} + \frac{2x^4}{4!} + \ldots\right)\right]$$

i.e. $\boldsymbol{\cosh x = 1 + \frac{x^2}{2!} + \frac{x^4}{4!} + \ldots}$ (which is valid for all values of x)

$\cosh x$ is an even function and contains only even powers of x in its expansion.

$$\sinh x = \frac{1}{2}(e^x - e^{-x}) = \frac{1}{2}\left[\left(1 + x + \frac{x^2}{2!} + \frac{x^3}{3!} + \frac{x^4}{4!} + \frac{x^5}{5!} + \ldots\right) - \left(1 - x + \frac{x^2}{2!} - \frac{x^3}{3!} + \frac{x^4}{4!} - \frac{x^5}{5!} + \ldots\right)\right]$$

$$= \frac{1}{2}\left[2x + \frac{2x^3}{3!} + \frac{2x^5}{5!} + \ldots\right]$$

i.e. $\boldsymbol{\sinh x = x + \frac{x^3}{3!} + \frac{x^5}{5!} + \ldots}$ (which is valid for all values of x)

$\sinh x$ is an odd function and contains only odd powers of x in its expansion.

---

14 Partial Fractions

By algebraic addition,

$$\frac{1}{x-2}+\frac{3}{x+1}=\frac{(x+1)+3(x-2)}{(x-2)(x+1)}=\frac{4x-5}{x^2-x-2}$$

The reverse process of moving from $\frac{4x-5}{x^2-x-2}$ to $\frac{1}{x-2}+\frac{3}{x+1}$ is called resolving into **partial fractions**.
In order to resolve an algebraic expression into partial fractions:

(i) the denominator must factorise (in the above example, x^2-x-2 factorises as $(x-2)(x+1)$), and
(ii) the numerator must be at least one degree less than the denominator (in the above example $(4x-5)$ is of degree 1 since the highest powered x term is x^1 and (x^2-x-2) is of degree 2).

When the degree of the numerator is equal to or higher than the degree of the denominator, the numerator must be divided by the denominator until the remainder is of less degree than the denominator.

There are basically three types of partial fraction and the form of partial fraction used is summarised in Table 14.1, where $f(x)$ is assumed to be of less degree than the relevant denominator and A, B and C are constants to be determined.
(In the latter type in Table 1.2, ax^2+bx+c is a quadratic expression which does not factorise without containing surds or imaginary terms.)
Resolving an algebraic expression into partial fractions is used as a preliminary to integrating certain functions (see chapter 60).

For example, to resolve $\frac{11-3x}{x^2+2x-3}$ into partial fractions:
The denominator factorises as $(x-1)(x+3)$ and the numerator is of less degree than the denominator. Thus $\frac{11-3x}{x^2+2x-3}$ may be resolved into partial fractions.

Let $\frac{11-3x}{x^2+2x-3}\equiv\frac{11-3x}{(x-1)(x+3)}\equiv\frac{A}{(x-1)}+\frac{B}{(x+3)}$, where A and B are constants to be determined, i.e. $\frac{11-3x}{(x-1)(x+3)}\equiv\frac{A(x+3)+B(x-1)}{(x-1)(x+3)}$, by

Table 14.1

Type	Denominator containing	Expression	Form of partial fraction
1	Linear factors	$\frac{f(x)}{(x+a)(x-b)(x+c)}$	$\frac{A}{(x+a)}+\frac{B}{(x-b)}+\frac{C}{(x+c)}$
2	Repeated linear factors	$\frac{f(x)}{(x+a)^3}$	$\frac{A}{(x+a)}+\frac{B}{(x+a)^2}+\frac{C}{(x+a)^3}$
3	Quadratic factors	$\frac{f(x)}{(ax^2+bx+c)(x+d)}$	$\frac{Ax+B}{(ax^2+bx+c)}+\frac{C}{(x+d)}$

algebraic addition. Since the denominators are the same on each side of the identity then the numerators are equal to each other.

Thus, $11 - 3x \equiv A(x+3) + B(x-1)$

To determine constants A and B, values of x are chosen to make the term in A or B equal to zero.

When $x = 1$, then $11 - 3(1) \equiv A(1+3) + B(0)$

i.e. $8 = 4A$

i.e. $\mathbf{A = 2}$

When $x = -3$, then $11-3(-3) \equiv A(0) + B(-3-1)$

i.e. $20 = -4B$

i.e. $\mathbf{B = -5}$

Thus $$\mathbf{\frac{11-3x}{x^2+2x-3} \equiv \frac{2}{(x-1)} + \frac{-5}{(x+3)} \equiv \frac{2}{(x-1)} - \frac{5}{(x+3)}}$$

$$\left[\text{Check: } \frac{2}{(x-1)} - \frac{5}{(x+3)} = \frac{2(x+3)-5(x-1)}{(x-1)(x+3)} = \frac{11-3x}{x^2+2x-3}\right]$$

In another example, to express $\dfrac{x^3-2x^2-4x-4}{x^2+x-2}$ in partial fractions:

The numerator is of higher degree than the denominator. Thus dividing out gives:

$$\begin{array}{r|l} & x - 3 \\ \hline x^2+x-2 & x^3 - 2x^2 - 4x - 4 \\ & \underline{x^3 + x^2 - 2x} \\ & -3x^2 - 2x - 4 \\ & \underline{-3x^2 - 3x + 6} \\ & x - 10 \end{array}$$

Thus $$\frac{x^3-2x^2-4x-4}{x^2+x-2} \equiv x - 3 + \frac{x-10}{x^2+x-2}$$

$$\equiv x - 3 + \frac{x-10}{(x+2)(x-1)}$$

Let $$\frac{x-10}{(x+2)(x-1)} \equiv \frac{A}{(x+2)} + \frac{B}{(x-1)} \equiv \frac{A(x-1)+B(x+2)}{(x+2)(x-1)}$$

Equating the numerators gives: $x - 10 \equiv A(x-1) + B(x+2)$

Let $x = -2$. Then $-12 = -3A$

i.e. $\mathbf{A = 4}$

Let $x = 1$. Then $-9 = 3B$

i.e. $\mathbf{B = -3}$

Hence $$\frac{x-10}{(x+2)(x-1)} \equiv \frac{4}{(x+2)} - \frac{3}{(x-1)}$$

Thus $$\frac{\mathbf{x^3-2x^2-4x-4}}{\mathbf{x^2+x-2}} \equiv \mathbf{x-3+\frac{4}{(x+2)}-\frac{3}{(x-1)}}$$

In another example to express $\frac{5x^2-2x-19}{(x+3)(x-1)^2}$ as the sum of three partial fractions:

The denominator is a combination of a linear factor and a repeated linear factor.

Let $$\frac{5x^2-2x-19}{(x+3)(x-1)^2} \equiv \frac{A}{(x+3)}+\frac{B}{(x-1)}+\frac{C}{(x-1)^2}$$

$$\equiv \frac{A(x-1)^2+B(x+3)(x-1)+C(x+3)}{(x+3)(x-1)^2},$$

by algebraic addition

Equating the numerators gives:

$$5x^2-2x-19 \equiv A(x-1)^2+B(x+3)(x-1)+C(x+3) \quad (1)$$

Let $x=-3$. Then

$$5(-3)^2-2(-3)-19 \equiv A(-4)^2+B(0)(-4)+C(0)$$

i.e. $32=16A$

i.e. $\mathbf{A=2}$

Let $x=1$. Then

$$5(1)^2-2(1)-19 \equiv A(0)^2+B(4)(0)+C(4)$$

i.e. $-16=4C$

i.e. $\mathbf{C=-4}$

Without expanding the RHS of equation (1) it can be seen that equating the coefficients of x^2 gives: $5=A+B$, and since $A=2$, $\mathbf{B=3}$

Hence $$\frac{\mathbf{5x^2-2x-19}}{\mathbf{(x+3)(x-1)^2}} \equiv \mathbf{\frac{2}{(x+3)}+\frac{3}{(x-1)}-\frac{4}{(x-1)^2}}$$

In another example, to resolve $\frac{3+6x+4x^2-2x^3}{x^2(x^2+3)}$ into partial fractions:

Terms such as x^2 may be treated as $(x+0)^2$, i.e. they are repeated linear factors. (x^2+3) is a quadratic factor which does not factorise without containing surds and imaginary terms.

Let $$\frac{3+6x+4x^2-2x^3}{x^2(x^2+3)} \equiv \frac{A}{x}+\frac{B}{x^2}+\frac{Cx+D}{(x^2+3)}$$

$$\equiv \frac{Ax(x^2+3)+B(x^2+3)+(Cx+D)x^2}{x^2(x^2+3)}$$

Equating the numerators gives:

$$3+6x+4x^2-2x^3 \equiv Ax(x^2+3)+B(x^2+3)+(Cx+D)x^2$$

$$\equiv Ax^3+3Ax+Bx^2+3B+Cx^3+Dx^2$$

Let $x=0$. Then $3=3B$

i.e. $\quad \mathbf{B = 1}$

Equating the coefficients of x^3 terms gives: $\quad -2 = A + C \quad (1)$

Equating the coefficients of x^2 terms gives: $\quad 4 = B + D$

Since $B = 1$, $\mathbf{D = 3}$

Equating the coefficients of x terms gives: $\quad 6 = 3A$

i.e. $\quad \mathbf{A = 2}$

From equation (1), since $A = 2$, $\mathbf{C = -4}$

Hence $\quad \dfrac{\mathbf{3 + 6x + 4x^2 - 2x^3}}{\mathbf{x^2(x^2 + 3)}} \equiv \dfrac{2}{x} + \dfrac{1}{x^2} + \dfrac{-4x + 3}{x^2 + 3} \equiv \dfrac{\mathbf{2}}{\mathbf{x}} + \dfrac{\mathbf{1}}{\mathbf{x^2}} + \dfrac{\mathbf{3 - 4x}}{\mathbf{x^2 + 3}}$

---

15 Number Sequences

Simple sequences

A set of numbers which are connected by a definite law is called a **series** or a **sequence of numbers**. Each of the numbers in the series is called a **term** of the series.

For example, 1, 3, 5, 7,· · is a series obtained by adding 2 to the previous term, and 2, 8, 32, 128,· · is a sequence obtained by multiplying the previous term by 4.

In another example, to find the next three terms in the series: 9, 5, 1,· · We notice that each term in the series 9, 5, 1,· · progressively decreases by 4, thus the next two terms will be $1 - 4$, i.e. $\mathbf{-3}$ and $-3 - 4$, i.e. $\mathbf{-7}$

In another example, to determine the next two terms in the series: 2, 6, 18, 54,· · We notice that the second term, 6, is three times the first term, the third term, 18, is three times the second term, and that the fourth term, 54, is three times the third term. Hence the fifth term will be $3 \times 54 = \mathbf{162}$ and the sixth term will be $3 \times 162 = \mathbf{486}$

The n'th term of a series

If a series is represented by a general expression, say, $2n + 1$, where n is an integer (i.e. a whole number), then by substituting $n = 1, 2, 3, \cdots$ the terms of the series can be determined; in this example, the first three terms will be:

$$2(1) + 1, 2(2) + 1, 2(3) + 1, \cdots, \quad \text{i.e. } 3, 5, 7, \cdots$$

What is the n'th term of the sequence 1, 3, 5, 7,· ·? Firstly, we notice that the gap between each term is 2, hence the law relating the numbers is:

'$2n$ + something'

The second term, $\quad 3 = 2n +$ something,

hence when $n = 2$ (i.e. the second term of the series), then $3 = 4+$ something and the 'something' must be -1. Thus **the n'th term of 1, 3, 5, 7,.. is $\mathbf{2n - 1}$**. Hence the fifth term is given by $2(5) - 1 = 9$, and the twentieth term is $2(20) - 1 = 39$, and so on.

Arithmetic progressions

When a sequence has a constant difference between successive terms it is called an **arithmetic progression** (often abbreviated to AP).
Examples include:

(i) 1, 4, 7, 10, 13, ... where the **common difference** is 3
and (ii) $a, a+d, a+2d, a+3d, \ldots$. where the common difference is d.

If the first term of an AP is 'a' and the common difference is 'd' then

$$\boxed{\text{the } n\text{'th term is : } \boldsymbol{a + (n - 1)d}}$$

In example (i) above, the 7th term is given by $1 + (7 - 1)3 = \mathbf{19}$, which may be readily checked.
The sum S of an AP can be obtained by multiplying the average of all the terms by the number of terms.
The average of all the terms $= \dfrac{a+l}{2}$, where 'a' is the first term and l is the last term, i.e. $l = a + (n - 1)d$, for n terms.
Hence the sum of n terms,

$$S_n = n\left(\frac{a+l}{2}\right) = \frac{n}{2}\{a + [a + (n - 1)d]\}$$

i.e.

$$\boxed{\boldsymbol{S_n = \frac{n}{2}[2a + (n - 1)d]}}$$

For example, the sum of the first 7 terms of the series 1, 4, 7, 10, 13, ... is given by

$$S_7 = \frac{7}{2}[2(1) + (7 - 1)3], \quad \text{since } a = 1 \text{ and } d = 3$$

$$= \frac{7}{2}[2 + 18] = \frac{7}{2}[20] = \mathbf{70}$$

In another example, to determine (a) the ninth, and (b) the sixteenth term of the series 2, 7, 12, 17, ...
2, 7, 12, 17, ... is an arithmetic progression with a common difference, d, of 5

(a) The n'th term of an AP is given by $a + (n - 1)d$
Since the first term $a = 2$, $d = 5$ and $n = 9$
then the 9th term is: $2 + (9 - 1)5 = 2 + (8)(5) = 2 + 40 = \mathbf{42}$
(b) The 16th term is: $2 + (16 - 1)5 = 2 + (15)(5) = 2 + 75 = \mathbf{77}$

Geometric progressions

When a sequence has a constant ratio between successive terms it is called a **geometric progression** (often abbreviated to GP). The constant is called the **common ratio, r**.
Examples include

(i) 1, 2, 4, 8, ... where the common ratio is 2

and (ii) $a, ar, ar^2, ar^3, \ldots.$ where the common ratio is r

If the first term of a GP is 'a' and the common ratio is r, then

$$\text{the } n\text{'th term is : } \mathbf{ar^{n-1}}$$

which can be readily checked from the above examples.

For example, the 8th term of the GP 1, 2, 4, 8,.... is $(1)(2)^7 = \mathbf{128}$, since $a = 1$ and $r = 2$

The sum of n terms, $\boldsymbol{S_n = \dfrac{a(1-r^n)}{(1-r)}}$ which is valid when $r < 1$

or $\boldsymbol{S_n = \dfrac{a(r^n - 1)}{(r-1)}}$ which is valid when $r > 1$

For example, the sum of the first 8 terms of the GP 1, 2, 4, 8, 16, ... is given by

$$S_8 = \frac{1(2^8 - 1)}{(2-1)}, \text{ since } a = 1 \text{ and } r = 2$$

i.e. $S_8 = \dfrac{1(256-1)}{1} = \mathbf{255}$

When the common ratio r of a GP is less than unity, the sum of n terms, $S_n = \dfrac{a(1-r^n)}{(1-r)}$, which may be written as $S_n = \dfrac{a}{(1-r)} - \dfrac{ar^n}{(1-r)}$

Since $r < 1$, r^n becomes less as n increases, i.e. $r^n \to 0$ as $n \to \infty$

Hence $\dfrac{ar^n}{(1-r)} \to 0$ as $n \to \infty$. Thus $S_n \to \dfrac{a}{(1-r)}$ as $n \to \infty$

The quantity $\dfrac{a}{(1-r)}$ is called the **sum to infinity**, S_∞, and is the limiting value of the sum of an infinite number of terms,

i.e. $\boldsymbol{S_\infty = \dfrac{a}{(1-r)}}$ which is valid when $-1 < r < 1$

For example, the sum to infinity of the GP $1 + \frac{1}{2} + \frac{1}{4} + \cdots\cdots$ is

$$S_\infty = \frac{1}{1 - \frac{1}{2}}, \text{ since } a = 1 \text{ and } r = \tfrac{1}{2},$$

i.e. $S_\infty = \mathbf{2}$

In another example, a hire tool firm finds that their net return from hiring tools is decreasing by 10% per annum. Their net gain on a certain tool this year is £400. To find the possible total of all future profits from this tool (assuming the tool lasts for ever):

The net gain forms a series: £400 + £400 × 0.9 + £400 × 0.9^2 + · · · · · , which is a GP with $a = 400$ and $r = 0.9$

The sum to infinity,

$$S_\infty = \frac{a}{(1-r)} = \frac{400}{(1-0.9)} = \mathbf{£4000 = total\ future\ profits}$$

In another example, a drilling machine is to have 6 speeds ranging from 50 rev/min to 750 rev/min. To determine their values, each correct to the nearest whole number, if the speeds form a geometric progression:
Let the GP of n terms be given by $a, ar, ar^2, \ldots. ar^{n-1}$
The first term $a = 50$ rev/min
The 6th term is given by ar^{6-1}, which is 750 rev/min, i.e. $ar^5 = 750$ from which $r^5 = \frac{750}{a} = \frac{750}{50} = 15$

Thus the common ratio, $r = \sqrt[5]{15} = 1.7188$

The first term is $a = 50$ rev/min

the second term is $ar = (50)(1.7188) = 85.94$,

the third term is $ar^2 = (50)(1.7188)^2 = 147.71$,

the fourth term is $ar^3 = (50)(1.7188)^3 = 253.89$,

the fifth term is $ar^4 = (50)(1.7188)^4 = 436.39$,

the sixth term is $ar^5 = (50)(1.7188)^5 = 750.06$

Hence, correct to the nearest whole number, the 6 speeds of the drilling machine are **50, 86, 148, 254, 436 and 750 rev/min**

---

16 The Binomial Series

Pascal's triangle

A **binomial expression** is one which contains two terms connected by a plus or minus sign. Thus $(p + q)$, $(a + x)^2$, $(2x + y)^3$ are examples of binomial expressions. Expanding $(a + x)^n$ for integer values of n from 0 to 6 gives the following results:

$$(a+x)^0 = 1$$
$$(a+x)^1 = a + x$$
$$(a+x)^2 = (a+x)(a+x) = a^2 + 2ax + x^2$$
$$(a+x)^3 = (a+x)^2(a+x) = a^3 + 3a^2x + 3ax^2 + x^3$$
$$(a+x)^4 = (a+x)^3(a+x) = a^4 + 4a^3x + 6a^2x^2 + 4ax^3 + x^4$$
$$(a+x)^5 = (a+x)^4(a+x) = a^5 + 5a^4x + 10a^3x^2 + 10a^2x^3 + 5ax^4 + x^5$$
$$(a+x)^6 = (a+x)^5(a+x) = a^6 + 6a^5x + 15a^4x^2 + 20a^3x^3 + 15a^2x^4 + 6ax^5 + x^6$$

From the above results the following patterns emerge:

(i) 'a' decreases in power moving from left to right.
(ii) 'x' increases in power moving from left to right.
(iii) The coefficients of each term of the expansions are symmetrical about the middle coefficient when n is even and symmetrical about the two middle coefficients when n is odd.
(iv) The coefficients are shown separately in Table 16.1 and this arrangement is known as **Pascal's triangle**. A coefficient of a term may be obtained by adding the two adjacent coefficients immediately above in the previous

Table 16.1

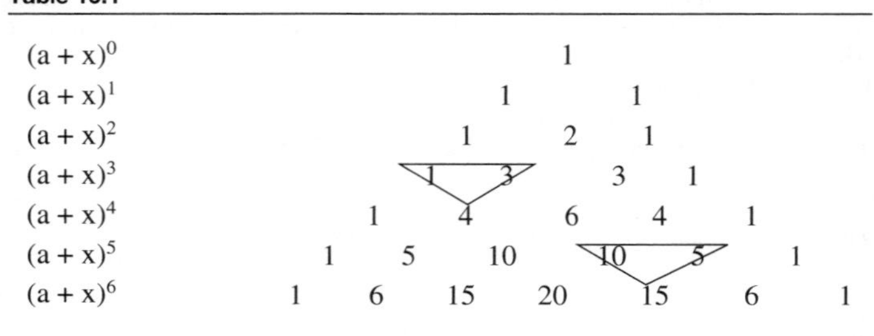

row. This is shown by the triangles in Table 16.1, where, for example, $1+3=4$, $10+5=15$, and so on.

(v) Pascal's triangle method is used for expansions of the form $(a+x)^n$ for integer values of n less than about 8

For example, using Pascal's triangle method to determine the expansion of $(a+x)^7$:

From Table 16.1, the row of Pascal's triangle corresponding to $(a+x)^6$ is as shown in (1) below. Adding adjacent coefficients gives the coefficients of $(a+x)^7$ as shown in (2) below.

	1 6		15	20 15		6	1	(1)
1	7	21	35	35	21	7	1	(2)

The first and last terms of the expansion of $(a+x)^7$ are a^7 and x^7 respectively.

The powers of 'a' decrease and the powers of 'x' increase moving from left to right.

Hence $$\mathbf{(a+x)^7 = a^7 + 7a^6x + 21a^5x^2 + 35a^4x^3 + 35a^3x^4 + 21a^2x^5 + 7ax^6 + x^7}$$

The binomial series

The **binomial series** or **binomial theorem** is a formula for raising a binomial expression to any power without lengthy multiplication. The general binomial expansion of $(a+x)^n$ is given by:

$$\mathbf{(a+x)^n = a^n + na^{n-1}x + \frac{n(n-1)}{2!}a^{n-2}x^2 + \frac{n(n-1)(n-2)}{3!}a^{n-3}x^3 + \ldots + x^n}$$

where 3! denotes $3 \times 2 \times 1$ and is termed 'factorial 3'.
With the binomial theorem n may be a fraction, a decimal fraction or a positive or negative integer.
In the general expansion of $(a + x)^n$ it is noted that the 4th term is: $\frac{n(n-1)(n-2)}{3!}a^{n-3}x^3$. The number 3 is very evident in this expression.
For any term in a binomial expansion, say the r'th term, $(r - 1)$ is very evident.
It may therefore be reasoned that **the r'th term of the expansion** $(a + x)^n$ is:

$$\frac{n(n-1)(n-2)\ldots \text{ to } (r-1) \text{ terms}}{(r-1)!}a^{n-(r-1)}x^{r-1}$$

If $a = 1$ in the binomial expansion of $(a + x)^n$ then:

$$\boldsymbol{(1+x)^n = 1 + nx + \frac{n(n-1)}{2!}x^2 + \frac{n(n-1)(n-2)}{3!}x^3 + \ldots\ldots}$$

which is valid for $-1 < x < 1$
When x is small compared with 1 then: $(1 + x)^n \approx 1 + nx$
For example, using the binomial series to determine the expansion of $(2 + x)^7$:
When $a = 2$ and $n = 7$ the binomial expansion is given by:

$$(2+x)^7 = 2^7 + 7(2)^6x + \frac{(7)(6)}{(2)(1)}(2)^5x^2 + \frac{(7)(6)(5)}{(3)(2)(1)}(2)^4x^3$$

$$+ \frac{(7)(6)(5)(4)}{(4)(3)(2)(1)}(2)^3x^4 + \frac{(7)(6)(5)(4)(3)}{(5)(4)(3)(2)(1)}(2)^2x^5$$

$$+ \frac{(7)(6)(5)(4)(3)(2)}{(6)(5)(4)(3)(2)(1)}(2)x^6 + \frac{(7)(6)(5)(4)(3)(2)(1)}{(7)(6)(5)(4)(3)(2)(1)}x^7$$

i.e. $\boldsymbol{(2+x)^7 = 128 + 448x + 672x^2 + 560x^3 + 280x^4 + 84x^5 + 14x^6 + x^7}$

In another example, to expand $\frac{1}{(1+2x)^3}$ in ascending powers of x as far as the term in x^3, using the binomial series:
Using the binomial expansion of $(1 + x)^n$, where $n = -3$ and x is replaced by $2x$ gives:

$$\frac{1}{(1+2x)^3} = (1+2x)^{-3} = 1 + (-3)(2x) + \frac{(-3)(-4)}{2!}(2x)^2$$

$$+ \frac{(-3)(-4)(-5)}{3!}(2x)^3 + \ldots$$

$$= \boldsymbol{1 - 6x + 24x^2 - 80x^3 +}$$

The expansion is valid provided $|2x| < 1$, i.e. $\boldsymbol{|x| < \frac{1}{2}}$ or $\boldsymbol{-\frac{1}{2} < x < \frac{1}{2}}$

In another example, to simplify $\dfrac{\sqrt[3]{(1-3x)}\sqrt{(1+x)}}{\left(1+\frac{x}{2}\right)^3}$ given that powers of x above the first may be neglected:

$$\frac{\sqrt[3]{(1-3x)}\sqrt{(1+x)}}{\left(1+\frac{x}{2}\right)^3} = (1-3x)^{1/3}(1+x)^{1/2}\left(1+\frac{x}{2}\right)^{-3}$$

$$\approx \left[1+\left(\frac{1}{3}\right)(-3x)\right]\left[1+\left(\frac{1}{2}\right)(x)\right]\left[1+(-3)\left(\frac{x}{2}\right)\right]$$

when expanded by the binomial theorem as far as the x term only,

$$= (1-x)\left(1+\frac{x}{2}\right)\left(1-\frac{3x}{2}\right)$$

$$= \left(1-x+\frac{x}{2}-\frac{3x}{2}\right)$$

when powers of x higher than unity are neglected

$$= \mathbf{(1-2x)}$$

Practical problems involving the binomial theorem

Binomial expansions may be used for numerical approximations, for calculations with small variations and in probability theory.

For example, the second moment of area of a rectangle through its centroid is given by $\dfrac{bl^3}{12}$. To determine the approximate change in the second moment of area if b is increased by 3.5% and l is reduced by 2.5%:

New values of b and l are $(1+0.035)b$ and $(1-0.025)l$ respectively.

$$\text{New second moment of area} = \frac{1}{12}[(1+0.035)b][(1-0.025)l]^3$$

$$= \frac{bl^3}{12}(1+0.035)(1-0.025)^3$$

$$\approx \frac{bl^3}{12}(1+0.035)(1-0.075),$$

neglecting powers of small terms

$$\approx \frac{bl^3}{12}(1+0.035-0.075),$$

neglecting products of small terms

$$\approx \frac{bl^3}{12}(1-0.040) \quad \text{or} \quad (0.96)\frac{bl^3}{12},$$

i.e. 96% of the original second moment of area

Hence the second moment of area is reduced by approximately 4%

---

17 Maclaurin's Series

Introduction

Some mathematical functions may be represented as power series, containing terms in ascending powers of the variable. For example,

$$e^x = 1 + x + \frac{x^2}{2!} + \frac{x^3}{3!} + \cdots$$

$$\sin x = x - \frac{x^3}{3!} + \frac{x^5}{5!} - \frac{x^7}{7!} + \cdots$$

and $$\cosh x = 1 + \frac{x^2}{2!} + \frac{x^4}{4!} + \cdots$$

Using a series, called **Maclaurin's series**, mixed functions containing, say, algebraic, trigonometric and exponential functions, may be expressed solely as algebraic functions, and differentiation and integration can often be more readily performed.
Maclaurin theorem or Maclaurin's series states:

$$\boldsymbol{f(x) = f(0) + xf'(0) + \frac{x^2}{2!}f''(0) + \frac{x^3}{3!}f'''(0) + \cdots} \qquad (1)$$

Conditions of Maclaurin's series

Maclaurin's series may be used to represent any function, say $f(x)$, as a power series provided that at $x = 0$ the following three conditions are met:

(a) $\boldsymbol{f(0) \neq \infty}$
For example, for the function $f(x) = \cos x$, $f(0) = \cos 0 = 1$, thus $\cos x$ meets the condition. However, if $f(x) = \ln x$, $f(0) = \ln 0 = -\infty$, thus $\ln x$ does not meet this condition.

(b) $\boldsymbol{f'(0), f''(0), f'''(0), \cdots \neq \infty}$
For example, for the function $f(x) = \cos x$, $f'(0) = -\sin 0 = 0$, $f''(0) = -\cos 0 = -1$, and so on; thus cos x meets this condition. However, if $f(x) = \ln x$, $f'(0) = \frac{1}{0} = \infty$, thus $\ln x$ does not meet this condition.

(c) **The resultant Maclaurin's series must be convergent**
In general, this means that the values of the terms, or groups of terms, must get progressively smaller and the sum of the terms must reach a limiting value. For example, the series $1 + \frac{1}{2} + \frac{1}{4} + \frac{1}{8} + \cdots$ is convergent since the value of the terms is getting smaller and the sum of the terms is approaching a limiting value of 2

Worked examples on Maclaurin's series

For example, to determine the first four terms of the power series for $\cos x$:

The values of $f(0)$, $f'(0)$, $f''(0)$, … in the Maclaurin's series are obtained as follows:

$$\begin{aligned}
f(x) &= \cos x & f(0) &= \cos 0 = 1\\
f'(x) &= -\sin x & f'(0) &= -\sin 0 = 0\\
f''(x) &= -\cos x & f''(0) &= -\cos 0 = -1\\
f'''(x) &= \sin x & f'''(0) &= \sin 0 = 0\\
f^{iv}(x) &= \cos x & f^{iv}(0) &= \cos 0 = 1\\
f^{v}(x) &= -\sin x & f^{v}(0) &= -\sin 0 = 0\\
f^{vi}(x) &= -\cos x & f^{vi}(0) &= -\cos 0 = -1
\end{aligned}$$

Substituting these values into equation (1) gives:

$$f(x) = \cos x = 1 + x(0) + \frac{x^2}{2!}(-1) + \frac{x^3}{3!}(0) + \frac{x^4}{4!}(1) + \frac{x^5}{5!}(0) + \frac{x^6}{6!}(-1) + \cdots$$

i.e. $\mathbf{\cos x = 1 - \dfrac{x^2}{2!} + \dfrac{x^4}{4!} - \dfrac{x^6}{6!} + \cdots}$

In another example, to determine the power series for $\cos 2\theta$:
Replacing x with 2θ in the series obtained in the previous example gives:

$$\cos 2\theta = 1 - \frac{(2\theta)^2}{2!} + \frac{(2\theta)^4}{4!} - \frac{(2\theta)^6}{6!} + \cdots$$

$$= 1 - \frac{4\theta^2}{2} + \frac{16\theta^4}{24} - \frac{64\,\theta^6}{720} + \cdots$$

i.e. $\mathbf{\cos 2\theta = 1 - 2\theta^2 + \dfrac{2}{3}\theta^4 - \dfrac{4}{45}\theta^6 + \cdots}$

In another example, to expand $\ln(1 + x)$ to five terms:

$$\begin{aligned}
f(x) &= \ln(1+x) & f(0) &= \ln(1+0) = 0\\
f'(x) &= \frac{1}{(1+x)} & f'(0) &= \frac{1}{1+0} = 1\\
f''(x) &= \frac{-1}{(1+x)^2} & f''(0) &= \frac{-1}{(1+0)^2} = -1\\
f'''(x) &= \frac{2}{(1+x)^3} & f'''(0) &= \frac{2}{(1+0)^3} = 2\\
f^{iv}(x) &= \frac{-6}{(1+x)^4} & f^{iv}(0) &= \frac{-6}{(1+0)^4} = -6\\
f^{v}(x) &= \frac{24}{(1+x)^5} & f^{v}(0) &= \frac{24}{(1+0)^5} = 24
\end{aligned}$$

Substituting these values into equation (1) gives:

$$f(x) = \ln(1+x) = 0 + x(1) + \frac{x^2}{2!}(-1) + \frac{x^3}{3!}(2) + \frac{x^4}{4!}(-6) + \frac{x^5}{5!}(24)$$

i.e. $\mathbf{\ln(1+x) = x - \frac{x^2}{2} + \frac{x^3}{3} - \frac{x^4}{4} + \frac{x^5}{5} - \cdots}$

Numerical integration using Maclaurin's series

The value of many integrals cannot be determined using the various analytical methods. In chapter 64, the trapezoidal, mid-ordinate and Simpson's rules are used to numerically evaluate such integrals. Another method of finding the approximate value of a definite integral is to express the function as a power series using Maclaurin's series, and then integrating each algebraic term in turn.

For example, to evaluate $\int_{0.1}^{0.4} 2e^{\sin\theta} d\theta$, correct to 3 significant figures:

A power series for $e^{\sin\theta}$ is firstly obtained using Maclaurin's series.

$f(\theta) = e^{\sin\theta}$ $\qquad f(0) = e^{\sin 0} = e^0 = 1$

$f'(\theta) = \cos\theta e^{\sin\theta}$ $\qquad f'(0) = \cos 0 e^{\sin 0} = (1)e^0 = 1$

$f''(\theta) = (\cos\theta)(\cos\theta e^{\sin\theta}) + (e^{\sin\theta})(-\sin\theta)$, by the product rule,

$= e^{\sin\theta}(\cos^2\theta - \sin\theta);$ $\qquad f''(0) = e^0(\cos^2 0 - \sin 0) = 1$

$f'''(\theta) = (e^{\sin\theta})[(2\cos\theta(-\sin\theta) - \cos\theta)] + (\cos^2\theta - \sin\theta)(\cos\theta e^{\sin\theta})$

$= e^{\sin\theta}\cos\theta[-2\sin\theta - 1 + \cos^2\theta - \sin\theta]$

$f'''(0) = e^0\cos 0[(0 - 1 + 1 - 0)] = 0$

Hence from equation (1):

$$e^{\sin\theta} = f(0) + \theta f'(0) + \frac{\theta^2}{2!}f''(0) + \frac{\theta^3}{3!}f'''(0) + \ldots$$

$$= 1 + \theta + \frac{\theta^2}{2} + 0$$

Thus $\int_{0.1}^{0.4} 2e^{\sin\theta} d\theta$

$$= \int_{0.1}^{0.4} 2\left(1 + \theta + \frac{\theta^2}{2}\right) d\theta = \int_{0.1}^{0.4} (2 + 2\theta + \theta^2) d\theta$$

$$= \left[2\theta + \frac{2\theta^2}{2} + \frac{\theta^3}{3}\right]_{0.1}^{0.4}$$

$$= \left(0.8 + (0.4)^2 + \frac{(0.4)^3}{3}\right) - \left(0.2 + (0.1)^2 + \frac{(0.1)^3}{3}\right)$$

$$= 0.98133 - 0.21033$$

$= \mathbf{0.771}$, correct to 3 significant figures

Limiting values

It is sometimes necessary to find limits of the form $\underset{\delta x \to a}{\text{limit}} \left\{ \frac{f(x)}{g(x)} \right\}$, where $f(a) = 0$ and $g(a) = 0$

For example, $\underset{\delta x \to 1}{\text{limit}} \left\{ \frac{x^2 + 3x - 4}{x^2 - 7x + 6} \right\} = \frac{1 + 3 - 4}{1 - 7 + 6} = \frac{0}{0}$, and $\frac{0}{0}$ is generally referred to as indeterminate.

L'Hopital's rule enables us to determine such limits when the differential coefficients of the numerator and denominator can be found.

L'Hopital's rule states: $\underset{\delta x \to a}{\text{limit}} \left\{ \frac{\boldsymbol{f(x)}}{\boldsymbol{g(x)}} \right\} = \underset{\delta x \to a}{\text{limit}} \left\{ \frac{\boldsymbol{f'(x)}}{\boldsymbol{g'(x)}} \right\}$ provided $g'(a) \neq 0$

It can happen that $\underset{\delta x \to 0}{\text{limit}} \left\{ \frac{f'(x)}{g'(x)} \right\}$ is still $\frac{0}{0}$; if so, the numerator and denominator are differentiated again (and again) until a non-zero value is obtained for the denominator.

For example, to determine $\underset{\delta x \to 1}{\text{limit}} \left\{ \frac{x^2 + 3x - 4}{x^2 - 7x + 6} \right\}$:

The first step is to substitute $x = 1$ into both numerator and denominator. In this case we obtain $\frac{0}{0}$. It is only when we obtain such a result that we then use L'Hopital's rule. Hence applying L'Hopital's rule,

$$\underset{\delta x \to 1}{\text{limit}} \left\{ \frac{x^2 + 3x - 4}{x^2 - 7x + 6} \right\} = \underset{\delta x \to 1}{\text{limit}} \left\{ \frac{2x + 3}{2x - 7} \right\}$$

i.e. both numerator and denominator have been differentiated

$$= \frac{5}{-5} = \mathbf{-1}$$

---

18 Solving Equations by Iterative Methods

Introduction to iterative methods

Many equations can only be solved graphically or by methods or successive approximations to the roots, called **iterative methods**. Three methods of successive approximations are (i) the bisection method, (ii) an algebraic method, and (iii) by using the Newton-Raphson formula.

Each successive approximation method relies on a reasonably good first estimate of the value of a root being made. One way of determining this is to sketch a graph of the function, say $y = f(x)$, and determine the approximate values of roots from the points where the graph cuts the x-axis. Another way is by using a functional notation method. This method uses the property that the value of the graph of $f(x) = 0$ changes sign for values of x just before and just after the value of a root. For example, one root of the equation $x^2 - x - 6 = 0$ is $x = 3$. Using functional notation:

$$f(x) = x^2 - x - 6$$

$$f(2) = 2^2 - 2 - 6 = -4$$

$$f(4) = 4^2 - 4 - 6 = +6$$

It can be seen from these results that the value of $f(x)$ changes from -4 at $f(2)$ to $+6$ at $f(4)$, indicating that a root lies between 2 and 4. This is shown more clearly in Figure 18.1.

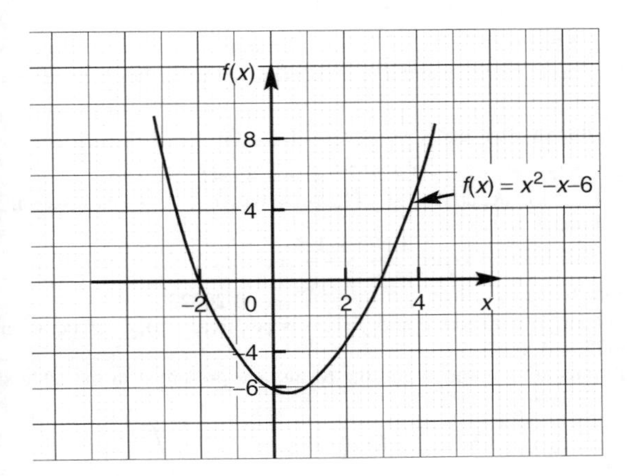

Figure 18.1

The bisection method

As shown above, by using functional notation it is possible to determine the vicinity of a root of an equation by the occurrence of a change of sign, i.e. if x_1 and x_2 are such that $f(x_1)$ and $f(x_2)$ have opposite signs, there is at least one root of the equation $f(x) = 0$ in the interval between x_1 and x_2 (provided $f(x)$ is a continuous function). In the **method of bisection** the mid-point of the interval, i.e. $x_3 = \dfrac{x_1 + x_2}{2}$, is taken, and from the sign of $f(x_3)$ it can be deduced whether a root lies in the half interval to the left or right of x_3. Whichever half interval is indicated, its mid-point is then taken and the procedure repeated. The method often requires many iterations and is therefore slow, but never fails to eventually produce the root. The procedure stops when two successive value of x are equal, to the required degree of accuracy.

For example, using the bisection method to determine the positive root of the equation $x + 3 = e^x$, correct to 3 decimal places:

Let $f(x) = x + 3 - e^x$ then, using functional notation:

$$\boldsymbol{f(0)} = 0 + 3 - e^0 = \mathbf{+2}$$

$$\boldsymbol{f(1)} = 1 + 3 - e^1 = \mathbf{+1.2817}\ldots$$

$$\boldsymbol{f(2)} = 2 + 3 - e^2 = \mathbf{-2.3890}\ldots$$

Since $f(1)$ is positive and $f(2)$ is negative, a root lies between $x = 1$ and $x = 2$. A sketch of $f(x) = x + 3 - e^x$, i.e. $x + 3 = e^x$ is shown in Figure 18.2.

Bisecting the interval between $x = 1$ and $x = 2$ gives $\dfrac{1+2}{2}$ i.e. 1.5

Hence $\boldsymbol{f(1.5)} = 1.5 + 3 - e^{1.5} = \mathbf{+0.01831}\ldots$

Since $f(1.5)$ is positive and $f(2)$ is negative, a root lies between $x = 1.5$ and $x = 2$. Bisecting this interval gives $\dfrac{1.5+2}{2}$ i.e. 1.75

Hence $\boldsymbol{f(1.75)} = 1.75 + 3 - e^{1.75} = \mathbf{-1.00460}\ldots$

Since $f(1.75)$ is negative and $f(1.5)$ is positive, a root lies between $x = 1.75$ and $x = 1.5$

Bisecting this interval gives $\dfrac{1.75+1.5}{2}$ i.e. 1.625

Hence $\boldsymbol{f(1.625)} = 1.625 + 3 - e^{1.625} = \mathbf{-0.45341}\ldots$

Since $f(1.625)$ is negative and $f(1.5)$ is positive, a root lies between $x = 1.625$ and $x = 1.5$

Bisecting this interval gives $\dfrac{1.625+1.5}{2}$ i.e. 1.5625

Hence $\boldsymbol{f(1.5625)} = 1.5625 + 3 - e^{1.5625} = \mathbf{-0.20823}\ldots$

Since $f(1.5625)$ is negative and $f(1.5)$ is positive, a root lies between $x = 1.5625$ and $x = 1.5$.

The iterations are continued and the results are presented in the table shown.

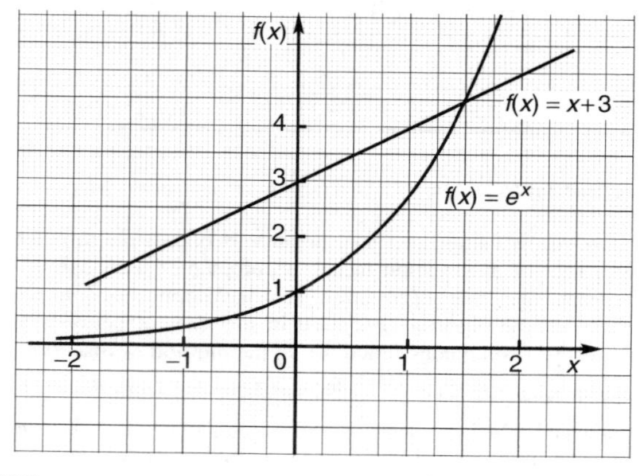

Figure 18.2

The last two values of x_3 in the table are 1.504882813 and 1.505388282, i.e. both are equal to 1.505, correct to 3 decimal places. The process therefore stops.

Hence the root of $x + 3 = e^x$ is $x = 1.505$, correct to 3 decimal places.

x_1	x_2	$x_3 = \frac{x_1 + x_2}{2}$	$f(x_3)$
		0	+2
		1	+1.2817..
		2	−2.3890..
1	2	1.5	+0.0183..
1.5	2	1.75	−1.0046..
1.5	1.75	1.625	−0.4534..
1.5	1.625	1.5625	−0.2082..
1.5	1.5625	1.53125	−0.0927..
1.5	1.53125	1.515625	−0.0366..
1.5	1.515625	1.5078125	−0.0090..
1.5	1.5078125	1.50390625	+0.0046..
1.50390625	1.5078125	1.505859375	−0.0021..
1.50390625	1.505859375	**1.504882813**	+0.0012..
1.504882813	1.505859375	**1.505388282**	

An algebraic method of successive approximations

This method can be used to solve equations of the form:
$a + bx + cx^2 + dx^3 + \ldots. = 0$, where $a, b, c, d, \ldots$ are constants.
Procedure:

First approximation

(a) Using a graphical or the functional notation method determine an approximate value of the root required, say x_1

Second approximation

(b) Let the true value of the root be $(x_1 + \delta_1)$
(c) Determine x_2 the approximate value of $(x_1 + \delta_1)$ by determining the value of $f(x_1 + \delta_1) = 0$, but neglecting terms containing products of δ_1

Third approximation

(d) Let the true value of the root be $(x_2 + \delta_2)$
(e) Determine x_3, the approximate value of $(x_2 + \delta_2)$ by determining the value of $f(x_2 + \delta_2) = 0$, but neglecting terms containing products of δ_2
(f) The fourth and higher approximations are obtained in a similar way.

Using the techniques given in paragraphs (b) to (f), it is possible to continue getting values nearer and nearer to the required root. The procedure is repeated until the value of the required root does not change on two consecutive approximations, when expressed to the required degree of accuracy.

[**Note on accuracy and errors**. Depending on the accuracy of evaluating the $f(x + \delta)$ terms, one or two iterations (i.e. successive approximations) might be saved. However, it is not usual to work to more than about 4 significant figures accuracy in this type of calculation. If a small error is made in calculations, the only likely effect is to increase the number of iterations.]

For example, to determine the value of the smallest positive root of the equation $3x^3 - 10x^2 + 4x + 7 = 0$, correct to 3 significant figures, using an algebraic method of successive approximations:

The functional notation method is used to find the value of the first approximation.

$$f(x) = 3x^3 - 10x^2 + 4x + 7$$

$$f(0) = 3(0)^3 - 10(0)^2 + 4(0) + 7 = 7$$

$$f(1) = 3(1)^3 - 10(1)^2 + 4(1) + 7 = 4$$

$$f(2) = 3(2)^3 - 10(2)^2 + 4(2) + 7 = -1$$

Following the above procedure:

First approximation

(a) Let the first approximation be such that it divides the interval 1 to 2 in the ratio of 4 to -1, i.e. let x_1 be 1.8

Second approximation

(b) Let the true value of the root, x_2, be $(x_1 + \delta_1)$

(c) Let $f(x_1 + \delta_1) = 0$, then since $x_1 = 1.8$,
$3(1.8 + \delta_1)^3 - 10(1.8 + \delta_1)^2 + 4(1.8 + \delta_1) + 7 = 0$
Neglecting terms containing products of δ_1 and using the binomial series gives:

$$3[1.8^3 + 3(1.8)^2\delta_1 - 10[1.8^2 + (2)(1.8)\delta_1] + 4(1.8 + \delta_1) + 7 \approx 0$$

$$3(5.832 + 9.720\delta_1) - 32.4 - 36\delta_1 + 7.2 + 4\delta_1 + 7 \approx 0$$

$$17.496 + 29.16\delta_1 - 32.4 - 36\delta_1 + 7.2 + 4\delta_1 + 7 \approx 0$$

$$\delta_1 \approx \frac{-17.496 + 32.4 - 7.2 - 7}{29.16 - 36 + 4} \approx -\frac{0.704}{2.84} \approx -0.2479$$

Thus $x_2 \approx 1.8 - 0.2479 = 1.5521$

Third approximation

(d) Let the true value of the root, x_3, be $(x_2 + \delta_2)$

(e) Let $f(x_2 + \delta_2) = 0$, then since $x_2 = 1.5521$,

$$3(1.5521 + \delta_2)^3 - 10(1.5521 + \delta_2)^2 + 4(1.5521 + \delta_2) + 7 = 0$$

Neglecting terms containing products of δ_2 gives:

$$11.217 + 21.681\delta_2 - 24.090 - 31.042\delta_2 + 6.2084 + 4\delta_2 + 7 \approx 0$$

$$\delta_2 \approx \frac{-11.217 + 24.090 - 6.2084 - 7}{21.681 - 31.042 + 4} \approx \frac{-0.3354}{-5.361} \approx 0.06256$$

Thus $x_3 \approx 1.5521 + 0.06256 \approx 1.6147$

(f) Values of x_4 and x_5 are found in a similar way.

$$f(x_3) + \delta_3 = 3(1.6147 + \delta_3)^3 - 10(1.6147 + \delta_3)^2 + 4(1.6147 + \delta_3) + 7 = 0$$

giving $\delta_3 \approx 0.003175$ and $x_4 \approx 1.618$, i.e. 1.62 correct to 3 significant figures

$$f(x_4 + \delta_4) = 3(1.618 + \delta_4)^3 - 10(1.618 + \delta_4)^2 + 4(1.618 + \delta_4) + 7 = 0$$

giving $\delta_4 \approx 0.0000417$, and $x_5 \approx 1.62$, correct to 3 significant figures. Since x_4 and x_5 are the same when expressed to the required degree of accuracy, then the required root is **1.62**, correct to 3 significant figures.

The Newton-Raphson method

The Newton-Raphson formula, often just referred to as **Newton's method**, may be stated as follows:

> *if r_1 is the approximate value of a real root of the equation $f(x) = 0$, then a closer approximation to the root r_2 is given by:*
>
> $$\boldsymbol{r_2 = r_1 - \frac{f(r_1)}{f'(r_1)}}$$

The advantages of Newton's method over the algebraic method of successive approximations is that it can be used for any type of mathematical equation (i.e. ones containing trigonometric, exponential, logarithmic, hyperbolic and algebraic functions), and it is usually easier to apply than the algebraic method. **For example**, using Newton's method to find the positive root of

$$(x+4)^3 - e^{1.92x} + 5\cos\frac{x}{3} = 9, \text{ correct to 3 significant figures:}$$

The functional notational method is used to determine the approximate value of the root.

$$f(x) = (x+4)^3 - e^{1.92x} + 5\cos\frac{x}{3} - 9$$

$$f(0) = (0+4)^3 - e^0 + 5\cos 0 - 9 = 59$$

$$f(1) = 5^3 - e^{1.92} + 5\cos\tfrac{1}{3} - 9 \approx 114$$

$$f(2) = 6^3 - e^{3.84} + 5\cos\tfrac{2}{3} - 9 \approx 164$$

$$f(3) = 7^3 - e^{5.76} + 5\cos 1 - 9 \approx 19$$

$$f(4) = 8^3 - e^{7.68} + 5\cos\tfrac{4}{3} - 9 \approx -1660$$

From these results, let a first approximation to the root be $r_1 = 3$
Newton's formula states that a better approximation to the root,

$$r_2 = r_1 - \frac{f(r_1)}{f'(r_1)}$$

$$f(r_1) = f(3) = 7^3 - e^{5.76} + 5\cos 1 - 9 = 19.35$$

$$f'(x) = 3(x+4)^2 - 1.92e^{1.92x} - \frac{5}{3}\sin\frac{x}{3}$$

$$f'(r_1) = f'(3) = 3(7)^2 - 1.92e^{5.76} - \frac{5}{3}\sin 1 = -463.7$$

Thus, $r_3 = 3 - \dfrac{19.35}{-463.7} = 3 + 0.042 = 3.042 = 3.04,$

correct to 3 significant figure

Similarly, $r_3 = 3.042 - \dfrac{f(3.042)}{f'(3.042)} = 3.042 - \dfrac{(-1.146)}{(-513.1)}$

$= 3.042 - 0.0022 = 3.0398$

$= 3.04,$ correct to 3 significant figures.

Since r_2 and r_3 are the same when expressed to the required degree of accuracy, then the required root is **3.04**, correct to 3 significant figures.

---

19 Computer Numbering Systems

Decimal and binary numbers

The system of numbers in everyday use is the **denary** or **decimal** system of numbers, using the digits 0 to 9. It has ten different digits (0, 1, 2, 3, 4, 5, 6, 7, 8 and 9) and is said to have a **radix** or **base** of 10.
The **binary** system of numbers has a radix of 2 and uses only the digits 0 and 1.

Conversion of binary to denary

The denary number 234.5 is equivalent to

$$2 \times 10^2 + 3 \times 10^1 + 4 \times 10^0 + 5 \times 10^{-1}$$

i.e. is the sum of terms comprising: (a digit) multiplied by (the base raised to some power).
In the binary system of numbers, the base is 2, so 1101.1 is equivalent to:

$$1 \times 2^3 + 1 \times 2^2 + 0 \times 2^1 + 1 \times 2^0 + 1 \times 2^{-1}$$

Thus the denary number equivalent to the binary number 1101.1 is $8 + 4 + 0 + 1 + \frac{1}{2}$, that is 13.5 i.e. $\mathbf{1101.1_2 = 13.5_{10}}$, the suffixes 2 and 10 denoting binary and denary systems of numbers respectively.
In another, to convert 101.0101_2 to a denary number:

$$101.0101_2 = 1 \times 2^2 + 0 \times 2^1 + 1 \times 2^0 + 0 \times 2^{-1} + 1 \times 2^{-2} + 0 \times 2^{-3} + 1 \times 2^{-4}$$

$$= 4 + 0 + 1 + 0 + 0.25 + 0 + 0.0625 = \mathbf{5.3125_{10}}$$

Conversion of denary to binary

An integer denary number can be converted to a corresponding binary number by repeatedly dividing by 2 and noting the remainder at each stage, as shown below for 39_{10}

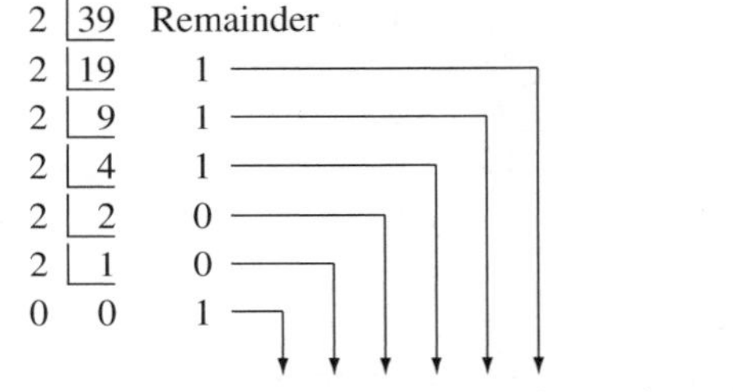

The result is obtained by writing the top digit of the remainder as the least significant bit, (a bit is a **b**inary dig**it** and the least significant bit is the one on the right). The bottom bit of the remainder is the most significant bit, i.e. the bit on the left. **Thus $39_{10} = 100111_2$**

The fractional part of a denary number can be converted to a binary number by repeatedly multiplying by 2, as shown below for the fraction 0.625

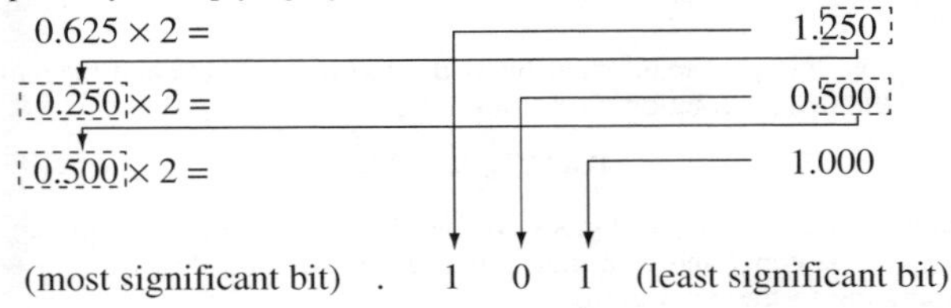

For fractions, the most significant bit of the result is the top bit obtained from the integer part of multiplication by 2. The least significant bit of the result is the bottom bit obtained from the integer part of multiplication by 2. **Thus $0.625_{10} = 0.101_2$**

Conversion of denary to binary via octal

For denary integers containing several digits, repeatedly dividing by 2 can be a lengthy process. In this case, it is usually easier to convert a denary number to a binary number via the **octal system** of numbers. This system has a radix of 8, using the digits 0, 1, 2, 3, 4, 5, 6 and 7. The denary number equivalent to the octal number 4317_8 is

$$4 \times 8^3 + 3 \times 8^2 + 1 \times 8^1 + 7 \times 8^0$$

i.e. $\quad 4 \times 512 + 3 \times 64 + 1 \times 8 + 7 \times 1$ or 2255_{10}

Thus $\mathbf{4317_8 = 2255_{10}}$

An integer denary number can be converted to a corresponding octal number by repeatedly dividing by 8 and noting the remainder at each stage, as shown below for 493_{10}

		Remainder
8	493	
8	61	5
8	7	5
	0	7

7 5 5

Thus $\mathbf{493_{10} = 755_8}$

The fractional part of a denary number can be converted to an octal number by repeatedly multiplying by 8, as shown below for the fraction 0.4375_{10}

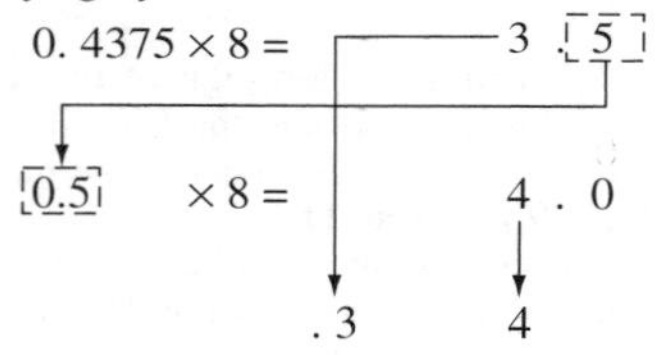

For fractions, the most significant bit is the top integer obtained by multiplication of the denary fraction by 8, thus

$$\mathbf{0.4375_{10} = 0.34_8}$$

The natural binary code for digits 0 to 7 is shown in Table 19.1, and an octal number can be converted to a binary number by writing down the three bits corresponding to the octal digit.

Thus $437_8 = 100\ 011\ 111_2$

and $26.35_8 = 010\ 110.011\ 101_2$

Table 19.1

Octal digit	*Natural binary number*
0	000
1	001
2	010
3	011
4	100
5	101
6	110
7	111

The '0' on the extreme left does not signify anything, thus

$$26.35_8 = 10\ 110.011\ 101_2$$

To convert 11 110 011.100 01_2 to a denary number via octal:
Grouping the binary number in three's from the binary point gives:

$$011\ 110\ 011.100\ 010_2$$

Using Table 19.1 to convert this binary number to an octal number gives: 363.42_8 and

$$\begin{aligned} 363.42_8 &= 3 \times 8^2 + 6 \times 8^1 + 3 \times 8^0 + 4 \times 8^{-1} + 2 \times 8^{-2} \\ &= 192 + 48 + 3 + 0.5 + 0.03125 \\ &= \mathbf{243.53125_{10}} \end{aligned}$$

Hence $\mathbf{11\ 110\ 011.100\ 01_2 = 363.42_8 = 243.53125_{10}}$

Hexadecimal numbers

The complexity of computers requires higher order numbering systems, such as octal (base 8) and hexadecimal (base 16), which are merely extensions of the binary system. A **hexadecimal numbering system** has a radix of 16 and uses the following 16 distinct digits:

0, 1, 2, 3, 4, 5, 6, 7, 8, 9, A, B, C, D, E and F

'A' corresponds to 10 in the denary system, B to 11, C to 12, and so on.

To convert from hexadecimal to decimal

For example, $1A_{16} = 1 \times 16^1 + A \times 16^0 = 1 \times 16^1 + 10 \times 1$

$= 16 + 10 = 26$

i.e. $\mathbf{1A_{16} = 26_{10}}$

Similarly, $\mathbf{2\mathit{E}_{16}} = 2 \times 16^1 + E \times 16^0 = 2 \times 16^1 + 14 \times 16^0$

$= 32 + 14 = 46_{10}$

and $\mathbf{1\mathit{BF}_{16}} = 1 \times 16^2 + B \times 16^1 + F \times 16^0$

$= 1 \times 16^2 + 11 \times 16^1 + 15 \times 16^0$

$= 256 + 176 + 15 = \mathbf{447_{10}}$

Table 19.2 compares decimal, binary, octal and hexadecimal numbers and shows, for example, that $\mathbf{23_{10} = 10111_2 = 27_8 = 17_{16}}$

To convert from decimal to hexadecimal

This is achieved by repeatedly dividing by 16 and noting the remainder at each stage, as shown below for 26_{10}

Table 19.2

Decimal	Binary	Octal	Hexadecimal
0	0000	0	0
1	0001	1	1
2	0010	2	2
3	0011	3	3
4	0100	4	4
5	0101	5	5
6	0110	6	6
7	0111	7	7
8	1000	10	8
9	1001	11	9
10	1010	12	A
11	1011	13	B
12	1100	14	C
13	1101	15	D
14	1110	16	E
15	1111	17	F
16	10000	20	10
17	10001	21	11
18	10010	22	12
19	10011	23	13
20	10100	24	14
21	10101	25	15
22	10110	26	16
23	10111	27	17
24	11000	30	18
25	11001	31	19
26	11010	32	1A
27	11011	33	1B
28	11100	34	1C
29	11101	35	1D
30	11110	36	1E
31	11111	37	1F
32	100000	40	20

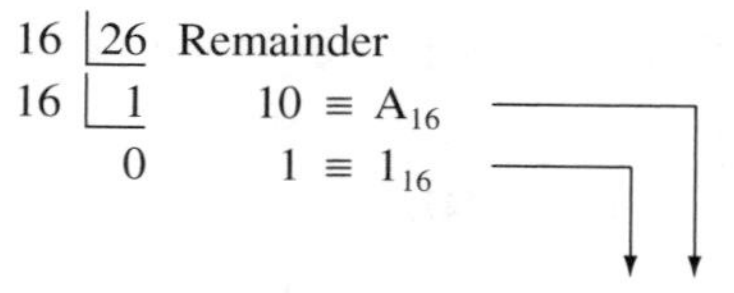

Hence $\mathbf{26_{10} = 1A_{16}}$

Similarly, for 447_{10}

$$\begin{array}{r|l l}
16 & 447 & \text{Remainder} \\
16 & 27 & 15 \equiv F_{16} \\
16 & 1 & 11 \equiv B_{16} \\
 & 0 & 1 \equiv 1_{16}
\end{array}$$

1 B F

Thus $\mathbf{447_{10} = 1BF_{16}}$

To convert from binary to hexadecimal

The binary bits are arranged in groups of four, starting from right to left, and a hexadecimal symbol is assigned to each group. For example, the binary number 1110011110101001 is initially grouped in fours as: 1110 0111 1010 1001 and a hexadecimal symbol assigned to each group as E 7 A 9 from Table 19.2
Hence $\mathbf{1110011110101001_2 = E7A9_{16}}$

To convert from hexadecimal to binary

The above procedure is reversed, thus, for example,

$6CF3_{16}$ = 0110 1100 1111 0011 from Table 19.2

i.e. $\mathbf{6CF3_{16} = 110110011110011_2}$

Mensuration

20 Areas of Plane Figures

Mensuration

Mensuration is a branch of mathematics concerned with the determination of lengths, areas and volumes.

Properties of quadrilaterals

A **polygon** is a closed plane figure bounded by straight lines. A polygon, which has:

(i) 3 sides is called a **triangle**
(ii) 4 sides is called a **quadrilateral**
(iii) 5 sides is called a **pentagon**
(iv) 6 sides is called a **hexagon**
(v) 7 sides is called a **heptagon**
(vi) 8 sides is called an **octagon**

There are five types of **quadrilateral**, these being

(i) rectangle
(ii) square
(iii) parallelogram
(iv) rhombus
(v) trapezium
(The properties of these are given below).

If the opposite corners of any quadrilateral are joined by a straight line, two triangles are produced. Since the sum of the angles of a triangle is 180°, the sum of the angles of a quadrilateral is 360°.
In a **rectangle**, shown in Figure 20.1:
(i) all four angles are right angles,
(ii) opposite sides are parallel and equal in length, and
(iii) diagonals AC and BD are equal in length and bisect one another.

In a **square**, shown in Figure 20.2:
(i) all four angles are right angles,
(ii) opposite sides are parallel,
(iii) all four sides are equal in length, and
(iv) diagonals PR and QS are equal in length and bisect one another at right angles.

In a **parallelogram**, shown in Figure 20.3:
(i) opposite angles are equal,
(ii) opposite sides are parallel and equal in length, and
(iii) diagonals WY and XZ bisect one another.

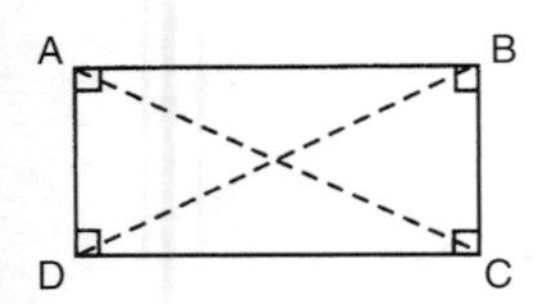

Figure 20.1

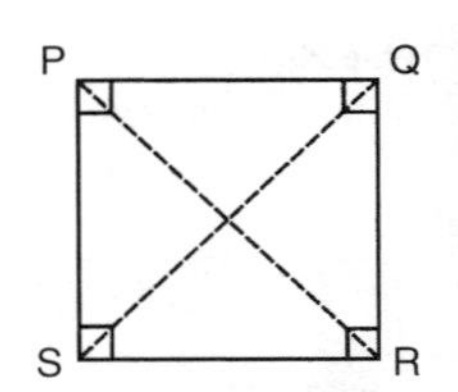

Figure 20.2

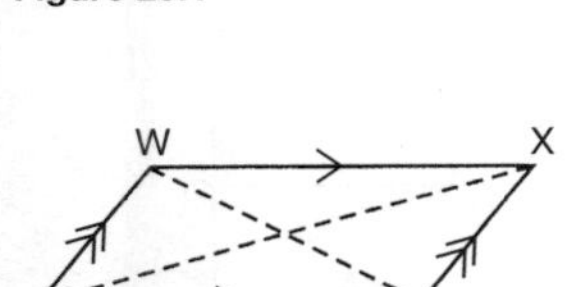

Figure 20.3

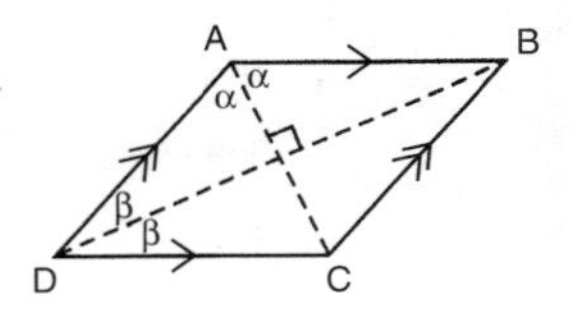

Figure 20.4

In a **rhombus**, shown in Figure 20.4:

(i) opposite angles are equal,
(ii) opposite angles are bisected by a diagonal,
(iii) opposite sides are parallel,
(iv) all four sides are equal in length, and
(v) diagonals AC and BD bisect one another at right angles.

In a **trapezium**, shown in Figure 20.5:
(i) only one pair of sides is parallel.

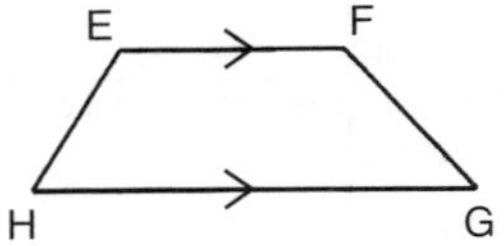

Figure 20.5

Areas of plane figures

A summary of areas of common shapes is shown in Table 20.1.
For example, a rectangular tray is 820 mm long and 400 mm wide. To find its area in (a) mm^2 (b) cm^2 (c) m^2:

(a) Area = length × width = 820 × 400 = **328 000 mm²**

(b) $1\ cm^2 = 100\ mm^2$, hence $328\,000\ mm^2 = \dfrac{328\,000}{100}\ cm^2$

$= \mathbf{3280\ cm^2}$

(c) $1\ m^2 = 10\,000\ cm^2$, hence $3280\ cm^2 = \dfrac{3280}{10\,000}\ m^2 = \mathbf{0.3280\ m^2}$

Table 20.1

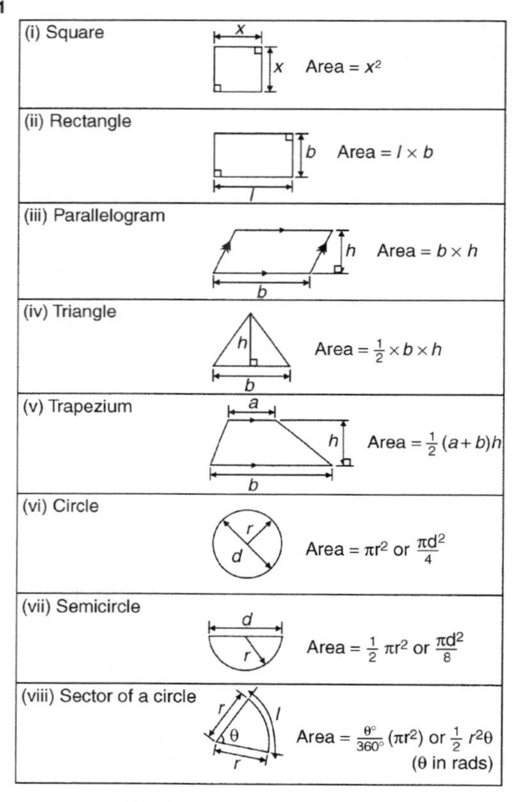

In another example, to find (a) the cross-sectional area of the girder shown in Figure 20.6(a) and (b) the area of the path shown in Figure 20.6(b):

(a) The girder may be divided into three separate rectangles as shown.

Area of rectangle A $= 50 \times 5 = 250\ \text{mm}^2$

Area of rectangle B $= (75 - 8 - 5) \times 6 = 62 \times 6 = 372\ \text{mm}^2$

Area of rectangle C $= 70 \times 8 = 560\ \text{mm}^2$

Total area of girder $= 250 + 372 + 560 = \mathbf{1182\ mm^2}$ or $\mathbf{11.82\ cm^2}$

(b) Area of path = area of large rectangle – area of small rectangle

$$= (25 \times 20) - (21 \times 16) = 500 - 336 = \mathbf{164\ m^2}$$

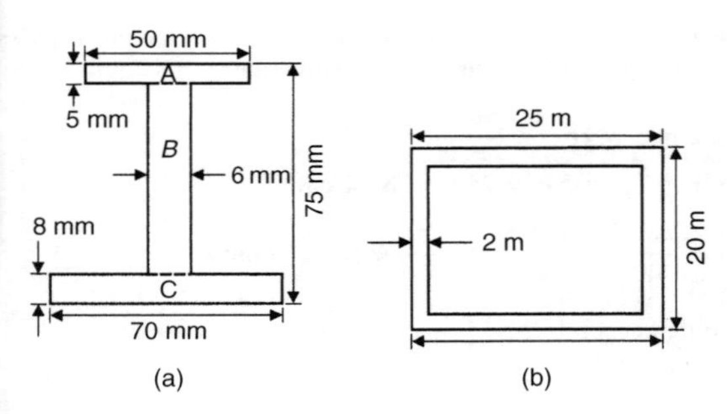

Figure 20.6

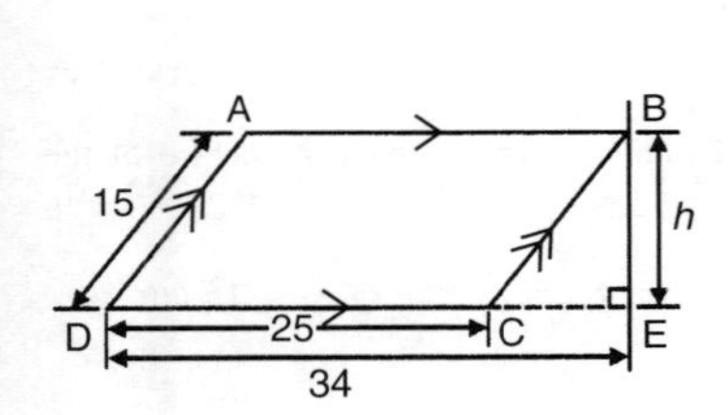

Figure 20.7

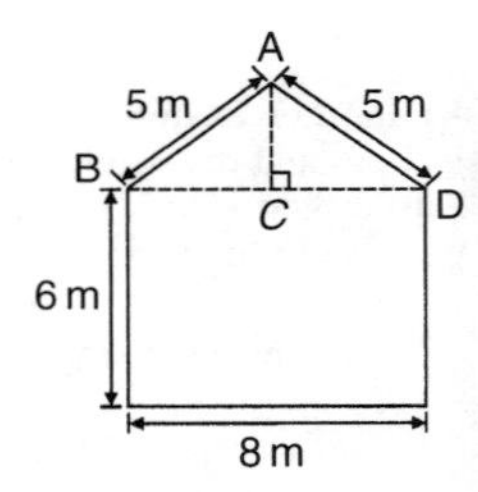

Figure 20.8

In another example, to find the area of the parallelogram shown in Figure 20.7 (dimensions are in mm):
Area of parallelogram = base × perpendicular height. The perpendicular height h is found using Pythagoras' theorem.

$$BC^2 = CE^2 + h^2$$

i.e. $$15^2 = (34 - 25)^2 + h^2$$

$$h^2 = 15^2 - 9^2 = 225 - 81 = 144$$

Hence, $h = \sqrt{144} = 12$ mm (-12 can be neglected).

Hence, area of ABCD = 25 × 12 = **300 mm²**

In another example, Figure 20.8 shows the gable end of a building. To determine the area of brickwork in the gable end:
The shape is that of a rectangle and a triangle.

$$\text{Area of rectangle} = 6 \times 8 = 48 \text{ m}^2$$

$$\text{Area of triangle} = \tfrac{1}{2} \times \text{base} \times \text{height}$$

CD = 4 m, AD = 5 m, hence AC = 3 m (since it is a 3, 4, 5 triangle)

$$\text{Hence, area of triangle ABD} = \tfrac{1}{2} \times 8 \times 3 = 12 \text{ m}^2$$

$$\text{Total area of brickwork} = 48 + 12 = \mathbf{60\ m^2}$$

In another example, to find the areas of the circles having (a) a radius of 5 cm, (b) a diameter of 15 mm, (c) a circumference of 70 mm:

Area of a circle $= \pi r^2$ or $\dfrac{\pi d^2}{4}$

(a) Area $= \pi r^2 = \pi(5)^2 = 25\pi =$ **78.54 cm²**

(b) Area $= \dfrac{\pi d^2}{4} = \dfrac{\pi(15)^2}{4} = \dfrac{225\pi}{4} =$ **176.7 mm²**

(c) Circumference, $c = 2\pi r$, hence $r = \dfrac{c}{2\pi} = \dfrac{70}{2\pi} = \dfrac{35}{\pi}$ mm

$$\text{Area of circle} = \pi r^2 = \pi\left(\frac{35}{\pi}\right)^2 = \frac{35^2}{\pi}$$

$$= \mathbf{389.9\ mm^2} \text{ or } \mathbf{3.899\ cm^2}$$

In another example, to calculate the area of a regular octagon, if each side is 5 cm and the width across the flats is 12 cm:

An octagon is an 8-sided polygon. If radii are drawn from the centre of the polygon to the vertices then 8 equal triangles are produced (see Figure 20.9).

$$\text{Area of one triangle} = \frac{1}{2} \times \text{base} \times \text{height} = \frac{1}{2} \times 5 \times \frac{12}{2} = 15\ \text{cm}^2$$

$$\text{Area of octagon} = 8 \times 15 = \mathbf{120\ cm^2}$$

In another example, to determine the area of a regular hexagon which has sides 8 cm long:

A hexagon is a 6-sided polygon that may be divided into 6 equal triangles as shown in Figure 20.10. The angle subtended at the centre of each triangle is $360°/6 = 60°$.

The other two angles in the triangle add up to 120° and are equal to each other.

Hence each of the triangles is equilateral with each angle 60° and each side 8 cm.

$$\text{Area of one triangle} = \tfrac{1}{2} \times \text{base} \times \text{height} = \tfrac{1}{2} \times 8 \times h$$

h is calculated using Pythagoras' theorem:

$$8^2 = h^2 + 4^2$$

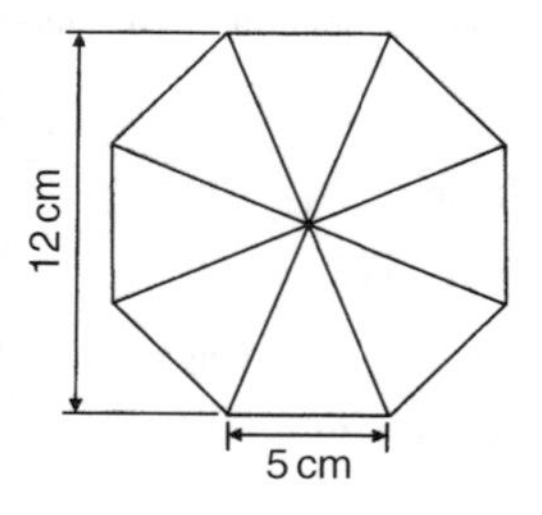

Figure 20.9

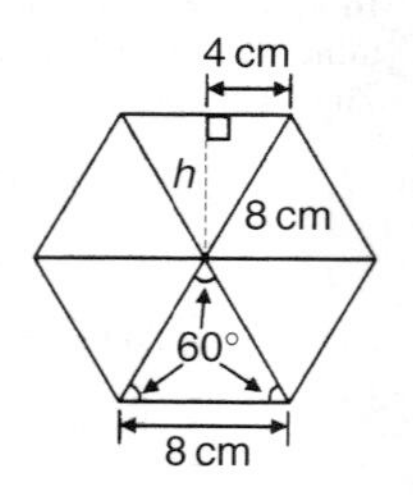

Figure 20.10

from which, $h = \sqrt{8^2 - 4^2} = 6.928$ cm

Hence area of one triangle $= \frac{1}{2} \times 8 \times 6.928 = 27.71$ cm^2

Area of hexagon $= 6 \times 27.71 =$ **166.3 cm^2**

Areas of similar shapes

The areas of similar shapes are proportional to the squares of corresponding linear dimensions. For example, Figure 20.11 shows two squares, one of which has sides three times as long as the other.

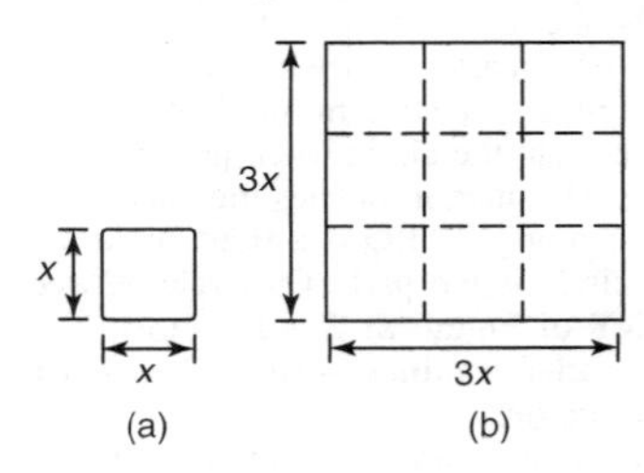

Figure 20.11

Area of Figure 20.11(a) $= (x)(x) = x^2$

Area of Figure 20.11(b) $= (3x)(3x) = 9x^2$

Hence Figure 20.11(b) has an area $(3)^2$, i.e. 9 times the area of Figure 20.11(a).

For example, a rectangular garage on a building plan has dimensions 10 mm by 20 mm. If the plan is drawn to a scale of 1 to 250, the true area of the garage in square metres, is determined as follows:
Area of garage on the plan = 10 mm × 20 mm = 200 mm^2
Since the areas of similar shapes are proportional to the squares of corresponding dimensions then:

True area of garage $= 200 \times (250)^2 = 12.5 \times 10^6$ mm^2

$= \dfrac{12.5 \times 10^6}{10^6}$ m^2 = **12.5 m^2**

---

21 The Circle and its Properties

Introduction

A **circle** is a plain figure enclosed by a curved line, every point on which is equidistant from a point within, called the **centre**.

Properties of circles

(i) The distance from the centre to the curve is called the **radius**, r, of the circle (see OP in Figure 21.1).

(ii) The boundary of a circle is called the **circumference**, c.

(iii) Any straight line passing through the centre and touching the circumference at each end is called the **diameter**, d (see QR in Figure 21.1). Thus $\boldsymbol{d = 2r}$.

(iv) The ratio $\dfrac{\text{circumference}}{\text{diameter}}$ = a constant for any circle.
This constant is denoted by the Greek letter π (pronounced 'pie'), where $\pi = 3.14159$, correct to 5 decimal places.
Hence $c/d = \pi$ or $\boldsymbol{c = \pi d}$ or $\boldsymbol{c = 2\pi r}$

(v) A **semicircle** is one half of the whole circle.

(vi) A **quadrant** is one quarter of a whole circle.

(vii) A **tangent** to a circle is a straight line that meets the circle in one point only and does not cut the circle when produced. AC in Figure 21.1 is a tangent to the circle since it touches the curve at point B only. If radius OB is drawn, then angle ABO is a right angle.

(viii) A **sector** of a circle is the part of a circle between radii (for example, the portion OXY of Figure 21.2 is a sector). If a sector is less than a semicircle it is called a **minor sector**, if greater than a semicircle it is called a **major sector**.

(ix) A **chord** of a circle is any straight line that divides the circle into two parts and is terminated at each end by the circumference. ST, in Figure 21.2 is a chord.

(x) A **segment** is the name given to the parts into which a circle is divided by a chord. If the segment is less than a semicircle it is called a **minor segment** (see shaded area in Figure 21.2). If the segment is greater than a semicircle it is called a **major segment** (see the unshaded area in Figure 21.2).

(xi) An **arc** is a portion of the circumference of a circle. The distance SRT in Figure 21.2 is called a **minor arc** and the distance SXYT is called a **major arc**.

(xii) The angle at the centre of a circle, subtended by an arc, is double the angle at the circumference subtended by the same arc. With reference to Figure 21.3: **Angle AOC = 2 × angle ABC**

(xiii) The angle in a semicircle is a right angle (see angle BQP in Figure 21.3).

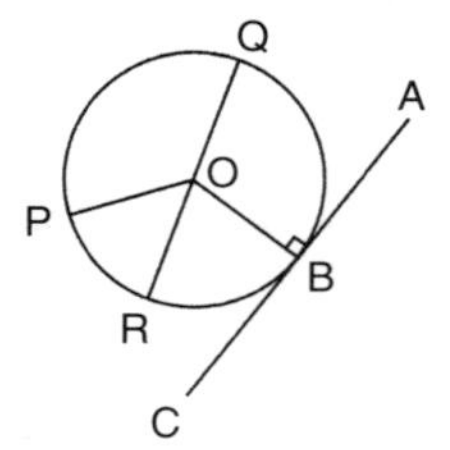

Figure 21.1

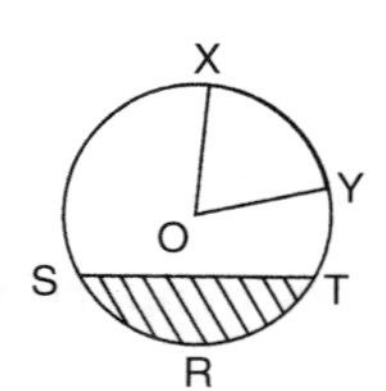

Figure 21.2

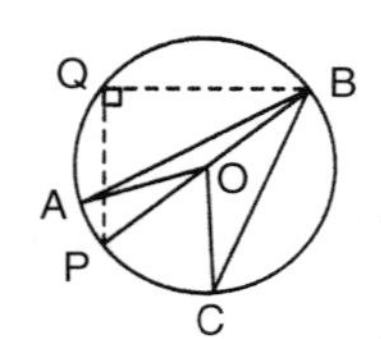

Figure 21.3

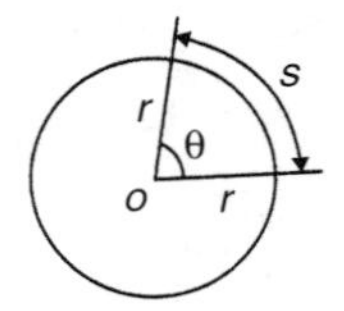

Figure 21.4

Arc length and area of a sector

One **radian** is defined as the angle subtended at the centre of a circle by an arc equal in length to the radius. With reference to Figure 21.4, for arc length s,

$$\theta \text{ radians} = s/r \quad \text{or} \quad \textbf{arc length, } \boxed{s = r\theta} \qquad (1)$$

where θ is in radians

When s = whole circumference ($= 2\pi r$) then $\theta = s/r = 2\pi r/r = 2\pi$

i.e. $\quad 2\pi$ radians $= 360°$ or $\boxed{\pi \textbf{ radians} = \mathbf{180°}}$

Thus 1 rad $= 180°/\pi = 57.30°$, correct to 2 decimal places.

Since π rad $= 180°$, then $\pi/2 = 90°$, $\pi/3 = 60°$, $\pi/4 = 45°$, and so on.

$$\textbf{Area of a sector} = \frac{\theta}{360}(\pi r^2) \quad \text{when } \theta \text{ is in degrees}$$

$$= \frac{\theta}{2\pi}(\pi r^2) = \frac{1}{2}r^2\theta \quad \text{when } \theta \text{ is in radians} \qquad (2)$$

For example, to convert (a) 125°, (b) 69°47', to radians:

(a) Since $180° = \pi$ rad then $1° = \pi/180$ rad, therefore

$$125° = 125\left(\frac{\pi}{180}\right) = \textbf{2.182 radians}$$

(b) $69°47' = 69\dfrac{47°}{60} = 69.783°$

$$69.783° = 69.783\left(\frac{\pi}{180}\right)^c = \textbf{1.218 radians}$$

In another example, to convert (a) 0.749 radians, (b) $3\pi/4$ radians, to degrees and minutes:

(a) Since π rad $= 180°$ then 1 rad $= 180°/\pi$, therefore

$$0.749 = 0.749\left(\frac{180}{\pi}\right)^\circ = 42.915°$$

$0.915° = (0.915 \times 60)' = 55'$, correct to the nearest minute, hence

0.749 radians = 42°55′

(b) Since $1 \text{ rad} = \left(\dfrac{180}{\pi}\right)^\circ$ then $\dfrac{3\pi}{4}$ rad $= \dfrac{3\pi}{4}\left(\dfrac{180}{\pi}\right)^\circ$

$$= \frac{3}{4}(180)^\circ = \mathbf{135°}$$

In another example, expressing (a) 150° (b) 270° (c) 37.5° in radians, in terms of π:
Since $180° = \pi$ rad then $1° = 180/\pi$, hence

$$\text{(a) } 150° = 150\left(\frac{\pi}{180}\right) \text{ rad} = \mathbf{\frac{5\pi}{6}\ rad}$$

$$\text{(b) } 270° = 270\left(\frac{\pi}{180}\right) \text{ rad} = \mathbf{\frac{3\pi}{2}\ rad}$$

$$\text{(c) } 37.5° = 37.5\left(\frac{\pi}{180}\right) \text{ rad} = \frac{75\pi}{360} \text{ rad} = \mathbf{\frac{5\pi}{24}\ rad}$$

In another example, to find the length of arc of a circle of radius 5.5 cm when the angle subtended at the centre is 1.20 radians:

From equation (1), length of arc, $s = r\theta$, where θ is in radians,

hence $s = (5.5)(1.20) = \mathbf{6.60\ cm}$

In another example, to determine the diameter and circumference of a circle if an arc of length 4.75 cm subtends an angle of 0.91 radians:

Since $s = r\theta$ then $r = \dfrac{s}{\theta} = \dfrac{4.75}{0.91} = 5.22$ cm

$$\text{Diameter} = 2 \times \text{ radius} = 2 \times 5.22 = \mathbf{10.44\ cm}$$

$$\text{Circumference, } c = \pi d = \pi(10.44) = \mathbf{32.80\ cm}$$

In another example, a football stadium floodlight can spread its illumination over an angle of 45° to a distance of 55 m. To determine the maximum area that is floodlit:

$$\textbf{Floodlit area} = \text{area of sector} = \frac{1}{2}r^2\theta$$

$$= \frac{1}{2}(55)^2\left(45 \times \frac{\pi}{180}\right) \text{ from equation (2)}$$

$$= \mathbf{1188\ m^2}$$

The equation of a circle

The simplest equation of a circle, centre at the origin, radius r, is given by:

$$x^2 + y^2 = r^2$$

For example, Figure 21.5 shows a circle $x^2 + y^2 = 9$
More generally, the equation of a circle, centre (a, b), radius r, is given by:

$$(x - a)^2 + (y - b)^2 = r^2 \qquad (1)$$

Figure 21.6 shows a circle $(x - 2)^2 + (y - 3)^2 = 4$

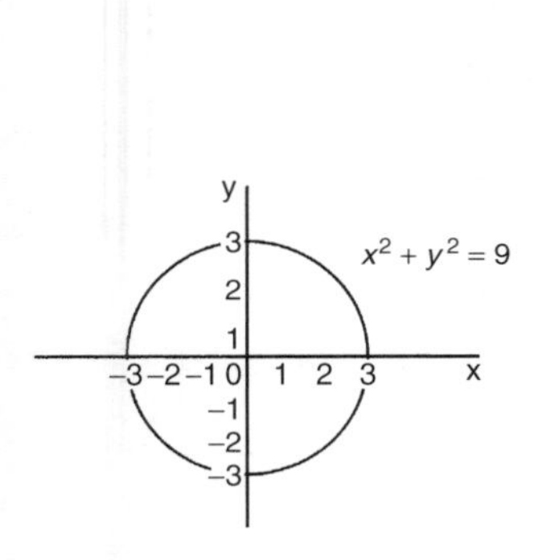

Figure 21.5

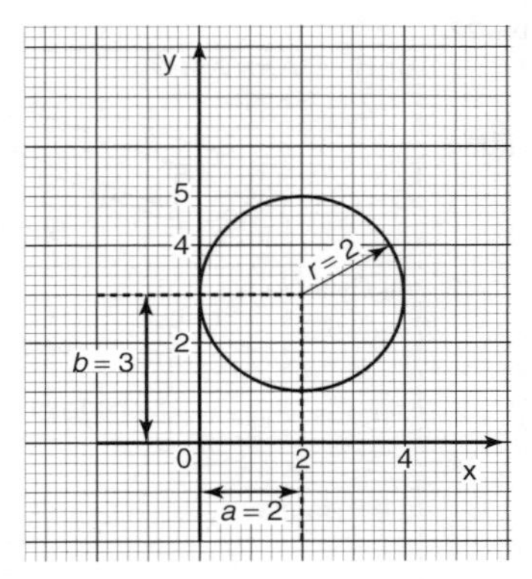

Figure 21.6

The general equation of a circle is:

$$x^2 + y^2 + 2ex + 2fy + c = 0 \qquad (2)$$

Multiplying out the bracketed terms in equation (1) gives:

$$x^2 - 2ax + a^2 + y^2 - 2by + b^2 = r^2$$

Comparing this with equation (2) gives:

$$2e = -2a, \text{ i.e. } \boldsymbol{a = -\frac{2e}{2}} \text{ and } 2f = -2b, \text{ i.e. } \boldsymbol{b = -\frac{2f}{2}}$$

and $c = a^2 + b^2 - r^2$, i.e. $\boldsymbol{r = \sqrt{a^2 + b^2 - c}}$
Thus, for example, the equation

$$x^2 + y^2 - 4x - 6y + 9 = 0$$

represents a circle with centre $a = -\left(\frac{-4}{2}\right)$, $b = -\left(\frac{-6}{2}\right)$, i.e. at (2, 3) and radius $r = \sqrt{2^2 + 3^2 - 9} = 2$
Hence $x^2 + y^2 - 4x - 6y + 9 = 0$ is the circle shown in Figure 21.6, which may be checked by multiplying out the brackets in the equation $(x - 2)^2 + (y - 3)^2 = 4$

---

22 Volumes of Common Solids

Volumes and surface areas of regular solids
A summary of volumes and surface areas of regular solids is shown in Table 22.1.
For example, a water tank is the shape of a rectangular prism having length 2 m, breadth 75 cm and height 50 cm. To determine the capacity of the tank in (a) m^3 (b) cm^3 (c) litres :

Table 22.1

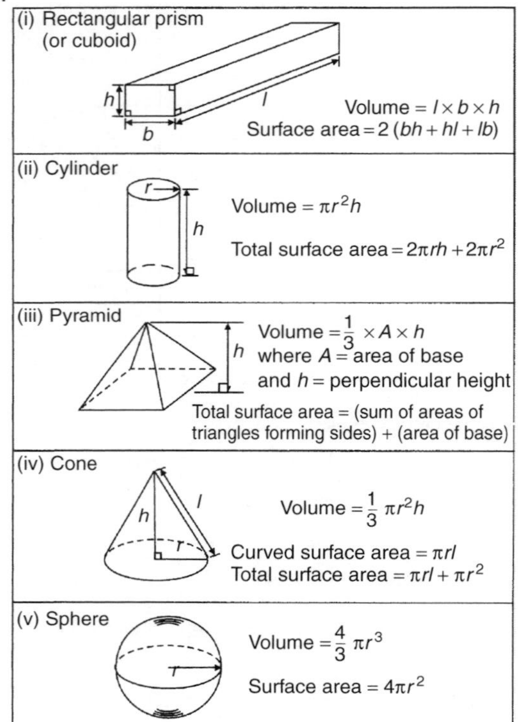

Volume of rectangular prism $= l \times b \times h$ (see Table 22.1)

(a) Volume of tank $= 2 \times 0.75 \times 0.5 = \mathbf{0.75\ m^3}$

(b) $1\ \text{m}^3 = 10^6\ \text{cm}^3$, hence $0.75\ \text{m}^3 = 0.75 \times 10^6\ \text{cm}^3 = \mathbf{750\,000\ cm^3}$

(c) $1\ \text{litre} = 1000\ \text{cm}^3$, hence $750\,000\ \text{cm}^3 = \dfrac{750\,000}{1000}$ litres $=$ **750 litres**

In another example, to calculate the volume and total surface area of the solid prism shown in Figure 22.1:
The solid shown in Figure 22.1 is a trapezoidal prism.

$$\text{Volume} = \text{cross-sectional area} \times \text{height}$$

$$= \tfrac{1}{2}(11 + 5)4 \times 15 = 32 \times 15 = \mathbf{480\ cm^3}$$

$$\text{Surface area} = \text{sum of two trapeziums} + 4 \text{ rectangles}$$

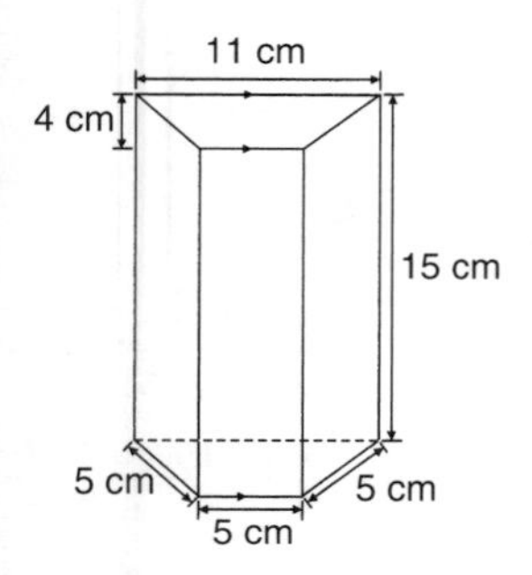

Figure 22.1

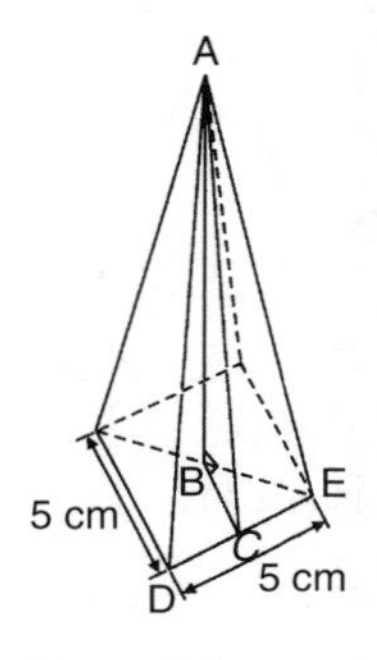

Figure 22.2

$$= (2 \times 32) + (5 \times 15) + (11 \times 15) + 2(5 \times 15)$$

$$= 64 + 75 + 165 + 150 = \mathbf{454\ cm^2}$$

In another example, to determine the volume and the total surface area of the square pyramid shown in Figure 22.2 if its perpendicular height is 12 cm :

$$\text{Volume of pyramid} = \tfrac{1}{3}\ (\text{area of base}) \times \text{perpendicular height}$$

$$= \tfrac{1}{3}(5 \times 5) \times 12 = \mathbf{100\ cm^3}$$

The total surface area consists of a square base and 4 equal triangles.
Area of triangle ADE $= \frac{1}{2} \times$ base $\times$ perpendicular height $= \frac{1}{2} \times 5 \times$ AC The length AC may be calculated using Pythagoras' theorem on triangle ABC, where AB = 12 cm, BC $= \frac{1}{2} \times 5 = 2.5$ cm, and AC $= \sqrt{AB^2 + BC^2}$ $= \sqrt{12^2 + 2.5^2} = 12.26$ cm

$$\text{Hence area of triangle ADE} = \tfrac{1}{2} \times 5 \times 12.26 = 30.65\ \text{cm}^2$$

$$\text{Total surface area of pyramid} = (5 \times 5) + 4(30.65) = \mathbf{147.6\ cm^2}$$

In another example, to determine the volume and total surface area of a cone of radius 5 cm and perpendicular height 12 cm:
The cone is shown in Figure 22.3.

$$\text{Volume of cone} = \tfrac{1}{3}\pi r^2 h = \tfrac{1}{3} \times \pi \times 5^2 \times 12 = \mathbf{314.2\ cm^3}$$

$$\text{Total surface area} = \text{curved surface area} + \text{area of base} = \pi rl + \pi r^2$$

From Figure 22.3, slant height l may be calculated using Pythagoras' theorem

$$l = \sqrt{12^2 + 5^2} = 13\ \text{cm},$$

Hence total surface area $= (\pi \times 5 \times 13) + (\pi \times 5^2) = \mathbf{282.7\ cm^2}$
In another example, a wooden section is shown in Figure 22.4. To find (a) its volume (in m^3), and (b) its total surface area:
The section of wood is a prism whose end comprises a rectangle and a semicircle. Since the radius of the semicircle is 8 cm, the diameter is 16 cm.

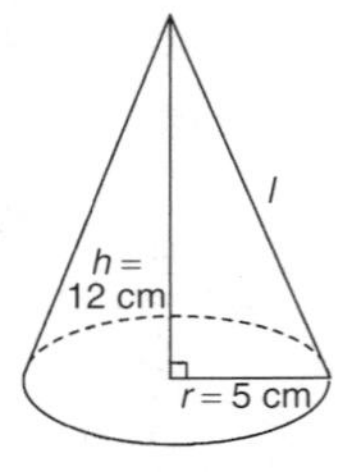

Figure 22.3

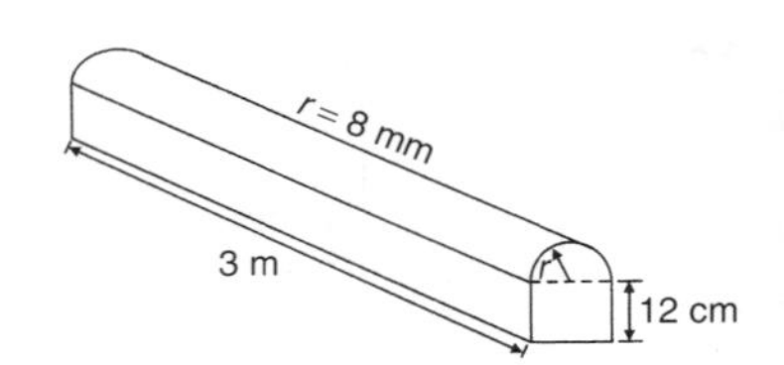

Figure 22.4

Hence the rectangle has dimensions 12 cm by 16 cm.

$$\text{Area of end} = (12 \times 16) + \tfrac{1}{2}\pi 8^2 = 292.5 \text{ cm}^2$$

$$\text{Volume of wooden section} = \text{area of end} \times \text{perpendicular height}$$

$$= 292.5 \times 300 = 87\,750 \text{ cm}^3$$

$$= \frac{87\,750 \text{ m}^3}{10^6} = \mathbf{0.08775 \text{ m}^3}$$

The total surface area comprises the two ends (each of area 292.5 cm^2), three rectangles and a curved surface (which is half a cylinder), hence

$$\text{total surface area} = (2 \times 292.5) + 2(12 \times 300) + (16 \times 300) + \tfrac{1}{2}(2\pi \times 8 \times 300)$$

$$= 585 + 7200 + 4800 + 2400\pi$$

$$= \mathbf{20\,125 \text{ cm}^2} \text{ or } \mathbf{2.0125 \text{ m}^2}$$

In another example, a boiler consists of a cylindrical section of length 8 m and diameter 6 m, on one end of which is surmounted a hemispherical section of diameter 6 m, and on the other end a conical section of height 4 m and base diameter 6 m. To calculate the volume of the boiler and the total surface area: The boiler is shown in Figure 22.5.

Volume of hemisphere, $P = \frac{2}{3}\pi r^3 = \frac{2}{3} \times \pi \times 3^3 = 18\pi \text{ m}^3$

Volume of cylinder, $Q = \pi r^2 h = \pi \times 3^2 \times 8 = 72\pi \text{ m}^3$

Volume of cone, $R = \frac{1}{3}\pi r^2 h = \frac{1}{3} \times \pi \times 3^2 \times 4 = 12\pi \text{ m}^3$

Total volume of boiler $= 18\pi + 72\pi + 12\pi = 102\pi = \mathbf{320.4 \text{ m}^3}$

Surface area of hemisphere, $P = \frac{1}{2}(4\pi r^2) = 2 \times \pi \times 3^2 = 18\pi \text{ m}^2$

Curved surface area of cylinder, $Q = 2\pi rh = 2 \times \pi \times 3 \times 8 = 48\pi \text{ m}^2$

The slant height of the cone, l, is obtained by Pythagoras' theorem on triangle ABC, i.e. $l = \sqrt{(4^2 + 3^2)} = 5$

Curved surface area of cone, $R = \pi rl = \pi \times 3 \times 5 = 15\pi \text{ m}^2$

Total surface area of boiler $= 18\pi + 48\pi + 15\pi = 81\pi = \mathbf{254.5 \text{ m}^2}$

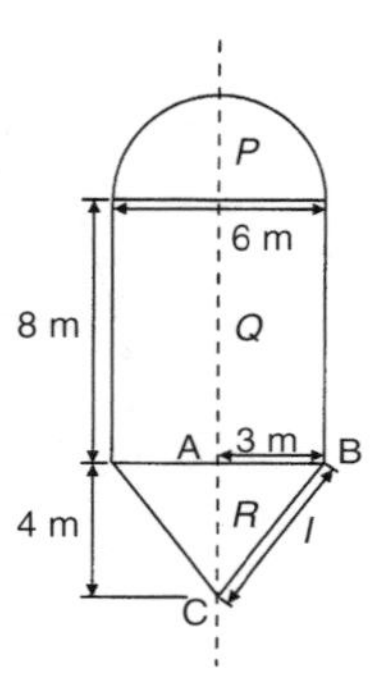

Figure 22.5

Volumes and surface areas of frusta of pyramids and cones

The **frustum** of a pyramid or cone is the portion remaining when a part containing the vertex is cut off by a plane parallel to the base.
The **volume of a frustum of a pyramid or cone** is given by the volume of the whole pyramid or cone minus the volume of the small pyramid or cone cut off.
The **surface area of the sides of a frustum of a pyramid or cone** is given by the surface area of the whole pyramid or cone minus the surface area of the small pyramid or cone cut off. This gives the lateral surface area of the frustum. If the total surface area of the frustum is required then the surface area of the two parallel ends are added to the lateral surface area.
There is an alternative method for finding the volume and surface area of a **frustum of a cone**. With reference to Figure 22.6:

$$\textbf{Volume} = \tfrac{1}{3}\pi h\,(R^2 + Rr + r^2)$$

$$\textbf{Curved surface area} = \pi l\,(R + r)$$

$$\textbf{Total surface area} = \pi l\,(R + r) + \pi r^2 + \pi R^2$$

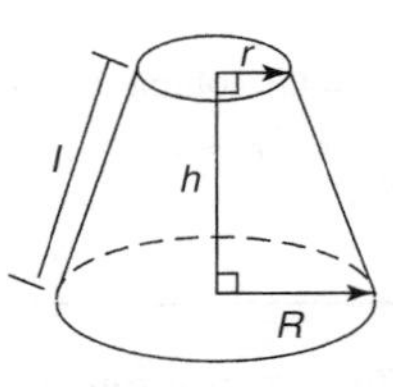

Figure 22.6

For example, a lampshade is in the shape of a frustum of a cone. The vertical height of the shade is 25.0 cm and the diameters of the ends are 20.0 cm and 10.0 cm, respectively. To determine the area of the material needed to form the lampshade, correct to 3 significant figures:
The curved surface area of a frustum of a cone $= \pi l(R + r)$ from above.
Since the diameters of the ends of the frustum are 20.0 cm and 10.0 cm, then from Figure 22.7,

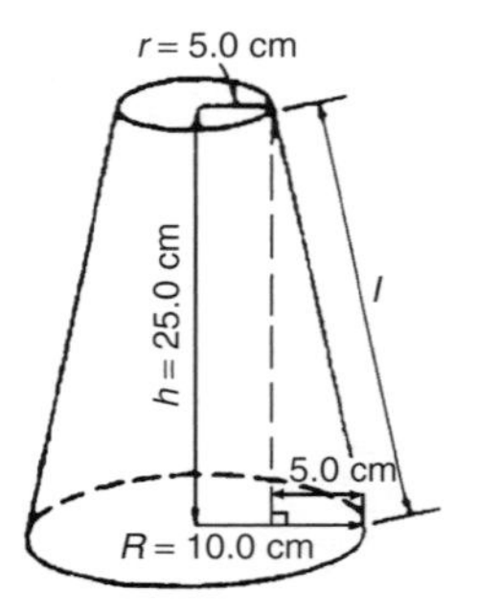

Figure 22.7

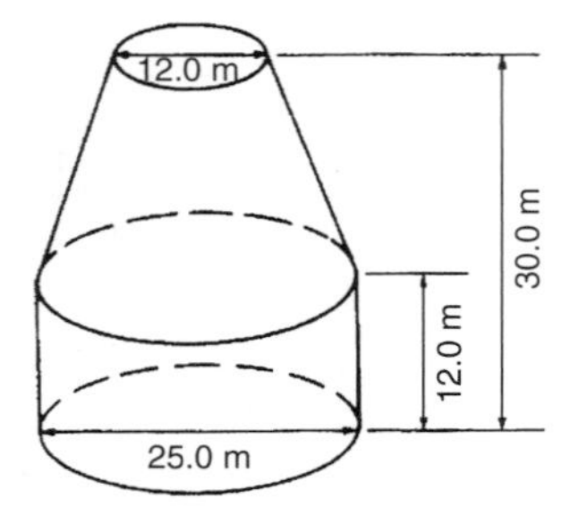

Figure 22.8

$r = 5.0$ cm, $R = 10.0$ cm and $l = \sqrt{[25.0^2 + 5.0^2]} = 25.50$ cm, from Pythagoras' theorem.
Hence curved surface area $= \pi(25.50)(10.0 + 5.0) = 1201.7\ \text{cm}^2$, i.e. the area of material needed to form the lampshade is **1200 cm^2**, correct to 3 significant figures.

In another example, a cooling tower is in the form of a cylinder surmounted by a frustum of a cone as shown in Figure 22.8. To determine the volume of air space in the tower if 40% of the space is used for pipes and other structures:

Volume of cylindrical portion $= \pi r^2 h = \pi \left(\frac{25.0}{2}\right)^2 (12.0) = 5890\ \text{m}^3$

Volume of frustum of cone $= \frac{1}{3}\pi h(R^2 + Rr + r^2)$ where $h = 30.0 - 12.0 = 18.0$ m, $R = 25.0/2 = 12.5$ m and $r = 12.0/2 = 6.0$ m.

Hence volume of frustum of cone $= \frac{1}{3}\pi(18.0)[(12.5)^2 + (12.5)(6.0) + (6.0)^2]$

$$= 5038\ \text{m}^3$$

Total volume of cooling tower $= 5890 + 5038 = 10\,928\ \text{m}^3$

If 40% of space is occupied then **volume of air space**

$$= 0.6 \times 10\,928 = \mathbf{6557\ m^3}$$

The frustum and zone of a sphere

Volume of sphere $= \frac{4}{3}\pi r^3$ and the surface area of sphere $= 4\pi r^2$.
A **frustum of a sphere** is the portion contained between two parallel planes. In Figure 22.9, PQRS is a frustum of the sphere. A **zone of a sphere** is the curved surface of a frustum. With reference to Figure 22.9:

Surface area of a zone of a sphere = $2\pi rh$

Volume of frustum of sphere = $\frac{\pi h}{6}(h^2 + 3r_1^2 + 3r_2^2)$

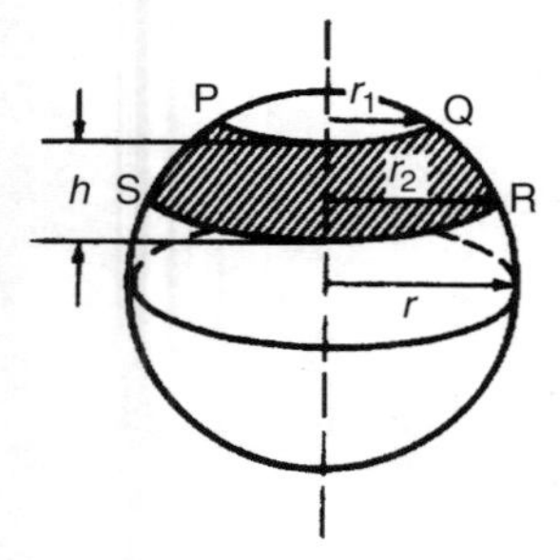

Figure 22.9

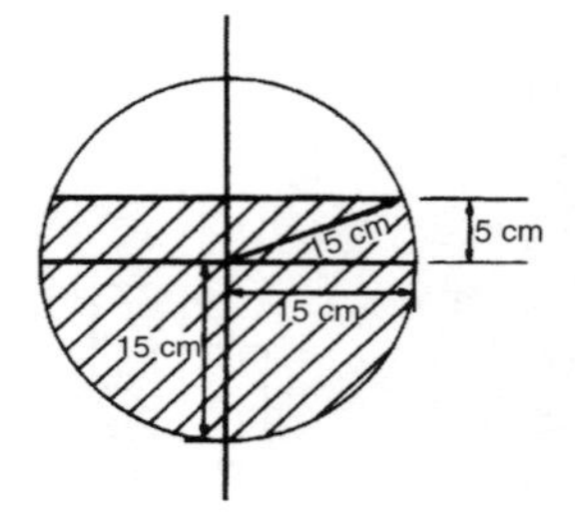

Figure 22.10

For example, to determine the volume of a frustum of a sphere of diameter 49.74 cm if the diameter of the ends of the frustum are 24.0 cm and 40.0 cm, and the height of the frustum is 7.00 cm:

From above, volume of frustum of a sphere $= \dfrac{\pi h}{6}(h^2 + 3r_1^2 + 3r_2^2)$

where $h = 7.00$ cm, $r_1 = 24.0/2 = 12.0$ cm and $r_2 = 40.0/2 = 20.0$ cm.

$$\text{Hence volume of frustum} = \frac{\pi(7.00)}{6}[(7.00)^2 + 3(12.0)^2 + 3(20.0)^2]$$

$$= \mathbf{6161\ cm^3}$$

In another example, to determine the curved surface area of the frustum in the previous example:

The curved surface area of the frustum = surface area of zone = $2\pi rh$ (from above), where r = radius of sphere $= 49.74/2 = 24.87$ cm and $h = 7.00$ cm.

Hence, surface area of zone $= 2\pi(24.87)(7.00) =$ **1094 cm²**

In another example, a spherical storage tank is filled with liquid to a depth of 20 cm. To determine the number of litres of liquid in the container (1 litre = 1000 cm³), if the internal diameter of the vessel is 30 cm:

The liquid is represented by the shaded area in the section shown in Figure 22.10. The volume of liquid comprises a hemisphere and a frustum of thickness 5 cm.

Hence volume of liquid $= \dfrac{2}{3}\pi r^3 + \dfrac{\pi h}{6}[h^2 + 3r_1^2 + 3r_2^2]$

where $r_2 = 30/2 = 15$ cm and $r_1 = \sqrt{15^2 - 5^2} = 14.14$ cm

$$\text{Volume of liquid} = \frac{2}{3}\pi(15)^3 + \frac{\pi(5)}{6}[5^2 + 3(14.14)^2 + 3(15)^2]$$

$$= 7069 + 3403 = 10\,470\ \text{cm}^3$$

Since 1 litre = 1000 cm³, the number of litres of liquid

$$= \frac{10\,470}{1000} = \mathbf{10.47\ litres}$$

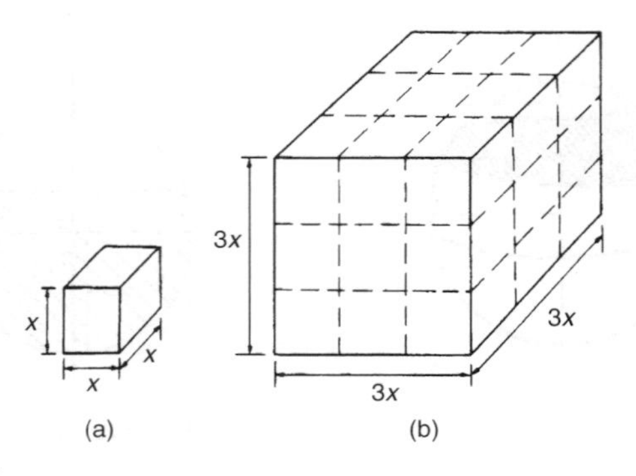

Figure 22.11

Volumes of similar shapes

The volumes of similar bodies are proportional to the cubes of corresponding linear dimensions. For example, Figure 22.11 shows two cubes, one of which has sides three times as long as those of the other.

$$\text{Volume of Figure 22.11(a)} = (x)(x)(x) = x^3$$

$$\text{Volume of Figure 22.11(b)} = (3x)(3x)(3x) = 27x^3$$

Hence Figure 22.11(b) has a volume $(3)^3$, i.e. 27 times the volume of Figure 22.11(a).

For example, a car has a mass of 1000 kg. A model of the car is made to a scale of 1 to 50. To determine the mass of the model if the car and its model are made of the same material:

$\dfrac{\text{Volume of model}}{\text{Volume of car}} = \left(\dfrac{1}{50}\right)^3$ since the volume of similar bodies are proportional to the cube of corresponding dimensions. Mass = density × volume, and since both car and model are made of the same material then:

$$\frac{\text{Mass of model}}{\text{Mass of car}} = \left(\frac{1}{50}\right)^3$$

$$\text{Hence mass of model} = (\text{mass of car})\left(\frac{1}{50}\right)^3$$

$$= \frac{1000}{50^3} = \mathbf{0.008\ kg\ or\ 8\ g}$$

23 Irregular Areas and Volumes and Mean Values

Areas of irregular figures

Areas of irregular plane surfaces may be approximately determined by using (a) a planimeter, (b) the trapezoidal rule, (c) the mid-ordinate rule, and

(d) Simpson's rule. Such methods may be used, for example, by engineers estimating areas of indicator diagrams of steam engines, surveyors estimating areas of plots of land or naval architects estimating areas of water planes or transverse sections of ships.

(a) **A planimeter** is an instrument for directly measuring small areas bounded by an irregular curve.

(b) **Trapezoidal rule**

To determine the areas PQRS in Figure 23.1:

(i) Divide base PS into any number of equal intervals, each of width d (the greater the number of intervals, the greater the accuracy)

(ii) Accurately measure ordinates y_1, y_2, y_3, etc

(iii) Area PQRS $= d\left[\dfrac{y_1 + y_7}{2} + y_2 + y_3 + y_4 + y_5 + y_6\right]$

In general, the trapezoidal rule states:

$$\textbf{Area} = \begin{pmatrix}\textbf{width of}\\ \textbf{interval}\end{pmatrix}\left[\frac{1}{2}\begin{pmatrix}\textbf{first + last}\\ \textbf{ordinate}\end{pmatrix} + \begin{pmatrix}\textbf{sum of remaining}\\ \textbf{ordinates}\end{pmatrix}\right]$$

(c) **Mid-ordinate rule**

To determine the area ABCD of Figure 23.2:

(i) Divide base AD into any number of equal intervals, each of width d (the greater the number of intervals, the greater the accuracy)

(ii) Erect ordinates in the middle of each interval (shown by broken lines in Figure 23.2)

(iii) Accurately measure ordinates y_1, y_2, y_3, etc.

(iv) Area ABCD $= d(y_1 + y_2 + y_3 + y_4 + y_5 + y_6)$

In general, the mid-ordinate rule states:

$$\textbf{Area = (width of interval)(sum of mid-ordinates)}$$

(d) **Simpson's rule**

To determine the area PQRS of Figure 23.1:

(i) Divide base PS into an **even** number of intervals, each of width d (the greater the number of intervals, the greater the accuracy)

(ii) Accurately measure ordinates y_1, y_2, y_3, etc.

(iii) Area PQRS $= \dfrac{d}{3}[(y_1 + y_7) + 4(y_2 + y_4 + y_6) + 2(y_3 + y_5)]$

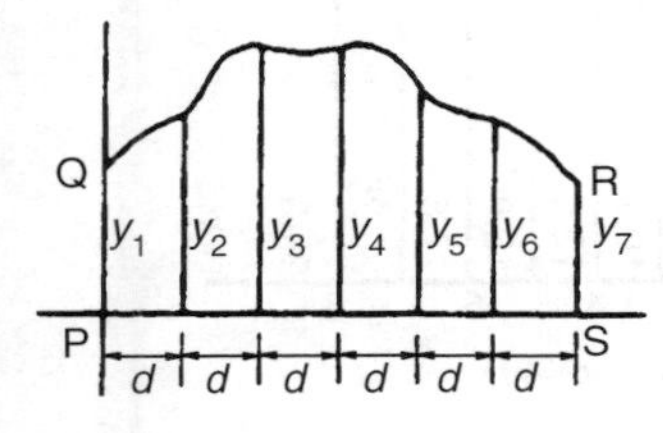

Figure 23.1

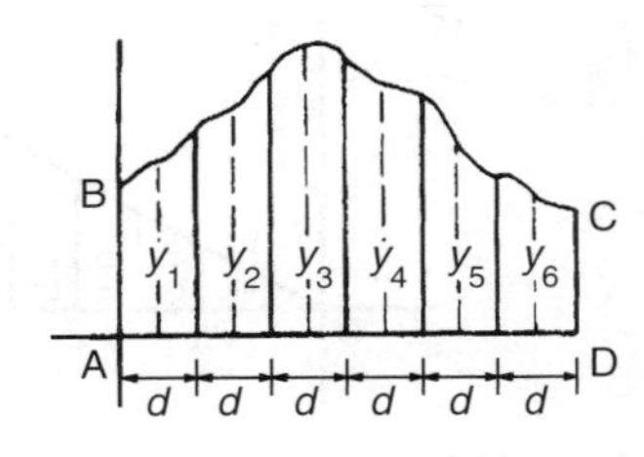

Figure 23.2

In general, Simpson's rule states:

$$\textbf{Area} = \frac{1}{3}\begin{pmatrix}\textbf{width of}\\ \textbf{interval}\end{pmatrix}\left[\begin{pmatrix}\textbf{first + last}\\ \textbf{ordinate}\end{pmatrix} + 4\begin{pmatrix}\textbf{sum of even}\\ \textbf{ordinates}\end{pmatrix} + 2\begin{pmatrix}\textbf{sum of remaining}\\ \textbf{odd ordinates}\end{pmatrix}\right]$$

For example, a car starts from rest and its speed is measured every second for 6 s:

Time t(s)	0	1	2	3	4	5	6
Speed v (m/s)	0	2.5	5.5	8.75	12.5	17.5	24.0

To determine the distance travelled in 6 seconds (i.e. the area under the v/t graph), by (a) the trapezoidal rule, (b) the mid-ordinate rule, and (c) Simpson's rule:
A graph of speed/time is shown in Figure 23.3.

(a) **Trapezoidal rule** (see para.(b) above)
The time base is divided into 6 strips each of width 1 s, and the length of the ordinates measured. Thus

$$\textbf{area} = (1)\left[\left(\frac{0 + 24.0}{2}\right) + 2.5 + 5.5 + 8.75 + 12.5 + 17.5\right] = \textbf{58.75 m}$$

(b) **Mid-ordinate rule** (see para.(c) above)
The time base is divided into 6 strips each of width 1 second.
Mid-ordinates are erected as shown in Figure 23.3 by the broken lines.
The length of each mid-ordinate is measured. Thus

$$\textbf{area} = (1)[1.25 + 4.0 + 7.0 + 10.75 + 15.0 + 20.25] = \textbf{58.25 m}$$

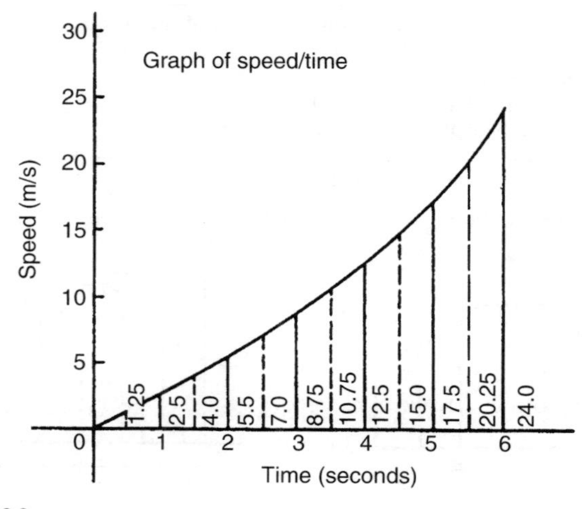

Figure 23.3

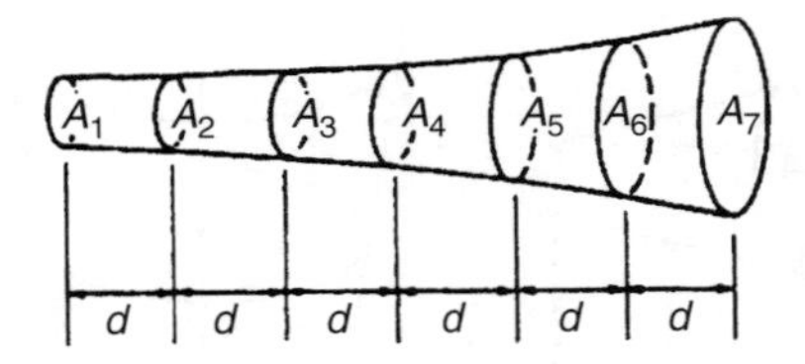

Figure 23.4

(c) **Simpson's rule** (see para.(d) above)
The time base is divided into 6 strips each of width 1 s, and the length of the ordinates measured. Thus

area $= \frac{1}{3}(1)[(0 + 24.0)+4(2.5 + 8.75 + 17.5)+2(5.5 + 12.5)] =$ **58.33 m**

Volumes of irregular solids using Simpson's rule

If the cross-sectional areas A_1, A_2, A_3, .. of an irregular solid bounded by two parallel planes are known at equal intervals of width d (as shown in Figure 23.4), then by Simpson's rule:

$$\textbf{Volume, } V = \frac{d}{3}[(A_1 + A_7) + 4(A_2 + A_4 + A_6) + 2(A_3 + A_5)]$$

For example, a tree trunk is 12 m in length and has a varying cross-section. The cross-sectional areas at intervals of 2 m measured from one end are:

0.52, 0.55, 0.59, 0.63, 0.72, 0.84, 0.97 m^2

To estimate the volume of the tree trunk:
A sketch of the tree trunk is similar to that shown in Figure 23.4 above, where $d = 2$ m, $A_1 = 0.52$ m^2, $A_2 = 0.55$ m^2, and so on.
Using Simpson's rule for volumes gives:

$$\text{Volume} = \frac{2}{3}[(0.52 + 0.97) + 4(0.55 + 0.63 + 0.84) + 2(0.59 + 0.72)]$$

$$= \frac{2}{3}[1.49 + 8.08 + 2.62] = \mathbf{8.13\ m^3}$$

Prismoidal rule for finding volumes

The prismoidal rule applies to a solid of length x divided by only three equidistant plane areas, A_1, A_2 and A_3 as shown in Figure 23.5 and is merely an extension of Simpson's rule — but for volumes.
With reference to Figure 23.5,

$$\textbf{Volume, } V = \frac{x}{6}[A_1 + 4A_2 + A_3]$$

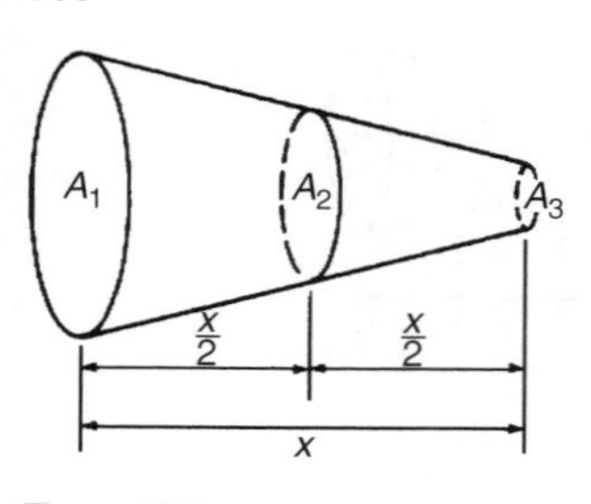

Figure 23.5

Figure 23.6

The prismoidal rule gives precise values of volume for regular solids such as pyramids, cones, spheres and prismoids.

For example, a container is in the shape of a frustum of a cone. Its diameter at the bottom is 18 cm and at the top 30 cm. To determine the capacity of the container, correct to the nearest litre, by the prismoidal rule, if the depth is 24 cm :

The container is shown in Figure 23.6. At the midpoint, i.e. at a distance of 12 cm from one end, the radius r_2 is $(9 + 15)/2 = 12$ cm, since the sloping sides change uniformly.

Volume of container by the prismoidal rule $= \frac{x}{6}[A_1 + 4A_2 + A_3]$, from above, where $x = 24$ cm, $A_1 = \pi(15)^2$ cm^2, $A_2 = \pi(12)^2$ cm^2 and $A_3 = \pi(9)^2$ cm^2

$$\text{Hence volume of container} = \frac{24}{6}[\pi(15)^2 + 4\pi(12)^2 + \pi(9)^2]$$

$$= 4[706.86 + 1809.56 + 254.47]$$

$$= 11\,080 \text{ cm}^3 = \frac{11\,080}{1000} \text{ litres}$$

$$= \textbf{11 litres, correct to the nearest litre}$$

The mean or average value of a waveform

The mean or average value, y, of the waveform shown in Figure 23.7 is given by:

$$y = \frac{\textbf{area under curve}}{\textbf{length of base, } b}$$

If the mid-ordinate rule is used to find the area under the curve, then:

$$y = \frac{\text{sum of mid-ordinates}}{\text{number of mid-ordinates}}$$

$$\left(= \frac{y_1 + y_2 + y_3 + y_4 + y_5 + y_6 + y_7}{7} \quad \text{for Figure 23.7}\right)$$

For a **sine wave**, the mean or average value:

(i) over one complete cycle is zero (see Figure 23.8(a)),

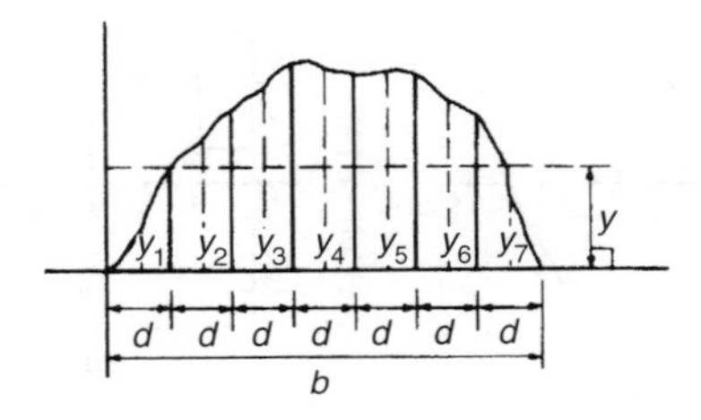

Figure 23.7

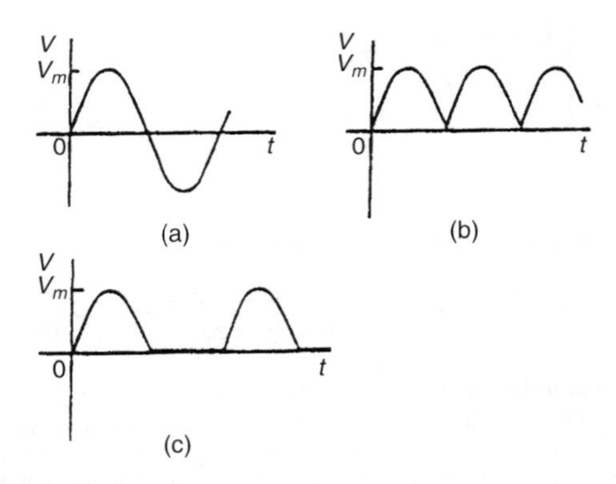

Figure 23.8

(ii) over half a cycle is **0.637 × maximum value**, or **$2/\pi$ × maximum value**,
(iii) of a full-wave rectified waveform (see Figure 23.8(b)) is **0.637 × maximum value**
(iv) of a half-wave rectified waveform (see Figure 23.8(c)) is **0.318 × maximum value**, or **$1/\pi$ × maximum value**

For example, to determine the average values over half a cycle of the periodic waveforms shown in Figure 23.9:

(a) Area under triangular waveform (a) for a half cycle is given by:

$$\text{Area} = \tfrac{1}{2}(\text{base})(\text{perpendicular height}) = \tfrac{1}{2}(2 \times 10^{-3})(20) = 20 \times 10^{-3}\ \text{Vs}$$

$$\text{Average value of waveform} = \frac{\text{area under curve}}{\text{length of base}} = \frac{20 \times 10^{-3}\ \text{Vs}}{2 \times 10^{-3}\ \text{s}} = \mathbf{10\ V}$$

(b) Area under waveform (b) for a half cycle $= (1 \times 1) + (3 \times 2) = 7$ As

$$\text{Average value of waveform} = \frac{\text{area under curve}}{\text{length of base}} = \frac{7\ \text{As}}{3\ \text{s}} = \mathbf{2.33\ A}$$

(c) A half cycle of the voltage waveform (c) is completed in 4 ms.

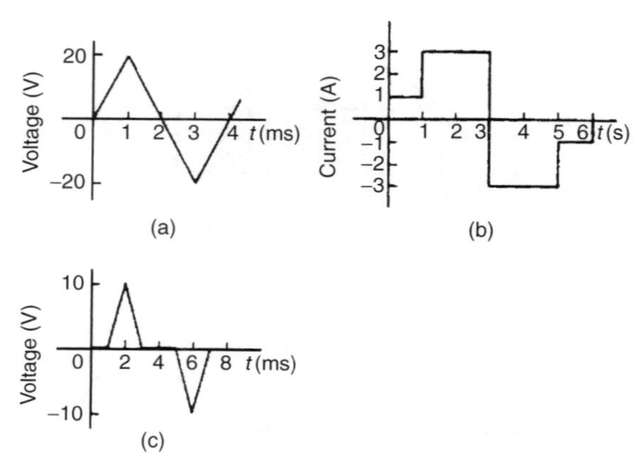

Figure 23.9

$$\text{Area under curve} = \frac{1}{2}\{(3-1)10^{-3}\}(10) = 10 \times 10^{-3}\ \text{Vs}$$

$$\text{Average value of waveform} = \frac{\text{area under curve}}{\text{length of base}} = \frac{10 \times 10^{-3}\ \text{Vs}}{4 \times 10^{-3}\ \text{s}} = \mathbf{2.5\ V}$$

In another example, an indicator diagram for a steam engine is shown in Figure 23.10. The base line has been divided into 6 equally spaced intervals and the lengths of the 7 ordinates measured with the results shown in centimetres. To determine (a) the area of the indicator diagram using Simpson's rule, and (b) the mean pressure in the cylinder given that 1 cm represents 100 kPa.

(a) The width of each interval is $\frac{12.0}{6}$ cm. Using Simpson's rule,

$$\text{area} = \frac{1}{3}(2.0)[(3.6+1.6)+4(4.0+2.9+1.7)+2(3.5+2.2)]$$

$$= \frac{2}{3}[5.2+34.4+11.4] = \mathbf{34\ cm^2}$$

(b) Mean height of ordinates $= \frac{\text{area of diagram}}{\text{length of base}} = \frac{34}{12} = 2.83$ cm

Since 1 cm represents 100 kPa,

the mean pressure in the cylinder = 2.83 cm × 100 kPa/cm = **283 kPa**

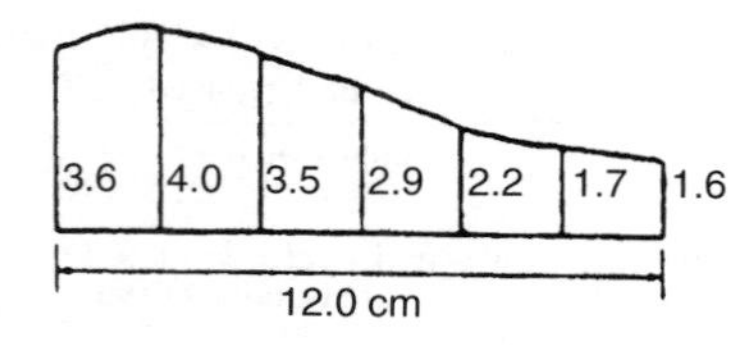

Figure 23.10

Geometry and Trigonometry

24 Geometry and Triangles

Angular measurement

Geometry is a part of mathematics in which the properties of points, lines, surfaces and solids are investigated.
An **angle** is the amount of rotation between two straight lines.
Angles may be measured in either **degrees** or **radians** (see Chapter 21).
1 revolution = 360 degrees, thus 1 degree = $\frac{1}{360}$ th of one revolution. Also 1 minute = $\frac{1}{60}$ th of a degree and 1 second = $\frac{1}{60}$ th of a minute. 1 minute is written as $1'$ and 1 second is written as $1''$ Thus $\mathbf{1^\circ = 60'}$ **and** $\mathbf{1' = 60''}$
For example, to determine (a) $13^\circ 42' 51'' + 48^\circ 22' 17''$ (b) $37^\circ 12' 8'' - 21^\circ 17' 25''$:

(a)
$$\begin{array}{lr} & 13^\circ 42' 51'' \\ & \underline{48^\circ 22' 17''} \\ \text{Adding:} & \underline{\mathbf{62^\circ\ 5'\ 8''}} \\ & 1^\circ 1' \end{array}$$

(b)
$$\begin{array}{lr} & 36^\circ 11' \\ & \not{3}\not{7}^\circ \not{1}\not{2}' 8'' \\ & \underline{21^\circ 17' 25''} \\ \text{Subtracting:} & \underline{\mathbf{15^\circ 54' 43''}} \end{array}$$

In another example, to convert $78^\circ 15' 26''$ to degrees:

Since 1 second = $\frac{1}{60}$th of a minute,

$$26'' = \left(\frac{26}{60}\right)' = 0.4333'$$

Hence $78^\circ 15' 26'' = 78^\circ 15.4\dot{3}'$

$$15.4333' = \left(\frac{15.4\dot{3}}{60}\right)^\circ = 0.2572^\circ, \text{ correct to 4 decimal places.}$$

Hence $\mathbf{78^\circ 15' 26'' = 78.26^\circ}$, correct to 4 significant places.

Types and properties of angles

(a) (i) Any angle between 0° and 90° is called an **acute angle**.
(ii) An angle equal to 90° is called a **right angle**.
(iii) Any angle between 90° and 180° is called an **obtuse angle**.

(iv) Any angle greater than 180° and less than 360° is called a **reflex angle**.

(b) (i) An angle of 180° lies on a straight line.

(ii) If two angles add up to 90° they are called **complementary angles**.

(iii) If two angles add up to 180° they are called **supplementary angles**.

(iv) **Parallel lines** are straight lines which are in the same plane and never meet. (Such lines are denoted by arrows, as in Figure 24.1).

(v) A straight line which crosses two parallel lines is called a **transversal** (see MN in Figure 24.1).

(c) With reference to Figure 24.1:

(i) $a = c$, $b = d$, $e = g$ and $f = h$. Such pairs of angles are called **vertically opposite angles**.

(ii) $a = e$, $b = f$, $c = g$ and $d = h$. Such pairs of angles are called **corresponding angles**.

(iii) $c = e$ and $b = h$. Such pairs of angles are called **alternate angles**.

(iv) $b + e = 180°$ and $c + h = 180°$. Such pairs of angles are called **interior angles**.

For example, the angle complementary to 58°39′ is (90° − 58°39′), i.e. **31°21′**

In another example, the angle supplementary to 111°11′ is (180° − 111°11′), i.e. **68°49′**

In another example, to determine angle $\boldsymbol{\beta}$ in Figure 24.2:

$$\alpha = 180° - 133° = 47° \text{ (i.e. supplementary angles)}$$

$$\alpha = \boldsymbol{\beta} = \mathbf{47°} \text{ (corresponding angles between parallel lines).}$$

In another example, to determine the value of angle $\boldsymbol{\theta}$ in Figure 24.3:
Let a straight line FG be drawn through E such that FG is parallel to AB and CD. ∠BAE = ∠AEF (alternate angles between parallel lines AB and FG), hence ∠AEF = 23°37′. ∠ECD = ∠FEC (alternate angles between parallel lines FG and CD), hence ∠FEC = 35°49′

Angle $\boldsymbol{\theta}$ = ∠AEF + ∠FEC = 23°37′ + 35°49′ = **59°26′**

In another example, to determine angles c and d in Figure 24.4:

$b = 46°$ (corresponding angles between parallel lines).

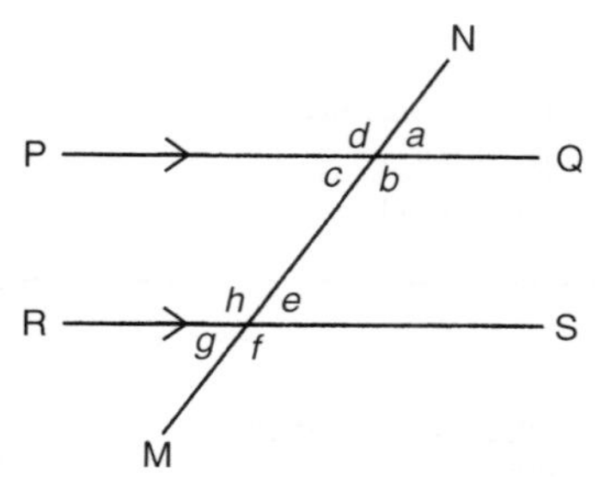

Figure 24.1

Figure 24.2

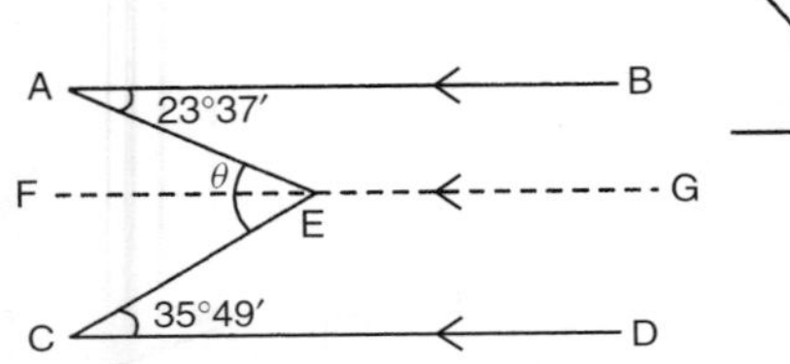

Figure 24.3

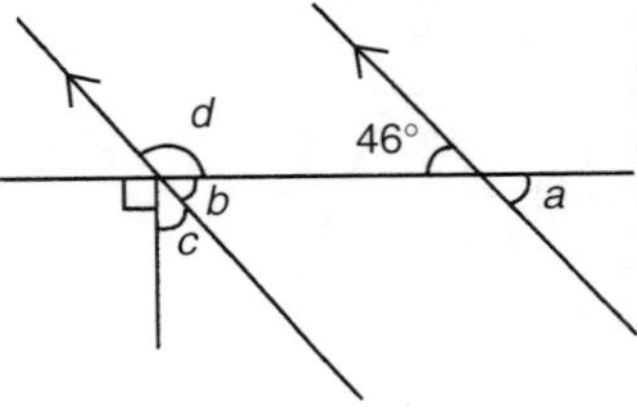

Figure 24.4

Also $b + c + 90° = 180°$ (angles on a straight line).

Hence $46° + c + 90° = 180°$, from which $\boldsymbol{c} = \mathbf{44°}$.

b and d are supplementary, hence $\boldsymbol{d} = 180° - 46° = \mathbf{134°}$

Alternatively, $90° + c = d$ (vertically opposite angles).

Properties of triangles

A triangle is a figure enclosed by three straight lines. The sum of the three angles of a triangle is equal to 180°. Types of triangles:

(i) An **acute-angled triangle** is one in which all the angles are acute, i.e. all the angles are less than 90°.
(ii) A **right-angled triangle** is one that contains a right angle.
(iii) An **obtuse-angled triangle** is one that contains an obtuse angle, i.e. one angle which lies between 90° and 180°.
(iv) An **equilateral triangle** is one in which all the sides and all the angles are equal (i.e. each 60°).
(v) An **isosceles triangle** is one in which two angles and two sides are equal.
(vi) A **scalene triangle** is one with unequal angles and therefore unequal sides.

With reference to Figure 24.5:

(i) Angles A, B and C are called **interior angles** of the triangle.
(ii) Angle θ is called an **exterior angle** of the triangle and is equal to the sum of the two opposite interior angles, i.e. $\theta = A + C$
(iii) $a + b + c$ is called the **perimeter** of the triangle.

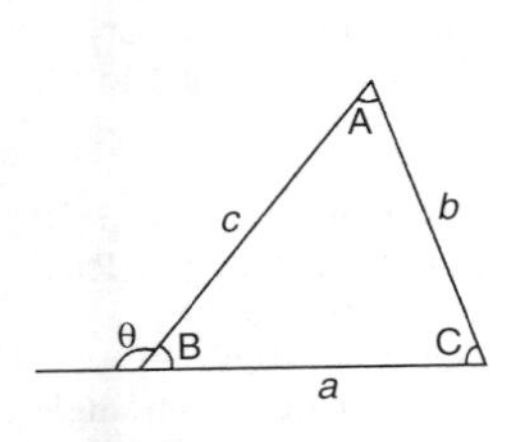

Figure 24.5

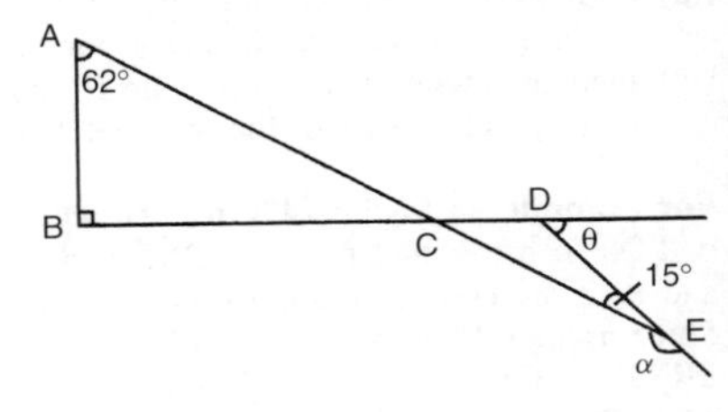

Figure 24.6

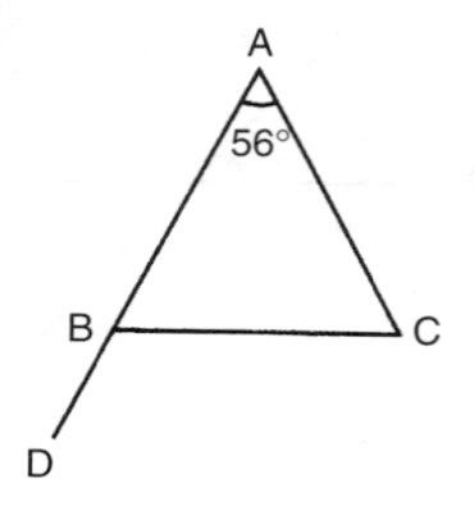

Figure 24.7

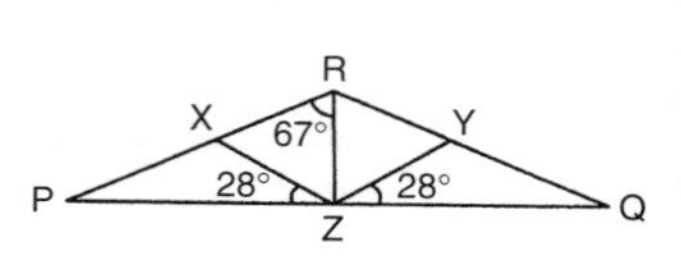

Figure 24.8

For example, to determine the value of θ and α in Figure 24.6:
In triangle ABC, $\angle A + \angle B + \angle C = 180°$ (angles in a triangle add up to 180°), hence $\angle C = 180° - 90° - 62° = 28°$. Thus $\angle DCE = 28°$ (vertically opposite angles). $\theta = \angle DCE + \angle DEC$ (exterior angle of a triangle is equal to the sum of the two opposite interior angles). Hence $\angle\theta = 28° + 15° = \mathbf{43°}$
$\angle\alpha$ and $\angle DEC$ are supplementary, thus $\alpha = 180° - 15° = \mathbf{165°}$
In another example, ABC is an isosceles triangle in which the unequal angle BAC is 56°. AB is extended to D as shown in Figure 24.7. To determine the angle DBC:
Since the three interior angles of a triangle add up to 180° then
$56° + \angle B + \angle C = 180°$, i.e. $\angle B + \angle C = 180° - 56° = 124°$

$$\text{Triangle ABC is isosceles hence } \angle B = \angle C = \frac{124°}{2} = 62°$$

$\angle DBC = \angle A + \angle C$ (exterior angle equals sum of two interior opposite angles), i.e. $\angle \mathbf{DBC} = 56° + 62° = \mathbf{118°}$

[Alternatively, $\angle DBC + \angle ABC = 180°$ (i.e. supplementary angles)]

Congruent triangles

Two triangles are said to be **congruent** if they are equal in all respects, i.e. three angles and three sides in one triangle are equal to three angles and three sides in the other triangle. Two triangles are congruent if:

(i) the three sides of one are equal to the three sides of the other (SSS),
(ii) they have two sides of the one equal to two sides of the other, and if the angles included by these sides are equal (SAS),
(iii) two angles of the one are equal to two angles of the other and any side of the first is equal to the corresponding side of the other (ASA), or
(iv) their hypotenuses are equal and if one other side of one is equal to the corresponding side of the other (RHS).

For example, in Figure 24.8, triangle PQR is isosceles with Z the mid-point of PQ. To prove that triangles PXZ and QYZ are congruent, triangles RXZ and RYZ are congruent and to find the values of angles RPZ and RXZ:
Since triangle PQR is isosceles PR = RQ and thus $\angle QPR = \angle RQP$
$\angle RXZ = \angle QPR + 28°$ and $\angle RYZ = \angle RQP + 28°$ (exterior angles of a triangle equal the sum of the two interior opposite angles). Hence $\angle RXZ = \angle RYZ$.

$\angle PXZ = 180° - \angle RXZ$ and $\angle QYZ = 180° - \angle RYZ$. Thus $\angle PXZ = \angle QYZ$.
Triangles PXZ and QYZ are congruent since $\angle XPZ = \angle YQZ$, PZ = ZQ and $\angle XZP = \angle YZQ$ (ASA). Hence XZ = YZ.
Triangles PRZ and QRZ are congruent since PR = RQ, $\angle RPZ = \angle RQZ$ and PZ = ZQ (SAS). Hence $\angle RZX = \angle RZY$.
Triangles RXZ and RYZ are congruent since $\angle RXZ = \angle RYZ$, XZ = YZ and $\angle RZX = \angle RZY$ (ASA). $\angle QRZ = 67°$ and thus $\angle PRQ = 67° + 67° = 134°$. Hence

$$\angle \mathbf{RPZ} = \angle RQZ = \frac{180° - 134°}{2} = \mathbf{23°} \quad \text{and} \quad \angle \mathbf{RXZ} = 23° + 28° = \mathbf{51°}$$

(external angle of a triangle equals the sum of the two interior opposite angles)

Similar triangles

Two triangles are said to be **similar** if the angles of one triangle are equal to the angles of the other triangle. With reference to Figure 24.9: Triangles ABC and PQR are similar and the corresponding sides are in proportion to each other,

i.e. $\quad \dfrac{\boldsymbol{p}}{\boldsymbol{a}} = \dfrac{\boldsymbol{q}}{\boldsymbol{b}} = \dfrac{\boldsymbol{r}}{\boldsymbol{c}}$

For example, to find the length of side a in Figure 24.10:
In triangle ABC, $50° + 70° + \angle C = 180°$, from which $\angle C = 60°$

In triangle DEF, $\angle E = 180° - 50° - 60° = 70°$. Hence triangles ABC and DEF are similar, since their angles are the same. Since corresponding sides are in proportion to each other then:

$$\frac{a}{d} = \frac{c}{f} \quad \text{i.e.} \quad \frac{a}{4.42} = \frac{12.0}{5.0}$$

Hence $\quad \boldsymbol{a} = \dfrac{12.0}{5.0}(4.42) = \mathbf{10.61\ cm}$

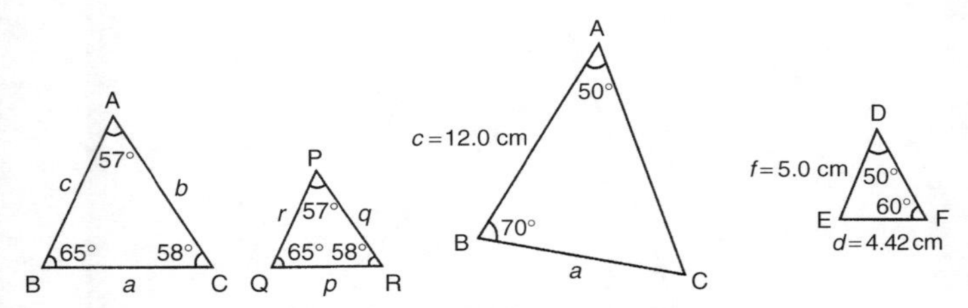

Figure 24.9

Figure 24.10

Construction of triangles

To construct any triangle the following drawing instruments are needed: (i) ruler and/or straight edge, (ii) compass, (iii) protractor, (iv) pencil.
For example, to construct a triangle whose sides are 6 cm, 5 cm and 3 cm: With reference to Figure 24.11:

(i) Draw a straight line of any length, and with a pair of compasses, mark out 6 cm length and label it AB.
(ii) Set compass to 5 cm and with centre at A describe arc DE.
(iii) Set compass to 3 cm and with centre at B describe arc FG.
(iv) The intersection of the two curves at C is the vertex of the required triangle. Join AC and BC by straight lines.

It may be proved by measurement that the ratio of the angles of a triangle is not equal to the ratio of the sides (i.e. in this problem, the angle opposite the 3 cm side is not equal to half the angle opposite the 6 cm side).
In another example, To construct a triangle ABC such that $a = 6$ cm, $b = 3$ cm and $\angle C = 60°$:

With reference to Figure 24.12:

(i) Draw a line BC, 6 cm long.
(ii) Using a protractor centred at C make an angle of 60° to BC.
(iii) From C measure a length of 3 cm and label A.
(iv) Join B to A by a straight line.

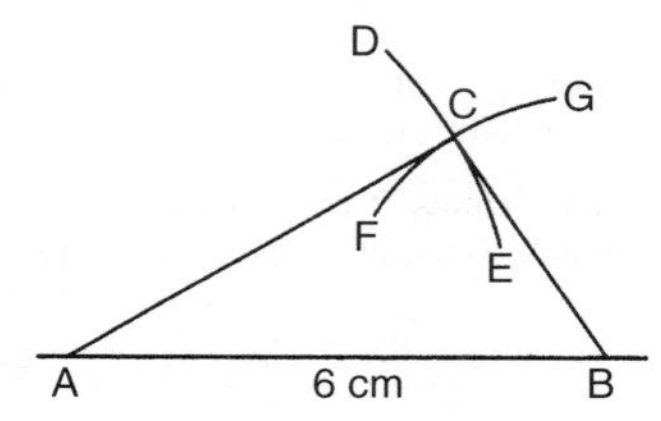

Figure 24.11

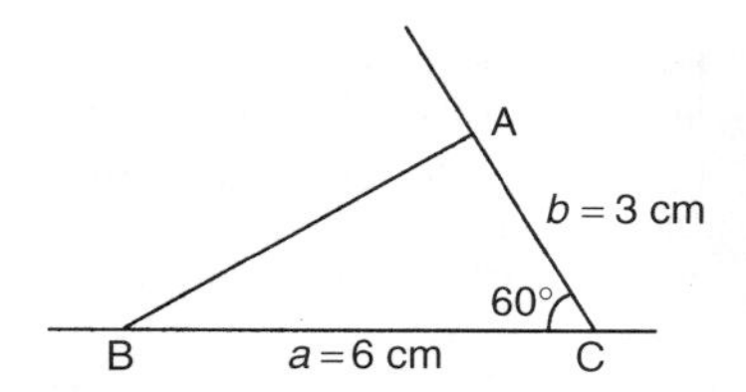

Figure 24.12

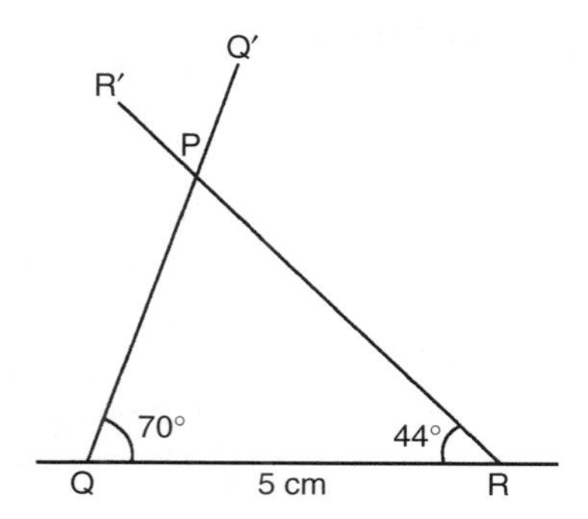

Figure 24.13

In another example, to construct a triangle PQR given that QR = 5 cm, $\angle Q = 70°$ and $\angle R = 44°$:
With reference to Figure 24.13:
(i) Draw a straight line 5 cm long and label it QR.
(ii) Use a protractor centred at Q and make an angle of 70°. Draw QQ′.
(iii) Use a protractor centred at R and make an angle of 44°. Draw RR′.
(iv) The intersection of QQ′ and RR′ forms the vertex P of the triangle.

---

25 Introduction to Trigonometry

Trigonometry is the branch of mathematics that deals with the measurement of sides and angles of triangles, and their relationship with each other. There are many applications in engineering where knowledge of trigonometry is needed.

The theorem of Pythagoras

With reference to Figure 25.1, the side opposite the right angle (i.e. side b) is called the **hypotenuse**. The **theorem of Pythagoras** states:
'In any right-angled triangle, the square on the hypotenuse is equal to the sum of the squares on the other two sides.'
Hence $$\boldsymbol{b^2 = a^2 + c^2}$$

For example, To find the length of EF in Figure 25.2:

By Pythagoras' theorem: $e^2 = d^2 + f^2$

Hence
$$13^2 = d^2 + 5^2$$
$$169 = d^2 + 25$$
$$d^2 = 169 - 25 = 144$$

Thus $d = \sqrt{144} = 12$ cm

i.e. **EF = 12 cm**

In another example, two aircraft leave an airfield at the same time. One travels due north at an average speed of 300 km/h and the other due west at an average speed of 220 km/h. To calculate their distance apart after 4 hours:

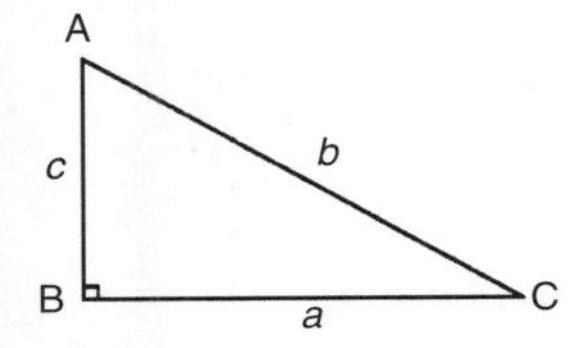

Figure 25.1

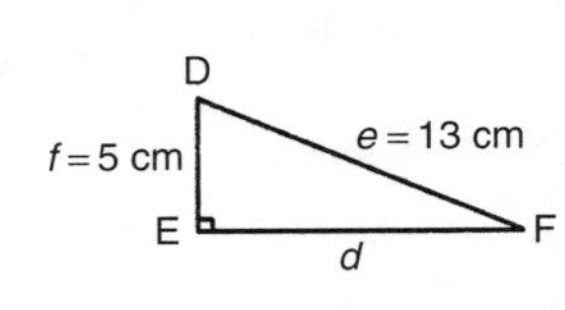

Figure 25.2

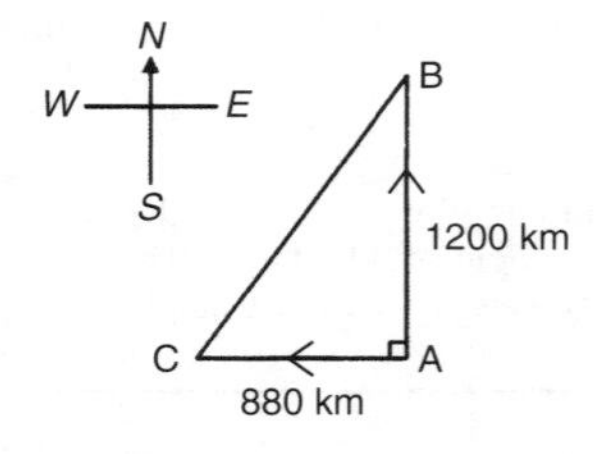

Figure 25.3

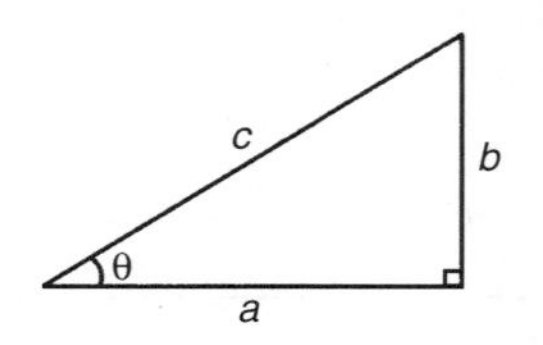

Figure 25.4

After 4 hours, the first aircraft has travelled $4 \times 300 = 1200$ km, due north, and the second aircraft has travelled $4 \times 220 = 880$ km due west, as shown in Figure 25.3. Distance apart after 4 hours = BC
From Pythagoras' theorem:

$$\begin{aligned} BC^2 &= 1200^2 + 880^2 \\ &= 1\,440\,000 + 774\,400 \quad \text{and} \quad BC = \sqrt{2\,214\,400} \end{aligned}$$

Hence distance apart after 4 hours = 1488 km

Trigonometric ratios of acute angles

(a) With reference to the right-angled triangle shown in Figure 25.4:

(i) sine $\theta = \dfrac{\text{opposite side}}{\text{hypotenuse}}$, i.e. $\boldsymbol{\sin\theta = \dfrac{b}{c}}$

(ii) cosine $\theta = \dfrac{\text{adjacent side}}{\text{hypotenuse}}$, i.e. $\boldsymbol{\cos\theta = \dfrac{a}{c}}$

(iii) tangent $\theta = \dfrac{\text{opposite side}}{\text{adjacent side}}$, i.e. $\boldsymbol{\tan\theta = \dfrac{b}{a}}$

(iv) secant $\theta = \dfrac{\text{hypotenuse}}{\text{adjacent side}}$, i.e. $\boldsymbol{\sec\theta = \dfrac{c}{a}}$

(v) cosecant $\theta = \dfrac{\text{hypotenuse}}{\text{opposite side}}$, i.e. $\boldsymbol{\text{cosec}\,\theta = \dfrac{c}{b}}$

(vi) cotangent $\theta = \dfrac{\text{adjacent side}}{\text{opposite side}}$, i.e. $\boldsymbol{\cot\theta = \dfrac{a}{b}}$

(b) From above,

(i) $\dfrac{\sin\theta}{\cos\theta} = \dfrac{\frac{b}{c}}{\frac{a}{c}} = \dfrac{b}{a} = \tan\theta$, i.e. $\boldsymbol{\tan\theta = \dfrac{\sin\theta}{\cos\theta}}$

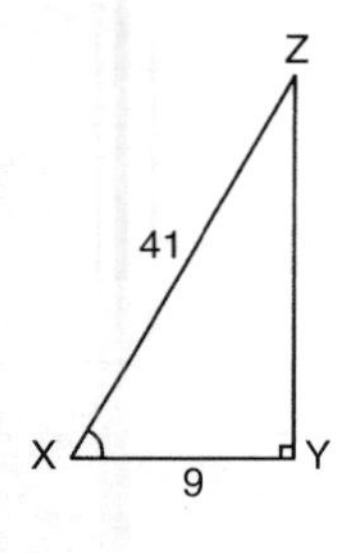

Figure 25.5

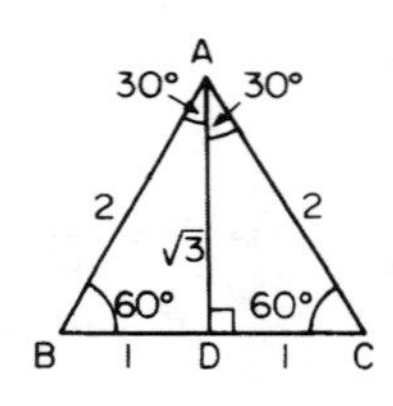

Figure 25.6

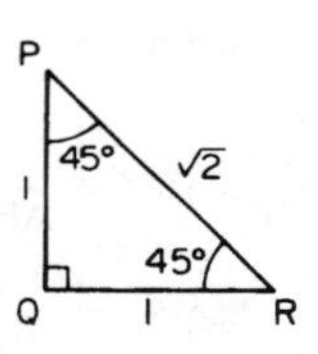

Figure 25.7

(ii) $\dfrac{\cos\theta}{\sin\theta} = \dfrac{\frac{a}{c}}{\frac{b}{c}} = \dfrac{a}{b} = \cot\theta$, i.e. $\boldsymbol{\cot\theta = \dfrac{\cos\theta}{\sin\theta}}$

(iii) $\boldsymbol{\sec\theta = \dfrac{1}{\cos\theta}}$

(iv) $\boldsymbol{\operatorname{cosec}\theta = \dfrac{1}{\sin\theta}}$ (Note '*s*' and '*c*' go together)

(v) $\boldsymbol{\cot\theta = \dfrac{1}{\tan\theta}}$

Secants, cosecants and cotangents are called the **reciprocal ratios**.
For example, to determine the value of the other five trigonometry ratios if $\cos X = \dfrac{9}{41}$:

Figure 25.5 shows a right-angled triangle XYZ.

Since $\cos X = \dfrac{9}{41}$, then XY = 9 units and XZ = 41 units.

Using Pythagoras' theorem: $41^2 = 9^2 + YZ^2$

from which $YZ = \sqrt{41^2 - 9^2} = 40$ units

Thus $\boldsymbol{\sin X = \dfrac{40}{41}}$, $\boldsymbol{\tan X = \dfrac{40}{9} = 4\dfrac{4}{9}}$, $\boldsymbol{\operatorname{cosec} X = \dfrac{41}{40} = 1\dfrac{1}{40}}$,

$\boldsymbol{\sec X = \dfrac{41}{9} = 4\dfrac{5}{9}}$ and $\boldsymbol{\cot X = \dfrac{9}{40}}$

Fractional and surd forms of trigonometric ratios

In Figure 25.6, ABC is an equilateral triangle of side 2 units. AD bisects angle A and bisects the side BC. Using Pythagoras' theorem on triangle ABC gives:

$AD = \sqrt{2^2 - 1^2} = \sqrt{3}$.

Hence, $\sin 30^\circ = \frac{BD}{AB} = \frac{1}{2}, \quad \cos 30^\circ = \frac{AD}{AB} = \frac{\sqrt{3}}{2}$

and $\tan 30^\circ = \frac{BD}{AD} = \frac{1}{\sqrt{3}}$

$$\sin 60^\circ = \frac{AD}{AB} = \frac{\sqrt{3}}{2}, \quad \cos 60^\circ = \frac{BD}{AB} = \frac{1}{2}$$

and $\tan 60^\circ = \frac{AD}{BD} = \sqrt{3}$

In Figure 25.7, PQR ia an isosceles triangle with $PQ = QR = 1$ unit. By Pythagoras' theorem, $PR = \sqrt{1^2 + 1^2} = \sqrt{2}$

Hence, $\sin 45^\circ = \frac{1}{\sqrt{2}}, \cos 45^\circ = \frac{1}{\sqrt{2}}$ and $\tan 45^\circ = 1$

A quantity that is not exactly expressible as a rational number is called a **surd**. For example, $\sqrt{2}$ and $\sqrt{3}$ are called surds because they cannot be expressed as a fraction and the decimal part may be continued indefinitely. For example,

$$\sqrt{2} = 1.4142135\ldots, \quad \text{and} \quad \sqrt{3} = 1.7320508\ldots$$

From above, $\sin 30^\circ = \cos 60^\circ$, $\sin 45^\circ = \cos 45^\circ$ and $\sin 60^\circ = \cos 30^\circ$.
In general, $\mathbf{\sin\theta = \cos(90^\circ - \theta)}$ and $\mathbf{\cos\theta = \sin(90^\circ - \theta)}$

For example, it may be checked by calculator that $\sin 25^\circ = \cos 65^\circ$, $\sin 42^\circ = \cos 48^\circ$ and $\cos 84^\circ 10' = \sin 5^\circ 50'$, and so on.

For example, to evaluate $\frac{3\tan 60^\circ - 2\cos 30^\circ}{\tan 30^\circ}$ using surd forms:

From above, $\tan 60^\circ = \sqrt{3}$, $\cos 30^\circ = \frac{\sqrt{3}}{2}$ and $\tan 30^\circ = \frac{1}{\sqrt{3}}$, hence

$$\frac{3\tan 60^\circ - 2\cos 30^\circ}{\tan 30^\circ} = \frac{3(\sqrt{3}) - 2\left(\frac{\sqrt{3}}{2}\right)}{\frac{1}{\sqrt{3}}} = \frac{3\sqrt{3} - \sqrt{3}}{\frac{1}{\sqrt{3}}} = \frac{2\sqrt{3}}{\frac{1}{\sqrt{3}}}$$

$$= 2\sqrt{3}\left(\frac{\sqrt{3}}{1}\right) = 2(3) = \mathbf{6}$$

Solution of right-angled triangles

To 'solve a right-angled triangle' means 'to find the unknown sides and angles'. This is achieved by using (i) the theorem of Pythagoras, and/or (ii) trigonometric ratios.

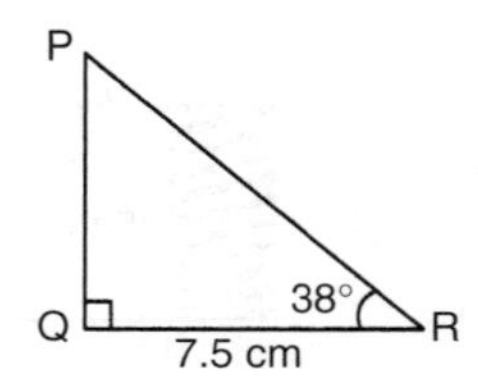

Figure 25.8

For example, to find the lengths of PQ and PR in triangle PQR shown in Figure 25.8:

$$\tan 38^\circ = \frac{PQ}{QR} = \frac{PQ}{7.5},$$

hence $\quad PQ = 7.5 \tan 38^\circ = 7.5(0.7813) = \mathbf{5.860\ cm}$

$$\cos 38^\circ = \frac{QR}{PR} = \frac{7.5}{PR},$$

hence $\quad PR = \dfrac{7.5}{\cos 38^\circ} = \dfrac{7.5}{0.7880} = \mathbf{9.518\ cm}$

[Check: Using Pythagoras' theorem $(7.5)^2 + (5.860)^2 = 90.59 = (9.518)^2$]

Angles of elevation and depression

(a) If, in Figure 25.9, BC represents horizontal ground and AB a vertical flagpole, then the **angle of elevation** of the top of the flagpole, A, from the point C is the angle that the imaginary straight line AC must be raised (or elevated) from the horizontal CB, i.e. angle $\boldsymbol{\theta}$.

(b) If, in Figure 25.10, PQ represents a vertical cliff and R a ship at sea, then the **angle of depression** of the ship from point P is the angle through which the imaginary straight line PR must be lowered (or depressed) from the horizontal to the ship, i.e. angle $\boldsymbol{\phi}$. (Note, $\angle$PRQ is also $\boldsymbol{\phi}$ — alternate angles between parallel lines.)

For example, an electricity pylon stands on horizontal ground. At a point 80 m from the base of the pylon, the angle of elevation of the top of the pylon is 23°. To calculate the height of the pylon to the nearest metre:

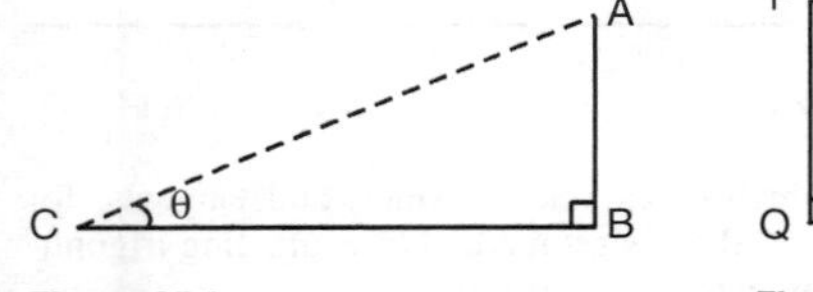

Figure 25.9

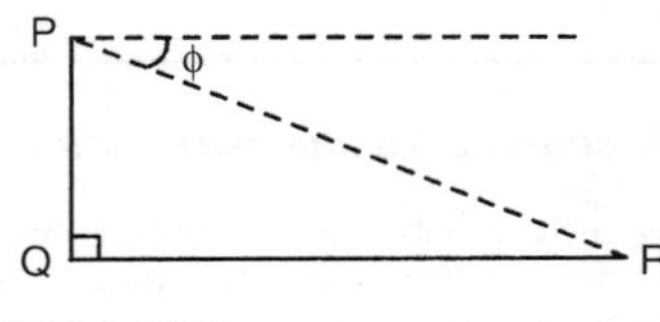

Figure 25.10

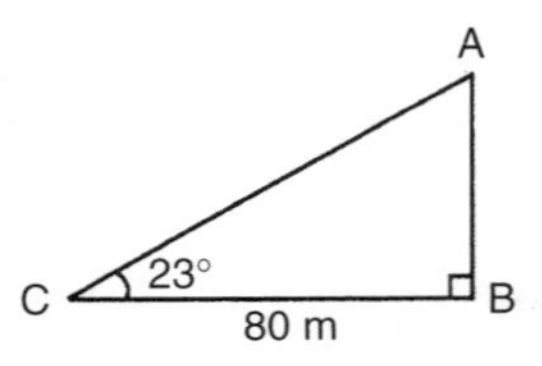

Figure 25.11

Figure 25.12

Figure 25.11 shows the pylon AB and the angle of elevation of A from point C is 23°. Now $\tan 23° = \dfrac{AB}{BC} = \dfrac{AB}{80}$.

Hence height of pylon $AB = 80 \tan 23° = 80(0.4245) = 33.96$ m

$= \mathbf{34\ m\ to\ the\ nearest\ metre}$

In another example, the angle of depression of a ship viewed at a particular instant from the top of a 75 m vertical cliff is 30°. The ship is sailing away from the cliff at constant speed and 1 minute later its angle of depression from the top of the cliff is 20°. To find (a) the distance of the ship from the base of the cliff at this instant, and (b) the speed of the ship in km/h:

(a) Figure 25.12 shows the cliff AB, the initial position of the ship at C and the final position at D. Since the angle of depression is initially 30° then $\angle ACB = 30°$ (alternate angles between parallel lines).

$\tan 30° = \dfrac{AB}{BC} = \dfrac{75}{BC}$ hence **initial position of ship from base of cliff**,

$BC = \dfrac{75}{\tan 30°} = \dfrac{75}{0.5774} = \mathbf{129.9\ m}$

(b) In triangle ABD, $\tan 20° = \dfrac{AB}{BD} = \dfrac{75}{BC + CD} = \dfrac{75}{129.9 + x}$

Hence $\quad 129.9 + x = \dfrac{75}{\tan 20°} = \dfrac{75}{0.3640} = 206.0$ m

from which $\quad x = 206.0 - 129.9 = 76.1$ m

Thus the ship sails 76.1 m in 1 minute, i.e. 60 s,

hence, **speed of ship** $= \dfrac{\text{distance}}{\text{time}} = \dfrac{76.1}{60}$ m/s

$= \dfrac{76.1 \times 60 \times 60}{60 \times 1000}$ km/h $= \mathbf{4.57\ km/h}$

Evaluating trigonometric ratios

Four-figure tables are available which gives sines, cosines, and tangents, for angles between 0° and 90°. However, the easiest method of evaluating trigonometric functions of any angle is by using a **calculator**.

The following values, correct to 4 decimal places, may be checked:

sine $18^\circ = 0.3090$,	cosine $56^\circ = 0.5592$
sine $172^\circ = 0.1392$	cosine $115^\circ = -0.4226$,
sine $241.63^\circ = -0.8799$,	cosine $331.78^\circ = 0.8811$
tangent $29^\circ = 0.5543$,	tangent $296.42^\circ = -2.0127$
tangent $178^\circ = -0.0349$	

To evaluate, say, sine $42^\circ 23'$ using a calculator means finding sine $42\dfrac{23^\circ}{60}$ since there are 60 minutes in 1 degree.

$$\frac{23}{60} = 0.383\dot{3}, \text{ thus } 42^\circ 23' = 42.383\dot{3}^\circ$$

Thus $\quad$ sine $42^\circ 23' = \text{sine } 42.383\dot{3}^\circ = 0.6741$,

correct to 4 decimal places.

Similarly, $\quad$ cosine $72^\circ 38' = \text{cosine } 72\dfrac{38^\circ}{60} = 0.2985$,

correct to 4 decimal places.

Most calculators contain only sine, cosine and tangent functions. Thus to evaluate secants, cosecants and cotangents, reciprocals need to be used. The following values, correct to 4 decimal places, may be checked:

$$\text{secant } 32^\circ = \frac{1}{\cos 32^\circ} = 1.1792 \qquad \text{secant } 215.12^\circ = \frac{1}{\cos 215.12^\circ} = -1.2226$$

$$\text{cosecant } 75^\circ = \frac{1}{\sin 75^\circ} = 1.0353 \qquad \text{cosecant } 321.62^\circ = \frac{1}{\sin 321.62^\circ} = -1.6106$$

$$\text{cotangent } 41^\circ = \frac{1}{\tan 41^\circ} = 1.1504 \qquad \text{cotangent } 263.59^\circ = \frac{1}{\tan 263.59^\circ} = 0.1123$$

For example, to evaluate, correct to 4 significant figures: (a) $\sin 1.481$ (b) $\tan 2.93$ (c) secant 5.37 (d) cosecant $\pi/4$

(a) $\sin 1.481$ means the sine of 1.481 radians. Hence a calculator needs to be on the radian function. Hence $\sin 1.481 = \mathbf{0.9960}$

(b) $\tan 2.93 = \mathbf{-0.2148}$

(c) $\sec 5.37 = \dfrac{1}{\cos 5.37} = \mathbf{1.6361}$

(d) $\text{cosec}(\pi/4) = \dfrac{1}{\sin(\pi/4)} = \dfrac{1}{\sin 0.785398\ldots} = \mathbf{1.4142}$

In another example, to determine the acute angles:
(a) $\sec^{-1} 2.3164$ (b) $\text{cosec}^{-1} 1.1784$

(a) $\sec^{-1} 2.3164 = \cos^{-1}\left(\dfrac{1}{2.3164}\right) = \cos^{-1} 0.4317\ldots$

$$= \mathbf{64.42^\circ} \text{ or } \mathbf{64^\circ 25'} \text{ or } \mathbf{1.124\ rad}$$

(b) $\text{cosec}^{-1} 1.1784 = \sin^{-1}\left(\frac{1}{1.1784}\right) = \sin^{-1} 0.8486\ldots$

$$= \mathbf{58.06^\circ} \text{ or } \mathbf{58^\circ 4'} \text{ or } \mathbf{1.013\ rad}$$

In another example, to evaluate correct to 4 decimal places: (a) $\sec(-115^\circ)$ (b) $\text{cosec}(-95^\circ 47')$

(a) Positive angles are considered by convention to be anticlockwise and negative angles as clockwise.
Hence -115° is actually the same as 245° (i.e. $360^\circ - 115^\circ$)

Hence $\quad \sec(-115^\circ) = \sec 245^\circ = \dfrac{1}{\cos 245^\circ} = \mathbf{-2.3662}$

(b) $\text{cosec}(-95^\circ 47') = \dfrac{1}{\sin\left(-95\frac{47^\circ}{60}\right)} = \mathbf{-1.0051}$

26 Cartesian and Polar Co-ordinates

Introduction

There are two ways in which the position of a point in a plane can be represented. These are (a) by **Cartesian co-ordinates**, i.e. (x, y), and (b) by **polar co-ordinates**, i.e. (r, θ), where r is a 'radius' from a fixed point and θ is an angle from a fixed point.

Changing from Cartesian into polar co-ordinates

In Figure 26.1, if lengths x and y are known, then the length of r can be obtained from Pythagoras' theorem (see Chapter 25) since OPQ is a right-angled triangle.

Hence $\quad r^2 = (x^2 + y^2)$

from which, $\quad \boxed{\mathbf{r = \sqrt{x^2 + y^2}}}$

From trigonometric ratios (see Chapter 25), $\tan\theta = \dfrac{y}{x}$

from which $\boxed{\boldsymbol{\theta = \tan^{-1}\frac{y}{x}}}$

$r = \sqrt{x^2 + y^2}$ and $\theta = \tan^{-1}\dfrac{y}{x}$ are the two formulae we need to change from Cartesian to polar co-ordinates. The angle θ, which may be expressed in degrees or radians, must **always** be measured from the positive x-axis, i.e.

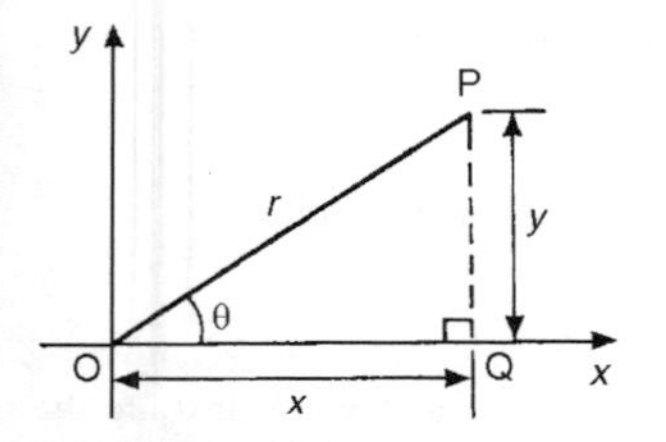

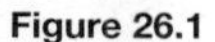
Figure 26.1

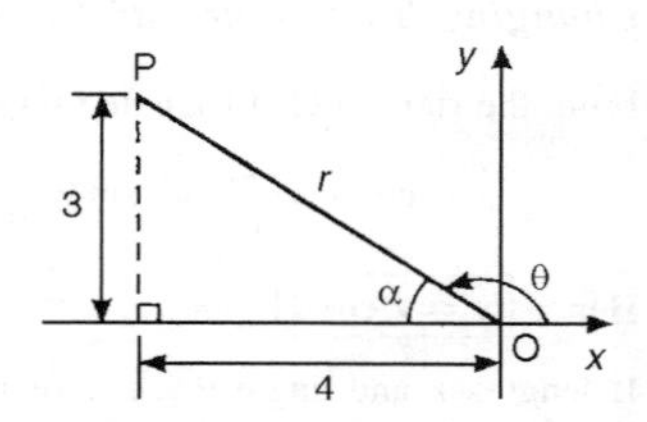

Figure 26.2

measured from the line OQ in Figure 26.1. It is suggested that when changing from Cartesian to polar co-ordinates a diagram should always be sketched.
For example, to express in polar co-ordinates the position (−4, 3):
A diagram representing the point using the Cartesian co-ordinates (−4, 3) is shown in Figure 26.2.

From Pythagoras' theorem, $r = \sqrt{4^2 + 3^2} = 5$

By trigonometric ratios, $\alpha = \tan^{-1}\frac{3}{4} = 36.87^\circ$ or 0.644 rad

Hence $\theta = 180^\circ - 36.87^\circ = 143.13^\circ$ or $\theta = \pi - 0.644 = 2.498$ rad
Hence the position of point P in polar coordinate form is (5, 143.13°) or (5, 2.498 rad)
In another example, to express (−5, −12) in polar co-ordinates:
A sketch showing the position (−5, −12) is shown in Figure 26.3.

$$r = \sqrt{5^2 + 12^2} = 13 \text{ and } \alpha = \tan^{-1}\frac{12}{5} = 67.38^\circ \text{ or } 1.176 \text{ rad}$$

Hence $\theta = 180^\circ + 67.38^\circ = 247.38^\circ$ or $\theta = \pi + 1.176 = 4.318$ rad
Thus (−5, −12) in Cartesian co-ordinates corresponds to (13, 247.38°) or (13, 4.318 rad) in polar co-ordinates.

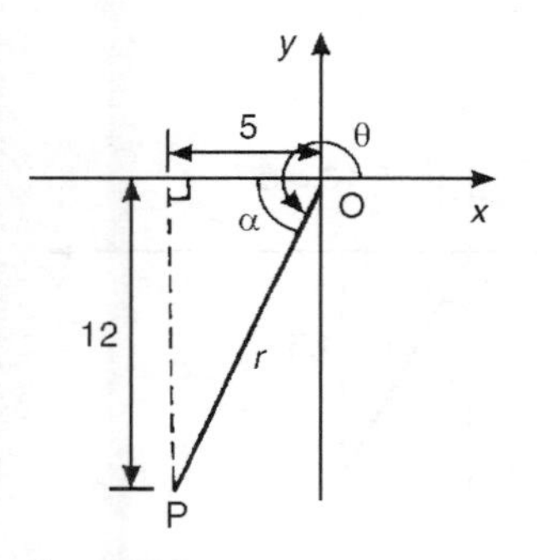

Figure 26.3

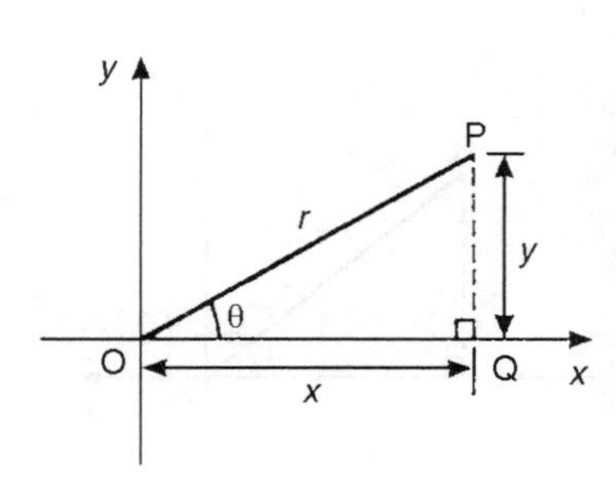

Figure 26.4

Changing from polar into Cartesian co-ordinates

From the right-angled triangle OPQ in Figure 26.4,

$$\cos\theta = \frac{x}{r} \text{ and } \sin\theta = \frac{y}{r}, \text{ from trigonometric ratios}$$

Hence $\boxed{x = r\cos\theta}$ and $\boxed{y = r\sin\theta}$

If lengths r and angle θ are known then $x = r\cos\theta$ and $y = r\sin\theta$ are the two formulae we need to change from polar to Cartesian co-ordinates.
For example, to express (6, 137°) in Cartesian co-ordinates:
A sketch showing the position (6, 137°) is shown in Figure 26.5.

$$x = r\cos\theta = 6\cos 137^\circ = -4.388$$

which corresponds to length OA in Figure 26.5.

$$y = r\sin\theta = 6\sin 137^\circ = 4.092$$

which corresponds to length AB in Figure 26.5.
Thus (6, 137°) in polar co-ordinates corresponds to (−4.388, 4.092) in Cartesian co-ordinates.
(Note that when changing from polar to Cartesian co-ordinates it is not quite so essential to draw a sketch. Use of $x = r\cos\theta$ and $y = r\sin\theta$ automatically produces the correct signs).
In another example, to express (4.5, 5.16 rad) in Cartesian co-ordinates:
A sketch showing the position (4.5, 5.16 rad) is shown in Figure 26.6.

$$x = r\cos\theta = 4.5\cos 5.16 = 1.948$$

which corresponds to length OA in Figure 26.6.

$$y = r\sin\theta = 4.5\sin 5.16 = -4.057$$

which corresponds to length AB in Figure 26.6.
Thus (1.948, −4.057) in Cartesian co-ordinates corresponds to (4.5, 5.16 rad) in polar co-ordinates.

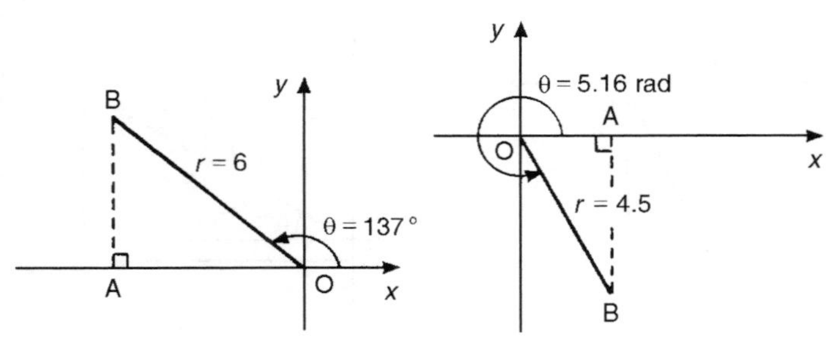

Figure 26.5

Figure 26.6

Use of R → P *and* P → R *functions on calculators*

Another name for Cartesian co-ordinates is **rectangular** co-ordinates. Many scientific notation calculators possess R → P and P → R functions. The R is the first letter of the word rectangular and the P is the first letter of the word polar. Check the operation manual for your particular calculator to determine how to use these two functions. They make changing from Cartesian to polar co-ordinates, and vice-versa, so much quicker and easier.

---

27 Triangles and Some Practical Applications

Sine and cosine rules

To **'solve a triangle'** means 'to find the values of unknown sides and angles'. If a triangle is **right-angled**, trigonometric ratios and the theorem of Pythagoras may be used for its solution, as shown in chapter 25. However, for a **non-right-angled triangle**, trigonometric ratios and Pythagoras' theorem **cannot** be used. Instead, two rules, called the **sine rule** and the**cosine rule**, are used.

Sine rule

With reference to triangle ABC of Figure 27.1, the **sine rule** states:

$$\frac{a}{\sin A} = \frac{b}{\sin B} = \frac{c}{\sin C}$$

The rule may be used only when:

(i) 1 side and any 2 angles are initially given, or

(ii) 2 sides and an angle (not the included angle) are initially given.

Cosine rule

With reference to triangle ABC of Figure 27.1, the **cosine rule** states:

$$a^2 = b^2 + c^2 - 2bc \cos A$$

$$\text{or } b^2 = a^2 + c^2 - 2ac \cos B$$

$$\text{or } c^2 = a^2 + b^2 - 2ab \cos C$$

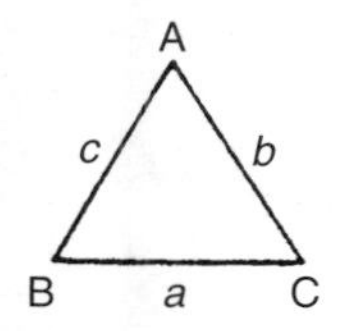

Figure 27.1

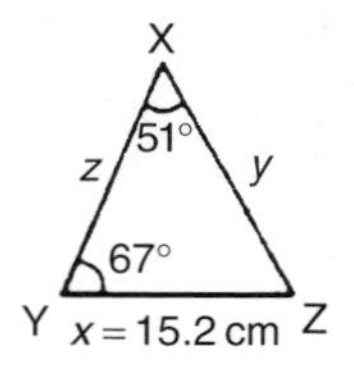

Figure 27.2

The rule may be used only when:
(i) 2 sides and the included angle are initially given, or
(ii) 3 sides are initially given.

Area of any triangle

The **area of any triangle** such as ABC of Figure 27.1 is given by:

(i) $\frac{1}{2}$ × **base** × **perpendicular height**, or
(ii) $\frac{1}{2}\boldsymbol{ab}$ **sin** $\boldsymbol{C}$ or $\frac{1}{2}\boldsymbol{ac}$ **sin** $\boldsymbol{B}$ or $\frac{1}{2}\boldsymbol{bc}$ **sin** $\boldsymbol{A}$, or
(iii) $\sqrt{[s(s-1)(s-b)(s-c)]}$, **where** $s = \dfrac{a+b+c}{2}$

For example, in a triangle XYZ, ∠X = 51°, ∠Y = 67° and YZ = 15.2 cm. To solve the triangle and find its area:
The triangle XYZ is shown in Figure 27.2. Since the angles in a triangle add up to 180°, then **Z** = 180° − 51° − 67° = **62°**. Applying the sine rule:

$$\frac{15.2}{\sin 51^\circ} = \frac{y}{\sin 67^\circ} = \frac{z}{\sin 62^\circ}$$

Using $\dfrac{15.2}{\sin 51^\circ} = \dfrac{y}{\sin 67^\circ}$ and transposing gives :

$$y = \frac{15.2 \sin 67^\circ}{\sin 51^\circ} = \mathbf{18.00\ cm = XZ}$$

Using $\dfrac{15.2}{\sin 51^\circ} = \dfrac{z}{\sin 62^\circ}$ and transposing gives :

$$z = \frac{15.2 \sin 62^\circ}{\sin 51^\circ} = \mathbf{17.27\ cm = XY}$$

$$\textbf{Area of triangle XYZ} = \tfrac{1}{2}xy \sin Z$$

$$= \tfrac{1}{2}(15.2)(18.00)\sin 62^\circ = \mathbf{120.8\ cm^2}$$

(or area = $\frac{1}{2}xz \sin Y = \frac{1}{2}(15.2)(17.27)\sin 67^\circ =$ **120.8 cm²**)
It is always worth checking with triangle problems that the longest side is opposite the largest angle, and vice-versa. In this problem, Y is the largest angle and XZ is the longest of the three sides.
In another example, to solve triangle DEF and find its area given that EF = 35.0 mm, DE = 25.0 mm and ∠E = 64°:
Triangle DEF is shown in Figure 27.3.
Applying the cosine rule: $e^2 = d^2 + f^2 - 2df \cos E$

i.e. $$e^2 = (35.0)^2 + (25.0)^2 - [2(35.0)(25.0)\cos 64^\circ]$$

$$= 1225 + 625 - 767.1 = 1083$$

from which, $$\boldsymbol{e} = \sqrt{1083} = \mathbf{32.91\ mm}$$

Applying the sine rule: $\frac{32.91}{\sin 64^\circ} = \frac{25.0}{\sin F}$

from which, $\sin F = \frac{25.0 \sin 64^\circ}{32.91} = 0.6828$

Thus $\angle F = \sin^{-1} 0.6828 = 43^\circ 4'$ or $136^\circ 56'$

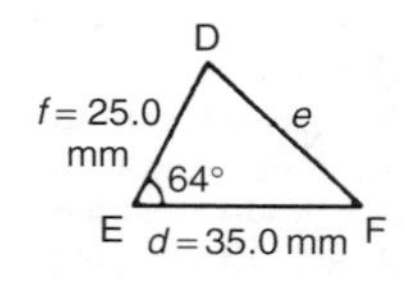

Figure 27.3

$F = 136^\circ 56'$ is not possible in this case since $136^\circ 56' + 64^\circ$ is greater than 180°. Thus only $\mathbf{F = 43^\circ 4'}$ is valid.

$$\angle \mathbf{D} = 180^\circ - 64^\circ - 43^\circ 4' = \mathbf{72^\circ 56'}$$

$$\text{Area of triangle DEF} = \tfrac{1}{2} d f \sin E$$

$$= \tfrac{1}{2}(35.0)(25.0)\sin 64^\circ = \mathbf{393.2\ mm^2}$$

Practical situations involving trigonometry

There are a number of **practical situations** where the use of trigonometry is needed to find unknown sides and angles of triangles.

For example, a room 8.0 m wide has a span roof that slopes at 33° on one side and 40° on the other. To find the length of the roof slopes, correct to the nearest centimetre:
A section of the roof is shown in Figure 27.4.
Angle at ridge, $B = 180^\circ - 33^\circ - 40^\circ = 107^\circ$

From the sine rule: $\frac{8.0}{\sin 107^\circ} = \frac{a}{\sin 33^\circ}$

from which, $a = \frac{8.0 \sin 33^\circ}{\sin 107^\circ} = 4.556$ m

Also from the sine rule: $\frac{8.0}{\sin 107^\circ} = \frac{c}{\sin 40^\circ}$

from which, $c = \frac{8.0 \sin 40^\circ}{\sin 107^\circ} = 5.377$ m

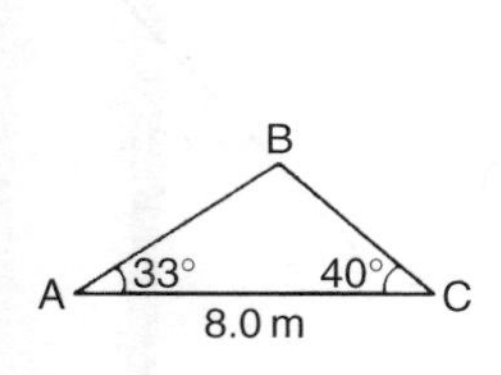

Figure 27.4

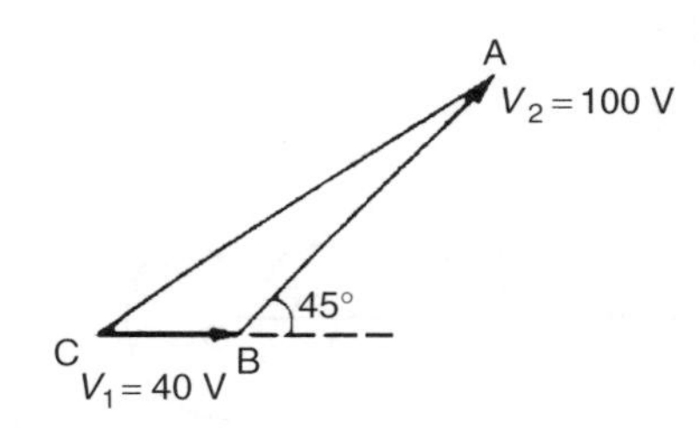

Figure 27.5

Hence the roof slopes are 4.56 m and 5.38 m, correct to the nearest centimetre.

In another example, two voltage phasors are shown in Figure 27.5 where $V_1 = 40$ V and $V_2 = 100$ V. To determine the value of their resultant (i.e. length OA) and the angle the resultant makes with V_1:

Angle OBA = 180° − 45° = 135°

Applying the cosine rule:

$$\begin{aligned}\text{OA}^2 &= V_1^2 + V_2^2 - 2V_1V_2\cos\text{OBA}\\ &= 40^2 + 100^2 - \{2(40)(100)\cos 135^\circ\}\\ &= 1600 + 10000 - \{-5657\}\\ &= 1600 + 10000 + 5657 = 17257\end{aligned}$$

The resultant $\text{OA} = \sqrt{17257} = 131.4$ V

Applying the sine rule:

$$\frac{131.4}{\sin 135^\circ} = \frac{100}{\sin\text{AOB}}$$

from which, $\sin\text{AOB} = \dfrac{100\sin 135^\circ}{131.4} = 0.5381$

Hence angle AOB = $\sin^{-1} 0.5381 = 32°33'$ (or $147°27'$, which is impossible in this case). Hence, **the resultant voltage is 131.4 volts at 32°33′ to V_1**

In another example, a crank mechanism of a petrol engine is shown in Figure 27.6. Arm OA is 10.0 cm long and rotates clockwise about 0. The connecting rod AB is 30.0 cm long and end B is constrained to move horizontally. To determine the angle between the connecting rod AB and the horizontal and the length of OB for the position shown in Figure 27.6:

Applying the sine rule: $\dfrac{\text{AB}}{\sin 50^\circ} = \dfrac{\text{AO}}{\sin B}$

from which, $\sin B = \dfrac{\text{AO}\sin 50^\circ}{\text{AB}} = \dfrac{10.0\sin 50^\circ}{30.0} = 0.2553$

Hence $B = \sin^{-1} 0.2553 = 14°47'$ (or $165°13'$, which is impossible in this case).

Hence the connecting rod AB makes an angle of 14°47′ with the horizontal.

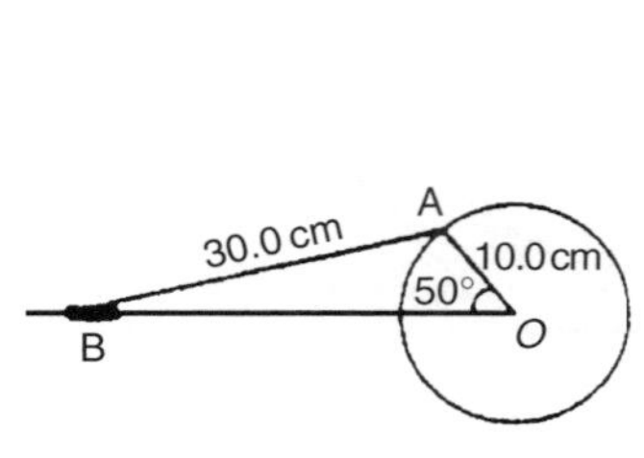

Figure 27.6

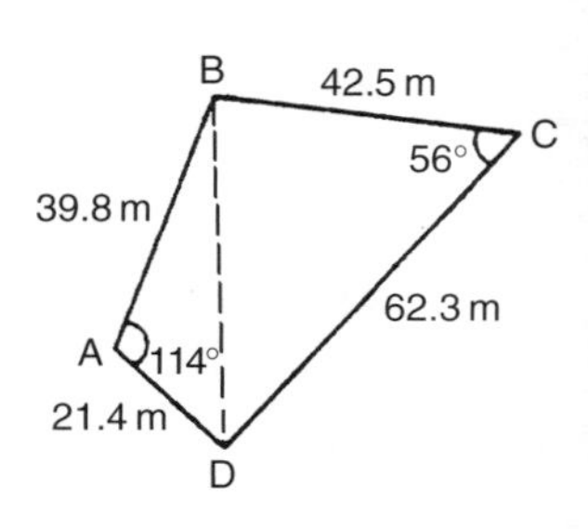

Figure 27.7

Angle OAB $= 180° - 50° - 14°47' = 115°13'$

Applying the sine rule: $\dfrac{30.0}{\sin 50°} = \dfrac{\text{OB}}{\sin 115°13'}$

from which, $\mathbf{OB} = \dfrac{30.0 \sin 115°13'}{\sin 50°} = \mathbf{35.43\ cm}$

In another example, the area of a field is in the form of a quadrilateral ABCD as shown in Figure 27.7. To determine its area:
A diagonal drawn from B to D divides the quadrilateral into two triangles.

Area of quadrilateral ABCD

$= $ area of triangle ABD + area of triangle BCD

$= \frac{1}{2}(39.8)(21.4)\sin 114° + \frac{1}{2}(42.5)(62.3)\sin 56°$

$= 389.04 + 1097.5 = \mathbf{1487\ m^2}$

---

28 Trigonometric Waveforms

Graphs of trigonometric functions

By drawing up tables of values from 0° to 360°, graphs of $y = \sin A$, $y = \cos A$ and $y = \tan A$ may be plotted as shown in Figure 28.1.
From the graphs it is seen that:

(i) Sine and cosine graphs oscillate between peak values of ± 1
(ii) The cosine curve is the same shape as the sine curve but displaced by 90°
(iii) The sine and cosine curves are continuous and they repeat at intervals of 360°; the tangent curve appears to be discontinuous and repeats at intervals of 180°.

Angles of any magnitude

Figure 28.2 shows rectangular axes XX′ and YY′ intersecting at origin 0. As with graphical work, measurements made to the right and above 0 are positive, while those to the left and downwards are negative. Let 0A be free to rotate about 0. By convention, when 0A moves anticlockwise angular measurement is considered positive, and vice versa.
Let OA be rotated anticlockwise so that θ_1 is any angle in the first quadrant and let perpendicular AB be constructed to form the right-angled triangle OAB in Figure 28.3. Since all three sides of the triangle are positive, the trigonometric ratios sine, cosine and tangent will all be positive in the first quadrant. (Note: OA is always positive since it is the radius of a circle).

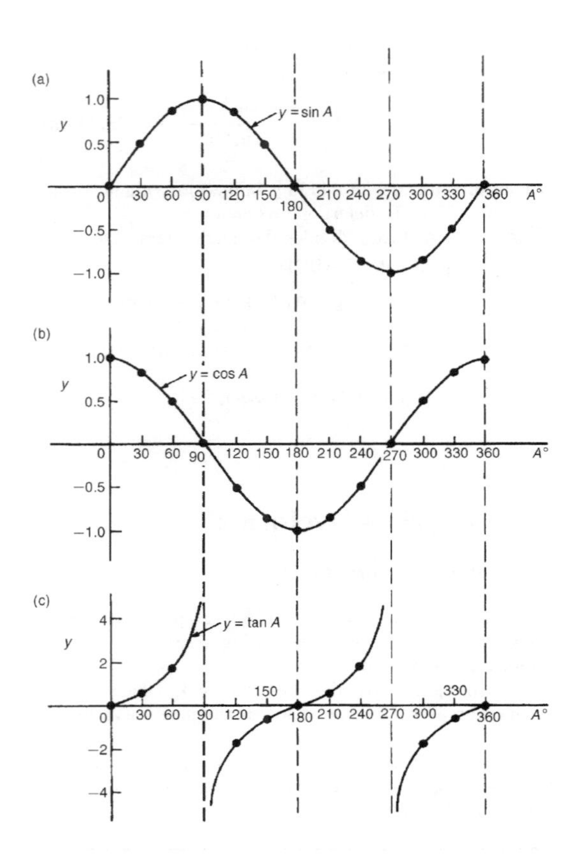

Figure 28.1

Let 0A be further rotated so that θ_2 is any angle in the second quadrant and let AC be constructed to form the right-angled triangle 0AC. Then

$$\sin\theta_2 = \frac{+}{+} = + \cos\theta_2 = \frac{-}{+} = - \tan\theta_2 = \frac{+}{-} = -$$

Let 0A be further rotated so that θ_3 is any angle in the third quadrant and let AD be constructed to form the right-angled triangle 0AD. Then

$$\sin\theta_3 = \frac{-}{+} = - \cos\theta_3 = \frac{-}{+} = - \tan\theta_3 = \frac{-}{-} = +$$

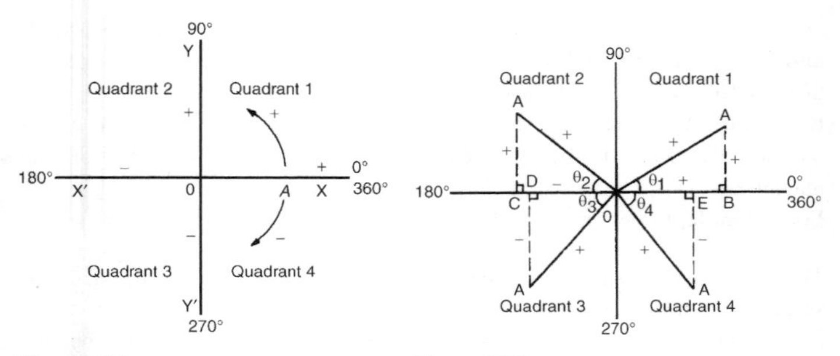

Figure 28.2 **Figure 28.3**

Let 0A be further rotated so that θ_4 is any angle in the fourth quadrant and let AE be constructed to form the right-angled triangle 0AE. Then

$$\sin\theta_4 = \frac{-}{+} = - \quad \cos\theta_4 = \frac{+}{+} = + \quad \tan\theta_4 = \frac{-}{+} = -$$

The above results are summarised in Figure 28.4. The letters underlined spell the word CAST when starting in the fourth quadrant and moving in an anti-clockwise direction.

In the first quadrant of Figure 28.1 all of the curves have positive values; in the second only sine is positive; in the third only tangent is positive; in the fourth only cosine is positive — exactly as summarised in Figure 28.4.

A knowledge of angles of any magnitude is needed when finding, for example, all the angles between 0° and 360° whose sine is, say, 0.3261. If 0.3261 is entered into a calculator and then the inverse sine key pressed (or $\sin^{-1}$ key) the answer 19.03° appears. However, there is a second angle between 0° and 360° which the calculator does not give. Sine is also positive in the second quadrant [either from CAST or from Figure 28.1(a)]. The other angle is shown

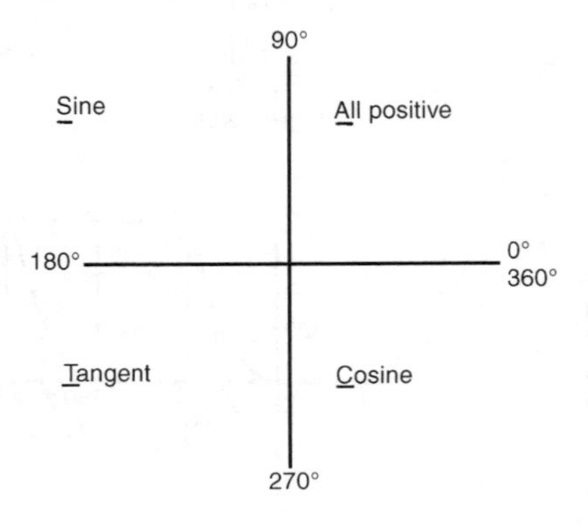

Figure 28.4

in Figure 28.5 as angle θ where $\theta = 180° - 19.03° = 160.97°$. Thus 19.03° **and** 160.97° are the angles between 0° and 360° whose sine is 0.3261 (check that sin 160.97° = 0.3261 on your calculator).

Be careful! Your calculator only gives you one of these answers. The second answer needs to be deduced from a knowledge of angles of any magnitude.

For example, to determine all the angles between 0° and 360° whose sine is −0.4638:

The angles whose sine is −0.4638 occurs in the third and fourth quadrants since sine is negative in these quadrants — see Figure 28.6.

From Figure 28.7, $\theta = \sin^{-1} 0.4638 = 27.63°$. Measured from 0°, the two angles between 0° and 360° whose sine is −0.4638 are 180° + 27.63°, i.e. **207.63°** and 360° − 27.63°, i.e. **332.37°**

(Note that a calculator only gives one answer, i.e. −27.632588°).

In another example, to determine all the angles between 0° and 360° whose tangent is 1.7629:

A tangent is positive in the first and third quadrants — see Figure 28.8.

From Figure 28.9, $\theta = \tan^{-1} 1.7629 = 60.44°$

Measured from 0°, the two angles between 0° and 360° whose tangent is 1.7629 are **60.44°** and 180° + 60.44°, i.e. **240.44°**

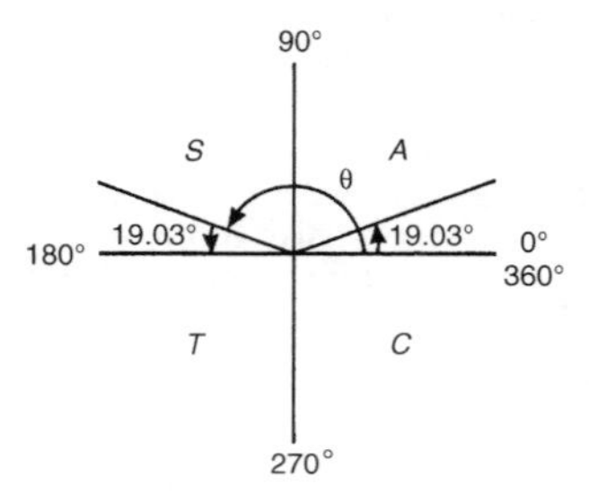

Figure 28.5

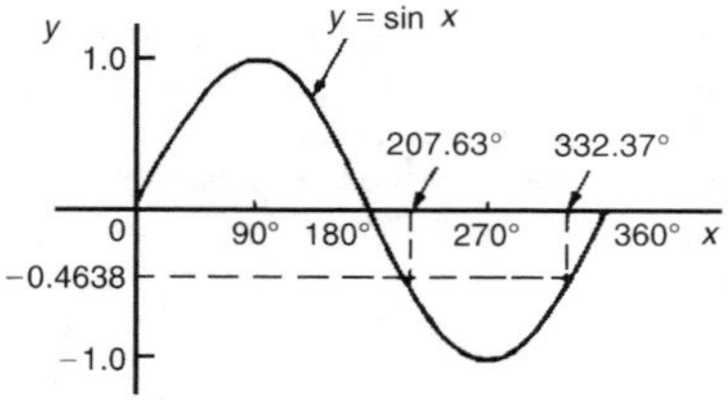

Figure 28.6

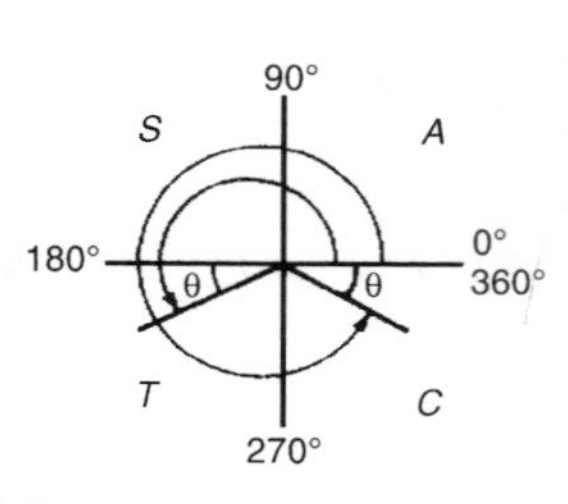

Figure 28.7

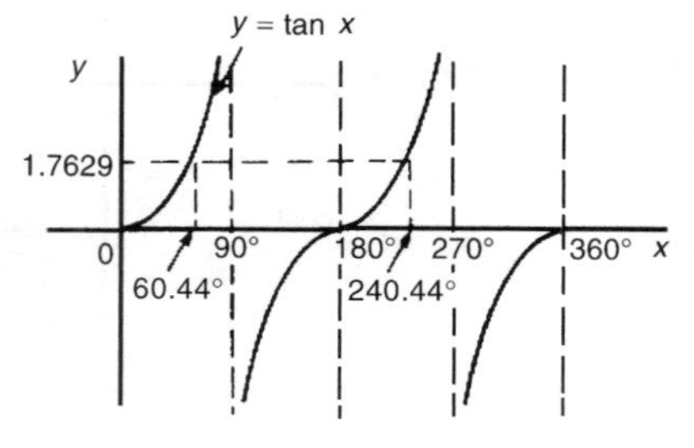

Figure 28.8

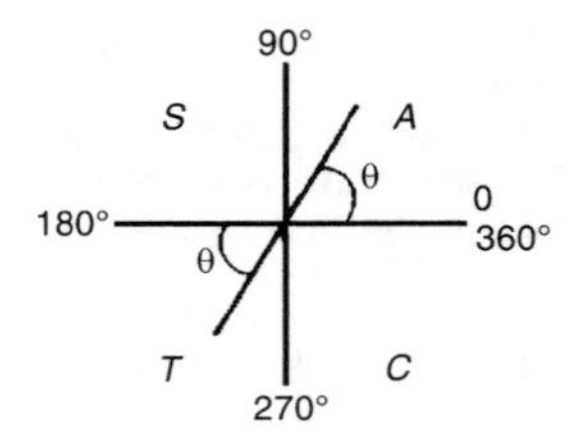

Figure 28.9

The production of a sine and cosine wave

In Figure 28.10, let OR be a vector 1 unit long and free to rotate anticlockwise about O. In one revolution a circle is produced and is shown with 15° sectors. Each radius arm has a vertical and a horizontal component. For example, at 30°, the vertical component is TS and the horizontal component is OS.
From trigonometric ratios,

$$\sin 30^\circ = \frac{TS}{TO} = \frac{TS}{1}, \quad \text{i.e.} \quad TS = \sin 30^\circ$$

and $$\cos 30^\circ = \frac{OS}{TO} = \frac{OS}{1}, \quad \text{i.e.} \quad OS = \cos 30^\circ$$

The vertical component TS may be projected across to T′S′, which is the corresponding value of 30° on the graph of y against angle x°. If all such

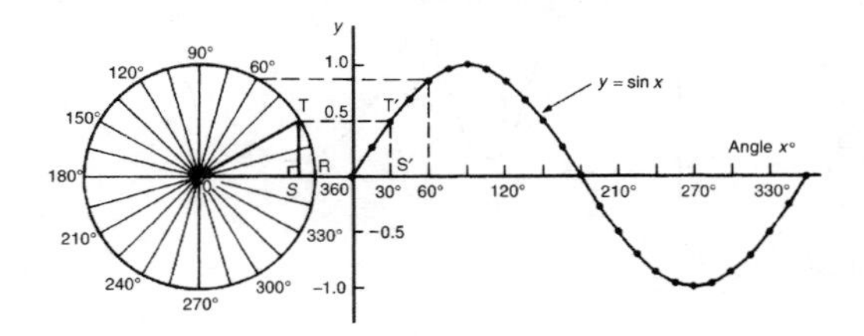

Figure 28.10

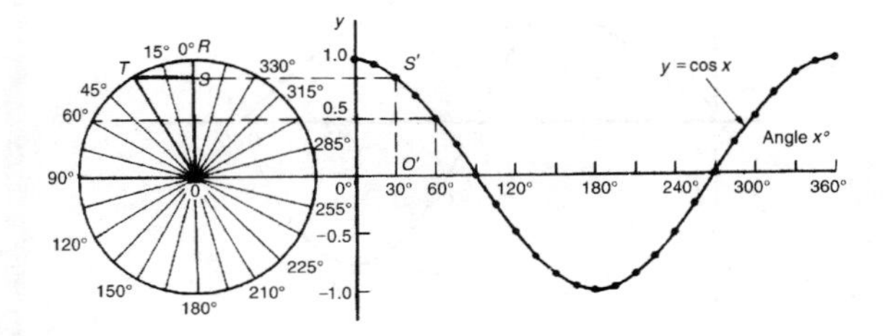

Figure 28.11

vertical components as TS are projected on to the graph, then a **sine wave** is produced as shown in Figure 28.10.

If all horizontal components such as OS are projected on to a graph of y against angle x°, then a **cosine wave** is produced. It is easier to visualise these projections by redrawing the circle with the radius arm OR initially in a vertical position as shown in Figure 28.11.

From Figures 28.10 and 28.11 it is seen that a cosine curve is of the same form as the sine curve but is displaced by 90° (or $\pi/2$ radians).

Sine and cosine curves

Graphs of sine and cosine waveforms

Graphs of $y = \sin A$ and $y = \sin 2A$ are shown in Figure 28.12

A graph of $y = \sin \frac{1}{2}A$ is shown in Figure 28.13.

Graph of $y = \cos A$ and $y = \cos 2A$ are shown in Figure 28.14.

A graph of $y = \cos \frac{1}{2}A$ is shown in Figure 28.15.

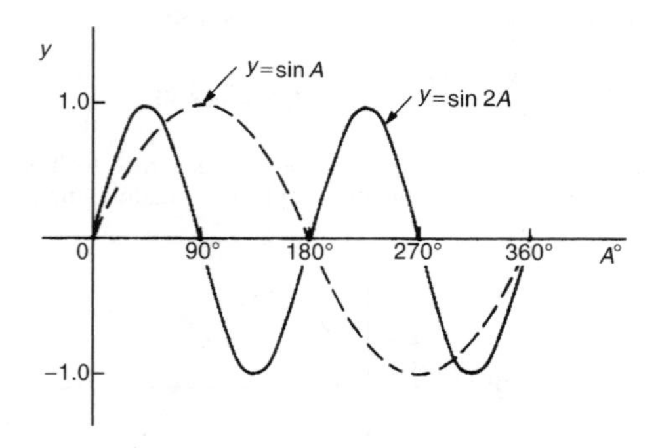

Figure 28.12

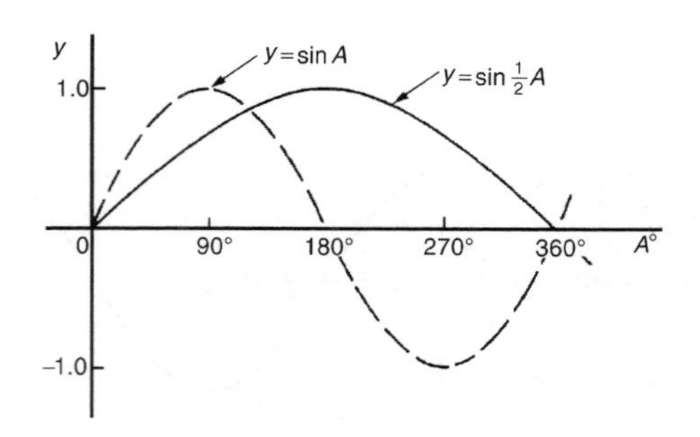

Figure 28.13

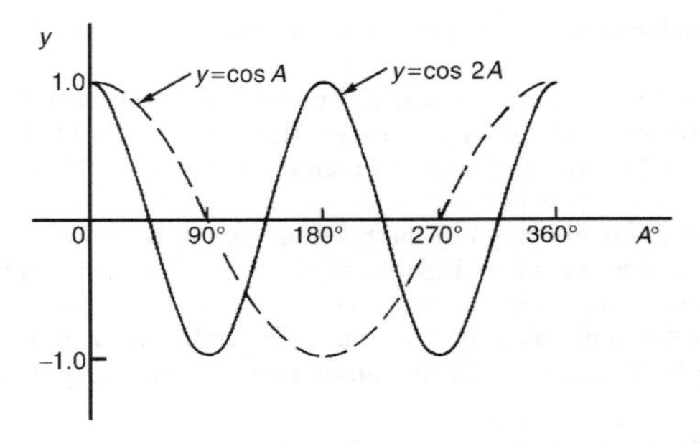

Figure 28.14

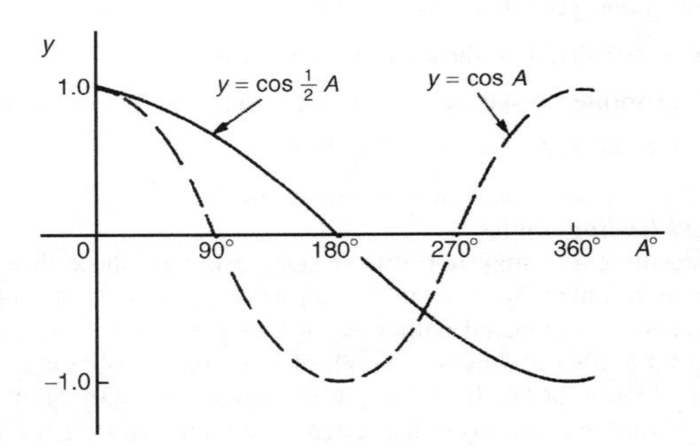

Figure 28.15

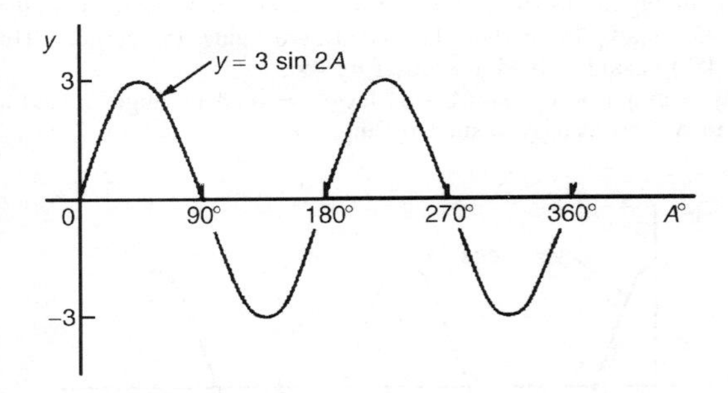

Figure 28.16

Periodic functions and period

Each of the graphs shown in Figures 28.12 to 28.15 will repeat themselves as angle A increases and are thus called **periodic functions**.

$y = \sin A$ and $y = \cos A$ repeat themselves every 360° (or 2π radians); thus 360° is called the **period** of these waveforms. $y = \sin 2A$ and $y = \cos 2A$

repeat themselves every 180° (or π radians); thus 180° is the period of these waveforms.
In general, if $y = \sin pA$ or $y = \cos pA$ (where p is a constant) then the period of the waveform is $360°/p$ (or $2\pi/p$ rad). Hence if $y = \sin 3A$ then the period is 360/3, i.e. 120°, and if $y = \cos 4A$ then the period is 360/4, i.e. 90°.

Amplitude

Amplitude is the name given to the maximum or peak value of a sine wave. Each of the graphs shown in Figures 28.12 to 28.15 has an amplitude of +1 (i.e. they oscillate between +1 and −1). However, if $y = 4\sin A$, each of the values of $\sin A$ is multiplied by 4 and the maximum value, and thus amplitude, is 4. Similarly, if $y = 5\cos 2A$, the amplitude is 5 and the period is $360°/2$, i.e. 180°.

For example, to sketch $y = 3\sin 2A$ from $A = 0$ to $A = 360°$:

Amplitude = 3 and period = 360/2 = 180°.

A sketch of $y = 3\sin 2A$ is shown in Figure 28.16.

In another example, to sketch $y = 4\cos 2x$ from $x = 0°$ to $x = 360°$:

Amplitude = 4 and period = $360°/2 = 180°$

A sketch of $y = 4\cos 2x$ is shown in Figure 28.17.

Lagging and leading angles

A sine or cosine curve may not always start at 0°. To show this a periodic function is represented by $y = \sin(A \pm \alpha)$ or $y = \cos(A \pm \alpha)$ where α is a phase displacement compared with $y = \sin A$ or $y = \cos A$.
By drawing up a table of values, a graph of $y = \sin(A - 60°)$ may be plotted as shown in Figure 28.18. If $y = \sin A$ is assumed to start at 0° then $y = \sin(A - 60°)$ starts 60° later (i.e. has a zero value 60° later). Thus $y = \sin(A - 60°)$ is said to **lag** $y = \sin A$ by 60°.
By drawing up a table of values, a graph of $y = \cos(A + 45°)$ may be plotted as shown in Figure 28.19. If $y = \cos A$ is assumed to start at 0° then $y = \cos(A + 45°)$ starts 45° earlier (i.e. has a zero value 45° earlier). Thus $y = \cos(A + 45°)$ is said to **lead** $y = \cos A$ by 45°.
Generally, a graph of $y = \sin(A - \alpha)$ lags $y = \sin A$ by angle α, and a graph of $y = \sin(A + \alpha)$ leads $y = \sin A$ by angle α.

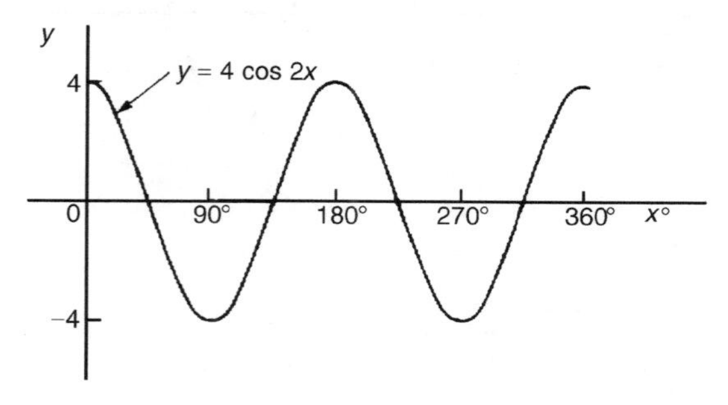

Figure 28.17

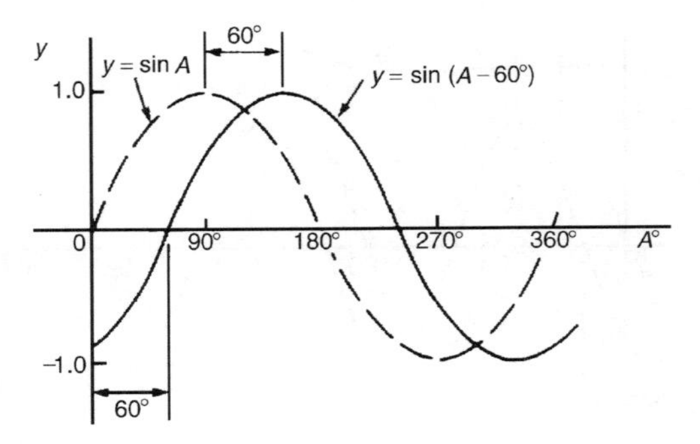

Figure 28.18

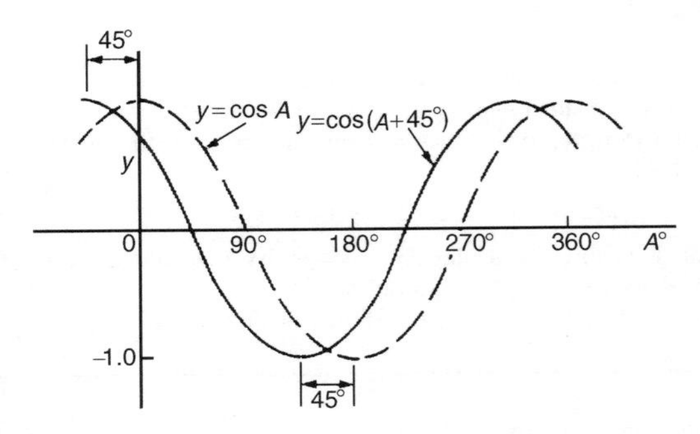

Figure 28.19

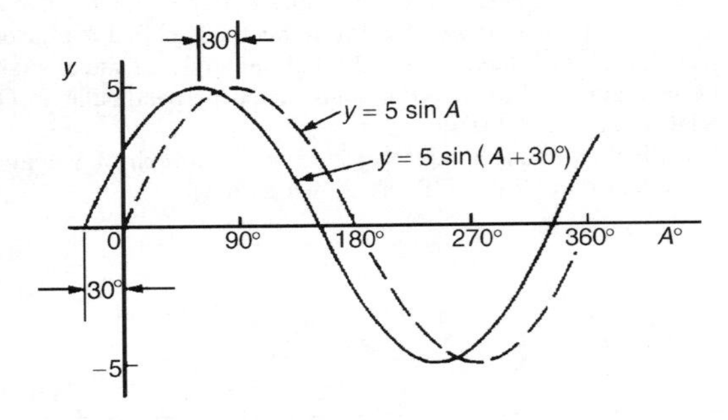

Figure 28.20

A cosine curve is the same shape as a sine curve but starts 90° earlier, i.e. leads by 90°. Hence $\cos A = \sin(A + 90^\circ)$.

For example, to sketch $y = 5\sin(A + 30^\circ)$ from $A = 0^\circ$ to $A = 360^\circ$:
Amplitude = 5 and period = $360^\circ/1 = 360^\circ$

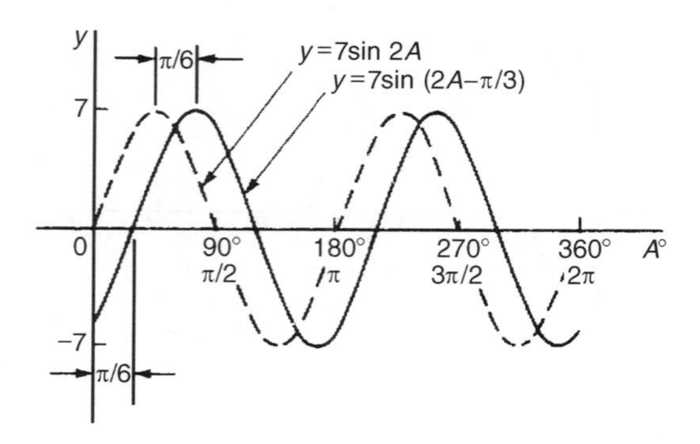

Figure 28.21

$5\sin(A + 30°)$ leads $5\sin A$ by 30° (i.e. starts 30° earlier)
A sketch of $y = 5\sin(A + 30°)$ is shown in Figure 28.20.
In another example, to sketch $y = 7\sin(2A - \pi/3)$ in the range $0 \leq A \leq 360°$:

Amplitude = 7 and period = $2\pi/2 = \pi$ radians

In general, $\boldsymbol{y = \sin(pt - \alpha)}$ **lags** $\boldsymbol{y = \sin pt}$ **by** $\boldsymbol{\alpha/p}$, hence $7\sin(2A - \pi/3)$ lags $7\sin 2A$ by $(\pi/3)/2$, i.e. $\pi/6$ rad or 30°.
A sketch of $y = 7\sin(2A - \pi/3)$ is shown in Figure 28.21.

Sinusoidal form A $\sin(\omega t \pm \alpha)$

In Figure 28.22, let OR represent a vector that is free to rotate anticlockwise about O at a velocity of ω rad/s. A rotating vector is called a **phasor**. After a time t seconds OR will have turned through an angle ωt radians (shown as angle TOR in Figure 28.22). If ST is constructed perpendicular to OR, then $\sin\omega t = \text{ST/OT}$, i.e. $\text{ST} = \text{OT}\sin\omega t$.
If all such vertical components are projected on to a graph of y against ωt, a sine wave results of amplitude OR (as shown earlier).

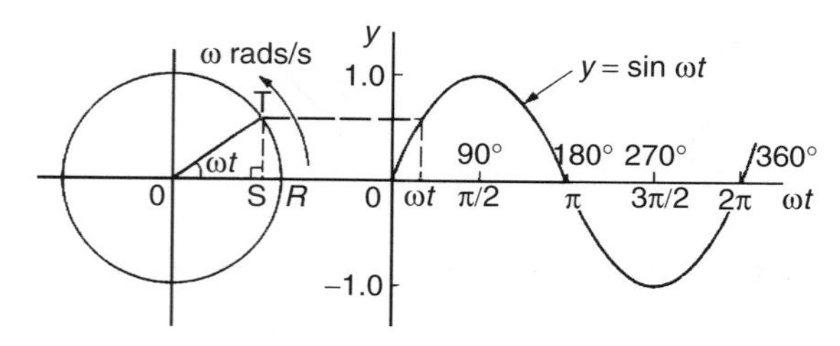

Figure 28.22

If phasor OR makes one revolution (i.e. 2π radians) in T seconds, then the angular velocity, $\omega = 2\pi/T$ rad/s, from which, $\boxed{T = 2\pi/\omega \text{ seconds.}}$
T is known as the **periodic time**.
The number of complete cycles occurring per second is called the **frequency, f**.

$$\text{Frequency} = \frac{\text{number of cycles}}{\text{second}} = \frac{1}{T} = \frac{\omega}{2\pi}\text{Hz} \quad \text{i.e.} \quad \boxed{f = \frac{\omega}{2\pi}\text{Hz}}$$

Hence **angular velocity**, $\boxed{\omega = 2\pi f \text{ rad/s}}$

Given a general sinusoidal function $\mathbf{y = A\sin(\omega t \pm \alpha)}$, then

(i) A = amplitude
(ii) ω = angular velocity $= 2\pi f$ rad/s
(iii) $\dfrac{2\pi}{\omega}$ = periodic time T seconds
(iv) $\dfrac{\omega}{2\pi}$ = frequency, f hertz
(v) α = angle of lead or lag (compared with $y = A\sin\omega t$), in radians.

For example, an alternating current is given by $i = 30\sin(100\pi t + 0.27)$ amperes. To find the amplitude, periodic time, frequency and phase angle (in degrees and minutes):

$i = 30\sin(100\pi t + 0.27)A$, hence **amplitude = 30 A**

Angular velocity $\omega = 100\pi$, hence

$$\textbf{periodic time, } T = \frac{2\pi}{\omega} = \frac{2\pi}{100\pi} = \frac{1}{50} = \mathbf{0.02 \text{ s or } 20 \text{ ms}}$$

$$\textbf{Frequency, } f = \frac{1}{T} = \frac{1}{0.02} = \mathbf{50 \text{ Hz}}$$

Phase angle, $\boldsymbol{\alpha} = 0.27 \text{ rad} = \left(0.27 \times \dfrac{180}{\pi}\right)^\circ = \mathbf{15.47^\circ}$ **or** $\mathbf{15^\circ 28'}$ **leading** $\mathbf{i = 30\sin(100\pi t)}$.

In another example, an oscillating mechanism has a maximum displacement of 2.5 m and a frequency of 60 Hz. At time $t = 0$ the displacement is 90 cm. To express the displacement in the general form $A\sin(\omega t \pm \alpha)$:

$$\text{Amplitude} = \text{maximum displacement} = 2.5 \text{ m}$$

$$\text{Angular velocity, } \omega = 2\pi f = 2\pi(60) = 120\pi \text{ rad/s}$$

Hence $\quad$ displacement $= 2.5\sin(120\pi t + \alpha)$ m

When $t = 0$, displacement = 90 cm = 0.90 m

Hence $\quad 0.90 = 2.5\sin(0 + \alpha)$ i.e. $\sin\alpha = \dfrac{0.90}{2.5} = 0.36$

Hence $\quad \alpha = \sin^{-1} 0.36 = 21.10^\circ = 21^\circ 6' = 0.368$ rad

Thus $\quad$ **displacement = 2.5 sin(120πt + 0.368) m**

In another example, the instantaneous value of voltage in an a.c. circuit at any time t seconds is given by $v = 340\sin(50\pi t - 0.541)$ volts. To determine
(a) the amplitude, periodic time, frequency and phase angle (in degrees),
(b) the value of the voltage when $t = 0$,
(c) the value of the voltage when $t = 10$ ms, and
(d) the time when the voltage first reaches 200 V

(a) **Amplitude = 340 V**
Angular velocity, $\omega = 50\pi$

Hence **periodic time**, $T = \frac{2\pi}{\omega} = \frac{2\pi}{50\pi} = \frac{1}{25} = \mathbf{0.04}$ **s** or **40 ms**

Frequency $f = \frac{1}{T} = \frac{1}{0.04} = \mathbf{25\ Hz}$

Phase angle $= 0.541$ rad $= \left(0.541 \times \frac{180}{\pi}\right)$

$= \mathbf{31^\circ}$ **lagging** $v = 340\sin(50\pi t)$

(b) **When $t = 0$**, $v = 340\sin(0 - 0.541) = 340\sin(-31^\circ) = \mathbf{-175.1\ V}$

(c) **When t = 10 ms** then $v = 340\sin\left(50\pi\frac{10}{10^3} - 0.541\right)$

$= 340\sin(1.0298) = 340\sin 59^\circ = \mathbf{291.4\ volts}$

(d) When $v = 200$ volts then $200 = 340\sin(50\pi t - 0.541)$

$$\frac{200}{340} = \sin(50\pi t - 0.541)$$

Hence $(50\pi t - 0.541) = \sin^{-1}\frac{200}{340}$

$= 36.03^\circ$ or 0.6288 rad

$50\pi t = 0.6288 + 0.541 = 1.1698$

Hence when $v = 200$ V, **time**, $t = \frac{1.1698}{50\pi} = \mathbf{7.447\ ms}$

A sketch of $v = 340\sin(50\pi t - 0.541)$ volts is shown in Figure 28.23.

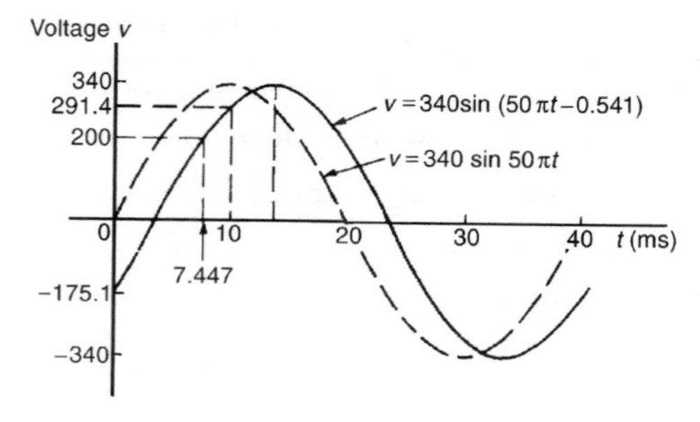

Figure 28.23

29 Trigonometric Identities and Equations

Trigonometric identities

A trigonometric identity is a relationship that is true for all values of the unknown variable.

$$\tan\theta = \frac{\sin\theta}{\cos\theta}, \quad \cot\theta = \frac{\cos\theta}{\sin\theta}, \quad \sec\theta = \frac{1}{\cos\theta}$$

$$\operatorname{cosec}\theta = \frac{1}{\sin\theta} \quad \text{and} \quad \cot\theta = \frac{1}{\tan\theta}$$

are examples of trigonometric identities from chapter 25. Applying Pythagoras' theorem to the right-angled triangle shown in Figure 29.1 gives:

$$a^2 + b^2 = c^2 \qquad (1)$$

Dividing each term of equation (1) by c^2 gives:

$$\frac{a^2}{c^2} + \frac{b^2}{c^2} = \frac{c^2}{c^2}, \text{ i.e. } \left(\frac{a}{c}\right)^2 + \left(\frac{b}{c}\right)^2 = 1$$

$$(\cos\theta)^2 + (\sin\theta)^2 = 1$$

Hence $\mathbf{\cos^2\theta + \sin^2\theta = 1}$ (2)

Dividing each term of equation (1) by a^2 gives:

$$\frac{a^2}{a^2} + \frac{b^2}{a^2} = \frac{c^2}{a^2}, \text{ i.e. } 1 + \left(\frac{b}{a}\right)^2 = \left(\frac{c}{a}\right)^2$$

Hence $\mathbf{1 + \tan^2\theta = \sec^2\theta}$ (3)

Dividing each term of equation (1) by b^2 gives:

$$\frac{a^2}{b^2} + \frac{b^2}{b^2} = \frac{c^2}{b^2}, \text{ i.e. } \left(\frac{a}{b}\right)^2 + 1 = \left(\frac{c}{b}\right)^2$$

Hence $\mathbf{\cot^2\theta + 1 = \operatorname{cosec}^2\theta}$ (4)

Equations (2), (3) and (4) are further examples of trigonometric identities.

For example, to prove the identity $\sin^2\theta\cot\theta\sec\theta = \sin\theta$:

With trigonometric identities it is necessary to start with the left-hand side (LHS) and attempt to make it equal to the right-hand side (RHS) or vice-versa. It is often useful to change all of the trigonometric ratios into sines and

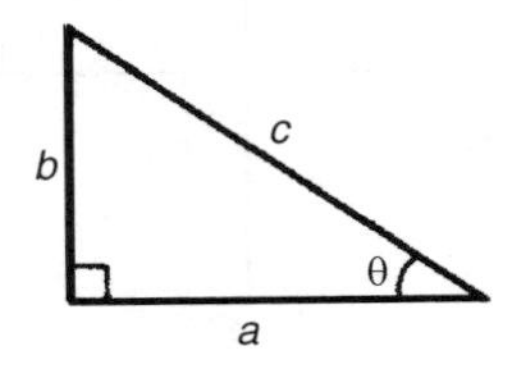

Figure 29.1

cosines where possible. Thus

$$\text{LHS} = \sin^2\theta\cot\theta\sec\theta = \sin^2\theta\left(\frac{\cos\theta}{\sin\theta}\right)\left(\frac{1}{\cos\theta}\right)$$

$$= \sin\theta \text{ (by cancelling)} = \text{RHS}$$

In another example, to prove that $\dfrac{1+\cot\theta}{1+\tan\theta} = \cot\theta$:

$$\text{LHS} = \frac{1+\cot\theta}{1+\tan\theta} = \frac{1+\dfrac{\cos\theta}{\sin\theta}}{1+\dfrac{\sin\theta}{\cos\theta}} = \frac{\dfrac{\sin\theta+\cos\theta}{\sin\theta}}{\dfrac{\cos\theta+\sin\theta}{\cos\theta}}$$

$$= \left(\frac{\sin\theta+\cos\theta}{\sin\theta}\right)\left(\frac{\cos\theta}{\cos\theta+\sin\theta}\right)$$

$$= \frac{\cos\theta}{\sin\theta} = \cot\theta = \text{RHS}$$

Trigonometric equations

Equations which contain trigonometric ratios are called **trigonometric equations**. There are usually an infinite number of solutions to such equations; however, solutions are often restricted to those between 0° and 360°.
A knowledge of angles of any magnitude is essential in the solution of trigonometric equations and calculators cannot be relied upon to give all the solutions (as shown in chapter 28). Figure 29.2 shows a summary for angles of any magnitude.

Equations of the type $a\sin^2 A + b\sin A + c = 0$

(i) **When $a = 0$,** $b\sin A + c = 0$, hence

$$\sin A = -\frac{c}{b} \quad \text{and} \quad \mathbf{A = \sin^{-1}\left(-\frac{c}{b}\right)}$$

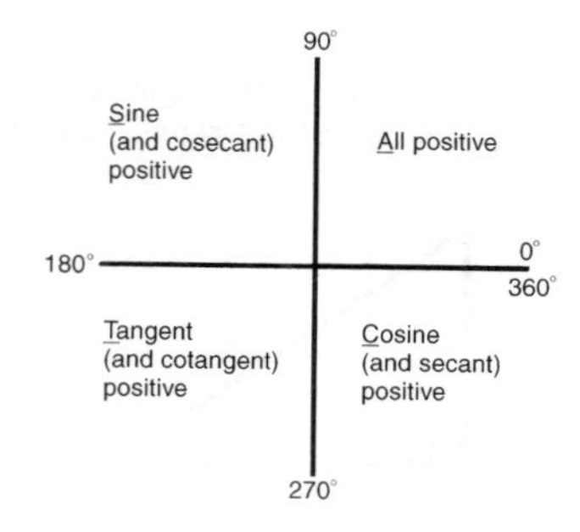

Figure 29.2

There are two values of A between 0° and 360° that satisfy such an equation, provided $-1 \le \frac{c}{b} \le 1$

(ii) **When $b = 0$,** $a \sin^2 A + c = 0$, hence

$$\sin^2 A = -\frac{c}{a}, \quad \sin A = \sqrt{\left(-\frac{c}{a}\right)} \quad \text{and} \quad \mathbf{A = \sin^{-1}\sqrt{\left(-\frac{c}{a}\right)}}$$

If either a or c is a negative number, then the value within the square root sign is positive. Since when a square root is taken there is a positive and negative answer there are four values of A between 0° and 360° which satisfy such an equation, provided $-1 \le \frac{c}{a} \le 1$

(iii) **When a, b and c are all non-zero**:
$a \sin^2 A + b \sin A + c = 0$ is a quadratic equation in which the unknown is $\sin A$. The solution of a quadratic equation is obtained either by factorising (if possible) or by using the quadratic formula:

$$\mathbf{\sin A = \frac{-b \pm \sqrt{(b^2 - 4ac)}}{2a}}$$

(iv) Often the trigonometric identities $\cos^2 A + \sin^2 A = 1$, $1 + \tan^2 A = \sec^2 A$ and $\cot^2 A + 1 = \text{cosec}^2$ A need to be used to reduce equations to one of the above forms.

For example, to solve the trigonometric equation $5 \sin\theta + 3 = 0$ for values of θ from 0° to 360°:

$$5 \sin\theta + 3 = 0, \text{ from which } \sin\theta = -3/5 = -0.6000$$

Hence $\theta = \sin^{-1}(-0.6000)$. Sine is negative in the third and fourth quadrants (see Figure 29.3). The acute angle $\sin^{-1}(0.6000) = 36.87°$ (shown as α in Figure 29.3(b)).

Hence $\theta = 180° + 36.87°$, i.e. **216.87°** or $\theta = 360° - 36.87°$, i.e. **323.13°**

In another example, to solve 4 $\sec t = 5$ for values of t between 0° and 360°:

$$4 \sec t = 5, \text{ from which } \sec t = \tfrac{5}{4} = 1.2500$$

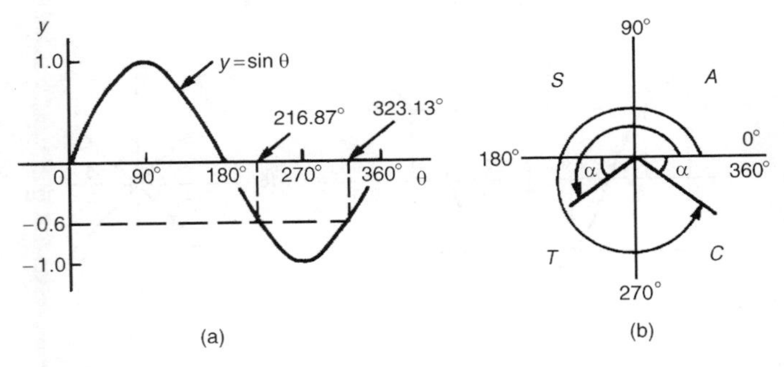

Figure 29.3

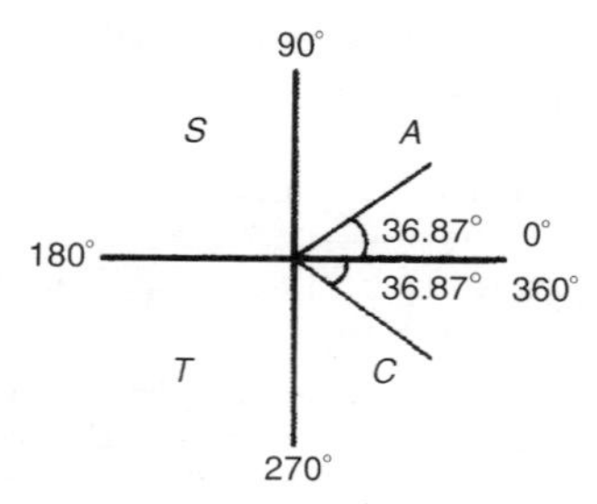

Figure 29.4

Hence $t = \sec^{-1} 1.2500$

Secant $= \dfrac{1}{\text{cosine}}$ is positive in the first and fourth quadrants (see Figure 29.4).

The acute angle $\sec^{-1} 1.2500 = 36.87°$. Hence

$$\mathbf{t = 36.87°} \text{ or } 360° - 36.87° = \mathbf{323.13°}$$

In another example, to solve $2 - 4\cos^2 A = 0$ for values of A in the range $0° < A < 360°$:

$$2 - 4\cos^2 A = 0, \text{ from which } \cos^2 A = \tfrac{2}{4} = 0.5000$$

Hence $\cos A = \sqrt{0.5000} = \pm 0.7071$ and $A = \cos^{-1}(\pm 0.7071)$
Cosine is positive in quadrants one and four and negative in quadrants two and three. Thus in this case there are four solutions, one in each quadrant (see Figure 29.5).

The acute angle $\cos^{-1} 0.7071 = 45°$.
Hence, $\mathbf{A = 45°,\ 135°,\ 225°}$ **or** $\mathbf{315°}$

In another example, to solve the equation $8\sin^2\theta + 2\sin\theta - 1 = 0$, for all values of θ between 0° and 360°:

Factorising $8\sin^2\theta + 2\sin\theta - 1 = 0$ gives $(4\sin\theta - 1)(2\sin\theta + 1) = 0$
Hence $4\sin\theta - 1 = 0$, from which, $\sin\theta = \frac{1}{4} = 0.2500$,
or $2\sin\theta + 1 = 0$, from which, $\sin\theta = -\frac{1}{2} = -0.5000$

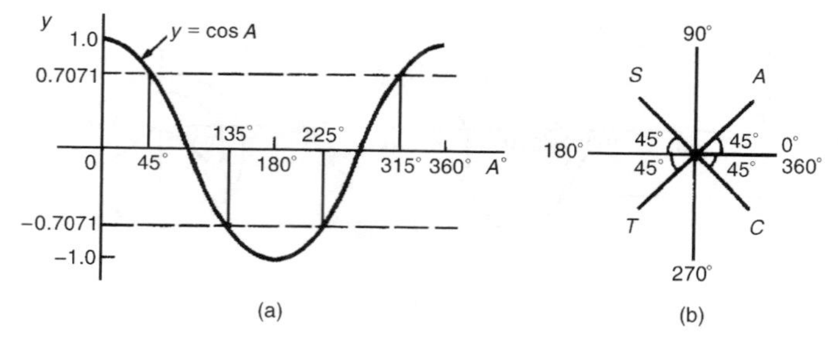

Figure 29.5

(Instead of factorising, the quadratic formula can, of course, be used).
$\theta = \sin^{-1} 0.250 = 14.48°$ or $165.52°$, since sine is positive in the first and second quadrants, or $\theta = \sin^{-1}(-0.5000) = 210°$ or $330°$, since sine is negative in the third and fourth quadrants. Hence $\theta =$ **14.48°, 165.52°, 210° or 330°**

In another example, to solve $18\sec^2 A - 3\tan A = 21$ for values of A between 0° and 360°:
$1 + \tan^2 A = \sec^2 A$. Substituting for $\sec^2 A$ in $18\sec^2 A - 3\tan A = 21$ gives

$$18(1 + \tan^2 A) - 3\tan A = 21$$

$$\text{i.e. } 18 + 18\tan^2 A - 3\tan A - 21 = 0$$

$$18\tan^2 A - 3\tan A - 3 = 0$$

$$\text{Factorising gives } (6\tan A - 3)(3\tan A + 1) = 0$$

Hence $6\tan A - 3 = 0$, from which, $\tan A = \frac{3}{6} = 0.5000$ or $3\tan A + 1 = 0$, from which, $\tan A = -\frac{1}{3} = -0.3333$. Thus $A = \tan^{-1}(0.5000) = 26.57°$ or $206.57°$, since tangent is positive in the first and third quadrants, or $A = \tan^{-1}(-0.3333) = 161.57°$ or $341.57°$, since tangent is negative in the second and fourth quadrants.

Hence, **A = 26.57°, 161.57°, 206.57° or 341.57°**

---

30 The Relationship Between Trigonometric and Hyperbolic Functions

In chapter 42, it is shown that

$$\cos\theta + j\sin\theta = e^{j\theta} \quad (1)$$

and
$$\cos\theta - j\sin\theta = e^{-j\theta} \quad (2)$$

Adding equations (1) and (2) gives:

$$\mathbf{\cos\theta = \tfrac{1}{2}(e^{j\theta} + e^{-j\theta})} \quad (3)$$

Subtracting equation (2) from equation (1) gives:

$$\mathbf{\sin\theta = \frac{1}{2j}(e^{j\theta} - e^{-j\theta})} \quad (4)$$

Substituting $j\theta$ for θ in equations (3) and (4) gives:

$$\cos j\theta = \tfrac{1}{2}(e^{j(j\theta)} + e^{-j(j\theta)})$$

and
$$\sin j\theta = \frac{1}{2j}(e^{j(j\theta)} - e^{-j(j\theta)})$$

Since $j^2 = -1$, $\cos j\theta = \frac{1}{2}(e^{-\theta} + e^{\theta}) = \frac{1}{2}(e^{\theta} + e^{-\theta})$

Hence from chapter 13, $\mathbf{\cos j\theta = \cosh\theta}$ (5)

Similarly, $$\sin j\theta = \frac{1}{2j}(e^{-\theta} - e^{\theta}) = -\frac{1}{2j}(e^{\theta} - e^{-\theta})$$
$$= \frac{-1}{j}\left[\frac{1}{2}(e^{\theta} - e^{-\theta})\right]$$
$$= -\frac{1}{j}\sinh\theta \text{ (see chapter 13)}$$

But $$-\frac{1}{j} = -\frac{1}{j} \times \frac{j}{j} = -\frac{j}{j^2} = j,$$

hence $$\mathbf{\sin j\theta = j \sinh\theta} \qquad (6)$$

Equations (5) and (6) may be used to verify that in all standard trigonometric identities, $j\theta$ may be written for θ and the identity still remains true.
For example, to verify that $\cos^2 j\theta + \sin^2 j\theta = 1$:
From equation (5), $\cos j\theta = \cosh\theta$, and from equation (6), $\sin j\theta = j\sinh\theta$
Thus, $\cos^2 j\theta + \sin^2 j\theta = \cosh^2\theta + j^2\sinh^2\theta$, and since $j^2 = -1$,
$\cos^2 j\theta + \sin^2 j\theta = \cosh^2\theta - \sinh^2\theta$
But, $\cosh^2\theta - \sinh^2\theta = 1$, from Chapter 13,

hence $\mathbf{\cos^2 j\theta + \sin^2 j\theta = 1}$

In another example, to verify that $\sin j2A = 2\sin jA\cos jA$:
From equation (6), writing $2A$ for θ, $\sin j2A = j\sinh 2A$, and from chapter 13, Table 13.1, page 59, $\sinh 2A = 2\sinh A\cosh A$
Hence, $\sin j2A = j(2\sinh A\cosh A)$
But, $\sinh A = \frac{1}{2}(e^A - e^{-A})$ and $\cosh A = \frac{1}{2}(e^A + e^{-A})$

Hence, $$\sin j2A = j2\left(\frac{e^A - e^{-A}}{2}\right)\left(\frac{e^A + e^{-A}}{2}\right)$$
$$= -\frac{2}{j}\left(\frac{e^A - e^{-A}}{2}\right)\left(\frac{e^A + e^{-A}}{2}\right)$$
$$= -\frac{2}{j}\left(\frac{\sin j\theta}{j}\right)(\cos j\theta)$$
$$= 2\sin jA\cos jA \quad \text{since } j^2 = -1$$

i.e. $\mathbf{\sin j2A = 2\sin jA\cos jA}$

Hyperbolic identities

From chapter 13, $\cosh\theta = \frac{1}{2}(e^{\theta} + e^{-\theta})$

Substituting $j\theta$ for θ gives:

$$\cosh j\theta = \frac{1}{2}(e^{j\theta} + e^{-j\theta}) = \cos\theta, \text{ from equation (3),}$$

i.e. $$\mathbf{\cosh j\theta = \cos\theta} \qquad (7)$$

Similarly, from chapter 13, $\sinh\theta = \frac{1}{2}(e^{\theta} - e^{-\theta})$

Substituting $j\theta$ for θ gives:

$$\sinh j\theta = \tfrac{1}{2}(e^{j\theta} - e^{-j\theta}) = j\sin\theta, \text{ from equation (4)}$$

Hence $\quad \mathbf{\sinh j\theta = j\sin\theta} \qquad (8)$

$$\tan j\theta = \frac{\sin j\theta}{\cos j\theta}$$

From equations (5) and (6), $\dfrac{\sin j\theta}{\cos j\theta} = \dfrac{j\sinh\theta}{\cosh\theta} = j\tanh\theta$

Hence $\quad \mathbf{\tan j\theta = j\tanh\theta} \qquad (9)$

Similarly, $\tanh j\theta = \dfrac{\sinh j\theta}{\cosh j\theta}$

From equations (7) and (8), $\dfrac{\sinh j\theta}{\cosh j\theta} = \dfrac{j\sin\theta}{\cos\theta} = j\tan\theta$

Hence $\quad \mathbf{\tanh j\theta = j\tan\theta} \qquad (10)$

Two methods are commonly used to verify hyperbolic identities. These are (a) by substituting $j\theta$ (and $j\phi$) in the corresponding trigonometric identity and using the relationships given in equations (5) to (10), and (b) by applying **Osborne's rule** given in chapter 13, page 58.

For example, to determine the corresponding hyperbolic identity by writing jA for θ in $\cot^2\theta + 1 = \text{cosec}^2\theta$:
Substituting jA for θ gives:

$$\cot^2 jA + 1 = \text{cosec}^2 jA, \text{ i.e. } \frac{\cos^2 jA}{\sin^2 jA} + 1 = \frac{1}{\sin^2 jA}$$

But from equation (5), $\cos jA = \cosh A$
and from equation (6), $\sin jA = j\sinh A$

Hence $$\frac{\cosh^2 A}{j^2\sinh^2 A} + 1 = \frac{1}{j^2\sinh^2 A}$$

and since $\quad j^2 = -1, \quad -\dfrac{\cosh^2 A}{\sinh^2 A} + 1 = -\dfrac{1}{\sinh^2 A}$

Multiplying throughout by -1, gives:

$$\frac{\cosh^2 A}{\sinh^2 A} - 1 = \frac{1}{\sinh^2 A} \text{ i.e. } \mathbf{\coth^2 A - 1 = \text{cosech}^2 A}$$

In another example, to show that
$\cosh A - \cosh B = 2\sinh\left(\dfrac{A+B}{2}\right)\sinh\left(\dfrac{A-B}{2}\right)$ by substituting jA and jB for θ and ϕ respectively in the trigonometric identity for $\cos\theta - \cos\phi$:

$$\cos\theta - \cos\phi = -2\sin\left(\frac{\theta+\phi}{2}\right)\sin\left(\frac{\theta-\phi}{2}\right) \text{ (see chapter 31)}$$

thus $\cos jA - \cos jB = -2\sin j\left(\frac{A+B}{2}\right)\sin j\left(\frac{A-B}{2}\right)$

But from equation (5), $\cos jA = \cosh A$
and from equation (6), $\sin jA = j\sinh A$

Hence, $\cosh A - \cosh B = -2j\sinh\left(\frac{A+B}{2}\right) j\sinh\left(\frac{A-B}{2}\right)$

$$= -2j^2\sinh\left(\frac{A+B}{2}\right)\sinh\left(\frac{A-B}{2}\right)$$

But $j^2 = -1$, hence

$$\mathbf{\cosh A - \cosh B = 2\sinh\left(\frac{A+B}{2}\right)\sinh\left(\frac{A-B}{2}\right)}$$

---

31 Compound Angles

Compound angle formulae

An electric current i may be expressed as $i = 5\sin(\omega t - 0.33)$ amperes. Similarly, the displacement x of a body from a fixed point can be expressed as $x = 10\sin(2t + 0.67)$ metres. The angles $(\omega t - 0.33)$ and $(2t + 0.67)$ are called **compound angles** because they are the sum or difference of two angles. The **compound angle formulae** for the sum and difference of two angles A and B are:

$$\sin(A+B) = \sin A\cos B + \cos A\sin B$$

$$\sin(A-B) = \sin A\cos B - \cos A\sin B$$

$$\cos(A+B) = \cos A\cos B - \sin A\sin B$$

$$\cos(A-B) = \cos A\cos B + \sin A\sin B$$

$$\tan(A+B) = \frac{\tan A + \tan B}{1 - \tan A\tan B}$$

$$\tan(A-B) = \frac{\tan A - \tan B}{1 + \tan A\tan B}$$

(Note, $\sin(A+B)$ is **not** equal to $(\sin A + \sin B)$, and so on.)
The compound-angle formulae are true for all values of A and B, and by substituting values of A and B into the formulae they may be shown to be true.
For example, to expand and simplify the following expressions (a) $\sin(\pi + \alpha)$ (b) $-\cos(90^\circ + \beta)$ (c) $\sin(A-B) - \sin(A+B)$:

(a) $\sin(\pi+\alpha) = \sin\pi\cos\alpha + \cos\pi\sin\alpha$ (from the formula for $\sin(A+B)$)

$$= (0)(\cos\alpha) + (-1)\sin\alpha = \mathbf{-\sin\alpha}$$

(b) $-\cos(90^\circ + \beta) = -[\cos 90^\circ \cos\beta - \sin 90^\circ \sin\beta]$

$$= [(0)(\cos\beta) - (1)\sin\beta] = \mathbf{\sin\beta}$$

(c) $\sin(A - B) - \sin(A + B)$

$$= [\sin A \cos B - \cos A \sin B] - [\sin A \cos B + \cos A \sin B]$$

$$= \mathbf{-2\cos A \sin B}$$

In another example, to prove that $\cos(y - \pi) + \sin\left(y + \frac{\pi}{2}\right) = 0$:

$$\cos(y - \pi) = \cos y \cos\pi + \sin y \sin\pi$$

$$= (\cos y)(-1) + (\sin y)(0) = -\cos y$$

$$\sin\left(y + \frac{\pi}{2}\right) = \sin y \cos\frac{\pi}{2} + \cos y \sin\frac{\pi}{2}$$

$$= (\sin y)(0) + (\cos y)(1) = \cos y$$

Hence $\cos(y - \pi) + \sin\left(y + \frac{\pi}{2}\right) = (-\cos y) + (\cos y) = \mathbf{0}$

In another example, to solve the equation $4\sin(x - 20^\circ) = 5\cos x$ for values of x between 0° and 90°:

$$4\sin(x - 20^\circ = 4[\sin x \cos 20^\circ - \cos x \sin 20^\circ],$$

from the formula for $\sin(A - B)$

$$= 4[\sin x(0.9397) - \cos x(0.3420)]$$

$$= 3.7588\sin x - 1.3680\cos x$$

Since $4\sin(x - 20^\circ) = 5\cos x$ then $3.7588\sin x - 1.3680\cos x = 5\cos x$. Rearranging gives:

$$3.7588\sin x = 5\cos x + 1.3680\cos x = 6.3680\cos x$$

and $$\frac{\sin x}{\cos x} = \frac{6.3680}{3.7588} = 1.6942$$

i.e. $\tan x = 1.6942$, and $x = \tan^{-1} 1.6942 = 59.449^\circ$ or $\mathbf{59^\circ 27'}$

[Check: LHS $= 4\sin(59.449^\circ - 20^\circ) = 4\sin 39.449^\circ = 2.542$

RHS $= 5\cos x = 5\cos 59.449^\circ = 2.542$]

Conversion of a $\sin\omega t + b\cos\omega t$ *into* $R\sin(\omega t + \alpha)$

(i) $R\sin(\omega t + \alpha)$ represents a sine wave of maximum value R, periodic time $2\pi/\omega$, frequency $\omega/2\pi$ and leading $R\sin\omega t$ by angle α (see Chapter 28).

(ii) $R\sin(\omega t+\alpha)$ may be expanded using the compound-angle formula for $\sin(A+B)$, where $A=\omega t$ and $B=\alpha$

Hence $$R\sin(\omega t+\alpha) = R[\sin\omega t\cos\alpha+\cos\omega t\sin\alpha]$$
$$= R\sin\omega t\cos\alpha + R\cos\omega t\sin\alpha$$
$$= (R\cos\alpha)\sin\omega t + (R\sin\alpha)\cos\omega t$$

(iii) If $a=R\cos\alpha$ and $b=R\sin\alpha$, where a and b are constants, then $R\sin(\omega t+\alpha)=a\sin\omega t+b\cos\omega t$, i.e. a sine and cosine function of the same frequency when added produce a sine wave of the same frequency (which is further demonstrated in Chapter 39).

(iv) Since $a=R\cos\alpha$, then $\cos\alpha=a/R$, and since $b=R\sin\alpha$, then $\sin\alpha=b/R$.

If the values of a and b are known then the values of R and α may be calculated. The relationship between constants a, b, R and α are shown in Figure 31.1.

From Figure 31.1, by Pythagoras' theorem: $\boldsymbol{R=\sqrt{a^2+b^2}}$ and from trigonometric ratios: $\boldsymbol{\alpha=\tan^{-1}b/a}$

For example, to find an expression for $3\sin\omega t+4\cos\omega t$ in the form $R\sin(\omega t+\alpha)$ and sketch graphs of $3\sin\omega t$, $4\cos\omega t$ and $R\sin(\omega t+\alpha)$ on the same axes:

Let $$3\sin\omega t+4\cos\omega t = R\sin(\omega t+\alpha)$$

then $$3\sin\omega t+4\cos\omega t = R[\sin\omega t\cos\alpha+\cos\omega t\sin\alpha]$$
$$= (R\cos\alpha)\sin\omega t + (R\sin\alpha)\cos\omega t$$

Equating coefficients of $\sin\omega t$ gives:

$$3=R\cos\alpha,\ \text{from which,}\ \cos\alpha=\frac{3}{R}$$

Equating coefficients of $\cos\omega t$ gives:

$$4=R\sin\alpha,\ \text{from which,}\ \sin\alpha=\frac{4}{R}$$

There is only one quadrant where both $\sin\alpha$ **and** $\cos\alpha$ are positive, and this is the first, as shown in Figure 31.2. From Figure 31.2, by Pythagoras' theorem:

$$R=\sqrt{3^2+4^2}=5$$

From trigonometric ratios: $\alpha=\tan^{-1}\frac{4}{3}=53.13°$ or 0.927 radians

Hence $\boldsymbol{3\sin\omega t+4\cos\omega t=5\sin(\omega t+0.927)}$

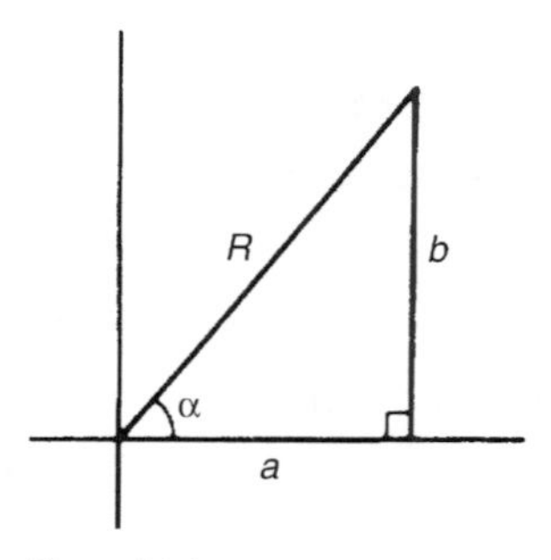

Figure 31.1

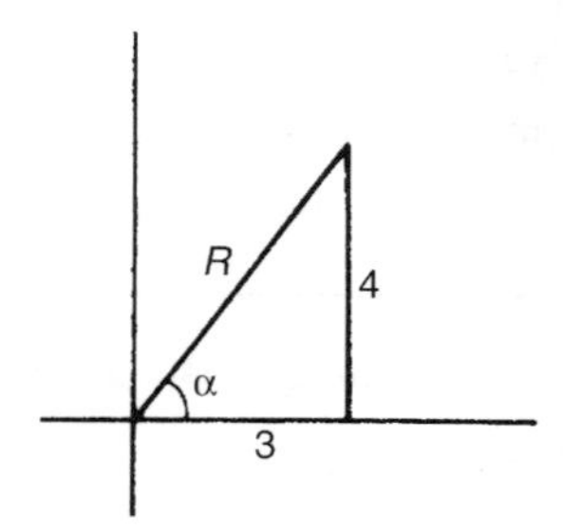

Figure 31.2

A sketch of $3\sin\omega t$, $4\cos\omega t$ and $5\sin(\omega t + 0.927)$ is shown in Figure 31.3.
Two periodic functions of the same frequency may be combined by
(a) plotting the functions graphically and combining ordinates at intervals, or
(b) by resolution of phasors by drawing or calculation.
The example below demonstrates a third method of combining waveforms.
For example, to express $4.6\sin\omega t - 7.3\cos\omega t$ in the form $R\sin(\omega t + \alpha)$:

Let $\quad 4.6\sin\omega t - 7.3\cos\omega t = R\sin(\omega t + \alpha)$

then $\quad 4.6\sin\omega t - 7.3\cos\omega t = R[\sin\omega t\cos\alpha + \cos\omega t\sin\alpha]$

$$= (R\cos\alpha)\sin\omega t + (R\sin\alpha)\cos\omega t$$

Equating coefficients of $\sin\omega t$ gives:

$$4.6 = R\cos\alpha, \text{ from which, } \cos\alpha = \frac{4.6}{R}$$

Equating coefficients of $\cos\omega t$ gives:

$$-7.3 = R\sin\alpha, \text{ from which } \sin\alpha = \frac{-7.3}{R}$$

There is only one quadrant where cosine is positive **and** sine is negative, i.e. the fourth quadrant, as shown in Figure 31.4. By Pythagoras' theorem:

$$R = \sqrt{4.6^2 + (-7.3)^2} = 8.628$$

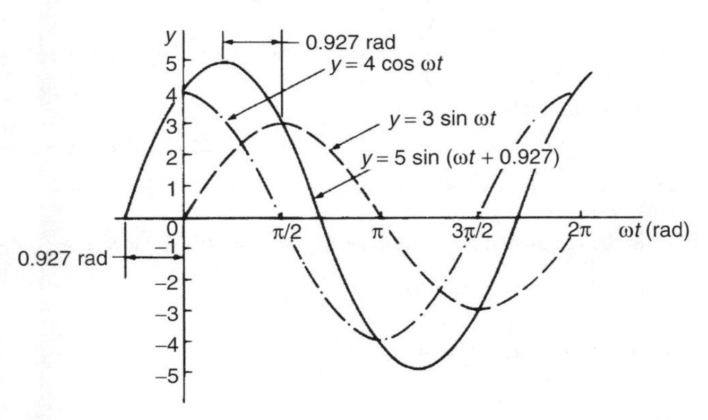

Figure 31.3

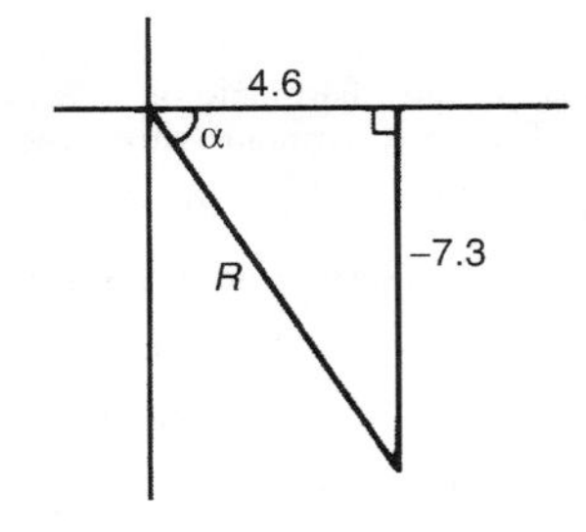

Figure 31.4

By trigonometric ratios:

$$\alpha = \tan^{-1}\left(\frac{-7.3}{4.6}\right) = -57.78^\circ \text{ or } -1.008 \text{ radians}$$

Hence 4.6 sin ωt − 7.3 cos ωt = 8.628 sin(ωt − 1.008)

Double Angles

(i) If, in the compound-angle formula for $\sin(A+B)$, we let $B = A$ then **sin 2A = 2 sin A cos A**

Also, for example, $\sin 4A = 2\sin 2A \cos 2A$

and $\sin 8A = 2\sin 4A \cos 4A$, and so on.

(ii) If, in the compound-angle formula for $\cos(A+B)$, we let $B = A$ then **cos 2A = cos²A − sin²A**
Since $\cos^2 A + \sin^2 A = 1$, then $\cos^2 A = 1 - \sin^2 A$, and
$\sin^2 A = 1 - \cos^2 A$, and two further formula for $\cos 2A$ can be produced.
Thus $\cos 2A = \cos^2 A - \sin^2 A = (1 - \sin^2 A) - \sin^2 A$
i.e. $\mathbf{\cos 2A = 1 - 2\sin^2 A}$
and $\cos 2A = \cos^2 A - \sin^2 A = \cos^2 A - (1 - \cos^2 A)$
i.e. $\mathbf{\cos 2A = 2\cos^2 A - 1}$

Also, for example,

$$\cos 4A = \cos^2 2A - \sin^2 2A \text{ or } 1 - 2\sin^2 2A \text{ or } 2\cos^2 2A - 1$$

and $\cos 6A = \cos^2 3A - \sin^2 3A$ or $1 - 2\sin^2 3A$ or $2\cos^2 3A - 1$, and so on.

(iii) If, in the compound-angle formula for $\tan(A+B)$, we let $B = A$

then $$\mathbf{\tan 2A = \frac{2\tan A}{1 - \tan^2 A}}$$

Also, for example, $$\tan 4A = \frac{2\tan 2A}{1 - \tan^2 2A}$$

and $$\tan 5A = \frac{2\tan \frac{5}{2}A}{1 - \tan^2 \frac{5}{2}A}$$

and so on.

For example, $I_3 \sin 3\theta$ is the third harmonic of a waveform. To express the third harmonic in terms of the first harmonic $\sin\theta$, when $I_3 = 1$:

When $I_3 = 1$, $I_3 \sin 3\theta = \sin 3\theta = \sin(2\theta + \theta)$

$$= \sin 2\theta \cos\theta + \cos 2\theta \sin\theta, \text{ from the } \sin(A+B) \text{ formula}$$

$$= (2\sin\theta\cos\theta)\cos\theta + (1 - 2\sin^2\theta)\sin\theta,$$

from the double angle expansions

$$= 2\sin\theta\cos^2\theta + \sin\theta - 2\sin^3\theta$$

$$= 2\sin\theta(1 - \sin^2\theta) + \sin\theta - 2\sin^3\theta, \text{ (since } \cos^2\theta = 1 - \sin^2\theta)$$

$$= 2\sin\theta - 2\sin^3\theta + \sin\theta - 2\sin^3\theta$$

i.e. $\mathbf{\sin 3\theta = 3\sin\theta - 4\sin^3\theta}$

In another example, to prove that $\cot 2x + \mathrm{cosec}\, 2x = \cot x$

$$\text{LHS} = \cot 2x + \mathrm{cosec}\, 2x = \frac{\cos 2x}{\sin 2x} + \frac{1}{\sin 2x} = \frac{\cos 2x + 1}{\sin 2x}$$

$$= \frac{(2\cos^2 x - 1) + 1}{\sin 2x} = \frac{2\cos^2 x}{\sin 2x}$$

$$= \frac{2\cos^2 x}{2\sin x\cos x} = \frac{\cos x}{\sin x} = \cot x = \text{RHS}$$

Changing products of sines and cosines into sums or differences

(i) $\sin(A+B) + \sin(A-B) = 2\sin A\cos B$ (from the earlier formulae)

i.e. $\mathbf{\sin A\cos B = \frac{1}{2}[\sin(A+B) + \sin(A-B)]}$ (1)

(ii) $\sin(A+B) - \sin(A-B) = 2\cos A\sin B$

i.e. $\mathbf{\cos A\sin B = \frac{1}{2}[\sin(A+B) - \sin(A-B)]}$ (2)

(iii) $\cos(A+B) + \cos(A-B) = 2\cos A\cos B$

i.e. $\mathbf{\cos A\cos B = \frac{1}{2}[\cos(A+B) + \cos(A-B)]}$ (3)

(iv) $\cos(A+B) - \cos(A-B) = -2\sin A\sin B$

i.e. $\mathbf{\sin A\sin B = -\frac{1}{2}[\cos(A+B) - \cos(A-B)]}$ (4)

For example, to express $\sin 4x\cos 3x$ as a sum or difference of sines and cosines:

From equation (1), $\sin 4x\cos 3x = \frac{1}{2}[\sin(4x+3x) + \sin(4x-3x)]$

$$= \mathbf{\frac{1}{2}(\sin 7x + \sin x)}$$

In another example, to express $2\cos 5\theta\sin 2\theta$ as a sum or difference of sines or cosines:
From equation (2),

$$2\cos 5\theta\sin 2\theta = 2\left\{\frac{1}{2}[\sin(5\theta+2\theta) - \sin(5\theta-2\theta)]\right\}$$

$$= \mathbf{\sin 7\theta - \sin 3\theta}$$

Changing sums or differences of sines and cosines into products

In the compound-angle formula let $(A+B) = X$ and $(A-B) = Y$

Solving the simultaneous equations gives

$$A = \frac{X+Y}{2} \text{ and } B = \frac{X-Y}{2}$$

Thus $\sin(A+B)+\sin(A-B)=2\sin A\cos B$

becomes $$\mathbf{\sin X + \sin Y = 2\sin\left(\frac{X+Y}{2}\right)\cos\left(\frac{X-Y}{2}\right)} \qquad (5)$$

Similarly, $$\mathbf{\sin X - \sin Y = 2\cos\left(\frac{X+Y}{2}\right)\sin\left(\frac{X-Y}{2}\right)} \qquad (6)$$

$$\mathbf{\cos X + \cos Y = 2\cos\left(\frac{X+Y}{2}\right)\cos\left(\frac{X-Y}{2}\right)} \qquad (7)$$

$$\mathbf{\cos X - \cos Y = -2\sin\left(\frac{X+Y}{2}\right)\sin\left(\frac{X-Y}{2}\right)} \qquad (8)$$

For example, to express $\sin 5\theta + \sin 3\theta$ as a product:
From equation (5),

$$\sin 5\theta + \sin 3\theta = 2\sin\left(\frac{5\theta+3\theta}{2}\right)\cos\left(\frac{5\theta-3\theta}{2}\right) = \mathbf{2\sin 4\theta\cos\theta}$$

In another example, to express $\sin 7x - \sin x$ as a product:
From equation (6),

$$\sin 7x - \sin x = 2\cos\left(\frac{7x+x}{2}\right)\sin\left(\frac{7x-x}{2}\right) = \mathbf{2\cos 4x\sin 3x}$$

Graphs

32 Straight Line Graphs

Introduction to graphs

A **graph** is a pictorial representation of information showing how one quantity varies with another related quantity.
The most common method of showing a relationship between two sets of data is to use **Cartesian** or **rectangular axes** as shown in Figure 32.1.
The points on a graph are called **co-ordinates**. Point A in Figure 32.1 has the co-ordinates (3, 2), i.e. 3 units in the x direction and 2 units in the y direction. Similarly, point B has co-ordinates (−4, 3) and C has co-ordinates (−3, −2). The origin has co-ordinates (0, 0).
The horizontal distance of a point from the vertical axis is called the **abscissa** and the vertical distance from the horizontal axis is called the **ordinate**.

The straight line graph

Let a relationship between two variables x and y be $y = 3x + 2$

When $x = 0$, $y = 3(0) + 2 = 2$. When $x = 1$, $y = 3(1) + 2 = 5$.
When $x = 2$, $y = 3(2) + 2 = 8$, and so on.
Thus co-ordinates (0, 2), (1, 5) and (2, 8) have been produced from the equation by selecting arbitrary values of x, and are shown plotted in Figure 32.2. When the points are joined together, a **straight-line graph** results.

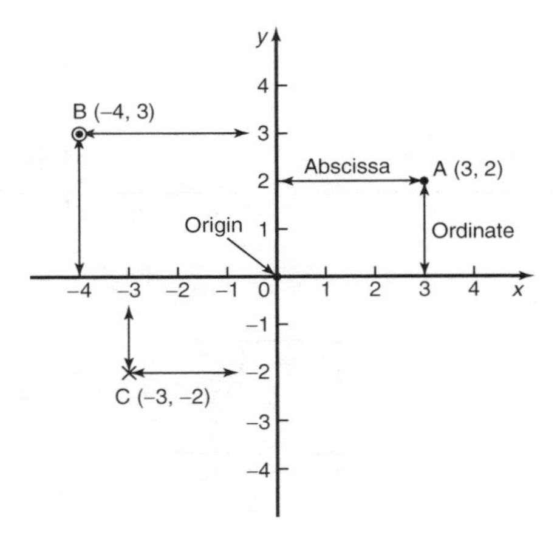

Figure 32.1

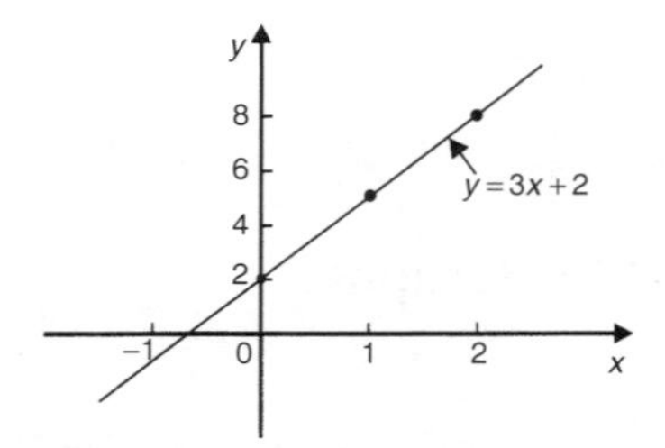

Figure 32.2

The **gradient** or **slope** of a straight line is the ratio of the change in the value of y to the change in the value of x between any two points on the line. If, as x increases, (→), y also increases (↑), then the gradient is positive.
In Figure 32.3(a),

$$\text{the gradient of AC} = \frac{\text{change in } y}{\text{change in } x} = \frac{\text{CB}}{\text{BA}} = \frac{7-3}{3-1} = \frac{4}{2} = 2$$

If as x increases (→), y decreases (↓), then the gradient is negative.
In Figure 32.3(b),

$$\text{the gradient of DF} = \frac{\text{change in } y}{\text{change in } x} = \frac{\text{FE}}{\text{ED}} = \frac{11-2}{-3-0} = \frac{9}{-3} = -3$$

Figure 32.3(c) shows a straight line graph $y = 3$. Since the straight line is horizontal the gradient is zero.

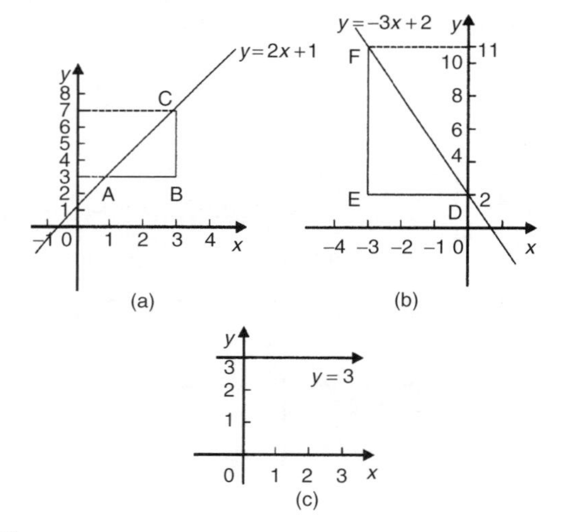

Figure 32.3

The value of y when $x = 0$ is called the **y-axis intercept**. In Figure 32.3(a) the y-axis intercept is 1 and in Figure 32.3(b) is 2.
If the equation of a graph is of the form $\boldsymbol{y = mx + c}$, where m and c are constants, **the graph will always be a straight line**, m representing the gradient and c the y-axis intercept. Thus $y = 5x + 2$ represents a straight line of gradient 5 and y-axis intercept 2. Similarly, $y = -3x - 4$ represents a straight line of gradient -3 and y-axis intercept -4.
In another example, to determine the gradient of the straight line graph passing through the co-ordinates $(-2, 5)$ and $(3, 4)$:
A straight line graph passing through co-ordinates (x_1, y_1) and (x_2, y_2) has a gradient given by:

$$m = \frac{y_2 - y_1}{x_2 - x_1} \qquad \text{(see Figure 32.4).}$$

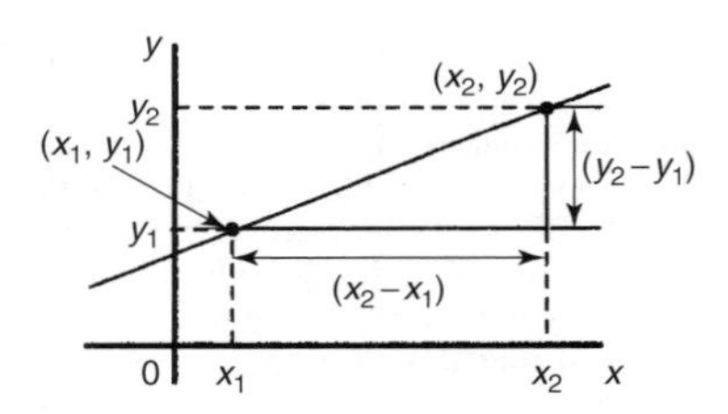

Figure 32.4

A straight line passes through $(-2, 5)$ and $(3, 4)$, from which, $x_1 = -2$, $y_1 = 5$, $x_2 = 3$ and $y_2 = 4$, hence gradient

$$m = \frac{y_2 - y_1}{x_2 - x_1} = \frac{4 - 5}{3 - (-2)} = -\frac{\mathbf{1}}{\mathbf{5}}$$

Summary of general rules to be applied when drawing graphs

(i) Give the graph a title clearly explaining what is being illustrated.
(ii) Choose scales such that the graph occupies as much space as possible on the graph paper being used.
(iii) Choose scales so that interpolation is made as easy as possible. Usually scales such as 1 cm = 1 unit, or 1 cm = 2 units, or 1 cm = 10 units are used. Awkward scales such as 1 cm = 3 units or 1 cm = 7 units should not be used.
(iv) The scales need not start at zero, particularly when starting at zero produces an accumulation of points within a small area of the graph paper.
(v) The co-ordinates, or points, should be clearly marked. This may be done either by a cross, or a dot and circle, or just by a dot (see Figure 32.1).
(vi) A statement should be made next to each axis explaining the numbers represented with their appropriate units.
(vii) Sufficient numbers should be written next to each axis without cramping.

Practical problems involving straight line graphs

When a set of co-ordinate values are given or are obtained experimentally and it is believed that they follow a law of the form $y = mx + c$, then if a straight line can be drawn reasonably close to most of the co-ordinate values when plotted, this verifies that a law of the form $y = mx + c$ exists. From the graph, constants m (i.e. gradient) and c (i.e. y-axis intercept) can be determined. This technique is called **determination of law** (see also Chapter 33).

For example, the temperature in degrees Celsius and the corresponding values in degrees Fahrenheit are shown in the table below.

°C	10	20	40	60	80	100
°F	50	68	104	140	176	212

Axes with suitable scales are shown in Figure 32.5. The co-ordinates (10, 50), (20, 68), (40, 104), and so on are plotted as shown. When the co-ordinates are joined, a straight line is produced. Since a straight line results there is a linear relationship between degrees Celsius and degrees Fahrenheit.
To find the Fahrenheit temperature at, say, 55°C, a vertical line AB is constructed from the horizontal axis to meet the straight line at B. The point where the horizontal line BD meets the vertical axis indicates the equivalent Fahrenheit temperature. **Hence 55°C is equivalent to 131°F**.
This process of finding an equivalent value in between the given information in the above table is called **interpolation**.
To find the Celsius temperature at, say, 167°F, a horizontal line EF is constructed as shown in Figure 32.5. The point where the vertical line FG cuts the horizontal axis indicates the equivalent Celsius temperature. **Hence 167°F is equivalent to 75°C**.

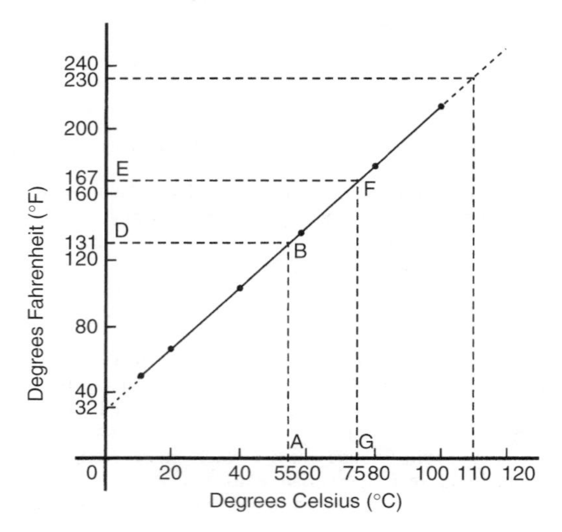

Figure 32.5

If the graph is assumed to be linear even outside of the given data, then the graph may be extended at both ends (shown by broken lines in Figure 32.5). From Figure 32.5, it is seen that **0°C corresponds to 32°F** and **230°F is seen to correspond to 110°C**.

The process of finding equivalent values outside of the given range is called **extrapolation**.

In another example, experimental tests to determine the breaking stress σ of rolled copper at various temperatures t gave the following results.

Stress σ N/cm^2	8.46	8.04	7.78	7.37	7.08	6.63
Temperature t°C	70	200	280	410	500	640

The co-ordinates (70, 8.46), (200, 8.04), and so on, are plotted as shown in Figure 32.6. Since the graph is a straight line then the values obey the law $\sigma = at + b$, and the gradient of the straight line, is

$$a = \frac{AB}{BC} = \frac{8.36 - 6.76}{100 - 600} = \frac{1.60}{-500} = \mathbf{-0.0032}$$

Vertical axis intercept, $\boldsymbol{b = 8.68}$

Hence the law of the graph is: $\boldsymbol{\sigma = -0.0032t + 8.68}$

When the temperature is, say, 250°C, stress σ is given by

$$\sigma = -0.0032(250) + 8.68 = \mathbf{7.88\ N/cm^2}$$

Rearranging $\quad \sigma = -0.0032t + 8.68$

gives: $\quad 0.0032\,t = 8.68 - \sigma,$

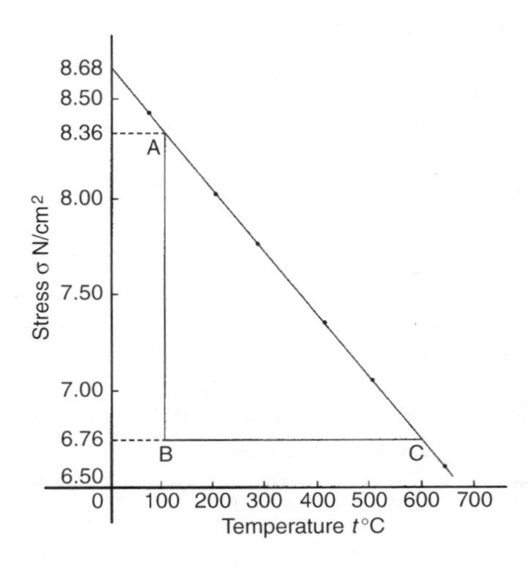

Figure 32.6

i.e. $$t = \frac{8.68 - \sigma}{0.0032}$$

Hence when the stress σ is, say, 7.54 N/cm^2,

$$\text{temperature } t = \frac{8.68 - 7.54}{0.0032} = \mathbf{356.3°C}$$

33 Reduction of Non-linear Laws to Linear Form

Determination of law

Frequently, the relationship between two variables, say x and y, is not a linear one, i.e. when x is plotted against y a curve results. In such cases the non-linear equation may be modified to the linear form, $y = mx + c$, so that the constants, and thus the law relating the variables can be determined. This technique is called **'determination of law'**.
Some examples of the reduction of equations to linear form include:

(i) $y = ax^2 + b$ compares with $Y = mX + c$, where $m = a$, $c = b$ and $X = x^2$.
Hence y is plotted vertically against x^2 horizontally to produce a straight line graph of gradient 'a' and y-axis intercept 'b'

(ii) $y = \frac{a}{x} + b$
y is plotted vertically against $\frac{1}{x}$ horizontally to produce a straight line graph of gradient 'a' and y-axis intercept 'b'

(iii) $y = ax^2 + bx$
Dividing both sides by x gives $\frac{y}{x} = ax + b$
Comparing with $Y = mX + c$ shows that $\frac{y}{x}$ is plotted vertically against x horizontally to produce a straight line graph of gradient 'a' and $\frac{y}{x}$ axis intercept 'b'.

For example, experimental values of x and y, shown below, are believed to be related by the law $y = ax^2 + b$.

x	1	2	3	4	5
y	9.8	15.2	24.2	36.5	53.0

If y is plotted against x a curve results and it is not possible to determine the values of constants a and b from the curve. Comparing $y = ax^2 + b$ with $Y = mX + c$ shows that y is to be plotted vertically against x^2 horizontally.

A table of values is drawn up as shown below.

x	1	2	3	4	5
x^2	1	4	9	16	25
y	9.8	15.2	24.2	36.5	53.0

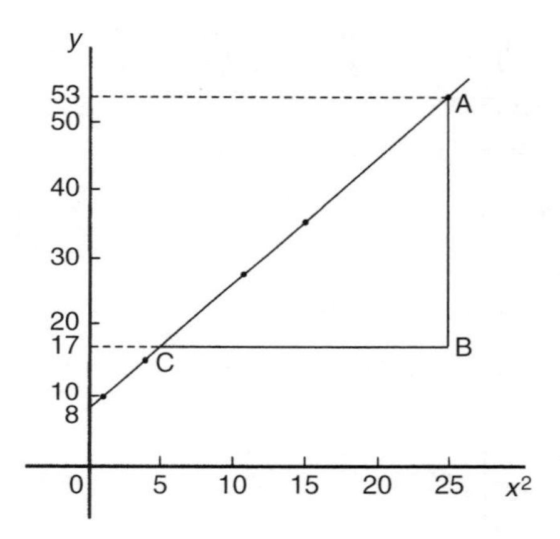

Figure 33.1

A graph of y against x^2 is shown in Figure 33.1, with the best straight line drawn through the points. Since a straight line graph results, the law is verified. From the graph, gradient

$$a = \frac{AB}{BC} = \frac{53 - 17}{25 - 5} = \frac{36}{20} = \mathbf{1.8}$$

and the y-axis intercept, $\boldsymbol{b} = \mathbf{8.0}$

Hence the law of the graph is $\boldsymbol{y} = \mathbf{1.8}\boldsymbol{x}^{\mathbf{2}} + \mathbf{8.0}$

In another example, values of load L newtons and distance d metres obtained experimentally are shown in the following table.

Load, L N	32.3	29.6	27.0	23.2	18.3	12.8	10.0	6.4
Distance, d m	0.75	0.37	0.24	0.17	0.12	0.09	0.08	0.07

Comparing $L = \frac{a}{d} + b$ i.e. $L = a\left(\frac{1}{d}\right) + b$ with $Y = mX + c$ shows that L is to be plotted vertically against $\frac{1}{d}$ horizontally. Another table of values is drawn up as shown below.

L	32.3	29.6	27.0	23.2	18.3	12.8	10.0	6.4
d	0.75	0.37	0.24	0.17	0.12	0.09	0.08	0.07
$\frac{1}{d}$	1.33	2.70	4.17	5.88	8.33	11.11	12.50	14.29

A graph of L against $\frac{1}{d}$ is shown in Figure 33.2. A straight line can be drawn through the points, which verifies that load and distance are related by a law of the form $L = \frac{a}{d} + b$

Gradient of straight line, $\boldsymbol{a} = \frac{AB}{BC} = \frac{31 - 11}{2 - 12} = \frac{20}{-10} = \mathbf{-2}$

L-axis intercept, $\boldsymbol{b} = \mathbf{35}$

Hence, the law of the graph is $\boldsymbol{L} = -\frac{\mathbf{2}}{\boldsymbol{d}} + \mathbf{35}$

When the distance d is, say, 0.20 m, load $L = \frac{-2}{0.20} + 35 = \mathbf{25.0\ N}$

Rearranging $L = -\frac{2}{d} + 35$ gives $\frac{2}{d} = 35 - L$ and $d = \frac{2}{35 - L}$

Hence, when the load L is, say, 20 N,

$$\text{distance } d = \frac{2}{35 - 20} = \frac{2}{15} = \mathbf{0.13\ m}$$

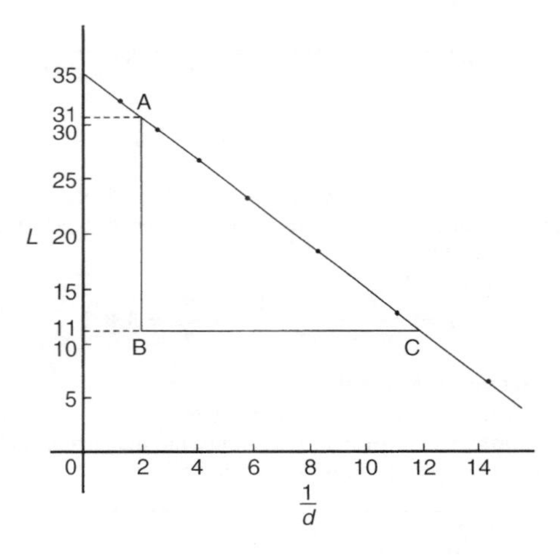

Figure 33.2

Determination of law involving logarithms

Examples of reduction of equations to linear form involving logarithms, include:

(i) $y = ax^n$

Taking logarithms to a base of 10 of both sides gives:

$$\lg y = \lg(ax^n) = \lg a + \lg x^n$$

i.e. $\quad \lg y = n \lg x + \lg a$

by the laws of logarithms which compares with $Y = mX + c$ and shows that $\lg y$ is plotted vertically against $\lg x$ horizontally to produce a straight line graph of gradient n and $\lg y$-axis intercept $\lg a$.

(ii) $y = ab^x$

Taking logarithms to a base of 10 of both sides gives:

$$\lg y = \lg(ab^x)$$

i.e. $\lg y = \lg a + \lg b^x$

i.e. $\lg y = x \lg b + \lg a$ by the laws of logarithms

or $\lg y = (\lg b)x + \lg a$ which compares with

$$Y = mX + c$$

and shows that $\lg y$ is plotted vertically against x horizontally to produce a straight line graph of gradient $\lg b$ and $\lg y$-axis intercept $\lg a$.

(iii) $y = ae^{bx}$

Taking logarithms to a base of e of both sides gives:

$$\ln y = \ln(ae^{bx})$$

i.e. $\ln y = \ln a + \ln e^{bx}$

i.e. $\ln y = \ln a + bx \ln e$

i.e. $\ln y = bx + \ln a$ (since $\ln e = 1$), which compares with

$$Y = mX + c$$

and shows that $\ln y$ is plotted vertically against x horizontally to produce a straight line graph of gradient b and $\ln y$-axis intercept $\ln a$.

For example, the current flowing in, and the power dissipated by a resistor are measured experimentally for various values and the results are as shown below.

Current, I amperes	2.2	3.6	4.1	5.6	6.8
Power, P watts	116	311	403	753	1110

To show that the law relating current and power is of the form $P = RI^n$, where R and n are constants, and determine the law:

Taking logarithms to a base of 10 of both sides of $P = RI^n$ gives:

$$\lg P = \lg(RI^n) = \lg R + \lg I^n$$

$$= \lg R + n \lg I \quad \text{by the laws of logarithms}$$

i.e. $\lg P = n \lg I + \lg R$, which is of the form $Y = mX + c$,

showing that $\lg P$ is to be plotted vertically against $\lg I$ horizontally.

A table of values for $\lg I$ and $\lg P$ is drawn up as shown below.

I	2.2	3.6	4.1	5.6	6.8
$\lg I$	0.342	0.556	0.613	0.748	0.833
P	116	311	403	753	1110
$\lg P$	2.064	2.493	2.605	2.877	3.045

A graph of $\lg P$ against $\lg I$ is shown in Figure 33.3 and since a straight line results the law $P = RI^n$ is verified.

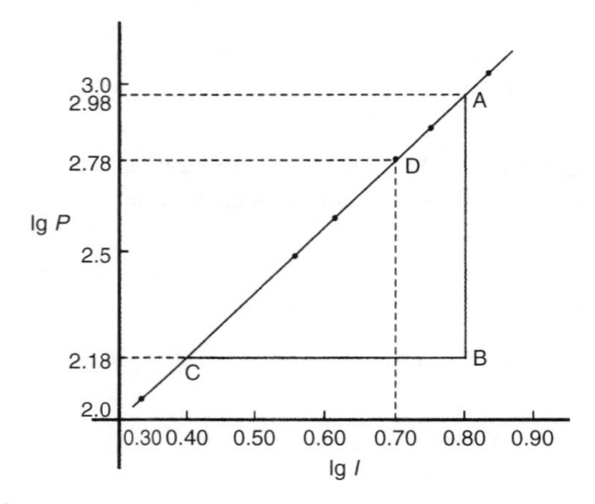

Figure 33.3

Gradient of straight line, $n = \dfrac{\text{AB}}{\text{BC}} = \dfrac{2.98 - 2.18}{0.8 - 0.4} = \dfrac{0.80}{0.4} = \mathbf{2}$

It is not possible to determine the vertical axis intercept on sight since the horizontal axis scale does not start at zero. Selecting any point from the graph, say point D, where $\lg I = 0.70$ and $\lg P = 2.78$, and substituting values into

$$\lg P = n \lg I + \lg R$$

gives: $\quad 2.78 = (2)(0.70) + \lg R$

from which $\quad \lg R = 2.78 - 1.40 = 1.38$

Hence $\quad R = \text{antilog } 1.38\ (= 10^{1.38}) = \mathbf{24.0}$

Hence the law of the graph is $P = 24.0I^2$

In another example, the current i mA flowing in a capacitor which is being discharged varies with time t ms as shown below.

i mA	203	61.14	22.49	6.13	2.49	0.615
t ms	100	160	210	275	320	390

To show that these results are related by a law of the form $i = Ie^{t/T}$, where I and T are constants:

Taking Napierian logarithms of both sides of $i = Ie^{t/T}$ gives

$$\ln i = \ln(Ie^{t/T}) = \ln I + \ln e^{t/T}$$

i.e. $\quad \ln i = \ln I + \dfrac{t}{T}$ (since $\ln e = 1$)

or $\quad \ln i = \left(\dfrac{1}{T}\right) t + \ln I$

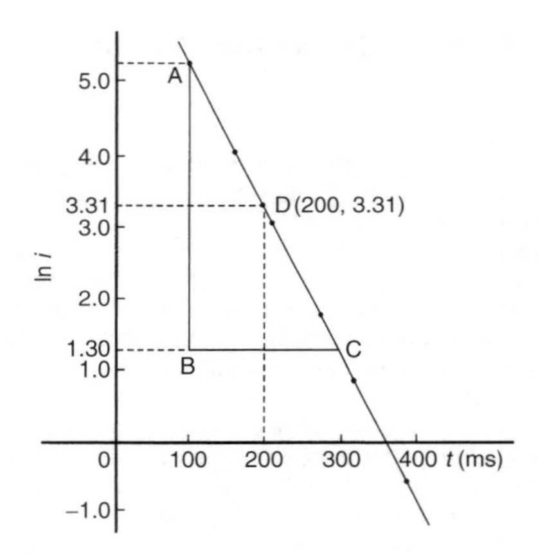

Figure 33.4

which compares with $y = mx + c$, showing that $\ln i$ is plotted vertically against t horizontally. (For methods of evaluating Napierian logarithms see Chapter 12).

Another table of values is drawn up as shown below.

t	100	160	210	275	320	390
i	203	61.14	22.49	6.13	2.49	0.615
$\ln i$	5.31	4.11	3.11	1.81	0.91	−0.49

A graph of $\ln i$ against t is shown in Figure 33.4 and since a straight line results the law $i = Ie^{t/T}$ is verified.

Gradient of straight line,

$$\frac{1}{T} = \frac{AB}{BC} = \frac{5.30 - 1.30}{100 - 300} = \frac{4.0}{-200} = -0.02$$

Hence $\quad T = \dfrac{1}{-0.02} = \mathbf{-50}$

Selecting any point on the graph, say point D, where $t = 200$ and $\ln i = 3.31$,

and substituting into $\quad \ln i = \left(\dfrac{1}{T}\right) t + \ln I$

gives: $\quad 3.31 = -\dfrac{1}{50}(200) + \ln I$

from which, $\quad \ln I = 3.31 + 4.0 = 7.31$

and I = antilog 7.31 $(= e^{7.31}) = 1495$ or **1500** correct to 3 significant figures

Hence the law of the graph is $i = 1500\, e^{-t/50}$

34 Graphs with Logarithmic Scales

Logarithmic scales

Graph paper is available where the scale markings along the horizontal and vertical axes are proportional to the logarithms of the numbers. Such graph paper is called **log-log graph paper**.
A **logarithmic scale** is shown in Figure 34.1 where the distance between, say 1 and 2, is proportional to $\lg 2 - \lg 1$, i.e. 0.3010 of the total distance from 1 to 10. Similarly, the distance between 7 and 8 is proportional to $\lg 8 - \lg 7$, i.e. 0.05799 of the total distance from 1 to 10. Thus the distance between markings progressively decreases as the numbers increase from 1 to 10.
With log-log graph paper the scale markings are from 1 to 9, and this pattern can be repeated several times. The number of times the pattern of markings is repeated on an axis signifies the number of **cycles**. When the vertical axis has, say, 3 sets of values from 1 to 9, and the horizontal axis has, say, 2 sets of values from 1 to 9, then this log-log graph paper is called 'log 3 cycle $\times$ 2 cycle' (see Figure 34.2). Many different arrangements are available ranging from 'log 1 cycle $\times$ 1 cycle' through to 'log 5 cycle $\times$ 5 cycle'.
To depict a set of values, say, from 0.4 to 161, on an axis of log-log graph paper, 4 cycles are required, from 0.1 to 1,1 to 10, 10 to 100 and 100 to 1000.

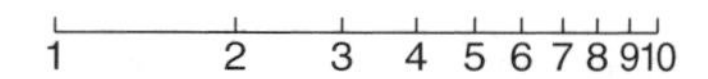

Figure 34.1

Graphs of the form $y = ax^n$

Taking logarithms to a base of 10 of both sides of $y = ax^n$ gives:

$$\lg y = \lg(ax^n) = \lg a + \lg x^n$$

i.e.

$$\lg y = n \lg x + \lg a$$

which compares with $\quad Y = mX + c$

Thus, by plotting $\lg y$ vertically against $\lg x$ horizontally, a straight line results, i.e. the equation $y = ax^n$ is reduced to linear form. With log-log graph paper available x and y may be plotted directly, without having first to determine their logarithms, as was the case in Chapter 33.

For example, experimental values of two related quantities x and y are shown below:

x	0.41	0.63	0.92	1.36	2.17	3.95
y	0.45	1.21	2.89	7.10	20.79	82.46

The law relating x and y is believed to be $y = ax^b$, where a and b are constants. To verify that this law is true and determine the approximate values of a and b: If $y = ax^b$ then $\lg y = b \lg x + \lg a$, from above, which is of the form $Y = mX + c$, showing that to produce a straight line graph $\lg y$ is plotted vertically against $\lg x$ horizontally. x and y may be plotted directly on to log-log graph paper as shown in Figure 34.2. The values of y range from 0.45

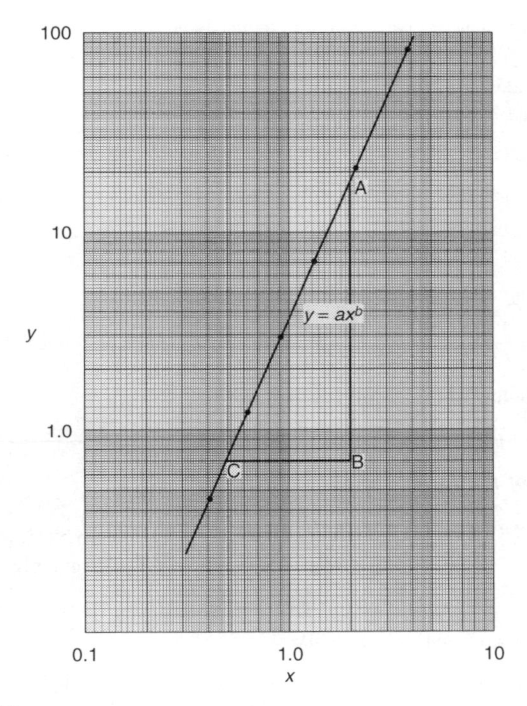

Figure 34.2

to 82.46 and 3 cycles are needed (i.e. 0.1 to 1, 1 to 10 and 10 to 100). The values of x range from 0.41 to 3.95 and 2 cycles are needed (i.e. 0.1 to 1 and 1 to 10). Hence 'log 3 cycle $\times$ 2 cycle' is used as shown in Figure 34.2 where the axes are marked and the points plotted. Since the points lie on a straight line the law $y = ax^b$ is verified.

To evaluate constants *a* and *b*:

Method 1. Any two points on the straight line, say points A and C, are selected, and AB and BC are measured (say in centimetres). Then, gradient,

$$\boldsymbol{b} = \frac{\text{AB}}{\text{BC}} = \frac{11.5 \text{ units}}{5 \text{ units}} = \mathbf{2.3}$$

Since $\lg y = b \lg x + \lg a$, when $x = 1$, $\lg x = 0$ and $\lg y = \lg a$.
The straight line crosses the ordinate $x = 1.0$ at $y = 3.5$. Hence $\lg a = \lg 3.5$, i.e. $\boldsymbol{a} = \mathbf{3.5}$

Method 2. Any two points on the straight line, say points A and C, are selected. A has co-ordinates (2, 17.25) and C has co-ordinates (0.5, 0.7).

Since $y = ax^b$ then $\quad 17.25 = a(2)^b \qquad (1)$

and $\quad 0.7 = a(0.5)^b \qquad (2)$

i.e. two simultaneous equations are produced and may be solved for a and b. Dividing equation (1) by equation (2) to eliminate a gives:

$$\frac{17.25}{0.7} = \frac{(2)^b}{(0.5)^b} = \left(\frac{2}{0.5}\right)^b$$

i.e. $24.643 = (4)^b$

Taking logarithms of both sides gives $\lg 24.643 = b \lg 4$,

i.e. $b = \dfrac{\lg 24.643}{\lg 4} = 2.3$, correct to 2 significant figures.

Substituting $b = 2.3$ in equation (1) gives: $17.25 = a(2)^{2.3}$,

i.e. $a = \dfrac{17.25}{(2)^{2.3}} = \dfrac{17.25}{4.925} = 3.5$, correct to 2 significant figures.

Hence the law of the graph is $y = 3.5x^{2.3}$

Graphs of the form $y = ab^x$

Taking logarithms to a base of 10 of both sides of $y = ab^x$ gives:

$$\lg y = \lg(ab^x) = \lg a + \lg b^x = \lg a + x \lg b$$

i.e. $\lg y = (\lg b)x + \lg a$

which compares with $Y = mX + c$

Thus, by plotting $\lg y$ vertically against x horizontally a straight line results, i.e. the graph $y = ab^x$ is reduced to linear form. In this case, graph paper having a linear horizontal scale and a logarithmic vertical scale may be used. This type of graph paper is called **log-linear graph paper**, and is specified by the number of cycles on the logarithmic scale. For example, graph paper having 3 cycles on the logarithmic scale is called 'log 3 cycle × linear' graph paper.

Graphs of the form $y = ae^{kx}$

Taking logarithms to a base of e of both sides of $y = ae^{kx}$ gives:

$$\ln y = \ln(ae^{kx}) = \ln a + \ln e^{kx} = \ln a + kx \ln e$$

i.e. $\ln y = kx + \ln a$ (since $\ln e = 1$)

which compares with $Y = mX + c$
Thus, by plotting $\ln y$ vertically against x horizontally, a straight line results, i.e. the equation $y = ae^{kx}$ is reduced to linear form. In this case, graph paper having a linear horizontal scale and a logarithmic vertical scale may be used.

For example, the voltage, v volts, across an inductor is believed to be related to time, t ms, by the law $v = Ve^{t/T}$, where V and T are constants.

Experimental results obtained are:

v volts	883	347	90	55.5	18.6	5.2
t ms	10.4	21.6	37.8	43.6	56.7	72.0

To show that the law relating voltage and time is as stated and determine the approximate values of V and T:

Since $v = Ve^{t/T}$ then $\ln v = \dfrac{1}{T}t + \ln V$, which is of the form $Y = mX + c$

Using 'log 3 cycle × linear' graph paper, the points are plotted as shown in Figure 34.3. Since the points are joined by a straight line the law $v = Ve^{t/T}$ is verified.

Gradient of straight line,

$$\frac{1}{T} = \frac{AB}{BC} = \frac{\ln 100 - \ln 10}{36.5 - 64.2} = \frac{2.3026}{-27.7}$$

Hence $\quad \boldsymbol{T} = \dfrac{-27.7}{2.3026} = \mathbf{-12.0}$, correct to 3 significant figures

Since the straight line does not cross the vertical axis at $t = 0$ in Figure 34.3, the value of V is determined by selecting any point, say A, having co-ordinates (36.5, 100) and substituting these values into $v = Ve^{t/T}$.

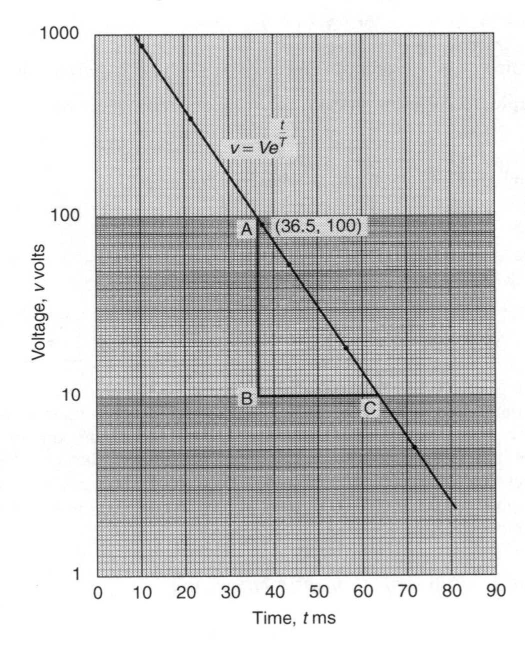

Figure 34.3

Thus $100 = Ve^{36.5/-12.0}$ i.e. $V = \dfrac{100}{e^{-36.5/12.0}}$ = **2090 volts**, correct to 3 significant figures.

Hence the law of the graph is $v = 2090e^{-t/12.0}$

When, say, time $t = 25$ ms, voltage $v = 2090e^{-25/12.0}$ = **260 V**

When, say, the voltage is 30.0 volts, $30.0 = 2090e^{-t/12.0}$,

hence $\quad e^{-t/12.0} = \dfrac{30.0}{2090} \quad$ and $\quad e^{t/12.0} = \dfrac{2090}{30.0} = 69.67$

Taking Napierian logarithms gives: $\dfrac{t}{12.0} = \ln 69.67 = 4.2438$
from which, time $t = (12.0)(4.2438)$ = **50.9 ms**

---

35 Graphical Solution of Equations

Graphical solution of simultaneous equations

Linear simultaneous equations in two unknowns may be solved graphically by:
(i) plotting the two straight lines on the same axes, and
(ii) noting their point of intersection.
The co-ordinates of the point of intersection give the required solution.

For example, to solve graphically the simultaneous equations

$$2x - y = 4$$

$$x + y = 5$$

Rearranging each equation into $y = mx + c$ form gives:

$$y = 2x - 4 \qquad (1)$$

$$y = -x + 5 \qquad (2)$$

Only three co-ordinates need be calculated for each graph since both are straight lines.

x	0	1	2
y = 2x − 4	−4	−2	0

x	0	1	2
y = −x + 5	5	4	3

Each of the graphs is plotted as shown in Figure 35.1. The point of intersection is at (3, 2) and since this is the only point which lies simultaneously on both lines then $x = 3$, $y = 2$ is the solution of the simultaneous equations.
(It is sometimes useful initially to sketch the two straight lines to determine the region where the point of intersection is. Then, if necessary, for greater accuracy, a graph having a smaller range of values can be drawn to 'magnify' the point of intersection).

Graphical solutions of quadratic equations

A general **quadratic equation** is of the form $y = ax^2 + bx + c$, where a, b and c are constants and a is not equal to zero.

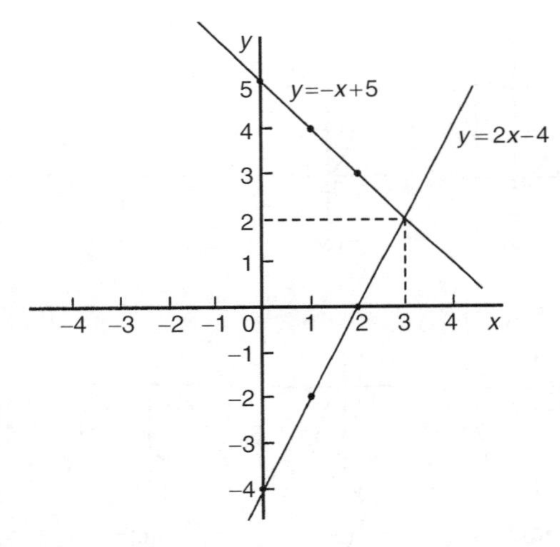

Figure 35.1

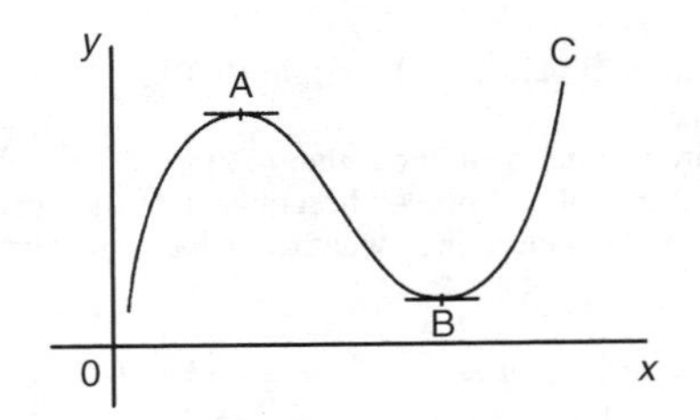

Figure 35.2

A graph of a quadratic equation always produces a shape called a **parabola**. The gradient of the curve between 0 and A and between B and C in Figure 35.2 is positive, whilst the gradient between A and B is negative. Points such as A and B are called **turning points**. At A the gradient is zero and, as x increases, the gradient of the curve changes from positive just before A to negative just after. Such a point is called a **maximum value**. At B the gradient is also zero, and, as x increases, the gradient of the curve changes from negative just before B to positive just after. Such a point is called a **minimum value**.

Quadratic graphs

(i) $\boldsymbol{y = ax^2}$
Graphs of $y = x^2$, $y = 3x^2$ and $y = \frac{1}{2}x^2$ are shown in Figure 35.3.
All have minimum values at the origin (0, 0).
Graphs of $y = -x^2$, $y = -3x^2$ and $y = -\frac{1}{2}x^2$ are shown in Figure 35.4.

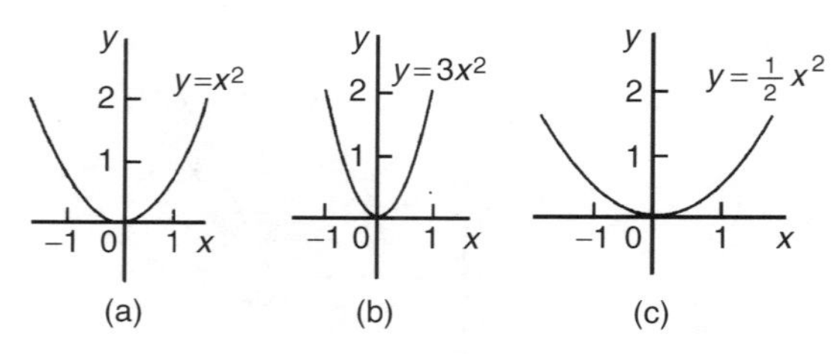

Figure 35.3

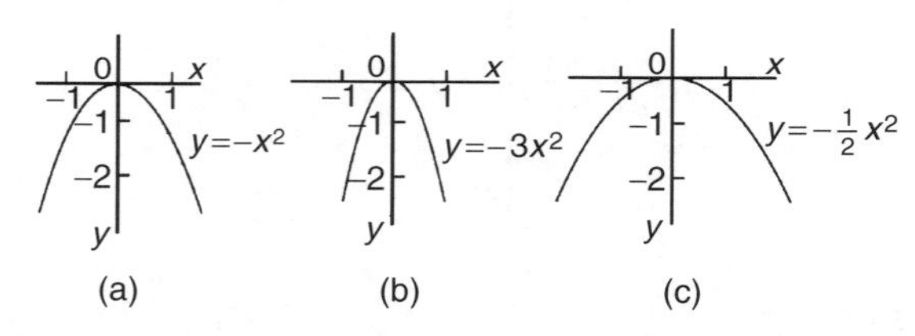

Figure 35.4

All have maximum values at the origin (0, 0).
When $y = ax^2$,
(a) curves are symmetrical about the y-axis,
(b) the magnitude of 'a' affects the gradient of the curve, and
(c) the sign of 'a' determines whether it has a maximum or minimum value

(ii) $\boldsymbol{y = ax^2 + c}$

Graphs of $y = x^2 + 3$, $y = x^2 - 2$, $y = -x^2 + 2$ and $y = -2x^2 - 1$ are shown in Figure 35.5.
When $y = ax^2 + c$:
(a) curves are symmetrical about the y-axis,
(b) the magnitude of 'a' affects the gradient of the curve, and
(c) the constant 'c' is the y-axis intercept

(iii) $\boldsymbol{y = ax^2 + bx + c}$

Whenever 'b' has a value other than zero the curve is displaced to the right or left of the y-axis. When b/a is positive, the curve is displaced $b/2a$ to the left of the y-axis, as shown in Figure 35.6(a). When b/a is negative the curve is displaced $b/2a$ to the right of the y-axis, as shown in Figure 35.6(b).

Quadratic equations of the form $ax^2 + bx + c = 0$ may be solved graphically by:

(i) plotting the graph $y = ax^2 + bx + c$, and
(ii) noting the points of intersection on the x-axis (i.e. where $y = 0$).

The x values of the points of intersection give the required solutions since at these points both $y = 0$ and $ax^2 + bx + c = 0$. The number of solutions, or roots of a quadratic equation, depends on how many times the curve cuts the

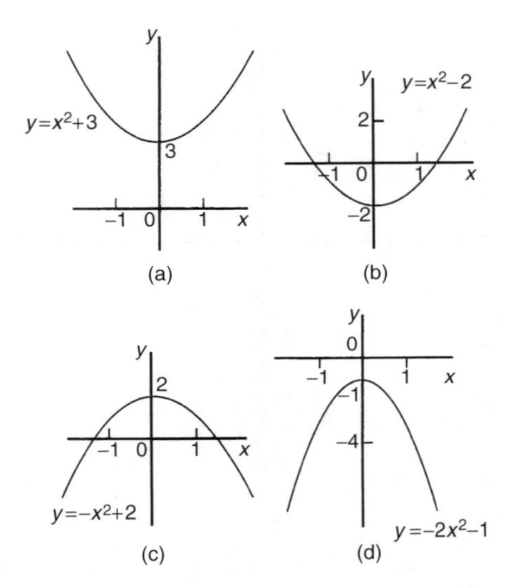

Figure 35.5

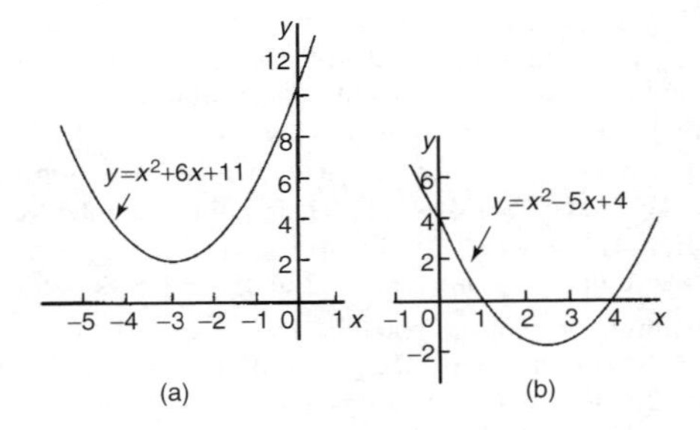

Figure 35.6

x-axis and there can be no real roots (as in Figure 35.6(a)) or one root (as in Figures 35.3 and 35.4) or two roots (as in Figure 35.6(b)).

For example, to solve the quadratic equation $4x^2 + 4x - 15 = 0$ graphically given that the solutions lie in the range $x = -3$ to $x = 2$:
Let $y = 4x^2 + 4x - 15$. A table of values is drawn up as shown below.

x	-3	-2	-1	0	1	2
$4x^2$	36	16	4	0	4	16
$4x$	-12	-8	-4	0	4	8
-15	-15	-15	-15	-15	-15	15
$y = 4x^2 + 4x - 15$	9	-7	-15	-15	-7	9

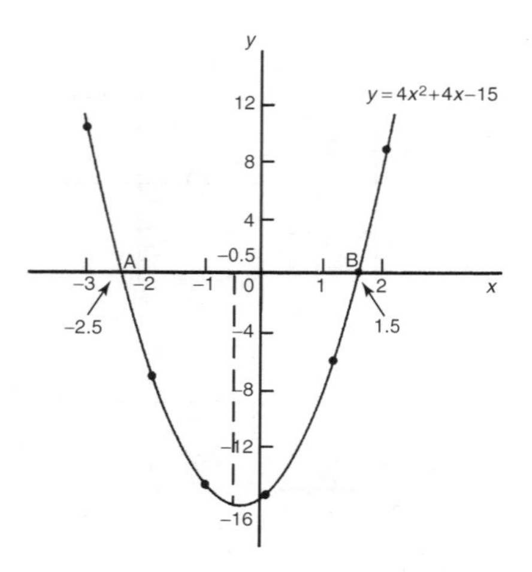

Figure 35.7

A graph of $y = 4x^2 + 4x - 15$ is shown in Figure 35.7. The only points where $y = 4x^2 + 4x - 15$ and $y = 0$ are the points marked A and B. This occurs at $\boldsymbol{x = -2.5}$ **and** $\boldsymbol{x = 1.5}$ and these are the solutions of the quadratic equation $4x^2 + 4x - 15 = 0$. (By substituting $x = -2.5$ and $x = 1.5$ into the original equation the solutions may be checked). The curve has a turning point at $(-0.5, -16)$ and the nature of the point is a **minimum**.

An alternative graphical method of solving $4x^2 + 4x - 15 = 0$ is to rearrange the equation as $4x^2 = -4x + 15$ and then plot two separate graphs-in this case $y = 4x^2$ and $y = -4x + 15$. Their points of intersection give the roots of equation $4x^2 = -4x + 15$, i.e. $4x^2 + 4x - 15 = 0$. This is shown in Figure 35.8, where the roots are $x = -2.5$ and $x = 1.5$ as before.

In another example, to plot the graph of $y = -2x^2 + 3x + 6$ for values of x from $x = -2$ to $x = 4$ and to use the graph to find the roots of the following equations (a) $-2x^2 + 3x + 6 = 0$ (b) $-2x^2 + 3x + 2 = 0$
(c) $-2x^2 + 3x + 9 = 0$ (d) $-2x^2 + x + 5 = 0$:
A table of values is drawn up as shown below.

x	−2	−1	0	1	2	3	4
$-2x^2$	−8	−2	0	−2	−8	−18	−32
$+3x$	−6	−3	0	3	6	9	12
$+6$	6	6	6	6	6	6	6
y	−8	1	6	7	4	−3	−14

A graph of $-2x^2 + 3x + 6$ is shown in Figure 35.9.

(a) The parabola $y = -2x^2 + 3x + 6$ and the straight line $y = 0$ intersect at A and B, where $\boldsymbol{x = -1.13}$ **and** $\boldsymbol{x = 2.63}$ and these are the roots of the equation $-2x^2 + 3x + 6 = 0$

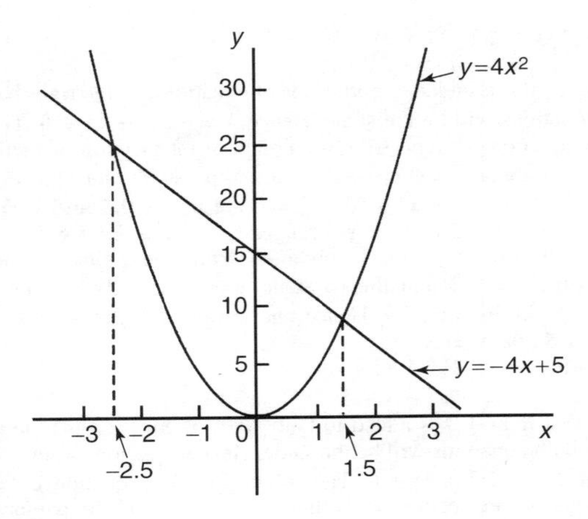

Figure 35.8

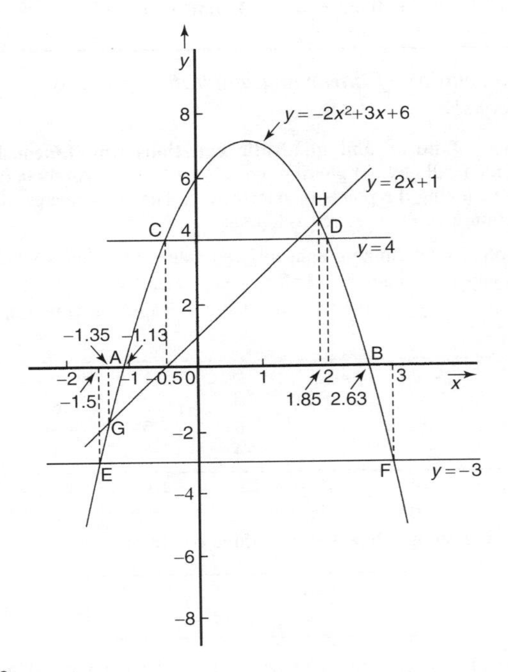

Figure 35.9

(b) Comparing $y = -2x^2 + 3x + 6$ (1)
with $0 = -2x^2 + 3x + 2$ (2)
shows that if 4 is added to both sides of equation (2), the right-hand side of both equations will be the same. Hence $4 = -2x^2 + 3x + 6$. The solution of this equation is found from the points of intersection of the line $y = 4$ and the parabola $y = -2x^2 + 3x + 6$, i.e. points C and D in Figure 35.9. Hence the roots of $-2x^2 + 3x + 2 = 0$ are $\boldsymbol{x = -0.5}$ **and** $\boldsymbol{x = 2}$

(c) $-2x^2 + 3x + 9 = 0$ may be rearranged as $-2x^2 + 3x + 6 = -3$, and the solution of this equation is obtained from the points of intersection of the line $y = -3$ and the parabola $y = -2x^2 + 3x + 6$, i.e. at points E and F in Figure 35.9 Hence the roots of $-2x^2 + 3x + 9 = 0$ are $\boldsymbol{x = -1.5}$ **and** $\boldsymbol{x = 3}$

(d) Comparing $y = -2x^2 + 3x + 6$ (3)
with $0 = -2x^2 + x + 5$ (4)
shows that if $2x + 1$ is added to both sides of equation (4) the right-hand side of both equations will be the same. Hence equation (4) may be written as $2x + 1 = -2x^2 + 3x + 6$. The solution of this equation is found from the points of intersection of the line $y = 2x + 1$ and the parabola $y = -2x^2 + 3x + 6$, i.e. points G and H in Figure 35.9. Hence the roots of $-2x^2 + x + 5 = 0$ are $\boldsymbol{x = -1.35}$ **and** $\boldsymbol{x = 1.85}$

Graphical solution of linear and quadratic equations simultaneously

The solution of **linear and quadratic equations simultaneously** may be achieved graphically by: (i) plotting the straight line and parabola on the same axes, and (ii) noting the points of intersection. The co-ordinates of the points of intersection give the required solutions.

For example, to determine graphically the values of x and y which simultaneously satisfy the equations $y = 2x^2 - 3x - 4$ and $y = 2 - 4x$:
$y = 2x^2 - 3x - 4$ is a parabola and a table of values is drawn up as shown below.

x	-2	-1	0	1	2	3
$2x^2$	8	2	0	2	8	18
$-3x$	6	3	0	-3	-6	-9
-4	-4	-4	-4	-4	-4	-4
y	10	1	-4	-5	-2	5

$y = 2 - 4x$ is a straight line and only three co-ordinates need be calculated.

x	0	1	2
y	2	-2	-6

The two graphs are plotted in Figure 35.10 and the points of intersection, shown as A and B, are at co-ordinates $(-2, 10)$ and $(1.5, -4)$. Hence the simultaneous solutions occur when $\boldsymbol{x = -2, y = 10}$ and when $\boldsymbol{x = 1.5, y = -4}$.

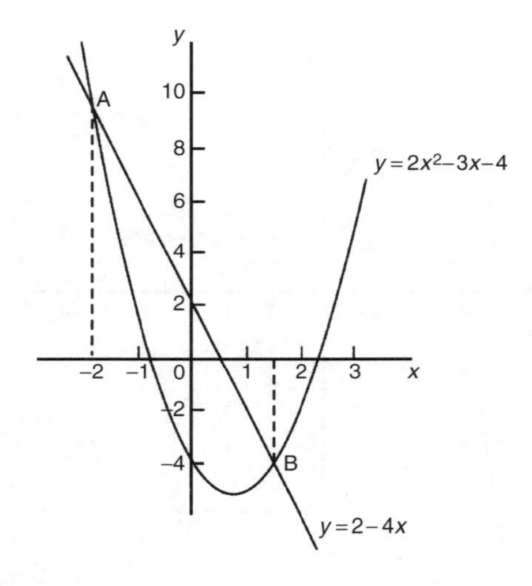

Figure 35.10

(These solutions may be checked by substituting into each of the original equations.)

Graphical solution of cubic equations

A **cubic equation** of the form $ax^3 + bx^2 + cx + d = 0$ may be solved graphically by:
(i) plotting the graph $y = ax^3 + bx^2 + cx + d$, and (ii) noting the points of intersection on the x-axis (i.e. where $y = 0$). The x-values of the points of intersection give the required solution since at these points both $y = 0$ and $ax^3 + bx^2 + cx + d = 0$.
The number of solutions, or roots of a cubic equation depends on how many times the curve cuts the x-axis and there can be one, two or three possible roots, as shown in Figure 35.11.

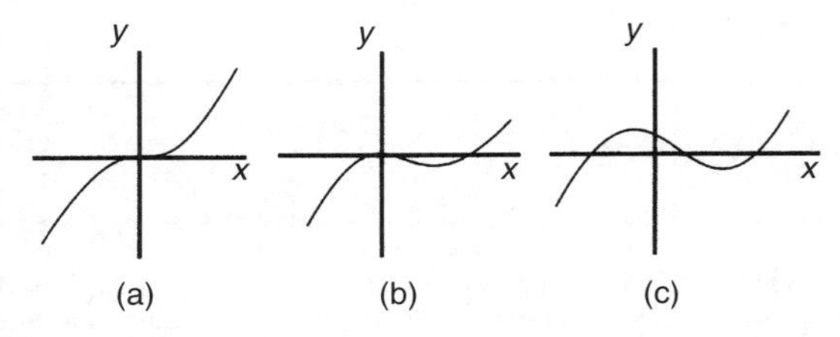

Figure 35.11

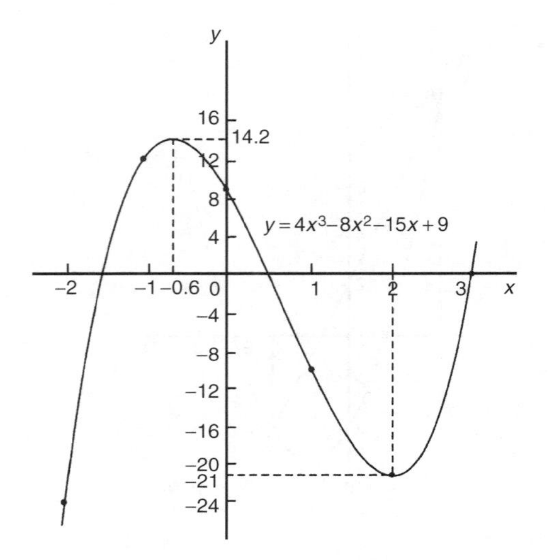

Figure 35.12

For example, to solve graphically the cubic equation $4x^3 - 8x^2 - 15x + 9 = 0$ given that the roots lie between $x = -2$ and $x = 3$:
Let $y = 4x^3 - 8x^2 - 15x + 9$. A table of values is drawn up as shown below.

x	-2	-1	0	1	2	3
$4x^3$	-32	-4	0	4	32	108
$-8x^2$	-32	-8	0	-8	-32	-72
$-15x$	30	15	0	-15	-30	-45
$+9$	9	9	9	9	9	9
y	-25	12	9	-10	-21	0

A graph of $y = 4x^3 - 8x^2 - 15x + 9$ is shown in Figure 35.12.
The graph crosses the x-axis (where $y = 0$) at $\boldsymbol{x = -1.5}$, $\boldsymbol{x = 0.5}$ **and** $\boldsymbol{x = 3}$ and these are the solutions to the cubic equation $4x^3 - 8x^2 - 15x + 9 = 0$.
The turning points occur at $\mathbf{(-0.6, 14.2)}$, which is a **maximum**, and $\mathbf{(2, -21)}$, which is a **minimum**.

---

36 Polar Curves

With Cartesian coordinates the equation of a curve is expressed as a general relationship between x and y, i.e. $y = f(x)$.
Similarly, with polar coordinates the equation of a curve is expressed in the form $r = f(\theta)$. When a graph of $r = f(\theta)$ is required a table of values needs to be drawn up and the coordinates, (r, θ) plotted.

For example, to plot the polar graph of $r = 5 \sin \theta$ between $\theta = 0°$ and $\theta = 360°$ using increments of 30°:
A table of values at 30° intervals is produced as shown below.

θ	0	30°	60°	90°	120°	150°	180°	210°
$r = 5 \sin \theta$	0	2.50	4.33	5.00	4.33	2.50	0	−2.50

θ	240°	270°	300°	330°	360°
$r = 5 \sin \theta$	−4.33	−5.00	−4.33	−2.50	0

The graph is plotted as shown in Figure 36.1.
Initially the zero line OA is constructed and then the broken lines in Figure 32.1 at 30° intervals are produced. The maximum value of r is 5.00 hence OA is scaled and circles drawn as shown with the largest at a radius of 5 units. The polar coordinates (0, 0°), (2.50, 30°), (4.33, 60°), (5.00, 90°)... are plotted and shown as points O, B, C, D, ... in Figure 32.1. When polar coordinate (0, 180°) is plotted and the points joined with a smooth curve a complete circle is seen to have been produced. When plotting the next point, (−2.50, 210°), since r is negative it is plotted in the opposite direction to 210°, i.e. 2.50 units long on the 30° axis. Hence the point (−2.50, 210°) is equivalent to the point (2.50, 30°).
Similarly, (−4.33, 240°) is the same point as (4.33, 60°).
When all the coordinates are plotted the graph $r = 5 \sin \theta$ appears as a single circle; it is, in fact, two circles, one on top of the other.

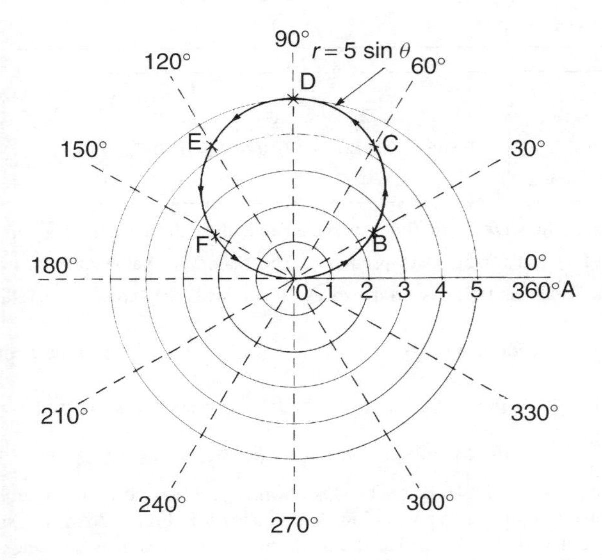

Figure 36.1

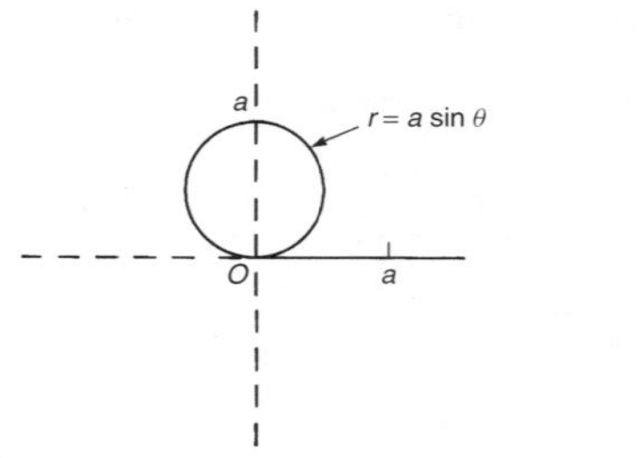

Figure 36.2

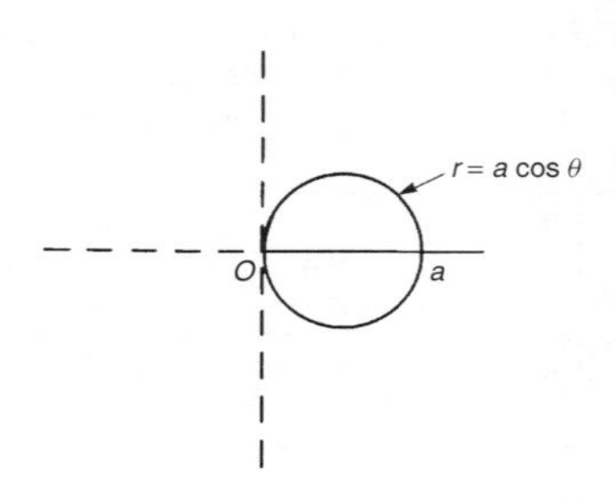

Figure 36.3

In general, a polar curve $\boldsymbol{r = a \sin\theta}$ is as shown in Figure 36.2.
In a similar manner to that explained above, it may be shown that the polar curve $\boldsymbol{r = a \cos\theta}$ is as sketched in Figure 36.3.

In another example, to plot the polar graph of $r = 4\sin^2\theta$ between $\theta = 0$ and $\theta = 2\pi$ radians using intervals of $\frac{\pi}{6}$:
A table of values is produced as shown below.

θ	0	$\frac{\pi}{6}$	$\frac{\pi}{3}$	$\frac{\pi}{2}$	$\frac{2\pi}{3}$	$\frac{5\pi}{6}$	π	$\frac{7\pi}{6}$
$\sin\theta$	0	0.50	0.866	1.00	0.866	0.50	0	−0.50
$r = 4\sin^2\theta$	0	1	3	4	3	1	0	1

θ	$\frac{4\pi}{3}$	$\frac{3\pi}{2}$	$\frac{5\pi}{3}$	$\frac{11\pi}{6}$	2π
$\sin\theta$	−0.866	−1.00	−0.866	−0.50	0
$r = 4\sin^2\theta$	3	4	3	1	0

The zero line *OA* is firstly constructed and then the broken lines at intervals of $\frac{\pi}{6}$ rad (or 30°) are produced. The maximum value of r is 4 hence *OA* is scaled and circles produced as shown with the largest at a radius of 4 units.
The polar coordinates (0, 0), $\left(1, \frac{\pi}{6}\right)$, $\left(3, \frac{\pi}{3}\right)$, ... (0, π) are plotted and shown as points O, B, C, D, E, F, O, respectively. Then $\left(1, \frac{7\pi}{6}\right)$, $\left(3, \frac{4\pi}{3}\right)$, ... (0, 0) are plotted as shown by points G, H, I, J, K, O respectively. Thus two distinct loops are produced as shown in Figure 36.4.
In general, a polar curve $r = a\sin^2\theta$ is as shown in Figure 36.5. In a similar manner it may be shown that the polar curve $r = a\cos^2\theta$ is as sketched in Figure 36.6.

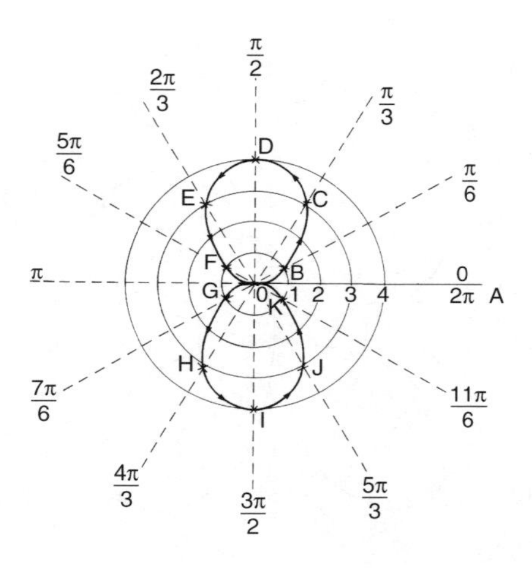

Figure 36.4

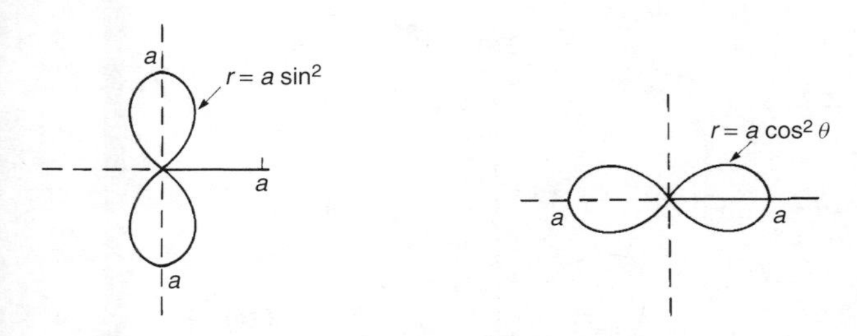

Figure 36.5 **Figure 36.6**

In another example, to plot the polar graph of $r = 3\sin 2\theta$ between $\theta = 0°$ and $\theta = 360°$, using 15° intervals:
A table of values is produced as shown below.

θ	0	15°	30°	45°	60°	75°	90°	105°	120°	135°
$r = 3\sin 2\theta$	0	1.5	2.6	3.0	2.6	1.5	0	−1.5	−2.6	−3.0

θ	150°	165°	180°	195°	210°	225°	240°
$r = 3\sin 2\theta$	−2.6	−1.5	0	1.5	2.6	3.0	2.6

θ	255°	270°	285°	300°	315°	330°	345°	360°
$r = 3\sin 2\theta$	1.5	0	−1.5	−2.6	−3.0	−2.6	−1.5	0

The polar graph $r = 3\sin 2\theta$ is plotted as shown in Figure 36.7 and is seen to contain four similar shaped loops displaced at 90° from each other.

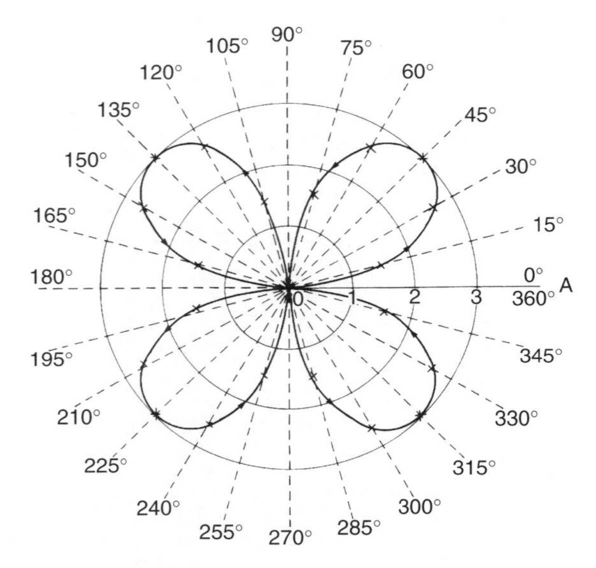

Figure 36.7

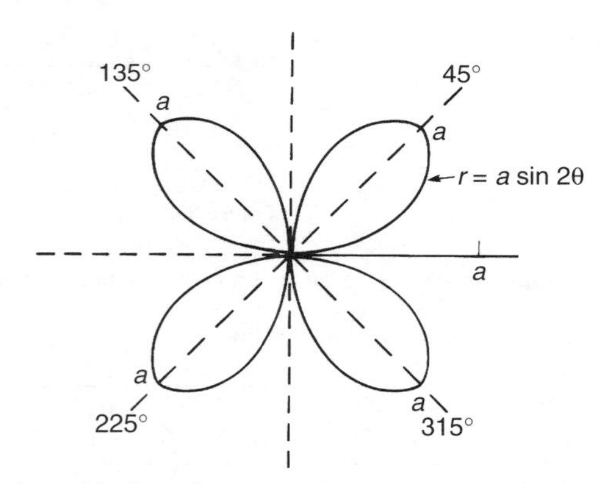

Figure 36.8

In general, a polar curve $r = a \sin 2\theta$ is as shown in Figure 36.8.
In a similar manner it may be shown that polar curves of $r = a \cos 2\theta$, $r = a \sin 3\theta$ and $r = a \cos 3\theta$ are as sketched in Figure 36.9.

In another example, to sketch the polar curve $r = 2\theta$ between $\theta = 0$ and $\theta = \dfrac{5\pi}{2}$ rad at intervals of $\dfrac{\pi}{6}$:

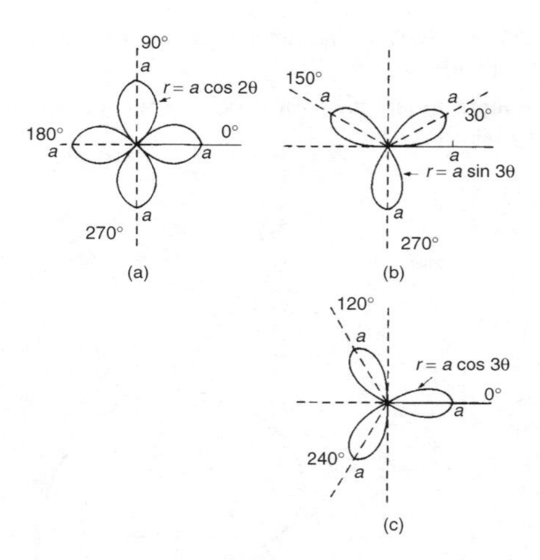

Figure 36.9

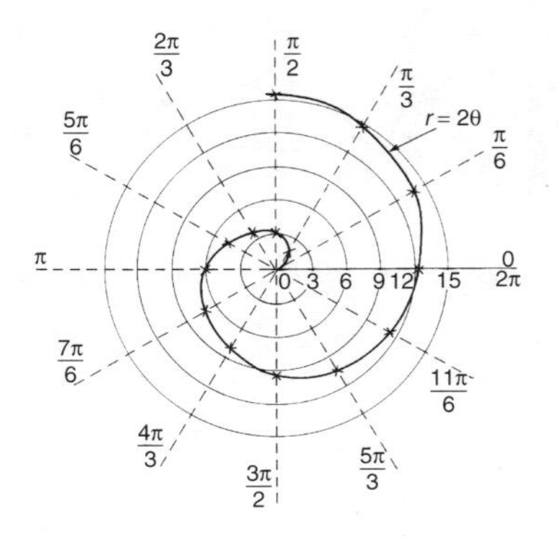

Figure 36.10

A table of values is produced as shown below.

θ	0	$\frac{\pi}{6}$	$\frac{\pi}{3}$	$\frac{\pi}{2}$	$\frac{2\pi}{3}$	$\frac{5\pi}{6}$	π	$\frac{7\pi}{6}$	$\frac{4\pi}{3}$
$r = 2\theta$	0	1.05	2.09	3.14	4.19	5.24	6.28	7.33	8.38

θ	$\frac{3\pi}{2}$	$\frac{5\pi}{3}$	$\frac{11\pi}{6}$	2π	$\frac{13\pi}{6}$	$\frac{7\pi}{3}$	$\frac{5\pi}{2}$
$r = 2\theta$	9.42	10.47	11.52	12.57	13.61	14.66	15.71

The polar graph of $r = 2\theta$ is shown in Figure 36.10 and is seen to be an ever-increasing spiral.

In another example, to plot the polar curve $r = 5(1 + \cos\theta)$ from $\theta = 0°$ to $\theta = 360°$, using 30° intervals:

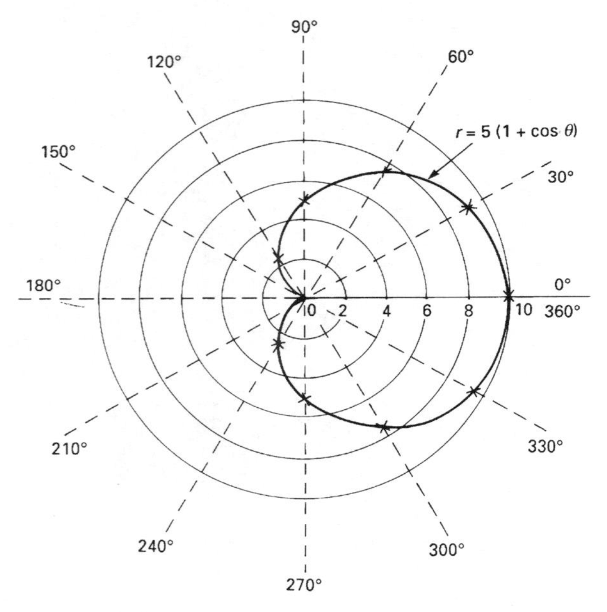

Figure 36.11

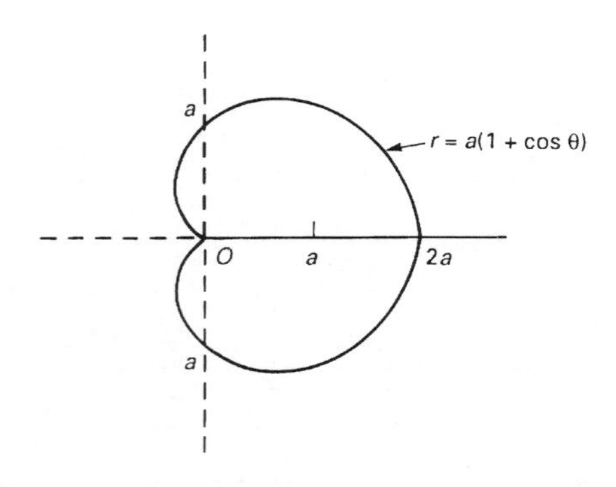

Figure 36.12

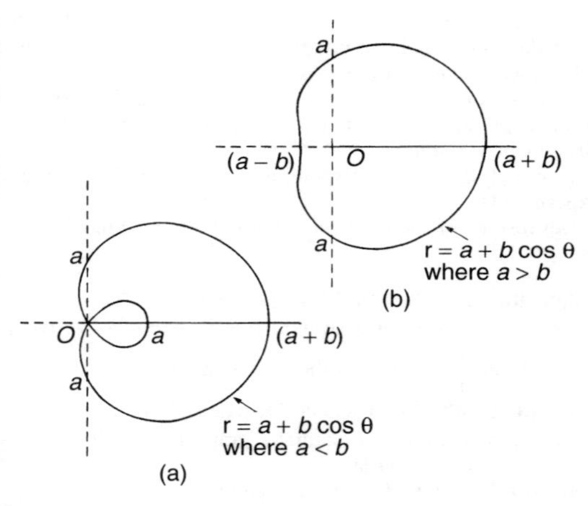

Figure 36.13

A table of values is shown below.

θ	0	30°	60°	90°	120°	150°
$r = 5(1 + \cos\theta)$	10.0	9.33	7.50	5.00	2.50	0.67

θ	180°	210°	240°	270°	300°	330°	360°
$R = 5(1 + \cos\theta)$	0	0.67	2.50	5.00	7.50	9.33	10.00

The polar curve $r = 5(1 + \cos\theta)$ is shown in Figure 36.11.
In general, a polar curve $r = a(1 + \cos\theta)$ is as shown in Figure 36.12 and the shape is called a **cardioid**.
In a similar manner it may be shown that the polar curve $r = a + b\cos\theta$ varies in shape according to the relative values of a and b. When $a = b$ the polar curve shown in Figure 36.12 results.
When $a < b$ the general shape shown in Figure 36.13(a) results and when $a > b$ the general shape shown in Figure 36.13(b) results.

37 Functions and their Curves

Standard curves

When a mathematical equation is known, co-ordinates may be calculated for a limited range of values, and the equation may be represented pictorially as a graph, within this range of calculated values. Sometimes it is useful to show all the characteristic features of an equation, and in this case a sketch depicting the equation can be drawn, in which all the important features are

shown, but the accurate plotting of points is less important. This technique is called 'curve sketching' and can involve the use of differential calculus, with, for example, calculations involving turning points.
If, say, y depends on, say, x, then y is said to be a function of x and the relationship is expressed as $y = f(x)$; x is called the independent variable and y is the dependent variable.
In engineering and science, corresponding values are obtained as a result of tests or experiments.
Here is a brief resumé of standard curves, some of which have been met earlier in this text.

(i) **Straight line** (see Chapter 32, page 155.)
The general equation of a straight line is $y = mx + c$, where m is the gradient $\left(\text{i.e. } \frac{dy}{dx}\right)$ and c is the y-axis intercept.

(ii) **Quadratic graphs** (see Chapter 35, page 171.)
The general equation of a quadratic graph is $y = ax^2 + bx + c$, and its shape is that of a parabola.

(iii) **Cubic equations** (see Chapter 35, page 177.)
The general equation of a cubic graph is
$y = ax^3 + bx^2 + cx + d$.

(iv) **Trigonometric functions** (see Chapter 28.)
Graphs of $y = \sin\theta$, $y = \cos\theta$ and $y = \tan\theta$ are shown in Figure 28.1, page 130.

(v) **Circle**
The simplest equation of a circle is $x^2 + y^2 = r^2$, with centre at the origin and radius r, as shown in Figure 21.5, page 95.
More generally, the equation of a circle, centre (a, b), radius r, is given by: $(x - a)^2 + (y - b)^2 = r^2$

(vi) **Ellipse**
The equation of an ellipse is $\frac{x^2}{a^2} + \frac{y^2}{b^2} = 1$ and the general shape is as shown in Figure 37.1.
The length AB is called the **major axis** and CD the **minor axis**. In the above equation, 'a' is the semi-major axis and 'b' is the semi-minor axis.
(Note that if $b = a$, the equation becomes $\frac{x^2}{a^2} + \frac{y^2}{a^2} = 1$,
i.e. $x^2 + y^2 = a^2$, which is a circle of radius a).

(vii) **Hyperbola**
The equation of a hyperbola is $\frac{x^2}{a^2} - \frac{y^2}{b^2} = 1$ and the general shape is shown in Figure 37.2. The curve is seen to be symmetrical about both the x- and y-axes.
The distance AB in Figure 37.2 is given by 2a.

(viii) **Rectangular hyperbola**
The equation of a rectangular hyperbola is $xy = c$ or $y = \frac{c}{x}$ and the general shape is shown in Figure 37.3.

(ix) **Logarithmic function**
$y = \ln x$ and $y = \lg x$ are both of the general shape shown in Figures 11.1 and 11.2, page 49.

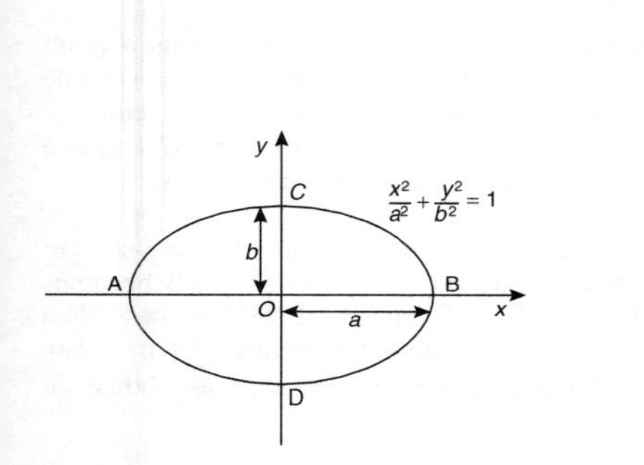

Figure 37.1

Figure 37.2

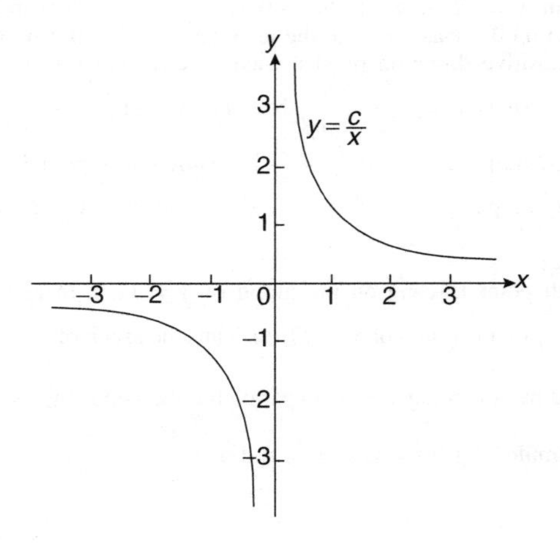

Figure 37.3

(x) Exponential functions
$y = e^x$ is of the general shape shown in Figure 12.1, page 52.

(xi) Polar curves
The equation of a polar curve is of the form $r = f(\theta)$ and examples of polar curves may be found on pages 178 to 185.

Simple transformations

From the graph of $y = f(x)$ it is possible to deduce the graphs of other functions which are transformations of $y = f(x)$. For example, knowing the graph of $y = f(x)$, can help us draw the graphs of $y = af(x)$, $y = f(x) + a$, $y = f(x + a)$, $y = f(ax)$, $y = -f(x)$ and $y = f(-x)$.

(i) $y = af(x)$

For each point (x_1, y_1) on the graph of $y = f(x)$ there exists a point (x_1, ay_1) on the graph of $y = af(x)$. Thus the graph of $y = af(x)$ can be obtained by stretching $y = f(x)$ parallel to the y-axis by a scale factor 'a'. Graphs of $y = x + 1$ and $y = 3(x + 1)$ are shown in Figure 37.4(a) and graphs of $y = \sin\theta$ and $y = 2\sin\theta$ are shown in Figure 37.4(b).

(ii) $y = f(x) + a$

The graph of $y = f(x)$ is translated by 'a' units parallel to the y-axis to obtain $y = f(x) + a$. For example, if $f(x) = x$, $y = f(x) + 3$ becomes $y = x + 3$, as shown in Figure 37.5(a). Similarly, if $f(\theta) = \cos\theta$, then $y = f(\theta) + 2$ becomes $y = \cos\theta + 2$, as shown in Figure 37.5(b). Also, if $f(x) = x^2$, then $y = f(x) + 3$ becomes $y = x^2 + 3$, as shown in Figure 37.5(c).

(iii) $y = f(x + a)$

The graph of $y = f(x)$ is translated by 'a' units parallel to the x-axis to obtain $y = f(x + a)$. If 'a' > 0 it moves $y = f(x)$ in the negative direction on the x-axis (i.e. to the left), and if 'a' < 0 it moves $y = f(x)$ in the positive direction on the x-axis (i.e. to the right). For example, if $f(x) = \sin x$, $y = f\left(x - \frac{\pi}{3}\right)$ becomes $y = \sin\left(x - \frac{\pi}{3}\right)$ as shown in Figure 37.6(a) and $y = \sin\left(x + \frac{\pi}{4}\right)$ is shown in Figure 37.6(b).

Similarly graphs of $y = x^2$, $y = (x - 1)^2$ and $y = (x + 2)^2$ are shown in Figure 37.7.

(iv) $y = f(ax)$

For each point (x_1, y_1) on the graph of $y = f(x)$, there exists a point $\left(\frac{x_1}{a}, y_1\right)$ on the graph of $y = f(ax)$. Thus the graph of $y = f(ax)$ can be obtained by stretching $y = f(x)$ parallel to the x-axis by a scale factor $\frac{1}{a}$.

For example, if $f(x) = (x - 1)^2$, and $a = \frac{1}{2}$,

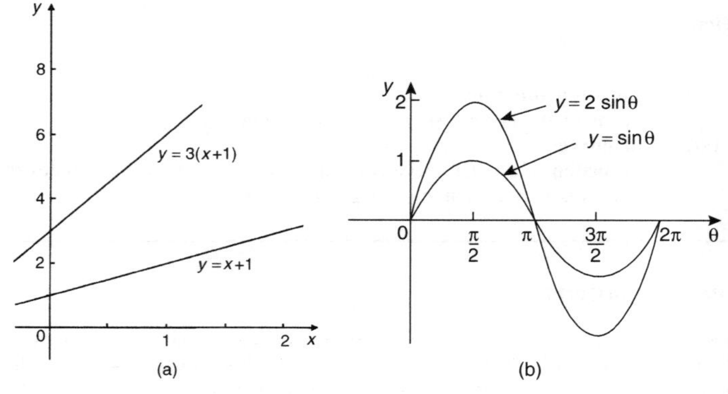

Figure 37.4

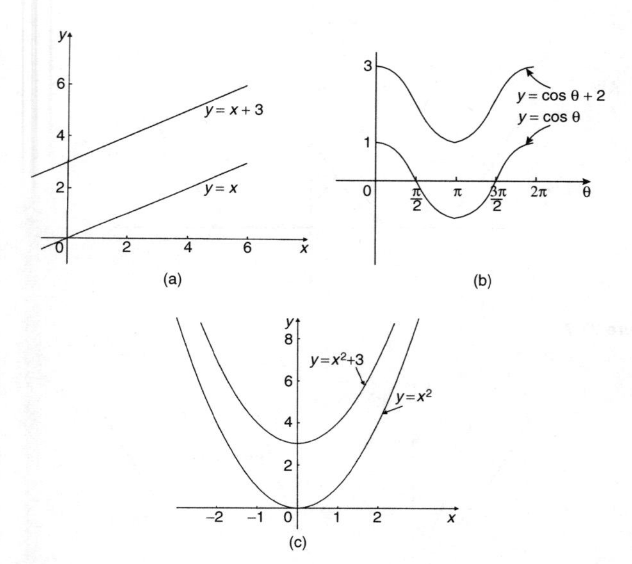

Figure 37.5

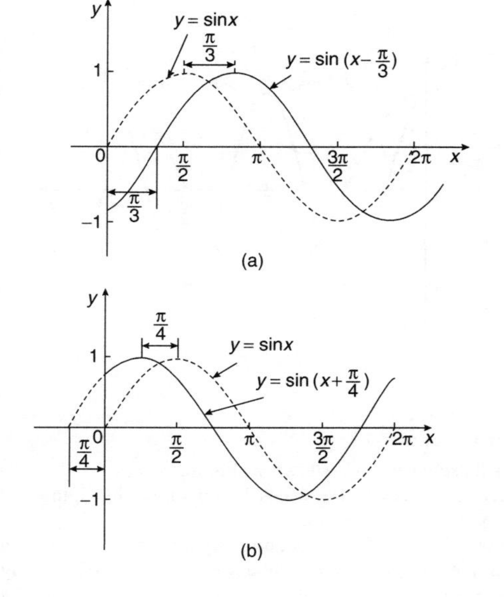

Figure 37.6

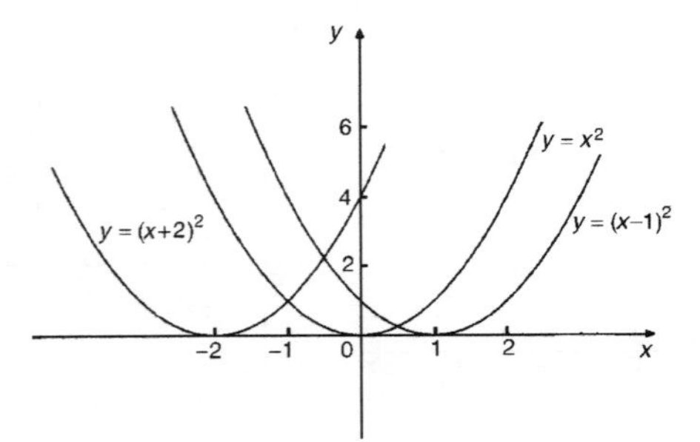

Figure 37.7

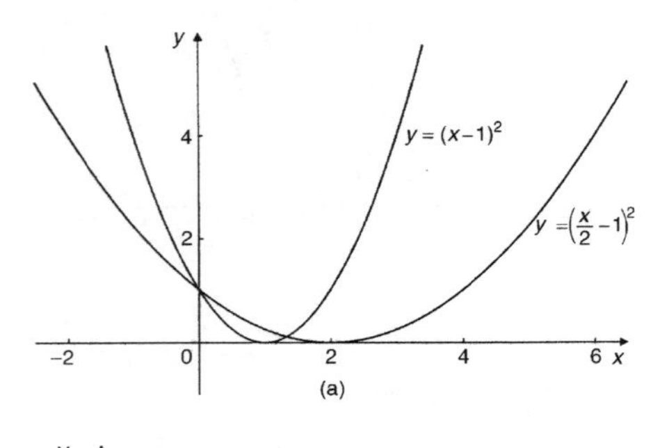

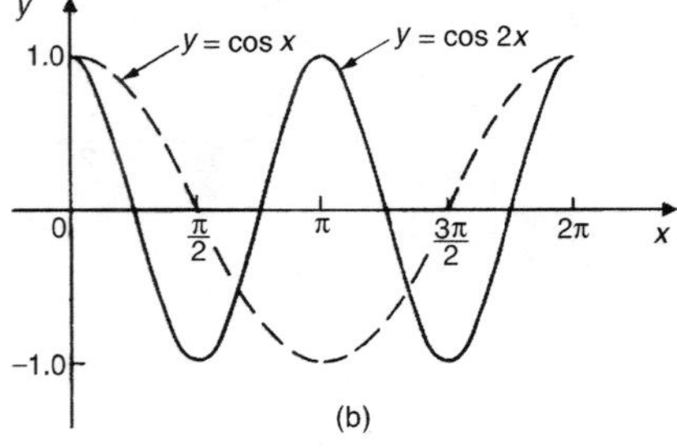

Figure 37.8

then $f(ax) = \left(\frac{x}{2} - 1\right)^2$.

Both of these curves are shown in Figure 37.8(a).

Similarly, $y = \cos x$ and $y = \cos 2x$ are shown in Figure 37.8(b).

(v) $\boldsymbol{y = -f(x)}$

The graph of $y = -f(x)$ is obtained by reflecting $y = f(x)$ in the x-axis. For example, graphs of $y = e^x$ and $y = -e^x$ are shown in Figure 37.9(a), and graphs of $y = x^2 + 2$ and $y = -(x^2 + 2)$ are shown in Figure 37.9(b).

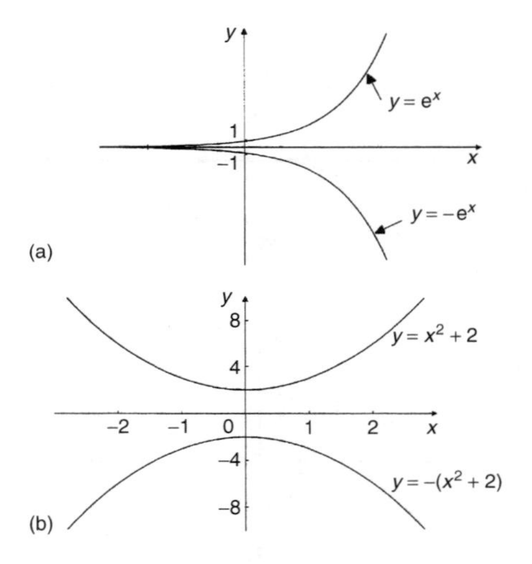

Figure 37.9

(vi) $\boldsymbol{y = f(-x)}$
The graph of $y = f(-x)$ is obtained by reflecting $y = f(x)$ in the y-axis. For example, graphs of $y = x^3$ and $y = (-x)^3 = -x^3$ are shown in Figure 37.10(a) and graphs of $y = \ln x$ and $y = -\ln x$ are shown in Figure 37.10(b).

Periodic functions

A function $f(x)$ is said to be **periodic** if $f(x + T) = f(x)$ for all values of x, where T is some positive number. T is the interval between two successive repetitions and is called the **period** of the function $f(x)$. For example, $y = \sin x$ is periodic in x with period 2π since $\sin x = \sin(x + 2\pi) = \sin(x + 4\pi)$, and so on. Similarly, $y = \cos x$ is a periodic function with period 2π since $\cos x = \cos(x + 2\pi) = \cos(x + 4\pi)$, and so on. In general,if $y = \sin \omega t$ or $y = \cos \omega t$ then the period of the waveform is $2\pi/\omega$. The function shown in Figure 37.11 is also periodic of period 2π and is defined by:

$$f(x) = \begin{cases} -1, & \text{when} \quad -\pi \le x \le 0 \\ 1. & \text{when} \quad 0 \le x \le \pi \end{cases}$$

Continuous and discontinuous functions

If a graph of a function has no sudden jumps or breaks it is called a **continuous function**, examples being the graphs of sine and cosine functions. However, other graphs make finite jumps at a point or points in the interval. The square wave shown in Figure 37.11 has **finite discontinuities** as $x = \pi$, 2π, 3π, and

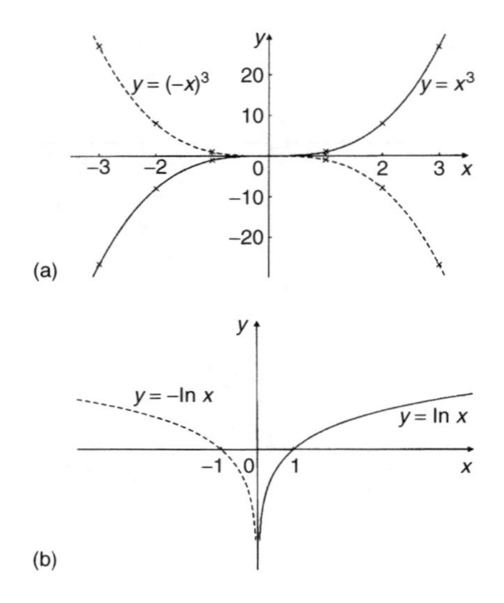

Figure 37.10

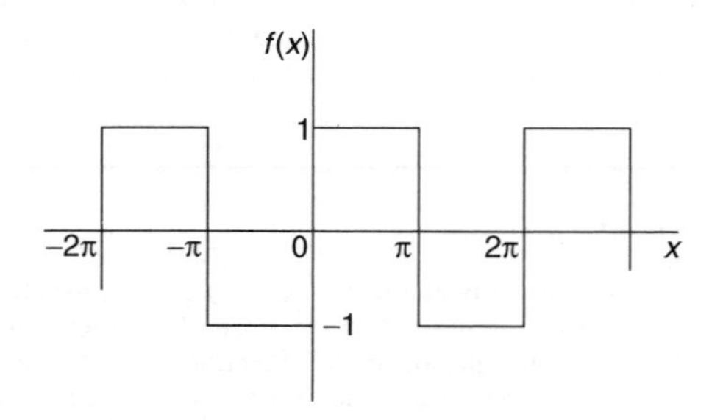

Figure 37.11

so on, and is therefore a **discontinuous function**. $y = \tan x$ is another example of a discontinuous function.

Even and odd functions

Even functions

A function $y = f(x)$ is said to be even if $f(-x) = f(x)$ for all values of x. Graphs of even functions are always **symmetrical about the y-axis** (i.e. is a mirror image). Two examples of even functions are $y = x^2$ and $y = \cos x$ as shown in Figure 33.12.

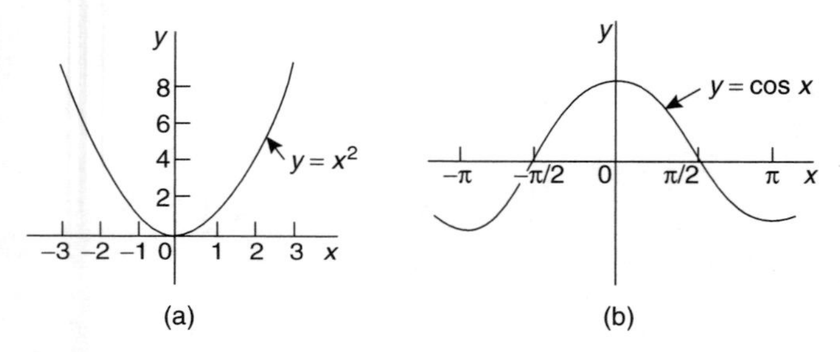

Figure 37.12

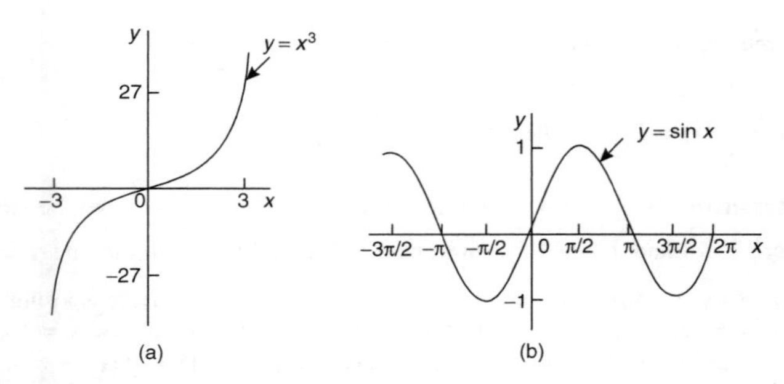

Figure 37.13

Odd functions

A function $y = f(x)$ is said to be odd if $f(-x) = -f(x)$ for all values of x. Graphs of odd functions are always **symmetrical about the origin**. Two examples of odd functions are $y = x^3$ and $y = \sin x$ as shown in Figure 37.13. Many functions are neither even nor odd, two such examples being $y = e^x$ and $y = \ln x$.

Inverse functions

If y is a function of x, the graph of y against x can be used to find x when any value of y is given. Thus the graph also expresses that x is a function of y. Two such functions are called **inverse functions**.

In general, given a function $y = f(x)$, its inverse may be obtained by interchanging the roles of x and y and then transposing for y. The inverse function is denoted by $y = f^{-1}(x)$.

For example, if $y = 2x + 1$, the inverse is obtained by

(i) transposing for x, i.e. $x = \dfrac{y-1}{2} = \dfrac{y}{2} - \dfrac{1}{2}$ and

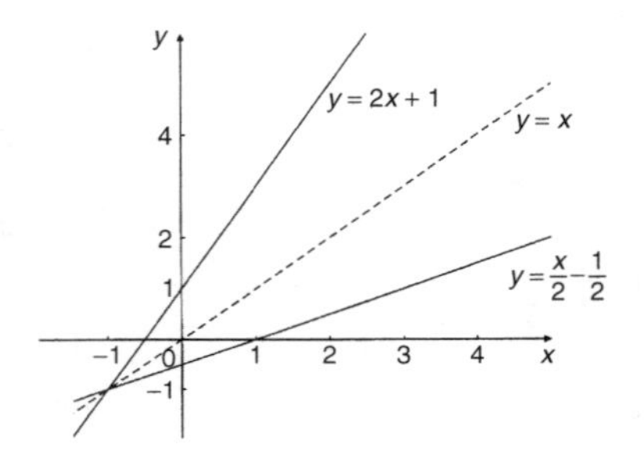

Figure 37.14

(ii) interchanging x and y, giving the inverse as $y = \frac{x}{2} - \frac{1}{2}$

Thus if $f(x) = 2x + 1$, then $f^{-1}(x) = \frac{x}{2} - \frac{1}{2}$

A graph of $f(x) = 2x + 1$ and its inverse $f^{-1}(x) = \frac{x}{2} - \frac{1}{2}$ is shown in Figure 37.14 and $f^{-1}(x)$ is seen to be a reflection of $f(x)$ in the line $y = x$.

In another example, if $y = x^2$, the inverse is obtained by (i) transposing for x, i.e. $x = \pm\sqrt{y}$ and (ii) interchanging x and y, giving the inverse $y = \pm\sqrt{x}$ Hence the inverse has two values for every value of x. Thus $f(x) = x^2$ does not have a single inverse. In such a case the domain of the original function may be restricted to $y = x^2$ for $x > 0$. Thus the inverse is then $y = +\sqrt{x}$.
A graph of $f(x) = x^2$ and its inverse $f^{-1}(x) = \sqrt{x}$ for $x > 0$ is shown in Figure 37.15 and, again, $f^{-1}(x)$ is seen to be a reflection of $f(x)$ in the line $y = x$.
It is noted from the latter example, that not all functions have a single inverse. An inverse, however, can be determined if the range is restricted.

Inverse trigonometric functions

If $y = \sin x$, then x is the angle whose sine is y. Inverse trigonometrical functions are denoted either by prefixing the function with 'arc' or by using^{-1}. Hence transposing $y = \sin x$ for x gives $x = \arcsin y$ or $\sin^{-1} y$. Interchanging x and y gives the inverse $y = \arcsin x$ or $\sin^{-1} x$.
Similarly, $y = \arccos x$, $y = \arctan x$, $y = \text{arcsec}\, x$, $y = \text{arccosec}\, x$ and $y = \text{arccot}\, x$ are all inverse trigonometric functions. The angle is always expressed in radians.
Inverse trigonometric functions are periodic so it is necessary to specify the smallest or principal value of the angle. For $\arcsin x$, $\arctan x$, $\text{arccosec}\, x$ and $\text{arccot}\, x$, the principal value is in the range $-\frac{\pi}{2} < y < \frac{\pi}{2}$.

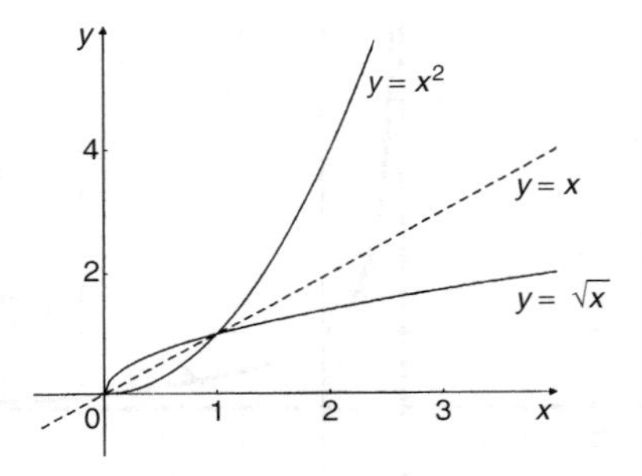

Figure 37.15

For arccos x and arcsec x the principal value is in the range $0 < y < \pi$. Graphs of the six inverse trigonometric functions are shown in Figure 53.1, page 290.

For example, to determine the principal values of

(a) $\arcsin 0.5$ (b) $\arctan(-1)$ (c) $\arccos\left(-\frac{\sqrt{3}}{2}\right)$

Using a calculator,

(a) $\arcsin 0.5 \equiv \sin^{-1} 0.5 = 30° = \frac{\pi}{6}$ **rad** or **0.5236 rad**

(b) $\arctan(-1) \equiv \tan^{-1}(-1) = -45° = -\frac{\pi}{4}$ **rad** or **−0.7854 rad**

(c) $\arccos\left(-\frac{\sqrt{3}}{2}\right) \equiv \cos^{-1}\left(-\frac{\sqrt{3}}{2}\right) = 150° = \frac{5\pi}{6}$ **rad** or **2.6180 rad**

Asymptotes

If a table of values for the function $y = \frac{x+2}{x+1}$ is drawn up for various values of x and then y plotted against x, the graph would be as shown in Figure 37.16. The straight lines AB, i.e. $x = -1$, and CD, i.e. $y = 1$, are known as **asymptotes**. An asymptote to a curve is defined as a straight line to which the curve approaches as the distance from the origin increases. Alternatively, an asymptote can be considered as a tangent to the curve at infinity.

Asymptotes parallel to the x- and y-axes

There is a simple rule that enables asymptotes parallel to the x- and y-axes to be determined. For a curve $y = f(x)$:

(i) the asymptotes parallel to the x-axis are found by equating the coefficient of the highest power of x to zero

(ii) the asymptotes parallel to the y-axis are found by equating the coefficient of the highest power of y to zero

With the above example $y = \frac{x+2}{x+1}$, rearranging gives:

$$y(x+1) = x+2$$

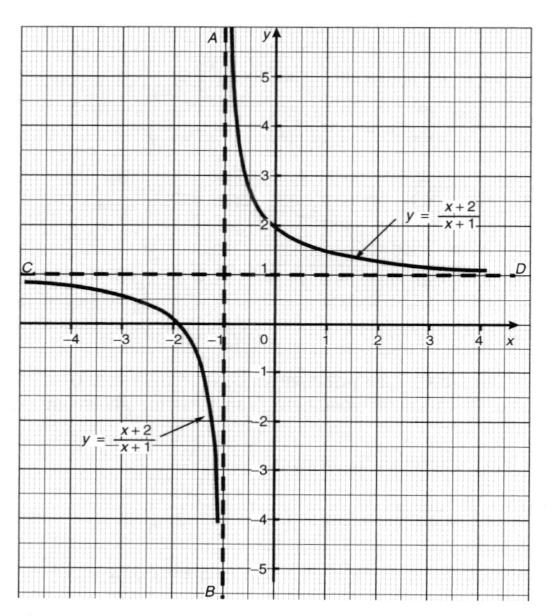

Figure 37.16

i.e. $\quad yx + y - x - 2 = 0 \quad (1)$

and $\quad x(y - 1) + y - 2 = 0$

The coefficient of the highest power of x (in this case x^1) is $(y - 1)$. Equating to zero gives: $y - 1 = 0$ from which, $\mathbf{y = 1}$, which is an asymptote of $y = \frac{x+2}{x+1}$ as shown in Figure 37.16.

Returning to equation (1): $\quad yx + y - x - 2 = 0$

from which, $\quad y(x + 1) - x - 2 = 0$

The coefficient of the highest power of y (in this case y^1 is) $(x + 1)$.

Equating to zero gives: $\quad x + 1 = 0$

from which, $\mathbf{x = -1}$, which is another asymptote of $y = \frac{x+2}{x+1}$ as shown in Figure 37.16.

Other asymptotes

To determine asymptotes other than those parallel to x- and y-axes a simple procedure is:

(i) substitute $y = mx + c$ in the given equation
(ii) simplify the expression
(iii) equate the coefficients of the two highest powers of x to zero and determine the values of m and c. $y = mx + c$ gives the asymptote.

For example, to determine the asymptotes for the function:

$$y(x + 1) = (x - 3)(x + 2)$$

Following the above procedure:

(i) Substituting $y = mx + c$ into $y(x + 1) = (x - 3)(x + 2)$

gives $(mx + c)(x + 1) = (x - 3)(x + 2)$

(ii) Simplifying gives $mx^2 + mx + cx + c = x^2 - x - 6$

and $(m - 1)x^2 + (m + c + 1)x + c + 6 = 0$

(iii) Equating the coefficient of the highest power of x to zero

gives $m - 1 = 0$ from which, $\boldsymbol{m = 1}$

Equating the coefficient of the next highest power of x to zero

gives $m + c + 1 = 0$

and since $m = 1$, $1 + c + 1 = 0$ from which, $\boldsymbol{c = -2}$

Hence $y = mx + c = 1x - 2$

i.e. $\boldsymbol{y = x - 2}$ **is an asymptote**

To determine any asymptotes parallel to the x-axis:

Rearranging $y(x + 1) = (x - 3)(x + 2)$

gives $yx + y = x^2 - x - 6$

The coefficient of the highest power of x (i.e. x^2) is 1. Equating this to zero gives $1 = 0$, which is not an equation of a line. Hence there is no asymptote parallel to the x-axis.
To determine any asymptotes parallel to the y-axis:
Since $y(x + 1) = (x - 3)(x + 2)$ the coefficient of the highest power of y is $x + 1$. Equating this to zero gives $x + 1 = 0$, from which, $x = -1$. Hence $\boldsymbol{x = -1}$ is an asymptote

When $\boldsymbol{x = 0}$, $y(1) = (-3)(2)$, i.e. $\boldsymbol{y = -6}$

When $\boldsymbol{y = 0}$, $0 = (x - 3)(x + 2)$, i.e. $\boldsymbol{x = 3}$ and $\boldsymbol{x = -2}$

A sketch of the function $y(x + 1) = (x - 3)(x + 2)$ is shown in Figure 37.17.

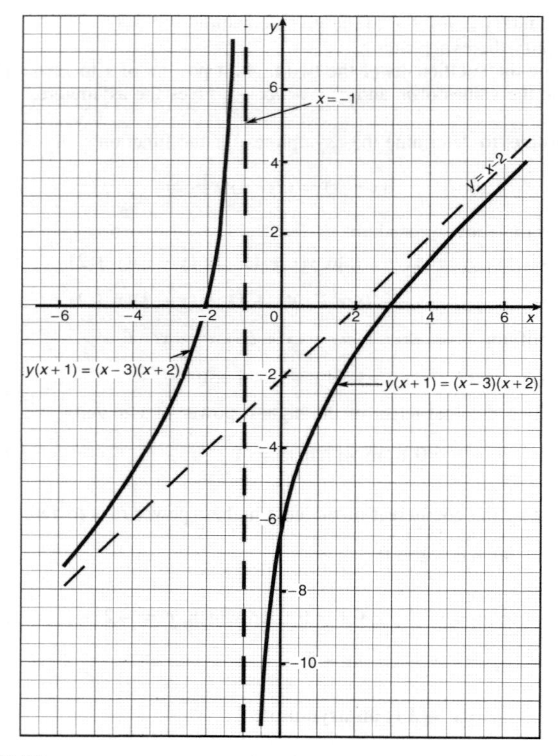

Figure 37.17

Brief guide to curve sketching

The following steps will give information from which the graphs of many types of functions $y = f(x)$ can be sketched.

(i) Use calculus to determine the location and nature of maximum and minimum points (see chapter 49)

(ii) Determine where the curve cuts the x- and y-axes

(iii) Inspect the equation for symmetry.

  (a) If the equation is unchanged when $-x$ is substituted for x, the graph will be symmetrical about the y-axis (i.e. it is an **even function**).

  (b) If the equation is unchanged when $-y$ is substituted for y, the graph will be symmetrical about the x-axis.

  (c) If $f(-x) = -f(x)$, the graph is symmetrical about the origin (i.e. it is an **odd function**).

(iv) Check for any asymptotes.

Vectors

38 Vectors

Introduction

Some physical quantities are entirely defined by a numerical value and are called **scalar quantities** or **scalars**. Examples of scalars include time, mass, temperature, energy and volume. Other physical quantities are defined by both a numerical value **and** a direction in space and these are called **vector quantities** or **vectors**. Examples of vectors include force, velocity, moment and displacement.

Vector addition

A vector may be represented by a straight line, the length of line being directly proportional to the magnitude of the quantity and the direction of the line being in the same direction as the line of action of the quantity. An arrow is used to denote the sense of the vector, that is, for a horizontal vector, say, whether it acts from left to right or vice-versa. The arrow is positioned at the end of the vector and this position is called the 'nose' of the vector. Figure 38.1 shows a velocity of 20 m/s at an angle of 45° to the horizontal and may be depicted by $\boldsymbol{oa} = 20$ m/s at 45° to the horizontal.
To distinguish between vector and scalar quantities, various ways are used. These include:

(i) **bold print**,
(ii) two capital letters with an arrow above them to denote the sense of direction, e.g. $\overrightarrow{\text{AB}}$, where A is the starting point and B the end point of the vector,
(iii) a line over the top of letters, e.g. $\overline{\text{AB}}$ or $\overline{a}$
(iv) letters with an arrow above, e.g. $\vec{a}$, $\vec{\text{A}}$
(v) underlined letters, e.g. $\underline{a}$
(vi) $xi + jy$, where i and j are axes at right-angles to each other; for example, $3i + 4j$ means 3 units in the i direction and 4 units in the j direction, as shown in Figure 38.2.
(vii) a column matrix $\begin{pmatrix} a \\ b \end{pmatrix}$; for example, the vector **OA** shown in Figure 38.2 could be represented by $\begin{pmatrix} 3 \\ 4 \end{pmatrix}$

Thus, in Figure 38.2, $\mathbf{OA} \equiv \overrightarrow{\text{OA}} \equiv \overline{\text{OA}} \equiv 3i + 4j \equiv \begin{pmatrix} 3 \\ 4 \end{pmatrix}$

The one adopted in this text is to denote vector quantities in **bold print**.
Thus, $\boldsymbol{oa}$ represents a vector quantity, but oa is the magnitude of the vector $\boldsymbol{oa}$. Also, positive angles are measured in an anticlockwise direction from a horizontal, right facing line and negative angles in a clockwise direction from this line — as with graphical work. Thus 90° is a line vertically upwards and −90° is a line vertically downwards.

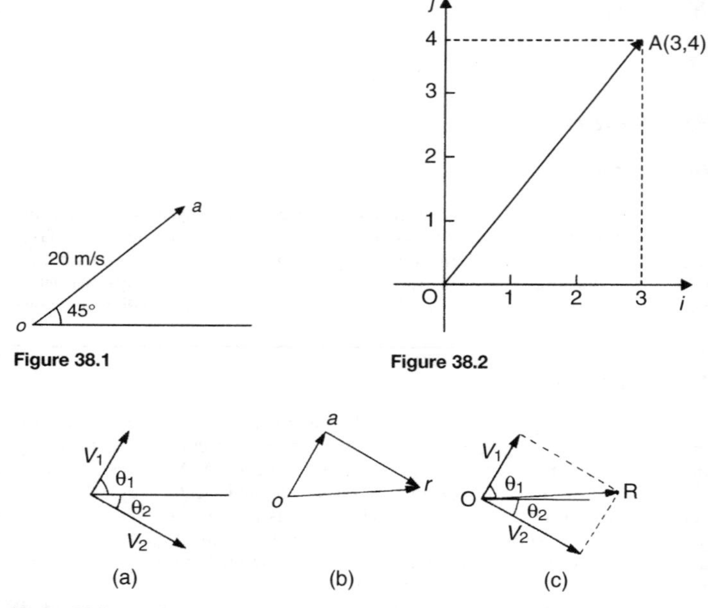

Figure 38.1

Figure 38.2

Figure 38.3

The resultant of adding two vectors together, say $\boldsymbol{V_1}$ at an angle θ_1 and $\boldsymbol{V_2}$ at angle $(-\theta_2)$, as shown in Figure 38.3(a), can be obtained by drawing ***oa*** to represent $\boldsymbol{V_1}$ and then drawing ***ar*** to represent $\boldsymbol{V_2}$. The resultant of $\boldsymbol{V_1} + \boldsymbol{V_2}$ is given by ***or***. This is shown in Figure 38.3(b), the vector equation being $\boldsymbol{oa} + \boldsymbol{ar} = \boldsymbol{or}$. This is called the **'nose-to-tail' method** of vector addition.
Alternatively, by drawing lines parallel to $\boldsymbol{V_1}$ and $\boldsymbol{V_2}$ from the noses of $\boldsymbol{V_2}$ and $\boldsymbol{V_1}$, respectively, and letting the point of intersection of these parallel lines be R, gives **OR** as the magnitude and direction of the resultant of adding $\boldsymbol{V_1}$ and $\boldsymbol{V_2}$, as shown in Figure 38.3(c). This is called the **'parallelogram' method** of vector addition.

For example, a force of 4 N is inclined at an angle of 45° to a second force of 7 N, both forces acting at a point. To find the magnitude of the resultant of these two forces and the direction of the resultant with respect to the 7 N force by both the 'triangle' and the 'parallelogram' methods:
The forces are shown in Figure 38.4(a). Although the 7 N force is shown as a horizontal line, it could have been drawn in any direction.
Using the **'nose-to-tail' method**, a line 7 units long is drawn horizontally to give vector ***oa*** in Figure 38.4(b). To the nose of this vector ***ar*** is drawn 4 units long at an angle of 45° to ***oa***. The resultant of vector addition is ***or*** and by measurement is **10.2 units long and at an angle of 16° to the 7 N force**.
Figure 38.4(c) uses the **'parallelogram' method** in which lines are drawn parallel to the 7 N and 4 N forces from the noses of the 4 N and 7 N forces,

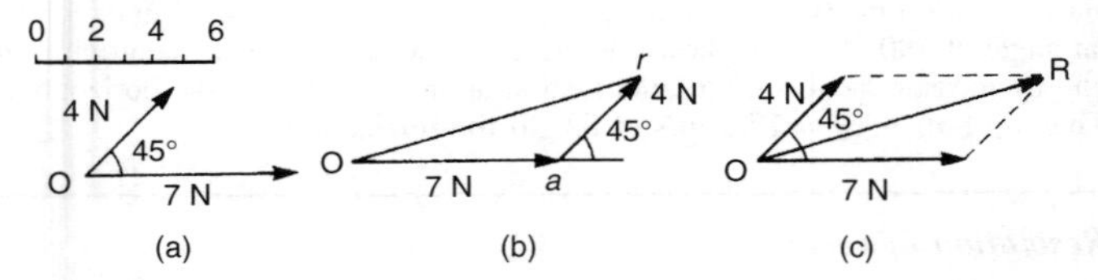

Figure 38.4

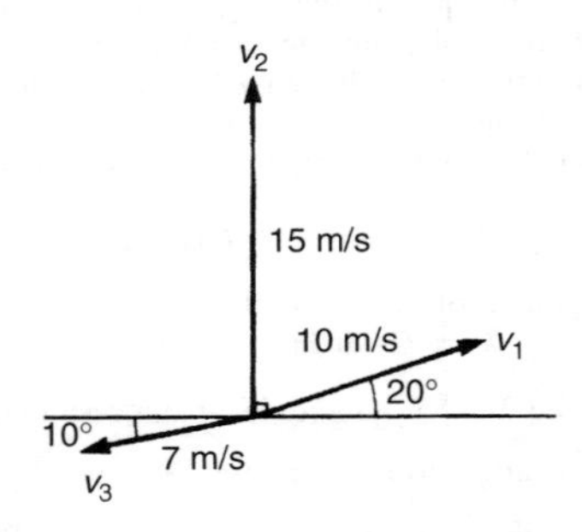

Figure 38.5

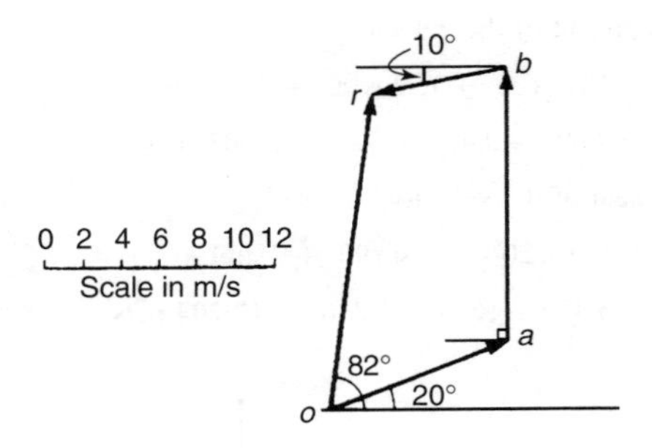

Figure 38.6

respectively. These intersect at R. Vector **OR** give the magnitude and direction of the resultant of vector addition and, as obtained by the 'nose-to-tail' method, is **10.2 units long at an angle of 16° to the 7 N force**.

In another example, to use a graphical method to determine the magnitude and direction of the resultant of the three velocities shown in Figure 38.5:
It is easier to use the 'nose-to-tail' method when more than two vectors are being added. The order in which the vectors are added is immaterial. In this case the order taken is v_1, then v_2, then v_3 but just the same result would have been obtained if the order had been, say, v_1, v_3 and finally v_2. v_1 is drawn 10 units long at an angle of 20° to the horizontal, shown by ***oa*** in Figure 38.6.

v_2 is added to v_1 by drawing a line 15 units long vertically upwards from a, shown as ***ab***. Finally, v_3 is added to $v_1 + v_2$ by drawing a line 7 units long at an angle at 190° from b, shown as ***br***. The resultant of vector addition is ***or*** and by measurement is 17.5 units long at an angle of 82° to the horizontal. Thus $\boldsymbol{v_1 + v_2 + v_3 =}$ **17.5 m/s at 82° to the horizontal**.

Resolution of vectors

A vector can be resolved into two component parts such that the vector addition of the component parts is equal to the original vector. The two components usually taken are a horizontal component and a vertical component. For the vector shown as $\boldsymbol{F}$ in Figure 38.7, the horizontal component is $F\cos\theta$ and the vertical component is $F\sin\theta$.
For the vectors $\boldsymbol{F_1}$ and $\boldsymbol{F_2}$ shown in Figure 38.8, the horizontal component of vector addition is:

$$H = F_1\cos\theta_1 + F_2\cos\theta_2$$

and the vertical component of vector addition is:

$$V = F_1\sin\theta_1 + F_2\sin\theta_2$$

Having obtained H and V, the magnitude of the resultant vector R is given by: $\boldsymbol{\sqrt{H^2 + V^2}}$ and its angle to the horizontal is given by $\mathbf{tan}^{-1}\dfrac{\boldsymbol{V}}{\boldsymbol{H}}$
For example, to calculate the resultant velocity of the three velocities shown in Figure 38.5:

Horizontal component of the velocity,

$$H = 10\cos 20^\circ + 15\cos 90^\circ + 7\cos 190^\circ$$
$$= 9.397 + 0 + (-6.894) = \mathbf{2.503\ m/s}$$

Vertical component of the velocity,

$$V = 10\sin 20^\circ + 15\sin 90^\circ + 7\sin 190^\circ$$
$$= 3.420 + 15 + (-1.216) = \mathbf{17.204\ m/s}$$

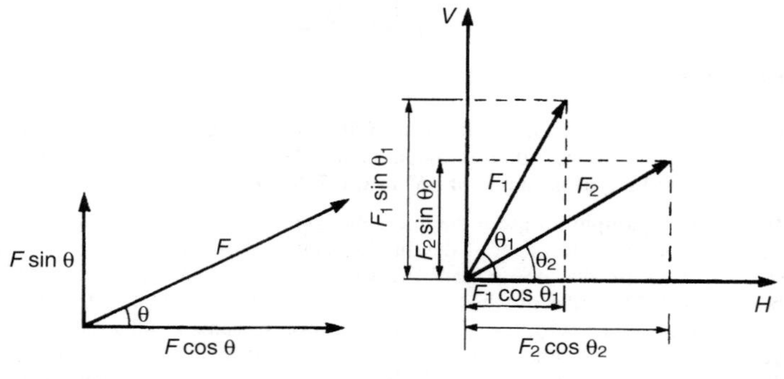

Figure 38.7

Figure 38.8

Magnitude of the resultant of vector addition

$$= \sqrt{H^2 + V^2} = \sqrt{2.503^2 + 17.204^2} = \sqrt{302.24} = \mathbf{17.39\ m/s}$$

Direction of the resultant of vector addition

$$= \tan^{-1}\left(\frac{V}{H}\right) = \tan^{-1}\left(\frac{17.204}{2.503}\right) = \tan^{-1} 6.8734 = 81.72^\circ$$

Thus, the resultant of the three velocities is a single vector of 17.39 m/s at 81.72° to the horizontal.

Vector subtraction

In Figure 38.9, a force vector $\boldsymbol{F}$ is represented by $\boldsymbol{oa}$. The vector $(-\boldsymbol{oa})$ can be obtained by drawing a vector from o in the opposite sense to $\boldsymbol{oa}$ but having the same magnitude, shown as $\boldsymbol{ob}$ in Figure 38.9, i.e. $\boldsymbol{ob} = (-\boldsymbol{oa})$
For two vectors acting at a point, as shown in Figure 38.10(a), the resultant of vector addition is $\boldsymbol{os} = \boldsymbol{oa} + \boldsymbol{ob}$. Figure 38.10(b) shows vectors $\boldsymbol{ob} + (-\boldsymbol{oa})$, that is, $\boldsymbol{ob} - \boldsymbol{oa}$ and the vector equation is $\boldsymbol{ob} - \boldsymbol{oa} = \boldsymbol{od}$. Comparing $\boldsymbol{od}$ in Figure 38.10(b) with the broken line ab in Figure 38.10(a) shows that the second diagonal of the 'parallelogram' method of vector addition gives the magnitude and direction of vector subtraction of $\boldsymbol{oa}$ from $\boldsymbol{ob}$.

For example, accelerations of $a_1 = 1.5$ m/s^2 at 90° and $a_2 = 2.6$ m/s^2 at 145° act at a point. To find $\boldsymbol{a_1} + \boldsymbol{a_2}$ and $\boldsymbol{a_1} - \boldsymbol{a_2}$ by (i) drawing a scale vector diagram and (ii) by calculation:

(i) The scale vector diagram is shown in Figure 38.11. By measurement,

$$\boldsymbol{a_1} + \boldsymbol{a_2} = \mathbf{3.7\ m/s^2\ at\ 126^\circ}$$

$$\boldsymbol{a_1} - \boldsymbol{a_2} = \mathbf{2.1\ m/s^2\ at\ 0^\circ}$$

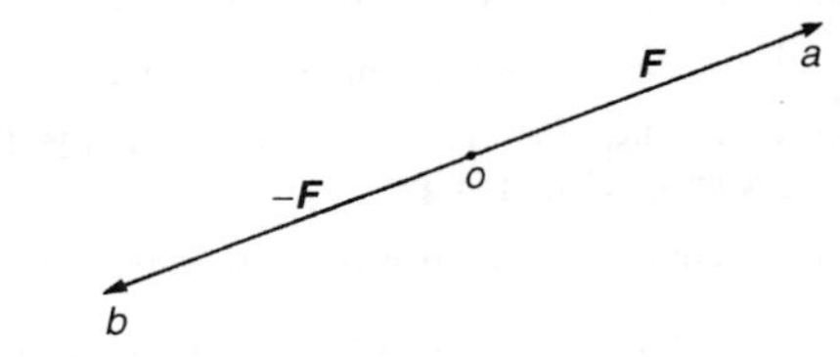

Figure 38.9

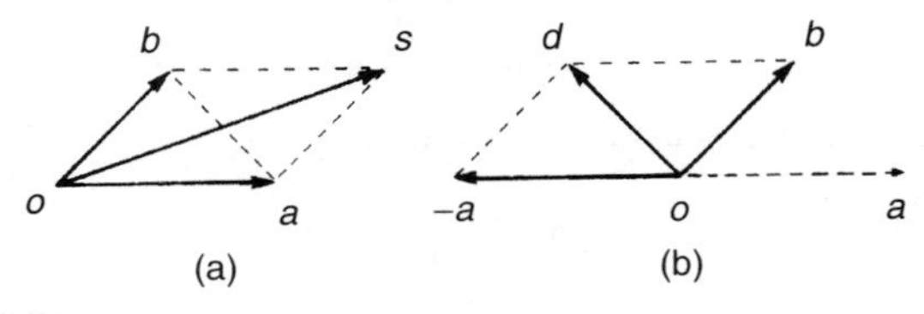

Figure 38.10

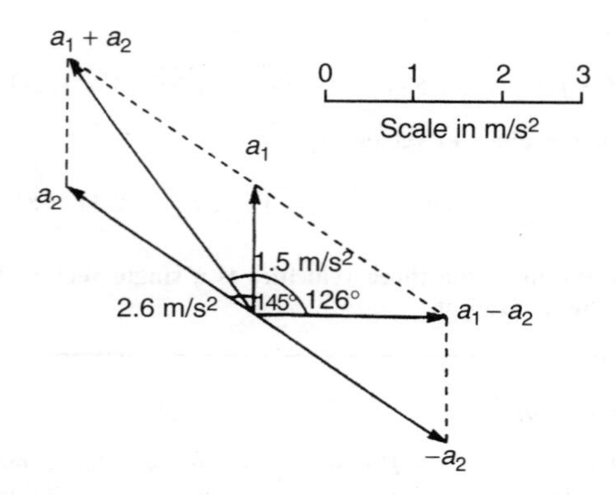

Figure 38.11

(ii) Resolving horizontally and vertically gives:

Horizontal component of $\boldsymbol{a_1} + \boldsymbol{a_2}$,

$$\boldsymbol{H} = 1.5\cos 90^\circ + 2.6\cos 145^\circ = -2.13$$

Vertical component of $\boldsymbol{a_1} + \boldsymbol{a_2}$,

$$\boldsymbol{V} = 1.5\sin 90^\circ + 2.6\sin 145^\circ = 2.99$$

Magnitude of $\boldsymbol{a_1} + \boldsymbol{a_2} = \sqrt{(-2.13)^2 + 2.99^2} = \mathbf{3.67\ m/s^2}$

Direction of $\boldsymbol{a_1} + \boldsymbol{a_2} = \tan^{-1}\left(\dfrac{2.99}{-2.13}\right)$ and must lie in the second quadrant since H is negative and V is positive.

$\tan^{-1}\left(\dfrac{2.99}{-2.13}\right) = -54.53^\circ$, and for this to be in the second quadrant, the true angle is 180° displaced, i.e. $180^\circ - 54.53^\circ$ or 125.47°.

Thus $\boldsymbol{a_1} + \boldsymbol{a_2} = \mathbf{3.67\ m/s^2\ at\ 125.47^\circ}$.

Horizontal component of $\boldsymbol{a_1} - \boldsymbol{a_2}$, that is, $\boldsymbol{a_1} + (-\boldsymbol{a_2})$

$$= 1.5\cos 90^\circ + 2.6\cos(145^\circ - 180^\circ) = 2.6\cos(-35^\circ) = 2.13$$

Vertical component of $\boldsymbol{a_1} - \boldsymbol{a_2}$, that is, $\boldsymbol{a_1} + (-\boldsymbol{a_2})$

$$= 1.5\sin 90^\circ + 2.6\sin(-35^\circ) = 0$$

Magnitude of $\boldsymbol{a_1} - \boldsymbol{a_2} = \sqrt{2.13^2 + 0^2} = 2.13\ \text{m/s}^2$

Direction of $\boldsymbol{a_1} - \boldsymbol{a_2} = \tan^{-1}\left(\dfrac{0}{2.13}\right) = 0^\circ$

Thus $\quad \boldsymbol{a_1} - \boldsymbol{a_2} = \mathbf{2.13\ m/s^2\ at\ 0^\circ}$

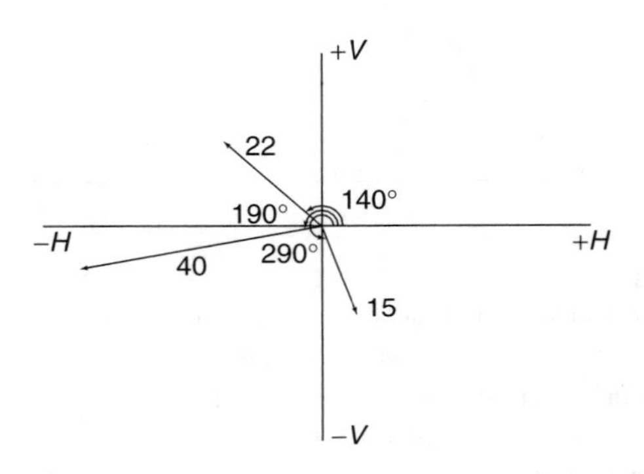

Figure 38.12

In another example, to calculate the resultant of $v_1 - v_2 + v_3$ when $v_1 =$ 22 units at 140°, $v_2 = 40$ units at 190° and $v_3 = 15$ units at 290°:

(i) The vectors are shown in Figure 38.12.

The horizontal component of $v_1 - v_2 + v_3$

$$= (22 \cos 140^\circ) - (40 \cos 190^\circ) + (15 \cos 290^\circ)$$

$$= (-16.85) - (-39.39) + (5.13) = \mathbf{27.67\ units}$$

The vertical component of $v_1 - v_2 + v_3$

$$= (22 \sin 140^\circ) - (40 \sin 190^\circ) + (15 \sin 290^\circ)$$

$$= (14.14) - (-6.95) + (-14.10) = \mathbf{6.99\ units}$$

The magnitude of the resultant, R, which can be represented by the mathematical symbol for 'the **modulus** of' as $|v_1 - v_2 + v_3|$ is given by:

$$|R| = \sqrt{27.67^2 + 6.99^2} = 28.54 \text{ units}$$

The direction of the resultant, $\boldsymbol{R}$, which can be represented by the mathematical symbol for 'the **argument** of' as $\arg(v_1 - v_2 + v_3)$ is given by:

$$\mathbf{arg\ R} = \tan^{-1}\left(\frac{6.99}{27.67}\right) = 14.18^\circ$$

Thus $v_1 - v_2 + v_3 =$ **28.54 units at 14.18°**

Relative velocity

For relative velocity problems, some fixed datum point needs to be selected. This is often a fixed point on the earth's surface. In any vector equation, only the start and finish points affect the resultant vector of a system. Two different systems are shown in Figure 38.13, but in each of the systems, the resultant vector is ***ad***.

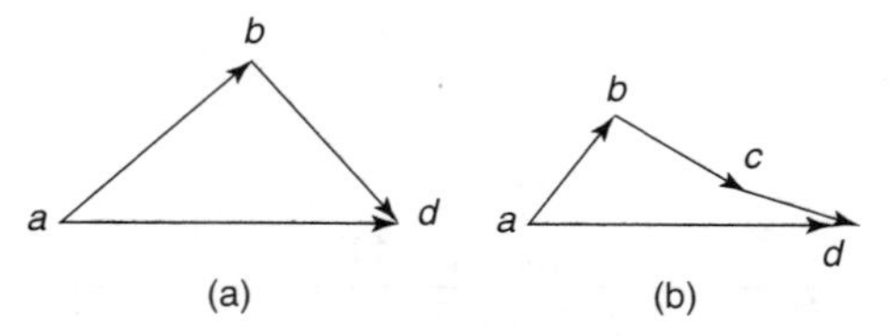

Figure 38.13

The vector equation of the system shown in Figure 38.13(a) is:

$$\boldsymbol{ad} = \boldsymbol{ab} + \boldsymbol{bd}$$

and that for the system shown in Figure 38.13(b) is:

$$\boldsymbol{ad} = \boldsymbol{ab} + \boldsymbol{bc} + \boldsymbol{cd}$$

Thus in vector equations of this form, only the first and last letters, a and d, respectively, fix the magnitude and direction of the resultant vector.

For example, two cars, P and Q, are travelling towards the junction of two roads which are at right angles to one another. Car P has a velocity of 45 km/h due east and car Q a velocity of 55 km/h due south. To calculate (i) the velocity of car P relative to car Q, and (ii) the velocity of car Q relative to car P:

(i) The directions of the cars are shown in Figure 38.14(a), called a **space diagram**. The velocity diagram is shown in Figure 38.14(b), in which $\boldsymbol{pe}$ is taken as the velocity of car P relative to point e on the earth's surface. The velocity of P relative to Q is vector $\boldsymbol{pq}$ and the vector equation is $\boldsymbol{pq} = \boldsymbol{pe} + \boldsymbol{eq}$. Hence the vector directions are as shown, $\boldsymbol{eq}$ being in the opposite direction to $\boldsymbol{qe}$. From the geometry of the vector triangle,

$$|\boldsymbol{pq}| = \sqrt{45^2 + 55^2} = 71.06 \text{ km/h} \quad \text{and}$$

$$\arg \boldsymbol{pq} = \tan^{-1}\left(\frac{55}{45}\right) = 50.71^\circ$$

i.e. **the velocity of car P relative to car Q is 71.06 km/h at 50.71°**

(ii) The velocity of car Q relative to car P is given by the vector equation $\boldsymbol{qp} = \boldsymbol{qe} + \boldsymbol{ep}$ and the vector diagram is as shown in Figure 38.14(c), having $\boldsymbol{ep}$ opposite in direction to $\boldsymbol{pe}$. From the geometry of this vector triangle:

$$|\boldsymbol{qp}| = \sqrt{45^2 + 55^2} = 71.06 \text{ m/s} \quad \text{and}$$

$$\arg \boldsymbol{qp} = \tan^{-1}\left(\frac{55}{45}\right) = 50.71^\circ$$

Figure 38.14

but must lie in the third quadrant, i.e. the required angle is $180^\circ + 50.71^\circ = 230.71^\circ$
Thus the velocity of car Q relative to car P is 71.06 m/s at 230.71°

39 Combination of Waveforms

Combination of two periodic functions

There are a number of instances in engineering and science where waveforms combine and where it is required to determine the single phasor (called the resultant) that could replace two or more separate phasors. (A phasor is a rotating vector). Uses are found in electrical alternating current theory, in mechanical vibrations, in the addition of forces and with sound waves. There are several methods of determining the resultant and two such methods — plotting/measuring, and resolution of phasors by calculation — are explained in this chapter.

Plotting periodic functions

This may be achieved by sketching the separate functions on the same axes and then adding (or subtracting) ordinates at regular intervals.

For example, the graphs of $y_1 = 3\sin A$ and $y_2 = 2\cos A$ are to be plotted from $A = 0^\circ$ to $A = 360^\circ$ on the same axes. To plot $y_R = 3\sin A + 2\cos A$ by adding ordinates, and obtain a sinusoidal expression for this resultant waveform:
$y_1 = 3\sin A$ and $y_2 = 2\cos A$ are shown plotted in Figure 39.1. Ordinates may be added at, say, 15° intervals. For example,

$$\text{at } 0^\circ,\ y_1 + y_2 = 0 + 2 = 2$$
$$\text{at } 15^\circ,\ y_1 + y_2 = 0.78 + 1.93 = 2.71$$
$$\text{at } 120^\circ,\ y_1 + y_2 = 2.60 + -1 = 1.6$$
$$\text{at } 210^\circ,\ y_1 + y_2 = -1.50 - 1.73 = -3.23, \text{ and so on}$$

The resultant waveform, shown by the broken line, has the same period, i.e. 360°, and thus the same frequency as the single phasors. The maximum value, or amplitude, of the resultant is 3.6. The resultant waveform **leads** $y_1 = 3\sin A$ by 34° or 0.593 rad. The sinusoidal expression for the resultant waveform is:

$$\mathbf{y_R = 3.6\sin(A + 34^\circ)} \quad \text{or} \quad \mathbf{y_R = 3.6\sin(A + 0.593)}$$

In another example, the graphs of $y_1 = 4\sin\omega t$ and $y_2 = 3\sin(\omega t - \pi/3)$ are to be plotted on the same axes, over one cycle. By adding ordinates at intervals plot $y_R = y_1 + y_2$. To obtain a sinusoidal expression for the resultant waveform:
$y_1 = 4\sin\omega t$ and $y_2 = 3\sin(\omega t - \pi/3)$ are shown plotted in Figure 39.2. Ordinates are added at 15° intervals and the resultant is shown by the broken line. The amplitude of the resultant is 6.1 and it **lags** y_1 by 25° or 0.436 rad. Hence the sinusoidal expression for the resultant waveform is:

$$\mathbf{y_R = 6.1\sin(\omega t - 0.436)}$$

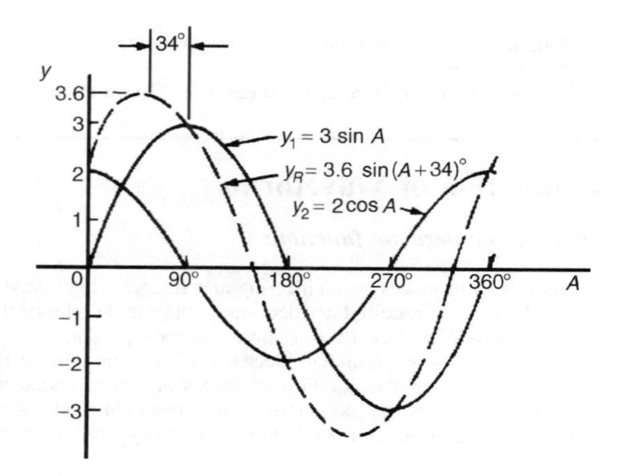

Figure 39.1

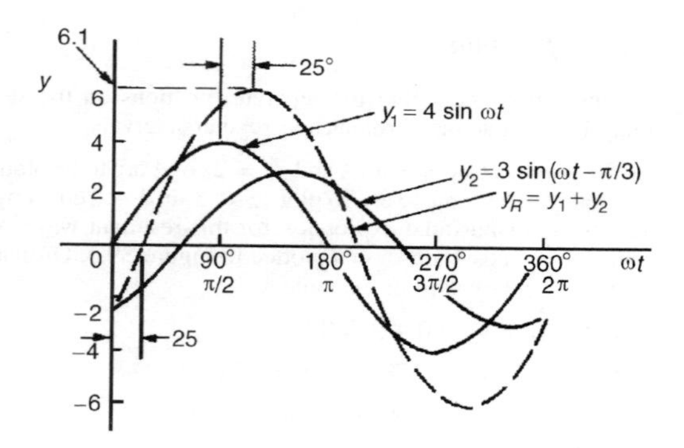

Figure 39.2

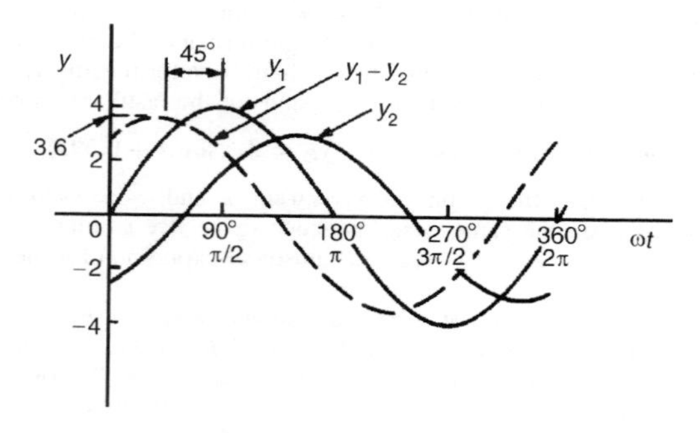

Figure 39.3

In another example, to determine a sinusoidal expression for $y_1 - y_2$ when $y_1 = 4\sin\omega t$ and $y_2 = 3\sin(\omega t - \pi/3)$:
y_1 and y_2 are shown plotted in Figure 39.3. At 15° intervals y_2 is subtracted from y_1. For example:

$$\text{at } 0^\circ,\ y_1 - y_2 = 0 - (-2.6) = +2.6$$
$$\text{at } 30^\circ,\ y_1 - y_2 = 2 - (-1.5) = +3.5$$
$$\text{at } 150^\circ,\ y_1 - y_2 = 2 - 3 = -1, \text{ and so on.}$$

The amplitude, or peak value of the resultant (shown by the broken line), is 3.6 and it leads y_1 by 45° or 0.79 rad. Hence

$$\mathbf{y_1 - y_2 = 3.6\sin(\omega t + 0.79)}$$

Resolution of phasors by calculation

The resultant of two periodic functions may be found from their relative positions when the time is zero. For example, if $y_1 = 4\sin\omega t$ and $y_2 = 3\sin(\omega t - \pi/3)$ then each may be represented as phasors as shown in Figure 39.4, y_1 being 4 units long and drawn horizontally and y_2 being 3 units long, lagging y_1 by $\pi/3$ radians or 60°. To determine the resultant of $y_1 + y_2$, y_1 is drawn horizontally as shown in Figure 39.5 and y_2 is joined to the end of y_1 at 60° to the horizontal. The resultant is given by y_R. This is the same as the diagonal of a parallelogram that is shown completed in Figure 39.6. Resultant y_R, in Figures 39.5 and 39.6, is determined either by:

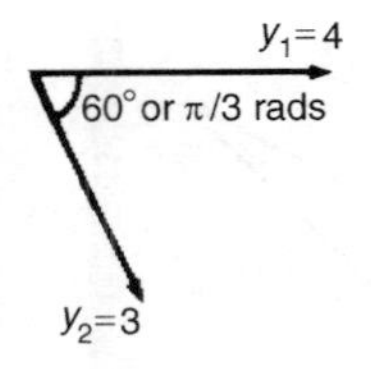

Figure 39.4

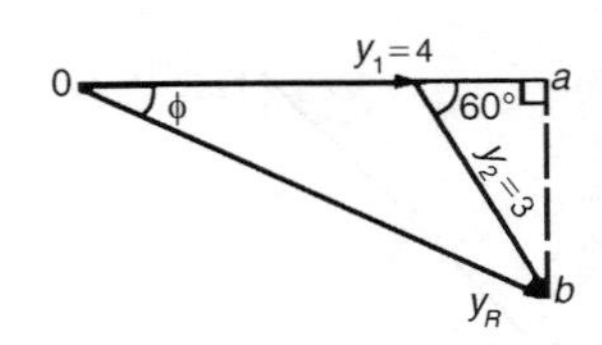

Figure 39.5

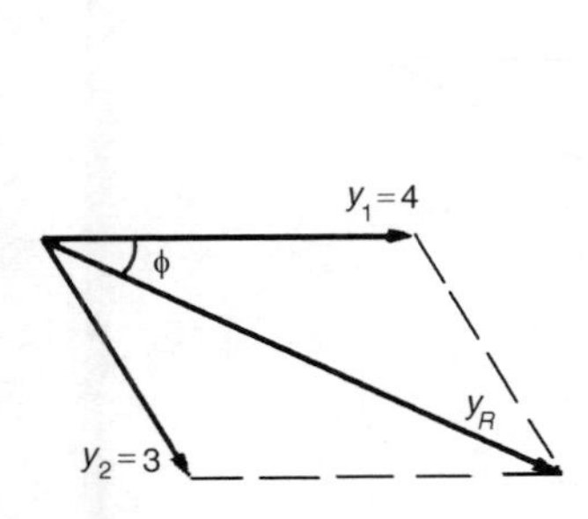

Figure 39.6

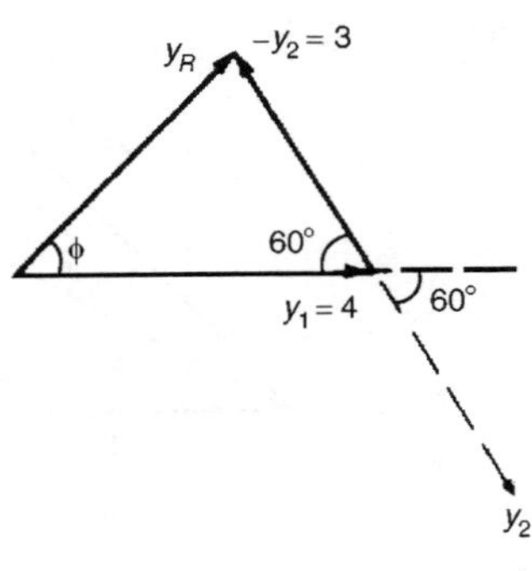

Figure 39.7

(a) use of the cosine rule (and then sine rule to calculate angle ϕ), or
(b) determining horizontal and vertical components of lengths oa and ab in Figure 39.5, and then using Pythagoras' theorem to calculate ob.

In the above example, by calculation, $y_R = 6.083$ and angle $\phi = 25.28°$ or 0.441 rad. Thus the resultant may be expressed in sinusoidal form as $y_R = 6.083 \sin(\omega t - 0.441)$. If the resultant phasor, $y_R = y_1 - y_2$ is required, then y_2 is still 3 units long but is drawn in the opposite direction, as shown in Figure 39.7, and y_R is determined by calculation.

For example, given $y_1 = 2 \sin \omega t$ and $y_2 = 3 \sin(\omega t + \pi/4)$, to obtain an expression for the resultant $y_R = y_1 + y_2$, (a) by drawing and (b) by calculation:

(a) When time $t = 0$ the position of phasors y_1 and y_2 are as shown in Figure 39.8(a). To obtain the resultant, y_1 is drawn horizontally, 2 units long, y_2 is drawn 3 units long at an angle of $\pi/4$ rads or 45° and joined to the end of y_1 as shown in Figure 39.8(b). y_R is measured as 4.6 units long and angle ϕ is measured as 27° or 0.47 rad. Alternatively, y_R is the diagonal of the parallelogram formed as shown in Figure 39.8(c).
Hence, by drawing, $\mathbf{y_R = 4.6 \sin(\omega t + 0.47)}$

(b) From Figure 39.8(b), and using the cosine rule:

$$y_R^2 = 2^2 + 3^2 - [2(2)(3) \cos 135° = 4 + 9 - [-8.485] = 21.49$$

Hence $y_R = \sqrt{21.49} = 4.64$

Using the sine rule: $$\frac{3}{\sin \phi} = \frac{4.64}{\sin 135°}$$

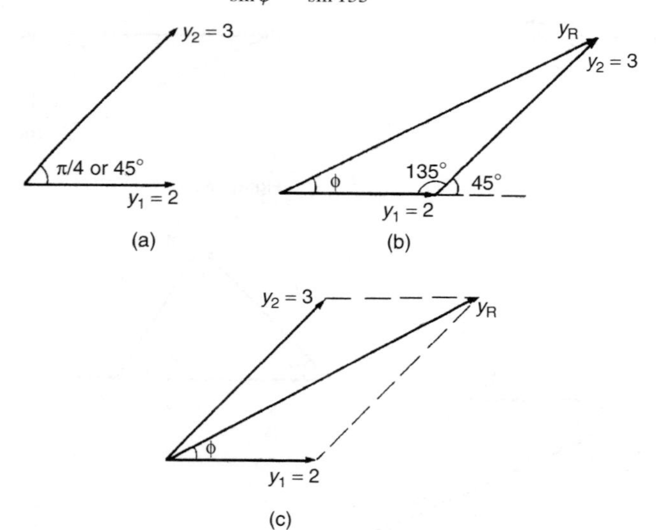

Figure 39.8

from which, $\sin\phi = \frac{3\sin 135^\circ}{4.64} = 0.4572$

Hence $\phi = \sin^{-1} 0.4572 = 27.21^\circ$ or 0.475 rad.

By calculation, $\boldsymbol{y_R = 4.64\sin(\omega t + 0.475)}$

40 Scalar and Vector Products

The unit triad

When a vector x of magnitude x units and direction θ° is divided by the magnitude of the vector, the result is a vector of unit length at angle θ°. The unit vector for a velocity of 10 m/s at 50° is $\frac{10 \text{ m/s at } 50^\circ}{10 \text{ m/s}}$, i.e. 1 at 50°. In general, the unit vector for $\boldsymbol{oa}$ is $\frac{\boldsymbol{oa}}{|\boldsymbol{oa}|}$, the $\boldsymbol{oa}$ being a vector and having both magnitude and direction and $|\boldsymbol{oa}|$ being the magnitude of the vector only.
One method of completely specifying the direction of a vector in space relative to some reference point is to use three unit vectors, mutually at right angles to each other, as shown in Figure 40.1. Such a system is called a **unit triad**. In Figure 40.2, one way to get from o to r is to move x units along $\boldsymbol{i}$ to point

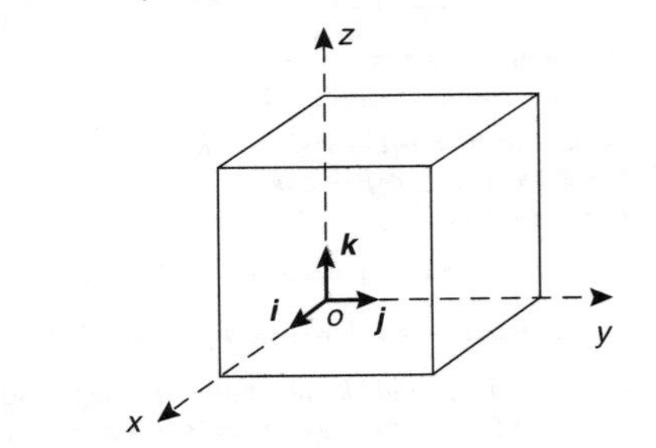

Figure 40.1

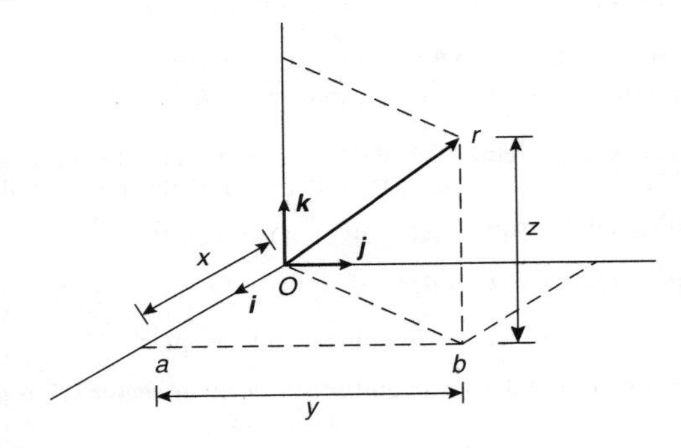

Figure 40.2

a, then y units in direction $\boldsymbol{j}$ to get to b and finally z units in direction $\boldsymbol{k}$ to get to r. The vector $\boldsymbol{or}$ is specified as

$$\boldsymbol{or} = \boldsymbol{xi} + \boldsymbol{yj} + \boldsymbol{zk}$$

The scalar product of two vectors

When vector $\boldsymbol{oa}$ is multiplied by a scalar quantity, say k, the magnitude of the resultant vector will be k times the magnitude of $\boldsymbol{oa}$ and its direction will remain the same. Thus $2 \times (5\text{ N at }20^\circ)$ results in a vector of magnitude 10 N at 20°.
One of the products of two vector quantities is called the **scalar** or **dot product** of two vectors and is defined as the product of their magnitudes multiplied by the cosine of the angle between them. The scalar product of $\boldsymbol{oa}$ and $\boldsymbol{ob}$ is shown as $\boldsymbol{oa} \bullet \boldsymbol{ob}$. For vectors $\boldsymbol{oa} = oa$ at θ_1, and $\boldsymbol{ob} = ob$ at θ_2 where $\theta_2 > \theta_1$, **the scalar product is**:

$$\boxed{\boldsymbol{oa} \bullet \boldsymbol{ob} = oa\ ob \cos(\theta_2 - \theta_1)}$$

It may be shown that $\boldsymbol{oa} \bullet \boldsymbol{ob} = \boldsymbol{ob} \bullet \boldsymbol{oa}$
The **angle between two vectors** can be expressed in terms of the vector constants as follows:

Since $\boldsymbol{a} \bullet \boldsymbol{b} = ab\ \cos\theta$, then $\boxed{\cos\theta = \dfrac{a \bullet b}{ab}}$ (1)

Let $\boldsymbol{a} = a_1\boldsymbol{i} + a_2\boldsymbol{j} + a_3\boldsymbol{k}$ and $\boldsymbol{b} = b_1\boldsymbol{i} + b_2\boldsymbol{j} + b_3\boldsymbol{k}$
$\boldsymbol{a} \bullet \boldsymbol{b} = (a_1\boldsymbol{i} + a_2\boldsymbol{j} + a_3\boldsymbol{k}) \bullet (b_1\boldsymbol{i} + b_2\boldsymbol{j} + b_3\boldsymbol{k})$
Multiplying out the brackets gives:

$$\boldsymbol{a} \bullet \boldsymbol{b} = a_1b_1\boldsymbol{i} \bullet \boldsymbol{i} + a_1b_2\boldsymbol{i} \bullet \boldsymbol{j} + a_1b_3\boldsymbol{i} \bullet \boldsymbol{k} + a_2b_1\boldsymbol{j} \bullet \boldsymbol{i} + a_2b_2\boldsymbol{j} \bullet \boldsymbol{j}$$
$$+ a_2b_3\boldsymbol{j} \bullet \boldsymbol{k} + a_3b_1\boldsymbol{k} \bullet \boldsymbol{i} + a_3b_2\boldsymbol{k} \bullet \boldsymbol{j} + a_3b_3\boldsymbol{k} \bullet \boldsymbol{k}$$

However, the unit vectors $\boldsymbol{i}$, $\boldsymbol{j}$ and $\boldsymbol{k}$ all have a magnitude of 1 and $\boldsymbol{i} \bullet \boldsymbol{i} = (1)(1)\cos 0^\circ = 1$, $\boldsymbol{i} \bullet \boldsymbol{j} = (1)(1)\cos 90^\circ = 0$, $\boldsymbol{i} \bullet \boldsymbol{k} = (1)(1)\cos 90^\circ = 0$ and similarly $\boldsymbol{j} \bullet \boldsymbol{j} = 1$, $\boldsymbol{j} \bullet \boldsymbol{k} = 0$ and $\boldsymbol{k} \bullet \boldsymbol{k} = 1$. Thus, only terms containing $\boldsymbol{i} \bullet \boldsymbol{i}$, $\boldsymbol{j} \bullet \boldsymbol{j}$ or $\boldsymbol{k} \bullet \boldsymbol{k}$ in the expansion above will not be zero.

Thus, the scalar product $\boxed{\boldsymbol{a} \bullet \boldsymbol{b} = a_1b_1 + a_2b_2 + a_3b_3}$ (2)
Both a and b in equation (1) can be expressed in terms of a_1, b_1, a_2, b_2, a_3 and b_3
From the geometry of Figure 40.3, the length of diagonal OP in terms of side lengths a, b and c can be obtained from Pythagoras' theorem as follows:

$$\text{OP}^2 = \text{OB}^2 + \text{BP}^2 \quad \text{and} \quad \text{OB}^2 = \text{OA}^2 + \text{AB}^2$$

$$\text{Thus,} \quad \text{OP}^2 = \text{OA}^2 + \text{AB}^2 + \text{BP}^2$$
$$= a^2 + b^2 + c^2, \quad \text{in terms of side lengths}$$

Thus, the **length** or **modulus** or **magnitude** or **norm of vector OP** is given by:

$$\boxed{\text{OP} = \sqrt{(a^2 + b^2 + c^2)}} \qquad (3)$$

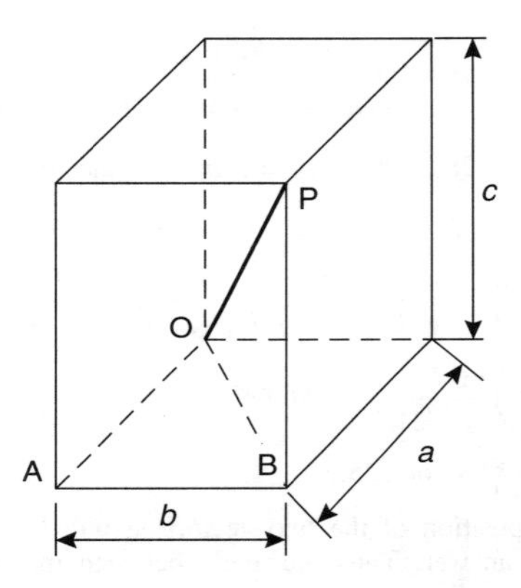

Figure 40.3

Relating this result to the two vectors $a_1\boldsymbol{i} + a_2\boldsymbol{j} + a_3\boldsymbol{k}$ and $b_1\boldsymbol{i} + b_2\boldsymbol{j} + b_3\boldsymbol{k}$, gives: $a = \sqrt{(a_1^2 + a_2^2 + a_3^2)}$ and $b = \sqrt{(b_1^2 + b_2^2 + b_3^2)}$
That is, from equation (1),

$$\boxed{\cos\theta = \frac{a_1b_1 + a_2b_2 + a_3b_3}{\sqrt{(a_1^2 + a_2^2 + a_3^2)}\sqrt{(b_1^2 + b_2^2 + b_3^2)}}} \qquad (4)$$

For example, to determine: (i) $\boldsymbol{p} \bullet \boldsymbol{q}$ (ii) $\boldsymbol{p} + \boldsymbol{q}$ (iii) $|\boldsymbol{p} + \boldsymbol{q}|$ and (iv) $|\boldsymbol{p}| + |\boldsymbol{q}|$ if $\boldsymbol{p} = 2\boldsymbol{i} + \boldsymbol{j} - \boldsymbol{k}$ and $\boldsymbol{q} = \boldsymbol{i} - 3\boldsymbol{j} + 2\boldsymbol{k}$:

(i) From equation (2), if $\boldsymbol{p} = a_1\boldsymbol{i} + a_2\boldsymbol{j} + a_3\boldsymbol{k}$ and $\boldsymbol{q} = b_1\boldsymbol{i} + b_2\boldsymbol{j} + b_3\boldsymbol{k}$

then $\quad \boldsymbol{p} \bullet \boldsymbol{q} = a_1b_1 + a_2b_2 + a_3b_3$

When $\quad \boldsymbol{p} = 2\boldsymbol{i} + \boldsymbol{j} - \boldsymbol{k}$, $a_1 = 2$, $a_2 = 1$ and $a_3 = -1$

and when $\quad \boldsymbol{q} = \boldsymbol{i} - 3\boldsymbol{j} + 2\boldsymbol{k}$, $b_1 = 1$, $b_2 = -3$ and $b_3 = 2$

Hence $\quad \boldsymbol{p} \bullet \boldsymbol{q} = (2)(1) + (1)(-3) + (-1)(2)$

i.e. $\quad \boldsymbol{p} \bullet \boldsymbol{q} = \mathbf{-3}$

(ii) $\boldsymbol{p} + \boldsymbol{q} = (2\boldsymbol{i} + \boldsymbol{j} - \boldsymbol{k}) + (\boldsymbol{i} - 3\boldsymbol{j} + 2\boldsymbol{k}) = \mathbf{3\boldsymbol{i} - 2\boldsymbol{j} + \boldsymbol{k}}$
(iii) $|\boldsymbol{p} + \boldsymbol{q}| = |3i - 2j + k|$
From equation (3), $|\boldsymbol{p} + \boldsymbol{q}| = \sqrt{[3^2 + (-2)^2 + 1^2)]} = \mathbf{\sqrt{14}}$
(iv) From equation (3), $|\boldsymbol{p}| = |2i + j - k| = \sqrt{[2^2 + 1^2 + (-1)^2]} = \sqrt{6}$
Similarly, $|\boldsymbol{q}| = |i - 3j + 2k| = \sqrt{[1^2 + (-3)^2 + 2^2]} = \sqrt{14}$
Hence $|\boldsymbol{p}| + |\boldsymbol{q}| = \sqrt{6} + \sqrt{14} = \mathbf{6.191}$,
correct to 3 decimal places

In another example, to determine the angle between vectors $\boldsymbol{oa}$ and $\boldsymbol{ob}$ when $\boldsymbol{oa} = \boldsymbol{i} + 2\boldsymbol{j} - 3\boldsymbol{k}$ and $\boldsymbol{ob} = 2\boldsymbol{i} - \boldsymbol{j} + 4\boldsymbol{k}$:

From equation (4), $\cos\theta = \dfrac{a_1b_1 + a_2b_2 + a_3b_3}{\sqrt{(a_1^2 + a_2^2 + a_3^2)}\sqrt{(b_1^2 + b_2^2 + b_3^2)}}$

Since $\quad \boldsymbol{oa} = \boldsymbol{i} + 2\boldsymbol{j} - 3\boldsymbol{k}, \quad a_1 = 1, a_2 = 2$ and $a_3 = -3$

Since $\quad \boldsymbol{ob} = 2\boldsymbol{i} - \boldsymbol{j} + 4\boldsymbol{k}, \quad b_1 = 2, b_2 = -1$ and $b_3 = 4$

Thus,
$$\cos\theta = \frac{(1\times 2) + (2\times -1) + (-3\times 4)}{\sqrt{(1^2 + 2^2 + (-3)^2)}\sqrt{(2^2 + (-1)^2 + 4^2)}}$$
$$= \frac{-12}{\sqrt{14}\sqrt{21}} = -0.6999$$

i.e. $\quad \theta = 134.4^\circ$ or 225.6°

By sketching the position of the two vectors, it will be seen that 225.6° is not an acceptable answer. Thus the angle between the vectors ***oa*** and ***ob***, $\boldsymbol{\theta = 134.4^\circ}$

Direction Cosines

From Figure 40.2, $\boldsymbol{or} = x\boldsymbol{i} + y\boldsymbol{j} + z\boldsymbol{k}$ and from equation (3), $|\boldsymbol{or}| = \sqrt{x^2 + y^2 + z^2}$.
If ***or*** makes angles of α, β and γ with the co-ordinate axes i, j and k respectively, then:

$$\boldsymbol{\cos\alpha = \frac{x}{\sqrt{x^2+y^2+z^2}}}, \quad \boldsymbol{\cos\beta = \frac{y}{\sqrt{x^2+y^2+z^2}}} \quad \text{and}$$
$$\boldsymbol{\cos\gamma = \frac{y}{\sqrt{x^2+y^2+z^2}}}$$

such that $\cos^2\alpha + \cos^2\beta + \cos^2\gamma = 1$
The values of $\cos\alpha$, $\cos\beta$ and $\cos\gamma$ are called the **direction cosines** of ***or***

Practical Application of Scalar Product

For example, a constant force of $\boldsymbol{F} = 10\boldsymbol{i} + 2\boldsymbol{j} - \boldsymbol{k}$ Newton's displaces an object from $\boldsymbol{A} = \boldsymbol{i} + \boldsymbol{j} + \boldsymbol{k}$ to $\boldsymbol{B} = 2\boldsymbol{i} - \boldsymbol{j} + 3\boldsymbol{k}$ (in metres). To find the work done in Newton metres:
The work done is the product of the applied force and the distance moved in the direction of the force,
i.e. $\quad$ **work done** $= \boldsymbol{F} \bullet \boldsymbol{d}$

The principles developed in the final example of chapter 39, apply equally to this example when determining the displacement. From the sketch shown in Figure 40.4,

$$\boldsymbol{AB} = \boldsymbol{AO} + \boldsymbol{OB} = \boldsymbol{OB} - \boldsymbol{OA}$$

that is $\quad \boldsymbol{AB} = (2\boldsymbol{i} - \boldsymbol{j} + 3\boldsymbol{k}) - (\boldsymbol{i} + \boldsymbol{j} + \boldsymbol{k}) = \boldsymbol{i} - 2\boldsymbol{j} + 2\boldsymbol{k}$

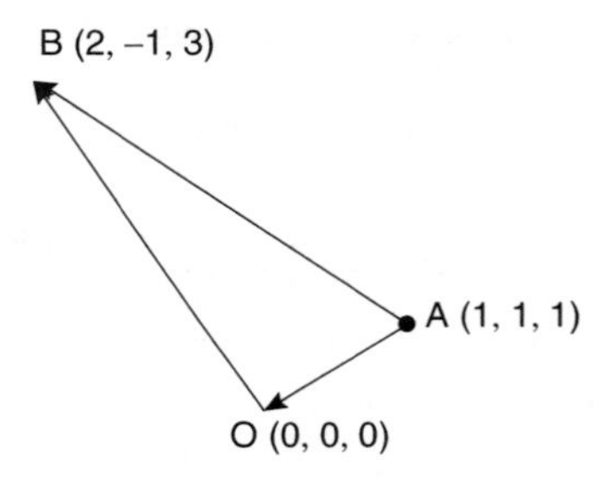

Figure 40.4

The work done is $\boldsymbol{F} \bullet \boldsymbol{d}$, that is $\boldsymbol{F} \bullet \boldsymbol{AB}$ in this case

i.e. **work done** $= (10\boldsymbol{i} + 2\boldsymbol{j} - \boldsymbol{k}) \bullet (\boldsymbol{i} - 2\boldsymbol{j} + 2\boldsymbol{k})$

But from equation (2), $\boldsymbol{a} \bullet \boldsymbol{b} = a_1b_1 + a_2b_2 + a_3b_3$

Hence **workdone** $= (10 \times 1) + (2 \times (-2)) + (-1) \times 2) =$ **4 Nm**

Vector Products

A second product of two vectors is called the **vector** or **cross product** and is defined in terms of its modulus and the magnitudes of the two vectors and the sine of the angle between them. The vector product of vectors ***oa*** and ***ob*** is written as ***oa*** × ***ob*** and is defined by:

$$|\boldsymbol{oa} \times \boldsymbol{ob}| = oa\ ob \sin\theta$$

where θ is the angle between the two vectors.

The direction of ***oa*** × ***ob*** is perpendicular to both ***oa*** and ***ob***, as shown in Figure 40.5

The direction is obtained by considering that a right-handed screw is screwed along ***oa*** × ***ob*** with its head at the origin and if the direction of ***oa*** × ***ob*** is correct, the head should rotate from ***oa*** to ***ob***, as shown in Figure 40.5(a). It follows that the direction of ***ob*** × ***oa*** is as shown in Figure 40.5(b). Thus

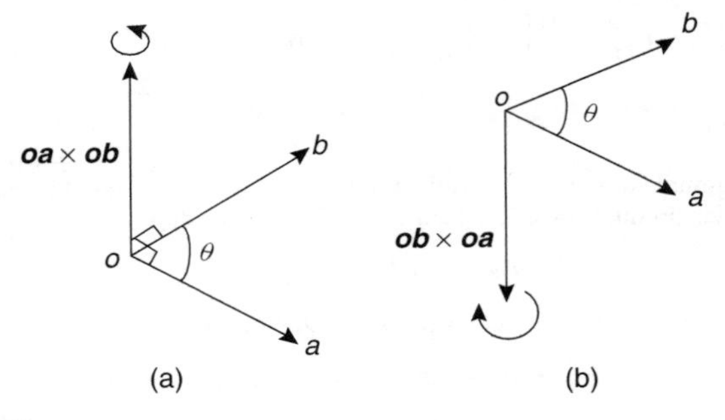

Figure 40.5

$\boldsymbol{oa} \times \boldsymbol{ob}$ is not equal to $\boldsymbol{ob} \times \boldsymbol{oa}$. The magnitudes of $oa\, ob \sin\theta$ are the same but their directions are 180° displaced, i.e.

$$\boldsymbol{oa} \times \boldsymbol{ob} = -\boldsymbol{ob} \times \boldsymbol{oa}$$

The vector product of two vectors may be expressed in terms of the unit vectors. Let two vectors, a and b, be such that:

$$\boldsymbol{a} = a_1\boldsymbol{i} + a_2\boldsymbol{j} + a_3\boldsymbol{k} \quad \text{and} \quad \boldsymbol{b} = b_1\boldsymbol{i} + b_2\boldsymbol{j} + b_3\boldsymbol{k}$$

$$\begin{aligned}\text{Then, } \boldsymbol{a} \times \boldsymbol{b} &= (a_1\boldsymbol{i} + a_2\boldsymbol{j} + a_3\boldsymbol{k}) \times (b_1\boldsymbol{i} + b_2\boldsymbol{j} + b_3\boldsymbol{k}) \\ &= a_1b_1\boldsymbol{i} \times \boldsymbol{i} + a_1b_2\boldsymbol{i} \times \boldsymbol{j} + a_1b_3\boldsymbol{i} \times \boldsymbol{k} + a_2b_1\boldsymbol{j} \times \boldsymbol{i} \\ &\quad + a_2b_2\boldsymbol{j} \times \boldsymbol{j} + a_2b_3\boldsymbol{j} \times \boldsymbol{k} + a_3b_1\boldsymbol{k} \times \boldsymbol{i} \\ &\quad + a_3b_2\boldsymbol{k} \times \boldsymbol{j} + a_3b_3\boldsymbol{k} \times \boldsymbol{k}\end{aligned}$$

But by the definition of a vector product,

$$\boldsymbol{i} \times \boldsymbol{j} = \boldsymbol{k}, \; \boldsymbol{j} \times \boldsymbol{k} = \boldsymbol{i} \quad \text{and} \quad \boldsymbol{k} \times \boldsymbol{i} = \boldsymbol{j}$$

Also $\quad \boldsymbol{i} \times \boldsymbol{i} = \boldsymbol{j} \times \boldsymbol{j} = \boldsymbol{k} \times \boldsymbol{k} = (1)(1)\sin 0° = 0$

Remembering that $\boldsymbol{a} \times \boldsymbol{b} = -\boldsymbol{b} \times \boldsymbol{a}$ gives:

$$\boldsymbol{a} \times \boldsymbol{b} = a_1b_2\boldsymbol{k} - a_1b_3\boldsymbol{j} - a_2b_1\boldsymbol{k} + a_2b_3\boldsymbol{i} + a_3b_1\boldsymbol{j} - a_3b_2\boldsymbol{i}$$

Grouping the $\boldsymbol{i}, \boldsymbol{j}$ and $\boldsymbol{k}$ terms together, gives

$$\boldsymbol{a} \times \boldsymbol{b} = (a_2b_3 - a_3b_2)\boldsymbol{i} + (a_3b_1 - a_1b_3)\boldsymbol{j} + (a_1b_2 - a_2b_1)\boldsymbol{k}$$

The vector product can be written in determinant form (see Chapter 43) as:

$$\boxed{\boldsymbol{a} \times \boldsymbol{b} = \begin{vmatrix} i & j & k \\ a_1 & a_2 & a_3 \\ b_1 & b_2 & b_3 \end{vmatrix}} \qquad (5)$$

The 3×3 determinant $\begin{vmatrix} i & j & k \\ a_1 & a_2 & a_3 \\ b_1 & b_2 & b_3 \end{vmatrix}$ is evaluated as:

$\boldsymbol{i}\begin{vmatrix} a_2 & a_3 \\ b_2 & b_3 \end{vmatrix} - \boldsymbol{j}\begin{vmatrix} a_1 & a_3 \\ b_1 & b_3 \end{vmatrix} + \boldsymbol{k}\begin{vmatrix} a_1 & a_2 \\ b_1 & b_2 \end{vmatrix}$ where

$\begin{vmatrix} a_2 & a_3 \\ b_2 & b_3 \end{vmatrix} = a_2b_3 - a_3b_2$, $\begin{vmatrix} a_1 & a_3 \\ b_1 & b_3 \end{vmatrix} = a_1b_3 - a_3b_1$ and $\begin{vmatrix} a_1 & a_2 \\ b_1 & b_2 \end{vmatrix} = a_1b_2 - a_2b_1$

The magnitude of the vector product of two vectors can be found by expressing it in scalar product form and then using the relationship

$$\boldsymbol{a} \bullet \boldsymbol{b} = a_1b_1 + a_2b_2 + a_3b_3$$

Squaring both sides of a vector product equation gives:

$$\begin{aligned}(|\boldsymbol{a} \times \boldsymbol{b}|)^2 &= a^2b^2\sin^2\theta = a^2b^2(1 - \cos^2\theta) \\ &= a^2b^2 - a^2b^2\cos^2\theta \qquad (6)\end{aligned}$$

It is stated earlier that $\boldsymbol{a} \bullet \boldsymbol{b} = ab\cos\theta$, hence

$$\boldsymbol{a} \bullet \boldsymbol{a} = a^2\cos\theta. \quad \text{But } \theta = 0°, \text{ thus}$$

$$\boldsymbol{a} \bullet \boldsymbol{a} = a^2$$

Also, $\cos\theta = \dfrac{a \bullet b}{ab}$

Multiplying both sides of this equation by a^2b^2 and squaring gives

$$a^2b^2\cos^2\theta = \frac{a^2b^2(a \bullet b)^2}{a^2b^2} = (\boldsymbol{a} \bullet \boldsymbol{b})^2$$

Substituting in equation (6) above for $a^2 = \boldsymbol{a} \bullet \boldsymbol{a}$, $b^2 = \boldsymbol{b} \bullet \boldsymbol{b}$ and $a^2b^2\cos^2\theta = (\boldsymbol{a} \bullet \boldsymbol{b})^2$ gives:

$$(|\boldsymbol{a} \times \boldsymbol{b}|)^2 = (\boldsymbol{a} \bullet \boldsymbol{a})(\boldsymbol{b} \bullet \boldsymbol{b}) - (\boldsymbol{a} \bullet \boldsymbol{b})^2$$

That is, $\boxed{|\boldsymbol{a} \times \boldsymbol{b}| = \sqrt{[(\boldsymbol{a} \bullet \boldsymbol{a})(\boldsymbol{b} \bullet \boldsymbol{b}) - (\boldsymbol{a} \bullet \boldsymbol{b})^2]}}$ (7)

For example, to find (i) $a \times b$ and (ii) $|\boldsymbol{a} \times \boldsymbol{b}|$ for the vectors $\boldsymbol{a} = \boldsymbol{i} + 4\boldsymbol{j} - 2\boldsymbol{k}$ and $\boldsymbol{b} = 2\boldsymbol{i} - \boldsymbol{j} + 3\boldsymbol{k}$:

(i) From equation 5,

$$\boldsymbol{a} \times \boldsymbol{b} = \begin{vmatrix} i & j & k \\ 1 & 4 & -2 \\ 2 & -1 & 3 \end{vmatrix} = \boldsymbol{i}\begin{vmatrix} 4 & -2 \\ -1 & 3 \end{vmatrix} - \boldsymbol{j}\begin{vmatrix} 1 & -2 \\ 2 & 3 \end{vmatrix} + \boldsymbol{k}\begin{vmatrix} 1 & 4 \\ 2 & -1 \end{vmatrix}$$

$$= \boldsymbol{i}(12 - 2) - \boldsymbol{j}(3 + 4) + \boldsymbol{k}(-1 - 8)$$

$$= \mathbf{10}\boldsymbol{i} - \mathbf{7}\boldsymbol{j} - \mathbf{9}\boldsymbol{k}$$

(ii) From equation (7), $|\boldsymbol{a} \times \boldsymbol{b}| = \sqrt{[(\boldsymbol{a} \bullet \boldsymbol{a})(\boldsymbol{b} \bullet \boldsymbol{b}) - (\boldsymbol{a} \bullet \boldsymbol{b})^2]}$

Now $\boldsymbol{a} \bullet \boldsymbol{a} = (1)(1) + (4 \times 4) + (-2)(-2) = 21$

$\boldsymbol{b} \bullet \boldsymbol{b} = (2)(2) + (-1)(-1) + (3)(3) = 14$

and $\boldsymbol{a} \bullet \boldsymbol{b} = (1)(2) + (4)(-1) + (-2)(3) = -8$

Thus $|\boldsymbol{a} \times \boldsymbol{b}| = \sqrt{(21 \times 14 - 64)} = \sqrt{230} = \mathbf{15.17}$

Practical application of vector products

For example, to find the moment and the magnitude of the moment of a force of $(\boldsymbol{i} + 2\boldsymbol{j} - 3\boldsymbol{k})$ Newton's about point B having co-ordinates (0, 1, 1), when the force acts on a line through A whose co-ordinates are (1, 3, 4):

The moment $\boldsymbol{M}$ about point B of a force vector $\boldsymbol{F}$ that has a position vector of $\boldsymbol{r}$ from A is given by: $\boldsymbol{M} = \boldsymbol{r} \times \boldsymbol{F}$

$\boldsymbol{r}$ is the vector from B to A, i.e. $\boldsymbol{r} = \boldsymbol{BA}$

But $\boldsymbol{BA} = \boldsymbol{BO} + \boldsymbol{OA} = \boldsymbol{OA} - \boldsymbol{OB}$ (see the final example in chapter 39), that is, $\boldsymbol{r} = (\boldsymbol{i} + 3\boldsymbol{j} + 4\boldsymbol{k}) - (\boldsymbol{j} + \boldsymbol{k}) = \boldsymbol{i} + 2\boldsymbol{j} + 3\boldsymbol{k}$

Moment, $\boldsymbol{M} = \boldsymbol{r} \times \boldsymbol{F} = (\boldsymbol{i} + 2\boldsymbol{j} + 3\boldsymbol{k}) \times (\boldsymbol{i} + 2\boldsymbol{j} - 3\boldsymbol{k})$

$$= \begin{vmatrix} i & j & k \\ 1 & 2 & 3 \\ 1 & 2 & -3 \end{vmatrix} = \boldsymbol{i}(-6-6) - \boldsymbol{j}(-3-3) + \boldsymbol{k}(2-2)$$

$$= \mathbf{-12}\boldsymbol{i} + \mathbf{6}\boldsymbol{j}\ \mathbf{Nm}$$

The magnitude of $\boldsymbol{M}$, $|\boldsymbol{M}| = |\boldsymbol{r} \times \boldsymbol{F}| = \sqrt{[(\boldsymbol{r} \bullet \boldsymbol{r})(\boldsymbol{F} \bullet \boldsymbol{F}) - (\boldsymbol{r} \bullet \boldsymbol{F})^2]}$

$$\boldsymbol{r} \bullet \boldsymbol{r} = (1)(1) + (2)(2) + (3)(3) = 14$$

$$\boldsymbol{F} \bullet \boldsymbol{F} = (1)(1) + (2)(2) + (-3)(-3) = 14$$

$$\boldsymbol{r} \bullet \boldsymbol{F} = (1)(1) + (2)(2) + (3)(-3) = -4$$

$$|\boldsymbol{M}| = \sqrt{[14 \times 14 - (-4)^2]} = \sqrt{180}\ \text{Nm} = \mathbf{13.42\ Nm}$$

Complex Numbers

41 Complex Numbers

Cartesian complex numbers

If the quadratic equation $x^2 + 2x + 5 = 0$ is solved using the quadratic formula then

$$x = \frac{-2 \pm \sqrt{[(2)^2 - (4)(1)(5)]}}{2(1)} = \frac{-2 \pm \sqrt{[-16]}}{2}$$

$$= \frac{-2 \pm \sqrt{[(16)(-1)]}}{2} = \frac{-2 \pm \sqrt{16}\sqrt{-1}}{2}$$

$$= \frac{-2 \pm 4\sqrt{-1}}{2} = -1 \pm 2\sqrt{-1}$$

It is not possible to evaluate $\sqrt{-1}$ in real terms. However, if an operator j is defined as $\boldsymbol{j = \sqrt{-1}}$ then the solution may be expressed as $x = -1 \pm j2$. $-1 + j2$ and $-1 - j2$ are known as **complex numbers**. Both solutions are of the form $a + jb$, 'a' being termed the **real part** and jb the **imaginary part**. A complex number of the form $a + jb$ is called a **Cartesian complex number**.

Since $\quad j = \sqrt{-1}$, then $j^2 = -1$,

$$j^3 = j^2 \times j = (-1) \times j = \boldsymbol{-j},$$

$$j^4 = j^2 \times j^2 = (-1) \times (-1) = \mathbf{1}$$

and $\quad j^{23} = j \times j^{22} = j \times (j^2)^{11} = j \times (-1)^{11} = j \times (-1) = \boldsymbol{-j}$

In pure mathematics the symbol i is used to indicate $\sqrt{-1}$ (i being the first letter of the word imaginary). However i is the symbol of electric current in engineering, and to avoid possible confusion the next letter in the alphabet, j, is used to represent $\sqrt{-1}$.

For example, the quadratic equation $2x^2 + 3x + 5 = 0$ is solved as follows:
Using the quadratic formula,

$$x = \frac{-3 \pm \sqrt{[(3)^2 - 4(2)(5)]}}{2(2)} = \frac{-3 \pm \sqrt{-31}}{4}$$

$$= \frac{-3 \pm \sqrt{(-1)}\sqrt{31}}{4} = \frac{-3 \pm j\sqrt{31}}{4}$$

Hence $\boldsymbol{x = -\frac{3}{4} + j\frac{\sqrt{31}}{4}}$ or $\boldsymbol{-0.750 \pm j1.392}$, correct to 3 decimal places.

(Note, a graph of $y = 2x^2 + 3x + 5$ does not cross the x-axis and hence $2x^2 + 3x + 5 = 0$ has no real roots).

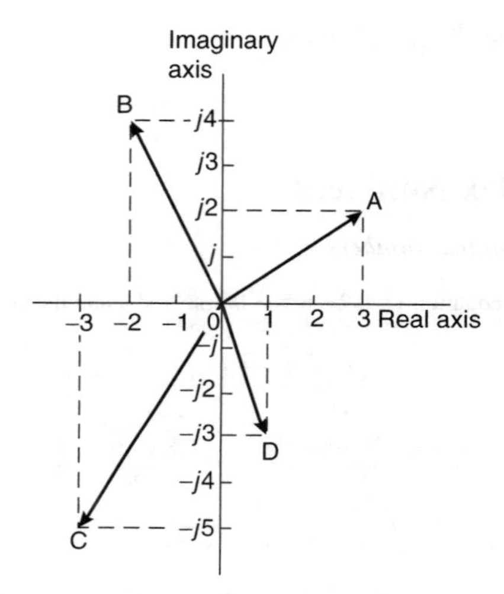

Figure 41.1

The Argand diagram

A complex number may be represented pictorially on rectangular or Cartesian axes. The horizontal (or x) axis is used to represent the real axis and the vertical (or y) axis is used to represent the imaginary axis. Such a diagram is called an **Argand diagram**. In Figure 41.1, the point A represents the complex number $(3 + j2)$ and is obtained by plotting the co-ordinates $(3, j2)$ as in graphical work. Figure 41.1 also shows the Argand points B, C and D representing the complex numbers $(-2 + j4)$, $(-3 - j5)$ and $(1 - j3)$ respectively.

Addition and subtraction of complex numbers

Two complex numbers are added/subtracted by adding/subtracting separately the two real parts and the two imaginary parts.
For example, if $Z_1 = a + jb$ and $Z_2 = c + jd$,

then $\quad Z_1 + Z_2 = (a + jb) + (c + jd) = (a + c) + j(b + d)$

and $\quad Z_1 - Z_2 = (a + jb) - (c + jd) = (a - c) + j(b - d)$

For example, $\quad (2 + j3) + (3 - j4) = 2 + j3 + 3 - j4 = \mathbf{5 - j1}$

and $\quad (2 + j3) - (3 - j4) = 2 + j3 - 3 + j4 = \mathbf{-1 + j7}$

Multiplication and division of complex numbers

Multiplication of complex numbers is achieved by assuming all quantities involved are real and then using $j^2 = -1$ to simplify.

Hence $(a + jb)(c + jd) = ac + a(jd) + (jb)c + (jb)(jd)$

$$= ac + jad + jbc + j^2bd$$

$$= (ac - bd) + j(ad + bc) \text{ since } j^2 = -1$$

For example, $(3 + j2)(4 - j5) = 12 - j15 + j8 - j^2 10$

$$= (12 - -10) + j(-15 + 8)$$

$$= \mathbf{22 - j7}$$

The **complex conjugate** of a complex number is obtained by changing the sign of the imaginary part. Hence the complex conjugate of $(a + jb)$ is $(a - jb)$. The product of a complex number and its complex conjugate is always a real number.
For example, $(3 + j4)(3 - j4) = 9 - j12 + j12 - j^2 16 = 9 + 16 = 25$
$[(a + jb)(a - jb)$ may be evaluated 'on sight' as $a^2 + b^2]$
Division of complex numbers is achieved by multiplying both numerator and denominator by the complex conjugate of the denominator.
For example,

$$\frac{2 - j5}{3 + j4} = \frac{2 - j5}{3 + j4} \times \frac{(3 - j4)}{(3 - j4)} = \frac{6 - j8 - j15 + j^2 20}{3^2 + 4^2}$$

$$= \frac{-14 - j23}{25} = \mathbf{\frac{-14}{25} - j\frac{23}{25}} \text{ or } \mathbf{-0.56 - j0.92}$$

Complex equations

If two complex numbers are equal, then their real parts are equal and their imaginary parts are equal. Hence if $a + jb = c + jd$, then $a = c$ and $b = d$
For example, solving the complex equation $(1 + j2)(-2 - j3) = a + jb$ gives:

$$(1 + j2)(-2 - j3) = a + jb$$

$$-2 - j3 - j4 - j^2 6 = a + jb$$

Hence $4 - j7 = a + jb$
Equating real and imaginary terms gives: $\boldsymbol{a = 4}$ and $\boldsymbol{b = -7}$

The polar form of a complex number

Let a complex number Z be $x + jy$ as shown in the Argand diagram of Figure 41.2.
Let distance OZ be r and the angle OZ makes with the positive real axis be θ.

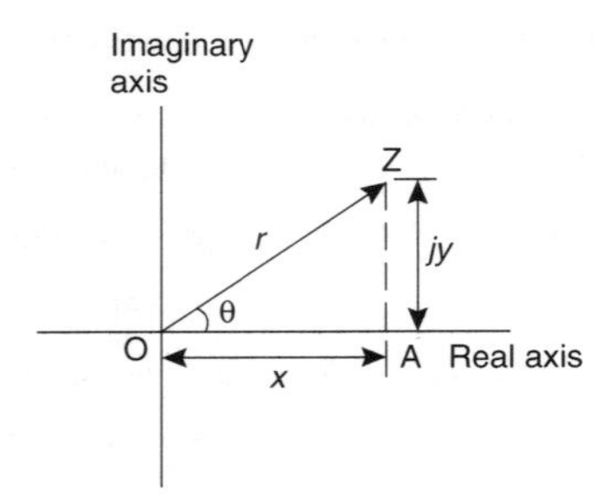

Figure 41.2

From trigonometry, $x = r\cos\theta$ and $y = r\sin\theta$
Hence $Z = x + jy = r\cos\theta + jr\sin\theta = r(\cos\theta + j\sin\theta)$
$Z = r(\cos\theta + j\sin\theta)$ is usually abbreviated to $Z = r\angle\theta$ which is known as the **polar form** of a complex number.
r is called the **modulus** (or magnitude) of Z and is written as mod Z or $|Z|$.
r is determined using Pythagoras' theorem on triangle OAZ in Figure 41.2,

i.e. $\boxed{\mathbf{r = \sqrt{x^2 + y^2}}}$

θ is called the **argument** (or amplitude) of Z and is written as arg Z.

By trigonometry on triangle OAZ, $\arg Z = \boxed{\boldsymbol{\theta = \tan^{-1}\frac{y}{x}}}$

Whenever changing from Cartesian form to polar form, or vice-versa, a sketch is invaluable for determining the quadrant in which the complex number occurs.
For example, expressing (a) $3 + j4$ and (b) $-3 + j4$ in polar form:

(a) $3 + j4$ is shown in Figure 41.3 and lies in the first quadrant.

$$\text{Modulus, } r = \sqrt{3^2 + 4^2} = 5$$

$$\text{and} \quad \text{argument } \theta = \tan^{-1}\tfrac{4}{3} = 53.13° = 53°8'$$

Hence $\mathbf{3 + j4 = 5\angle 53.13°}$

(b) $-3 + j4$ is shown in Figure 41.3 and lies in the second quadrant.
Modulus, $r = 5$ and angle $\alpha = 53.13°$, from part (a).
Argument $= 180° - 53.13° = 126.87°$ (i.e. the argument must be measured from the positive real axis)
Hence $\mathbf{-3 + j4 = 5\angle 126.87°}$

Similarly it may be shown that $\mathbf{(-3 - j4) = 5\angle 233.13°}$ **or** $\mathbf{5\angle -126.87°}$, (by convention the **principal value** is normally used, i.e. the numerically least value, such that $-\pi < \theta < \pi$), and $\mathbf{(3 - j4) = 5\angle -53.13°}$.

In another example, $7\angle -145°$ into $a + jb$ form:
$7\angle -145°$ is shown in Figure 41.4 and lies in the third quadrant.

$$\mathbf{7\angle -145°} = 7\cos(-145°) + j7\sin(-145°) = \mathbf{-5.734 - j4.015}$$

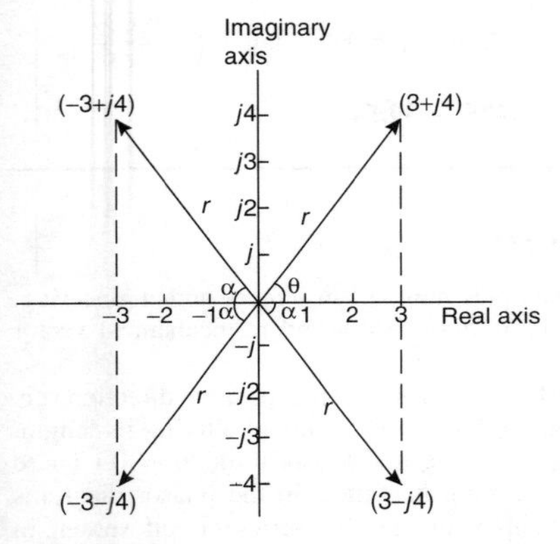

Figure 41.3

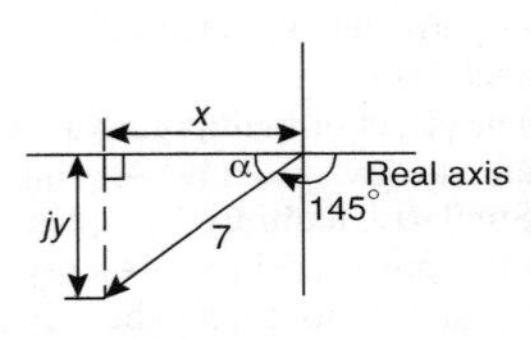

Figure 41.4

Multiplication and division in polar form

If $\quad Z_1 = r_1\angle\theta_1$ and $Z_2 = r_2\angle\theta_2$ then:

$$Z_1Z_2 = r_1r_2\angle(\theta_1 + \theta_2) \text{ and } \frac{Z_1}{Z_2} = \frac{r_1}{r_2}\angle(\theta_1 - \theta_2)$$

For example,

$$3\angle 16^\circ \times 5\angle{-44^\circ} \times 2\angle 80^\circ = (3 \times 5 \times 2)\angle[16^\circ + (-44^\circ) + 80^\circ]$$

$$= \mathbf{30\angle 52^\circ}$$

In another example, $\dfrac{16\angle 75^\circ}{2\angle 15^\circ} = \dfrac{16}{2}\angle(75^\circ - 15^\circ) = \mathbf{8\angle 60^\circ}$

In another example, to evaluate, in polar form $2\angle 30^\circ + 5\angle{-45^\circ} - 4\angle 120^\circ$:

$$2\angle 30^\circ = 2(\cos 30^\circ + j\sin 30^\circ) = 2\cos 30^\circ + j2\sin 30^\circ$$

$$= 1.732 + j1.000$$

$$5\angle{-45^\circ} = 5(\cos(-45^\circ) + j\sin(-45^\circ))$$

$$= 5\cos(-45^\circ) + j5\sin(-45^\circ) = 3.536 - j3.536$$

$$4\angle 120^\circ = 4(\cos 120^\circ + j\sin 120^\circ) = 4\cos 120^\circ + j4\sin 120^\circ$$

$$= -2.000 + j3.464$$

Hence

$$2\angle 30^\circ + 5\angle{-45^\circ} - 4\angle 120^\circ = (1.732 + j1.000) + (3.536 - j3.536) - (-2.000 + j3.464)$$

$$= 7.268 - j6.000,$$

which lies in the fourth quadrant

$$= \sqrt{7.268^2 + 6.000^2} \angle \tan^{-1}\left(\frac{-6.000}{7.268}\right)$$

$$= \mathbf{9.425\angle{-39.54^\circ}}$$

Applications of complex numbers

There are several applications of complex numbers in science and engineering, in particular in electrical alternating current theory and in mechanical vector analysis.

The effect of multiplying a phasor by j is to rotate it in a positive direction (i.e. anticlockwise) on an Argand diagram through 90° without altering its length. Similarly, multiplying a phasor by $-j$ rotates the phasor through $-90°$. These facts are used in a.c. theory since certain quantities in the phasor diagrams lie at 90° to each other. For example, in the *R-L* series circuit shown in Figure 41.5(a), V_L leads I by 90° (i.e. I lags V_L by 90°) and may be written as jV_L, the vertical axis being regarded as the imaginary axis of an Argand diagram. Thus $V_R + jV_L = V$ and since $V_R = IR$, $V = IX_L$ (where X_L is the inductive reactance, $2\pi fL$ ohms) and $V = IZ$ (where Z is the impedance) then $R + jX_L = Z$.

For example, $Z = (4 + j7)\ \Omega$ represents an impedance consisting of a 4 Ω resistance in series with an inductance of inductive reactance 7 Ω.

Similarly, for the *R-C* circuit shown in Figure 41.5(b), V_C lags I by 90° (i.e. I leads V_C by 90°) and $V_R - jV_C = V$, from which $R - jX_C = Z$ (where X_C is the capacitive reactance $\dfrac{1}{2\pi fC}$ ohms).

For example, $Z = (5 - j3)\ \Omega$ represents an impedance consisting of a 5 Ω resistance in series with a capacitance of capacitive reactance 3 Ω.

In another example, to determine the value of current I and its phase relative to the 240 V supply for the parallel circuit shown in Figure 41.6:

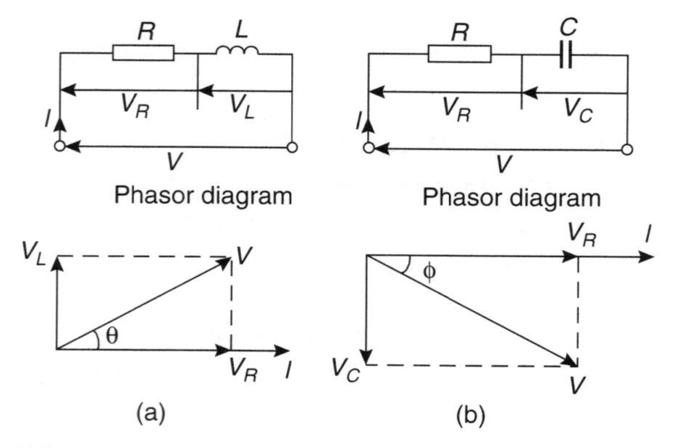

Figure 41.5

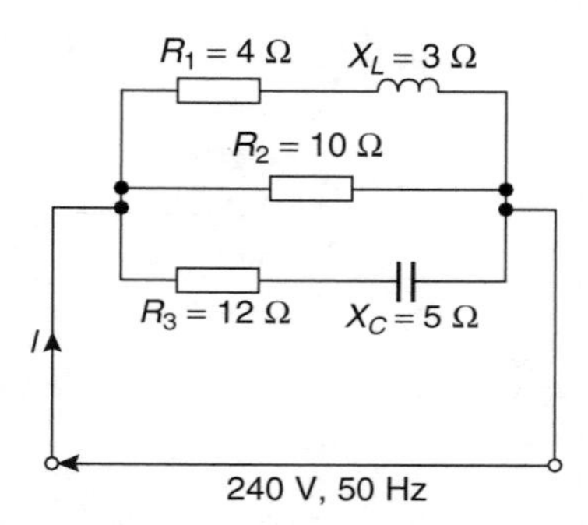

Figure 41.6

Current $I = \dfrac{V}{Z}$. Impedance Z for the three-branch parallel circuit is given by:
$\dfrac{1}{Z} = \dfrac{1}{Z_1}, + \dfrac{1}{Z_2} + \dfrac{1}{Z_3}$, where $Z_1 = 4 + j3$, $Z_2 = 10$ and $Z_3 = 12 - j5$

$$\text{Admittance, } Y_1 = \frac{1}{Z_1} = \frac{1}{4 + j3} = \frac{1}{4 + j3} \times \frac{4 - j3}{4 - j3}$$

$$= \frac{4 - j3}{4^2 + 3^2} = 0.160 - j0.120 \text{ siemens}$$

$$\text{Admittance, } Y_2 = \frac{1}{Z_2} = \frac{1}{10} = 0.10 \text{ siemens}$$

$$\text{Admittance, } Y_3 = \frac{1}{Z_3} = \frac{1}{12 - j5} = \frac{1}{12 - j5} \times \frac{12 + j5}{12 + j5}$$

$$= \frac{12 + j5}{12^2 + 5^2} = 0.0710 + j0.0296 \text{ siemens}$$

$$\text{Total admittance,} \quad Y = Y_1 + Y_2 + Y_3$$

$$= (0.160 - j0.120) + (0.10) + (0.0710 + j0.0296)$$

$$= 0.331 - j0.0904 = 0.343\angle -15.28^\circ \text{ siemens}$$

Current $I = \dfrac{V}{Z} = VY = (240\angle 0^\circ)(0.343\angle -15.28^\circ) = \mathbf{82.32\angle -15.28^\circ\ A}$

In another example, to determine the magnitude and direction of the resultant of the three coplanar forces shown in Figure 41.7:
Force A, $f_A = 10\angle 45^\circ$, force B, $f_B = 8\angle 120^\circ$ and force C, $f_C = 15\angle 210^\circ$

$$\text{The resultant force} \quad = f_A + f_B + f_C = 10\angle 45^\circ + 8\angle 120^\circ + 15\angle 210^\circ$$

$$= 10(\cos 45^\circ + j\sin 45^\circ) + 8(\cos 120^\circ + j\sin 120^\circ)$$

$$+ 15(\cos 210^\circ + j\sin 210^\circ)$$

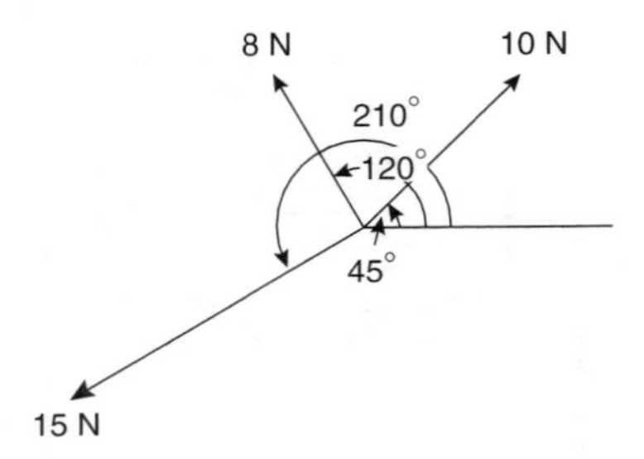

Figure 41.7

$$= (7.071 + j7.071) + (-4.00 + j6.928) + (-12.99 - j7.50)$$

$$= -9.919 + j6.499$$

Magnitude of resultant force $= \sqrt{(-9.919)^2 + 6.499^2} = \mathbf{11.86\ N}$

Direction of resultant force $= \tan^{-1}\left(\dfrac{6.499}{-9.919}\right) = \mathbf{146.77°}$

(since $-9.919 + j6.499$ lies in the second quadrant).

---

42 De Moivre's Theorem

Introduction

From multiplication of complex numbers in polar form,

$$(r\angle\theta) \times (r\angle\theta) = r^2\angle 2\theta$$

Similarly, $(r\angle\theta) \times (r\angle\theta) \times (r\angle\theta) = r^3\angle 3\theta$, and so on.
In general, **de Moivre's theorem** states:

$$\boxed{[r\angle\theta]^n = r^n\angle n\theta}$$

The theorem is true for all positive, negative and fractional values of n. The theorem is used to determine powers and roots of complex numbers.

Powers of complex numbers

For example, $[3\angle 20°]^4 = 3^4\angle(4 \times 20°) = 81\angle 80°$ by de Moivre's theorem.

In another example, to determine $(-2 + j3)^6$ in polar form:

$$(-2 + j3) = \sqrt{(-2)^2 + 3^2}\angle\tan^{-1}\frac{3}{-2} = \sqrt{13}\angle 123.69°,$$

since $-2 + j3$ lies in the second quadrant

$$
\begin{aligned}
(-2+j3)^6 &= [\sqrt{13}\angle 123.69^\circ]^6 \\
&= (\sqrt{13})^6\angle(6\times 123.69^\circ), \text{ by de Moivre's theorem} \\
&= 2197\angle 742.14^\circ \\
&= 2197\angle 382.14^\circ \quad (\text{since } 742.14 \equiv 742.14^\circ - 360^\circ = 382.14^\circ) \\
&= \mathbf{2197\angle 22.14^\circ} \quad (\text{since } 382.14^\circ \equiv 382.14^\circ - 360^\circ = 22.14^\circ)
\end{aligned}
$$

Roots of complex numbers

The **square root** of a complex number is determined by letting $n = \frac{1}{2}$ in de Moivre's theorem,

i.e. $\sqrt{r\angle\theta} = [r\angle\theta]^{1/2} = r^{1/2}\angle\frac{1}{2}\theta = \sqrt{r}\angle\frac{\theta}{2}$

There are two square roots of a real number, equal in size but opposite in sign.

For example, to determine the two square roots of the complex number $(5 + j12)$ in polar and Cartesian forms:

$$(5+j12) = \sqrt{5^2+12^2}\angle\tan^{-1}\frac{12}{5} = 13\angle 67.38^\circ$$

When determining square roots two solutions result. To obtain the second solution one way is to express $13\angle 67.38^\circ$ also as $13\angle(67.38^\circ + 360^\circ)$, i.e. $13\angle 427.38^\circ$. When the angle is divided by 2 an angle less than 360° is obtained.

Hence

$$
\begin{aligned}
\sqrt{5^2+12^2} &= \sqrt{13\angle 67.38^\circ} \text{ and } \sqrt{13\angle 427.38^\circ} \\
&= [13\angle 67.38^\circ]^{1/2} \text{ and } [13\angle 427.38^\circ]^{1/2} \\
&= 13^{1/2}\angle\left(\tfrac{1}{2}\times 67.38^\circ\right) \text{ and } 13^{1/2}\angle\left(\tfrac{1}{2}\times 427.38^\circ\right) \\
&= \sqrt{13}\angle 33.69^\circ \text{ and } \sqrt{13}\angle 213.69^\circ \\
&= 3.61\angle 33.69^\circ \text{ and } 3.61\angle 213.69^\circ
\end{aligned}
$$

Thus, in polar form, the two roots are $3.61\angle 33.69^\circ$ and $3.61\angle -146.69^\circ$

$$\sqrt{13}\angle 33.69^\circ = \sqrt{13}(\cos 33.69^\circ + j\sin 33.69^\circ) = 3.0 + j2.0$$

$$\sqrt{13}\angle 213.69^\circ = \sqrt{13}(\cos 213.69^\circ + j\sin 213.69^\circ) = -3.0 - j2.0$$

Thus, in Cartesian form, the two roots are $\pm(3.0 + j2.0)$.

From the Argand diagram shown in Figure 42.1 the two roots are seen to be 180° apart, which is always true when finding square roots of complex numbers.

In general, **when finding the nth root of a complex number, there are n solutions**. For example, there are three solutions to a cube root, five solutions to a fifth root, and so on. In the solutions to the roots of a complex number, the modulus, r, is always the same, but the arguments, θ, are different. Arguments are symmetrically spaced on an Argand diagram and are $\frac{360^\circ}{n}$ apart, where n

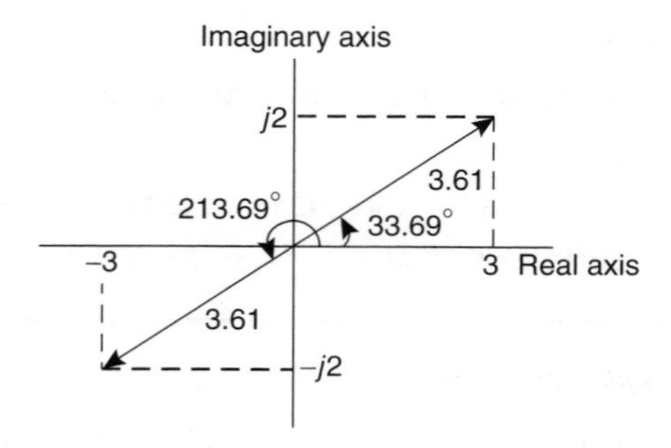

Figure 42.1

is the number of the roots required. Thus if one of the solutions to the cube root of a complex number is, say, $5\angle 20°$, the other two roots are symmetrically spaced $\frac{360°}{3}$, i.e. 120° from this root, and the three roots are $5\angle 20°$, $5\angle 140°$ and $5\angle 260°$.

The exponential form of a complex number

Certain mathematical functions may be expressed as power series, three examples being:

$$\text{(i) } e^x = 1 + x + \frac{x^2}{2!} + \frac{x^3}{3!} + \frac{x^4}{4!} + \frac{x^5}{5!} + \ldots \quad (1)$$

$$\text{(ii) } \sin x = x - \frac{x^3}{3!} + \frac{x^5}{5!} - \frac{x^7}{7!} + \ldots \quad (2)$$

$$\text{(iii) } \cos x = 1 - \frac{x^2}{2!} + \frac{x^4}{4!} - \frac{x^6}{6!} + \ldots \quad (3)$$

Replacing x in equation (1) by the imaginary number $j\theta$ gives:

$$e^{j\theta} = 1 + j\theta + \frac{(j\theta)^2}{2!} + \frac{(j\theta)^3}{3!} + \frac{(j\theta)^4}{4!} + \frac{(j\theta)^5}{5!} + \ldots$$

$$= 1 + j\theta + \frac{j^2\theta^2}{2!} + \frac{j^3\theta^3}{3!} + \frac{j^4\theta^4}{4!} + \frac{j^5\theta^5}{5!} + \ldots$$

By definition, $j = \sqrt{-1}$, hence $j^2 = -1$, $j^3 = -j$, $j^4 = 1$, $j^5 = j$, and so on.

Thus $e^{j\theta} = 1 + j\theta - \frac{\theta^2}{2!} - j\frac{\theta^3}{3!} + \frac{\theta^4}{4!} + j\frac{\theta^5}{5!} - \ldots$

Grouping real and imaginary terms gives:

$$e^{j\theta} = \left(1 - \frac{\theta^2}{2!} + \frac{\theta^4}{4!} - \ldots\right) + j\left(\theta - \frac{\theta^3}{3!} + \frac{\theta^5}{5!} - \ldots\right)$$

However, from equations (2) and (3):

$$\left(1 - \frac{\theta^2}{2!} + \frac{\theta^4}{4!} - \dots\right) = \cos\theta \quad \text{and} \quad \left(\theta - \frac{\theta^3}{3!} + \frac{\theta^5}{5!} - \dots\right) = \sin\theta$$

Thus $$\boxed{\boldsymbol{e^{j\theta} = \cos\theta + j\sin\theta}} \qquad (4)$$

Writing $-\theta$ for θ in equation (4), gives:

$$e^{j(-\theta)} = \cos(-\theta) + j\sin(-\theta)$$

However, $\cos(-\theta) = \cos\theta$ and $\sin(-\theta) = -\sin\theta$

Thus $$\boxed{\boldsymbol{e^{-j\theta} = \cos\theta - j\sin\theta}} \qquad (5)$$

The polar form of a complex number z is: $z = r(\cos\theta + j\sin\theta)$. But, from equation (4), $\cos\theta + j\sin\theta = e^{j\theta}$.

Therefore $\boxed{\boldsymbol{z = re^{j\theta}}}$

When a complex number is written in this way, it is said to be expressed in **exponential form**.

There are therefore three ways of expressing a complex number:

1. $z = (a + jb)$, called **Cartesian** or **rectangular form**,
2. $z = r(\cos\theta + j\sin\theta)$ or $r\angle\theta$, called **polar form**, and
3. $z = re^{j\theta}$ called **exponential form**.

The exponential form is obtained from the polar form. For example, $4\angle 30°$ becomes $4e^{j\pi/6}$ in exponential form. (Note that in $re^{j\theta}$, θ must be in radians).

For example, $(3 - j4) = \mathbf{5\angle{-53.13°}} = \mathbf{5\angle{-0.927}}$ in polar form

$$= \mathbf{5e^{-j0.927}} \text{ in exponential form}$$

In another example,

$$7.2e^{j1.5} = 7.2\angle 1.5 \text{ rad } (= 7.2\angle 85.94°) \text{ in polar form}$$
$$= 7.2\cos 1.5 + j7.2\sin 1.5$$
$$= \mathbf{(0.509 + j7.182)} \text{ in rectangular form}$$

In another example,

$$z = 2e^{1+j\pi/3} = (2e^1)(e^{j\pi/3}) \text{ by the laws of indices}$$
$$= (2e^1)\angle\frac{\pi}{3} \text{ (or } 2e\angle 60°\text{) in polar form}$$
$$= 2e\left(\cos\frac{\pi}{3} + j\sin\frac{\pi}{3}\right) = \mathbf{(2.718 + j4.708)} \text{ in Cartesian form}$$

In another example, if

$$z = 4e^{j1.3} \text{ then } \ln z = \ln(4e^{j1.3})$$

$$= \mathbf{\ln 4 + j1.3} \text{ (or } \mathbf{1.386 + j1.300}\text{) in Cartesian form.}$$

$$= \mathbf{1.90\angle 43.17°} \text{ or } \mathbf{1.90\angle 0.753} \text{ in polar form.}$$

In another example,

$$\ln(3 + j4) = \ln[5\angle 0.927] = \ln[5e^{j0.927}] = \ln 5 + \ln(e^{j0.927})$$

$$= \ln 5 + j0.927 = 1.609 + j0.927$$

$$= \mathbf{1.857\angle 29.95°} \text{ or } \mathbf{1.857\angle 0.523}$$

Matrices and Determinants

43 The Theory of Matrices and Determinants

Matrix notation

Matrices and determinants are mainly used for the solution of linear simultaneous equations. The theory of matrices and determinants is dealt with in this chapter and this theory is then used in chapter 44 to solve simultaneous equations. The coefficients of the variables for linear simultaneous equations may be shown in matrix form. The coefficients of x and y in the simultaneous equations

$$x + 2y = 3$$
$$4x - 5y = 6$$

become $\begin{pmatrix} 1 & 2 \\ 4 & -5 \end{pmatrix}$ in matrix notation

Similarly, the coefficients of p, q and r in the equations

$$1.3p - 2.0q + r = 7$$
$$3.7p + 4.8q - 7r = 3$$
$$4.1p + 3.8q + 12r = -6$$

become $\begin{pmatrix} 1.3 & -2.0 & 1 \\ 3.7 & 4.8 & -7 \\ 4.1 & 3.8 & 12 \end{pmatrix}$ in matrix form

The numbers within a matrix are called an **array** and the coefficients forming the array are called the **elements** of the matrix. The number of rows in a matrix is usually specified by m and the number of columns by n and a matrix referred to as an 'm by n' matrix. Thus, $\begin{pmatrix} 2 & 3 & 6 \\ 4 & 5 & 7 \end{pmatrix}$ is a '2 by 3' matrix.

Matrices cannot be expressed as a single numerical value, but they can often be simplified or combined, and unknown element values can be determined by comparison methods. Just as there are rules for addition, subtraction, multiplication and division of numbers in arithmetic, rules for these operations can be applied to matrices and the rules of matrices are such that they obey most of those governing the algebra of numbers.

Addition, Subtraction and Multiplication of Matrices

Addition of matrices

Corresponding elements in two matrices may be added to form a single matrix. **For example**,

$$\begin{pmatrix} 2 & -1 \\ -7 & 4 \end{pmatrix} + \begin{pmatrix} -3 & 0 \\ 7 & -4 \end{pmatrix} = \begin{pmatrix} 2+(-3) & -1+0 \\ -7+7 & 4+(-4) \end{pmatrix}$$

$$= \begin{pmatrix} \mathbf{-1} & \mathbf{-1} \\ \mathbf{0} & \mathbf{0} \end{pmatrix}$$

Subtraction of matrices

If A is a matrix and B is another matrix, then $(A - B)$ is a single matrix formed by subtracting the elements of B from the corresponding elements of A.
For example,

$$\begin{pmatrix} 2 & -1 \\ -7 & 4 \end{pmatrix} - \begin{pmatrix} -3 & 0 \\ 7 & -4 \end{pmatrix} = \begin{pmatrix} 2-(-3) & -1-0 \\ -7-7 & 4-(-4) \end{pmatrix}$$

$$= \begin{pmatrix} \mathbf{5} & \mathbf{-1} \\ \mathbf{-14} & \mathbf{8} \end{pmatrix}$$

Multiplication

When a matrix is multiplied by a number, called **scalar multiplication**, a single matrix results in which each element of the original matrix has been multiplied by the number.
For example, if $A = \begin{pmatrix} -3 & 0 \\ 7 & -4 \end{pmatrix}$, $B = \begin{pmatrix} 2 & -1 \\ -7 & 4 \end{pmatrix}$ and $C = \begin{pmatrix} 1 & 0 \\ -2 & -4 \end{pmatrix}$

then $\quad 2A - 3B + 4C = 2\begin{pmatrix} -3 & 0 \\ 7 & -4 \end{pmatrix} - 3\begin{pmatrix} 2 & -1 \\ -7 & 4 \end{pmatrix} + 4\begin{pmatrix} 1 & 0 \\ -2 & -4 \end{pmatrix}$

$$= \begin{pmatrix} -6 & 0 \\ 14 & -8 \end{pmatrix} - \begin{pmatrix} 6 & -3 \\ -21 & 12 \end{pmatrix} + \begin{pmatrix} 4 & 0 \\ -8 & -16 \end{pmatrix}$$

$$= \begin{pmatrix} -6-6+4 & 0-(-3)+0 \\ 14-(-21)+(-8) & -8-12+(-16) \end{pmatrix}$$

$$= \begin{pmatrix} \mathbf{-8} & \mathbf{3} \\ \mathbf{27} & \mathbf{-36} \end{pmatrix}$$

When a matrix A is multiplied by another matrix B, a single matrix results in which elements are obtained from the sum of the products of the corresponding rows of A and the corresponding columns of B.
Two matrices A and B may be multiplied together, provided the number of elements in the rows of matrix A are equal to the number of elements in the columns of matrix B. In general terms, when multiplying a matrix of dimensions (m by n) by a matrix of dimensions (n by r), the resulting matrix has dimensions (m by r). Thus a 2 by 3 matrix multiplied by a 3 by 1 matrix gives a matrix of dimensions 2 by 1.
For example, let $A = \begin{pmatrix} 2 & 3 \\ 1 & -4 \end{pmatrix}$ and $B = \begin{pmatrix} -5 & 7 \\ -3 & 4 \end{pmatrix}$

Let $A \times B = C$ where $C = \begin{pmatrix} c_{11} & c_{12} \\ c_{21} & c_{22} \end{pmatrix}$
C_{11} is the sum of the products of the first row elements of A and the first column elements of B taken one at a time, i.e. $C_{11} = (2 \times (-5)) + (3 \times (-3)) = -19$. C_{12} is the sum of the products of the first row elements of A and the second column elements of B, taken one at a time, i.e. $C_{12} = (2 \times 7) + (3 \times 4) = 26$. C_{21} is the sum of the products of the second row elements of A and the first column elements of B, taken one at a time, i.e.

$C_{21} = (1 \times (-5)) + (-4) \times (-3)) = 7$. Finally, C_{22} is the sum of the products of the second row elements of A and the second column elements of B, taken one at a time, i.e. $C_{22} = (1 \times 7) + (-4) \times 4) = -9$

Thus, $\boldsymbol{A} \times \boldsymbol{B} = \begin{pmatrix} \mathbf{-19} & \mathbf{26} \\ \mathbf{7} & \mathbf{-9} \end{pmatrix}$

In another example

$$\begin{pmatrix} 3 & 4 & 0 \\ -2 & 6 & -3 \\ 7 & -4 & 1 \end{pmatrix} \times \begin{pmatrix} 2 \\ 5 \\ -1 \end{pmatrix}$$

$$= \begin{pmatrix} (3 \times 2) + (4 \times 5) + (0 \times (-1)) \\ (-2 \times 2) + (6 \times 5) + (-3 \times (-1)) \\ (7 \times 2) + (-4 \times 5) + (1 \times (-1)) \end{pmatrix} = \begin{pmatrix} \mathbf{26} \\ \mathbf{29} \\ \mathbf{-7} \end{pmatrix}$$

In algebra, the commutative law of multiplication states that $a \times b = b \times a$. For matrices, this law is only true in a few special cases, and in general $A \times B$ is **not** equal to $B \times A$

The unit matrix

A **unit matrix**, I, is one in which all elements of the leading diagonal (\) have a value of 1 and all other elements have a value of 0. Multiplication of a matrix by I is the equivalent of multiplying by 1 in arithmetic.

The determinant of a 2 by 2 matrix

The **determinant** of a 2 by 2 matrix, $\begin{pmatrix} a & b \\ c & d \end{pmatrix}$ is defined as $(ad - bc)$.

The elements of the determinant of a matrix are written between vertical lines. Thus, the determinant of $\begin{pmatrix} 3 & -4 \\ 1 & 6 \end{pmatrix}$ is written as $\begin{vmatrix} 3 & -4 \\ 1 & 6 \end{vmatrix}$ and is equal to $(3 \times 6) - (-4 \times 1)$, i.e. $18 - (-4) = 22$. Hence the determinant of a matrix can be expressed as a single numerical value, i.e. $\begin{vmatrix} 3 & -4 \\ 1 & 6 \end{vmatrix} = 22$

The inverse or reciprocal of a 2 by 2 matrix

The inverse of matrix A is A^{-1} such that $A \times A^{-1} = I$, the unit matrix.

For any matrix $\begin{pmatrix} p & q \\ r & s \end{pmatrix}$ the inverse may be obtained by:

(i) interchanging the positions of p and s,
(ii) changing the signs of q and r, and
(iii) multiplying this new matrix by the reciprocal of the determinant of $\begin{pmatrix} p & q \\ r & s \end{pmatrix}$

Thus the inverse of matrix $\begin{pmatrix} 1 & 2 \\ 3 & 4 \end{pmatrix}$ is

$$\frac{1}{4-6}\begin{pmatrix} 4 & -2 \\ -3 & 1 \end{pmatrix} = \begin{pmatrix} -2 & 1 \\ \frac{3}{2} & -\frac{1}{2} \end{pmatrix}$$

The determinant of a 3 by 3 matrix

(i) The **minor** of an element of a 3 by 3 matrix is the value of the 2 by 2 determinant obtained by covering up the row and column containing that element.

Thus for the matrix $\begin{pmatrix} 1 & 2 & 3 \\ 4 & 5 & 6 \\ 7 & 8 & 9 \end{pmatrix}$ the minor of element 4 is obtained by covering the row (4 5 6) and the column $\begin{pmatrix} 1 \\ 4 \\ 7 \end{pmatrix}$, leaving the 2 by 2 determinant $\begin{vmatrix} 2 & 3 \\ 8 & 9 \end{vmatrix}$, i.e. the minor of element 4 is $(2 \times 9) - (3 \times 8) = -6$

(ii) The sign of a minor depends on its position within the matrix, the sign pattern being $\begin{pmatrix} + & - & + \\ - & + & - \\ + & - & + \end{pmatrix}$. Thus the signed-minor of element 4 in the matrix $\begin{pmatrix} 1 & 2 & 3 \\ 4 & 5 & 6 \\ 7 & 8 & 9 \end{pmatrix}$ is $-\begin{vmatrix} 2 & 3 \\ 8 & 9 \end{vmatrix} = -(-6) = 6$

The signed-minor of an element is called the **cofactor** of the element.

(iii) **The value of a 3 by 3 determinant is the sum of the products of the elements and their cofactors of any row or any column of the corresponding 3 by 3 matrix.**

There are thus six different ways of evaluating a 3×3 determinant-and all should give the same value.

For example, to evaluate $\begin{vmatrix} 1 & 4 & -3 \\ -5 & 2 & 6 \\ -1 & -4 & 2 \end{vmatrix}$:

Using the first row:

$$\begin{vmatrix} 1 & 4 & -3 \\ -5 & 2 & 6 \\ -1 & -4 & 2 \end{vmatrix} = 1\begin{vmatrix} 2 & 6 \\ -4 & 2 \end{vmatrix} - 4\begin{vmatrix} -5 & 6 \\ -1 & 2 \end{vmatrix} + (-3)\begin{vmatrix} -5 & 2 \\ -1 & -4 \end{vmatrix}$$

$$= (4 + 24) - 4(-10 + 6) - 3(20 + 2)$$

$$= 28 + 16 - 66 = \mathbf{-22}$$

Using the second column:

$$\begin{vmatrix} 1 & 4 & -3 \\ -5 & 2 & 6 \\ -1 & -4 & 2 \end{vmatrix} = -4\begin{vmatrix} -5 & 6 \\ -1 & 2 \end{vmatrix} + 2\begin{vmatrix} 1 & -3 \\ -1 & 2 \end{vmatrix} - (-4)\begin{vmatrix} 1 & -3 \\ -5 & 6 \end{vmatrix}$$

$$= -4(-10 + 6) + 2(2 - 3) + 4(6 - 15)$$

$$= 16 - 2 - 36 = \mathbf{-22}$$

The inverse or reciprocal of a 3 by 3 matrix

The **adjoint** of a matrix A is obtained by:

(i) forming a matrix B of the cofactors of A, and

(ii) **transposing** matrix B to give B^T, where B^T is the matrix obtained by writing the rows of B as the columns of B^T. Then $\mathbf{adj}\ \boldsymbol{A} = \boldsymbol{B}^T$

The **inverse of matrix** A, A^{-1} is given by $\boldsymbol{A}^{-1} = \dfrac{\mathbf{adj}\ \boldsymbol{A}}{|\boldsymbol{A}|}$ where adj A is the adjoint of matrix A and $|A|$ is the determinant of matrix A.

For example, to find the inverse of $\begin{pmatrix} 1 & 5 & -2 \\ 3 & -1 & 4 \\ -3 & 6 & -7 \end{pmatrix}$

$$\text{Inverse} = \frac{\text{adjoint}}{\text{determinant}}$$

The matrix of cofactors is $\begin{pmatrix} -17 & 9 & 15 \\ 23 & -13 & -21 \\ 18 & -10 & -16 \end{pmatrix}$

The transpose of the matrix of cofactors (i.e. the adjoint) is

$$\begin{pmatrix} -17 & 23 & 18 \\ 9 & -13 & -10 \\ 15 & -21 & -16 \end{pmatrix}$$

The determinant of

$$\begin{pmatrix} 1 & 5 & -2 \\ 3 & -1 & 4 \\ -3 & 6 & -7 \end{pmatrix} = 1(7 - 24) - 5(-21 + 12) - 2(18 - 3)$$

$$= -17 + 45 - 30 = -2$$

Hence the inverse of

$$\begin{pmatrix} 1 & 5 & -2 \\ 3 & -1 & 4 \\ -3 & 6 & -7 \end{pmatrix} = \frac{\begin{pmatrix} -17 & 23 & 18 \\ 9 & -13 & -10 \\ 15 & -21 & -16 \end{pmatrix}}{-2} = \begin{pmatrix} \mathbf{8.5} & \mathbf{-11.5} & \mathbf{-9} \\ \mathbf{-4.5} & \mathbf{6.5} & \mathbf{5} \\ \mathbf{-7.5} & \mathbf{10.5} & \mathbf{8} \end{pmatrix}$$

---

44 The Solution of Simultaneous Equations by Matrices and Determinants

Solution of simultaneous equations by matrices

Two unknowns

The procedure for solving linear simultaneous equations in **two unknowns using matrices** is:

(i) write the equations in the form

$$a_1x + b_1y = c_1$$
$$a_2x + b_2y = c_2$$

(ii) write the matrix equation corresponding to these equations,

i.e. $$\begin{pmatrix} a_1 & b_1 \\ a_2 & b_2 \end{pmatrix} \times \begin{pmatrix} x \\ y \end{pmatrix} = \begin{pmatrix} c_1 \\ c_2 \end{pmatrix}$$

(iii) determine the inverse matrix of $\begin{pmatrix} a_1 & b_1 \\ a_2 & b_2 \end{pmatrix}$,

i.e. $$\frac{1}{a_1b_2 - b_1a_2}\begin{pmatrix} b_2 & -b_1 \\ -a_2 & a_1 \end{pmatrix}, \text{ (from chapter 43)}$$

(iv) multiply each side of (ii) by the inverse matrix, and
(v) solve for x and y by equating corresponding elements.

For example, using matrices to solve the simultaneous equations:

$$3x + 5y - 7 = 0 \quad (1)$$
$$4x - 3y - 19 = 0 \quad (2)$$

(i) Writing the equations in the $a_1x + b_1y = c$ form gives:

$$3x + 5y = 7$$
$$4x - 3y = 19$$

(ii) The matrix equation is $\begin{pmatrix} 3 & 5 \\ 4 & -3 \end{pmatrix} \times \begin{pmatrix} x \\ y \end{pmatrix} = \begin{pmatrix} 7 \\ 19 \end{pmatrix}$

(iii) The inverse of matrix $\begin{pmatrix} 3 & 5 \\ 4 & -3 \end{pmatrix}$ is

$$\frac{1}{3 \times (-3) - 5 \times 4}\begin{pmatrix} -3 & -5 \\ -4 & 3 \end{pmatrix} = \begin{pmatrix} \frac{3}{29} & \frac{5}{29} \\ \frac{4}{29} & \frac{-3}{29} \end{pmatrix}$$

(iv) Multiplying each side of (ii) by (iii) and remembering that $A \times A^{-1} = I$, the unit matrix, gives:

$$\begin{pmatrix} 1 & 0 \\ 0 & 1 \end{pmatrix}\begin{pmatrix} x \\ y \end{pmatrix} = \begin{pmatrix} \frac{3}{29} & \frac{5}{29} \\ \frac{4}{29} & \frac{-3}{29} \end{pmatrix} \times \begin{pmatrix} 7 \\ 19 \end{pmatrix}$$

Thus $$\begin{pmatrix} x \\ y \end{pmatrix} = \begin{pmatrix} \frac{21}{29} + \frac{95}{29} \\ \frac{28}{29} - \frac{57}{29} \end{pmatrix} \quad \text{i.e.} \quad \begin{pmatrix} x \\ y \end{pmatrix} = \begin{pmatrix} 4 \\ -1 \end{pmatrix}$$

(v) By comparing corresponding elements: $\mathbf{x = 4}$ **and** $\mathbf{y = -1}$**,** which can be checked in the original equations.

Three unknowns

The procedure for solving linear simultaneous equations in **three unknowns using matrices** is:

(i) write the equations in the form

$$a_1x + b_1y + c_1z = d_1$$

$$a_2x + b_2y + c_2z = d_2$$

$$a_3x + b_3y + c_3z = d_3$$

(ii) write the matrix equation corresponding to these equations, i.e.

$$\begin{pmatrix} a_1 & b_1 & c_1 \\ a_2 & b_2 & c_2 \\ a_3 & b_3 & c_3 \end{pmatrix} \times \begin{pmatrix} x \\ y \\ z \end{pmatrix} = \begin{pmatrix} d_1 \\ d_2 \\ d_3 \end{pmatrix}$$

(iii) determine the inverse matrix of $\begin{pmatrix} a_1 & b_1 & c_1 \\ a_2 & b_2 & c_2 \\ a_3 & b_3 & c_3 \end{pmatrix}$ (see chapter 43)

(iv) multiply each side of (ii) by the inverse matrix, and

(v) solve for x, y and z by equating the corresponding elements.

For example, using matrices to solve the simultaneous equations:

$$x + y + z - 4 = 0 \quad (1)$$

$$2x - 3y + 4z - 33 = 0 \quad (2)$$

$$3x - 2y - 2z - 2 = 0 \quad (3)$$

(i) Writing the equations in the $a_1x + b_1y + c_1z = d_1$ form gives:

$$x + y + z = 4$$

$$2x - 3y + 4z = 33$$

$$3x - 2y - 2z = 2$$

(ii) The matrix equation is $\begin{pmatrix} 1 & 1 & 1 \\ 2 & -3 & 4 \\ 3 & -2 & -2 \end{pmatrix} \times \begin{pmatrix} x \\ y \\ z \end{pmatrix} = \begin{pmatrix} 4 \\ 33 \\ 2 \end{pmatrix}$

(iii) The inverse matrix of $A = \begin{pmatrix} 1 & 1 & 1 \\ 2 & -3 & 4 \\ 3 & -2 & -2 \end{pmatrix}$ is given by $A^{-1} = \dfrac{\text{adj } A}{|A|}$

The adjoint of A is the transpose of the matrix of the cofactors of the elements (see chapter 43). The matrix of cofactors is $\begin{pmatrix} 14 & 16 & 5 \\ 0 & -5 & 5 \\ 7 & -2 & -5 \end{pmatrix}$

and the transpose of this matrix gives: $\text{adj}\, A = \begin{pmatrix} 14 & 0 & 7 \\ 16 & -5 & -2 \\ 5 & 5 & -5 \end{pmatrix}$

The determinant of A, i.e. the sum of the products of elements and their cofactors, using a first row expansion is

$$1\begin{vmatrix}-3 & 4\\ -2 & -2\end{vmatrix} - 1\begin{vmatrix}2 & 4\\ 3 & -2\end{vmatrix} + 1\begin{vmatrix}2 & -3\\ 3 & -2\end{vmatrix}$$

$$= (1 \times 14) - (1 \times -16) + (1 \times 5) = 35$$

Hence the inverse of A, $A^{-1} = \dfrac{1}{35}\begin{pmatrix}14 & 0 & 7\\ 16 & -5 & -2\\ 5 & 5 & -5\end{pmatrix}$

(iv) Multiplying each side of (ii) by (iii), and remembering that $A \times A^{-1} = I$, the unit matrix, gives:

$$\begin{pmatrix}1 & 0 & 0\\ 0 & 1 & 0\\ 0 & 0 & 1\end{pmatrix} \times \begin{pmatrix}x\\ y\\ z\end{pmatrix} = \frac{1}{35}\begin{pmatrix}14 & 0 & 7\\ 16 & -5 & -2\\ 5 & 5 & -5\end{pmatrix} \times \begin{pmatrix}4\\ 33\\ 2\end{pmatrix}$$

$$\begin{pmatrix}x\\ y\\ z\end{pmatrix} = \frac{1}{35}\begin{pmatrix}(14 \times 4) + (0 \times 33) + (7 \times 2)\\ (16 \times 4) + (-5 \times 33) + ((-2) \times 2)\\ (5 \times 4) + (5 \times 33) + ((-5) \times 2)\end{pmatrix}$$

$$= \frac{1}{35}\begin{pmatrix}70\\ -105\\ 175\end{pmatrix} = \begin{pmatrix}2\\ -3\\ 5\end{pmatrix}$$

(v) By comparing corresponding elements, $\boldsymbol{x = 2, y = -3, z = 5}$, which can be checked in the original equations.

Solution of simultaneous equations by determinants

Two unknowns

When solving linear simultaneous equations in **two unknowns using determinants**:

(i) write the equations in the form

$$a_1x + b_1y + c_1 = 0$$

$$a_2x + b_2y + c_2 = 0$$

(ii) the solution is given by $\dfrac{x}{D_x} = \dfrac{-y}{D_y} = \dfrac{1}{D}$, where

$D_x = \begin{vmatrix}b_1 & c_1\\ b_2 & c_2\end{vmatrix}$ i.e. the determinant of the coefficients left when the x-column is covered up,

$D_y = \begin{vmatrix}a_1 & c_1\\ a_2 & c_2\end{vmatrix}$ i.e. the determinant of the coefficients left when the y-column is covered up,

and

$D = \begin{vmatrix}a_1 & b_1\\ a_2 & b_2\end{vmatrix}$ i.e. the determinant of the coefficients left when the constants-column is covered up.

For example, to solve the following simultaneous equations using determinants:

$$3x - 4y = 12$$
$$7x + 5y = 6.5$$

Following the above procedure:

(i) $3x - 4y - 12 = 0$

$7x + 5y - 6.5 = 0$

(ii) $$\frac{x}{\begin{vmatrix} -4 & -12 \\ 5 & -6.5 \end{vmatrix}} = \frac{-y}{\begin{vmatrix} 3 & -12 \\ 7 & -6.5 \end{vmatrix}} = \frac{1}{\begin{vmatrix} 3 & -4 \\ 7 & 5 \end{vmatrix}}$$

i.e. $$\frac{x}{(-4)(-6.5) - (-12)(5)} = \frac{-y}{(3)(-6.5) - (-12)(7)}$$
$$= \frac{1}{(3)(5) - (-4)(7)}$$

i.e. $$\frac{x}{26 + 60} = \frac{-y}{-19.5 + 84} = \frac{1}{15 + 28}$$

i.e. $$\frac{x}{86} = \frac{-y}{64.5} = \frac{1}{43}$$

Since $\dfrac{x}{86} = \dfrac{1}{43}$ then $\boldsymbol{x} = \dfrac{86}{43} = \mathbf{2}$

and since $\dfrac{-y}{64.5} = \dfrac{1}{43}$ then $\boldsymbol{y} = -\dfrac{64.5}{43} = \mathbf{-1.5}$

Three unknowns

When solving simultaneous equations in **three unknowns using determinants**:

(i) write the equations in the form

$$a_1x + b_1y + c_1z + d_1 = 0$$
$$a_2x + b_2y + c_2z + d_2 = 0$$
$$a_3x + b_3y + c_3z + d_3 = 0$$

(ii) the solution is given by: $\dfrac{x}{D_x} = \dfrac{-y}{D_y} = \dfrac{z}{D_z} = \dfrac{-1}{D}$

where $D_x = \begin{vmatrix} b_1 & c_1 & d_1 \\ b_2 & c_2 & d_2 \\ b_3 & c_3 & d_3 \end{vmatrix}$ i.e. the determinant of the coefficients obtained by covering up the x column

$D_y = \begin{vmatrix} a_1 & c_1 & d_1 \\ a_2 & c_2 & d_2 \\ a_3 & c_3 & d_3 \end{vmatrix}$ i.e. the determinant of the coefficients obtained by covering up the y column

$$D_z = \begin{vmatrix} a_1 & b_1 & d_1 \\ a_2 & b_2 & d_2 \\ a_3 & b_2 & d_3 \end{vmatrix}$$ i.e. the determinant of the coefficients obtained by covering up the z column

and $$D = \begin{vmatrix} a_1 & b_1 & c_1 \\ a_2 & b_2 & c_2 \\ a_3 & b_3 & c_3 \end{vmatrix}$$ i.e. the determinant of the coefficients obtained by covering up the constants column.

For example, a d.c. circuit comprises three closed loops. Applying Kirchhoff's laws to the closed loops gives the following equations for current flow in milliamperes:

$$2I_1 + 3I_2 - 4I_3 = 26$$
$$I_1 - 5I_2 - 3I_3 = -87$$
$$-7I_1 + 2I_2 + 6I_3 = 12$$

Using determinants to solve for I_1, I_2 and I_3:
Following the above procedure:

(i) $2I_1 + 3I_2 - 4I_3 - 26 \quad = 0$

$I_1 - 5I_2 - 3I_3 + 87 \quad = 0$

$-7I_1 + 2I_2 + 6I_3 - 12 = 0$

(ii) The solution is given by: $\dfrac{I_1}{D_{I_1}} = \dfrac{-I_2}{D_{I_2}} = \dfrac{I_3}{D_{I_3}} = \dfrac{-1}{D}$, where

$$D_{I_1} = \begin{vmatrix} 3 & -4 & -26 \\ -5 & -3 & 87 \\ 2 & 6 & -12 \end{vmatrix}$$

$$= (3)\begin{vmatrix} -3 & 87 \\ 6 & -12 \end{vmatrix} - (-4)\begin{vmatrix} -5 & 87 \\ 2 & -12 \end{vmatrix} + (-26)\begin{vmatrix} -5 & -3 \\ 2 & 6 \end{vmatrix}$$

$$= 3(-486) + 4(-114) - 26(-24) = \mathbf{-1290}$$

$$D_{I_2} = \begin{vmatrix} 2 & -4 & -26 \\ 1 & -3 & 87 \\ -7 & 6 & -12 \end{vmatrix}$$

$$= (2)(36 - 522) - (-4)(-12 + 609) + (-26)(6 - 21)$$

$$= -972 + 2388 + 390 = \mathbf{1806}$$

$$D_{I_3} = \begin{vmatrix} 2 & 3 & -26 \\ 1 & -5 & 87 \\ -7 & 2 & -12 \end{vmatrix}$$

$$= (2)(60 - 174) - (3)(-12 + 609) + (-26)(2 - 35)$$

$$= -228 - 1791 + 858 = \mathbf{-1161}$$

and $$D = \begin{vmatrix} 2 & 3 & -4 \\ 1 & -5 & -3 \\ -7 & 2 & 6 \end{vmatrix}$$

$$= (2)(-30 + 6) - (3)(6 - 21) + (-4)(2 - 35)$$

$$= -48 + 45 + 132 = \mathbf{129}$$

Thus $$\frac{I_1}{-1290} = \frac{-I_2}{1806} = \frac{I_3}{-1161} = \frac{-1}{129}$$

giving $$\boldsymbol{I_1} = -\frac{-1290}{129} = \mathbf{10\ mA} \quad \boldsymbol{I_2} = \frac{1806}{129} = \mathbf{14\ mA}$$

and $$\boldsymbol{I_3} = -\frac{-1161}{129} = \mathbf{9\ mA}$$

Solution of simultaneous equations using Cramer's rule

Cramer's rule states that if

$$a_{11}x + a_{12}y + a_{13}z = b_1$$

$$a_{21}x + a_{22}y + a_{23}z = b_2$$

$$a_{31}x + a_{32}y + a_{33}z = b_3$$

then $\boldsymbol{x} = \dfrac{\boldsymbol{D_x}}{\boldsymbol{D}}$, $\boldsymbol{y} = \dfrac{\boldsymbol{D_y}}{\boldsymbol{D}}$ and $\boldsymbol{z} = \dfrac{\boldsymbol{D_z}}{\boldsymbol{D}}$, where $D = \begin{vmatrix} a_{11} & a_{12} & a_{13} \\ a_{21} & a_{22} & a_{23} \\ a_{31} & a_{32} & a_{33} \end{vmatrix}$

$$D_x = \begin{vmatrix} b_1 & a_{12} & a_{13} \\ b_2 & a_{22} & a_{23} \\ b_3 & a_{32} & a_{33} \end{vmatrix}$$ i.e. the x-column has been replaced by the R.H.S. b column

$$D_y = \begin{vmatrix} a_{11} & b_1 & a_{13} \\ a_{21} & b_2 & a_{23} \\ a_{31} & b_3 & a_{33} \end{vmatrix}$$ i.e. the y-column has been replaced by the R.H.S. b column

$$D_z = \begin{vmatrix} a_{11} & a_{12} & b_1 \\ a_{21} & a_{22} & b_2 \\ a_{31} & a_{32} & b_3 \end{vmatrix}$$ i.e. the z-column has been replaced by the R.H.S. b column

For example, to solve the following simultaneous equations using Cramer's rule

$$x + y + z = 4$$

$$2x - 3y + 4z = 33$$

$$3x - 2y - 2z = 2$$

Following the above method:

$$\boldsymbol{D} = \begin{vmatrix} 1 & 1 & 1 \\ 2 & -3 & 4 \\ 3 & -2 & -2 \end{vmatrix} = 1(6 - -8) - 1(-4 - 12) + 1(-4 - -9)$$

$$= 14 + 16 + 5 = \mathbf{35}$$

$$\boldsymbol{D_x} = \begin{vmatrix} 4 & 1 & 1 \\ 33 & -3 & 4 \\ 2 & -2 & -2 \end{vmatrix} = 4(6 - -8) - 1(-66 - 8) + 1(-66 - -6)$$

$$= 56 + 74 - 60 = \mathbf{70}$$

$$D_y = \begin{vmatrix} 1 & 4 & 1 \\ 2 & 33 & 4 \\ 3 & 2 & -2 \end{vmatrix} = 1(-66-8) - 4(-4-12) + 1(4-99)$$

$$= -74 + 64 - 95 = \mathbf{-105}$$

$$D_z = \begin{vmatrix} 1 & 1 & 4 \\ 2 & -3 & 33 \\ 3 & -2 & 2 \end{vmatrix} = 1(-6 - -66) - 1(4-99) + 4(-4 - -9)$$

$$= 60 + 95 + 20 = \mathbf{175}$$

Hence $x = \dfrac{D_x}{D} = \dfrac{70}{35} = \mathbf{2}$, $y = \dfrac{D_y}{D} = \dfrac{-105}{35} = \mathbf{-3}$

and $z = \dfrac{D_z}{D} = \dfrac{175}{35} = \mathbf{5}$

Solution of simultaneous equations using the Gaussian elimination method

Consider the following simultaneous equations:

$$x + y + z = 4 \quad (1)$$

$$2x - 3y + 4z = 33 \quad (2)$$

$$3x - 2y - 2z = 2 \quad (3)$$

Leaving equation (1) as it is gives:

$$x + y + z = 4 \quad (1)$$

Equation (2) − 2 × equation (1) gives:

$$0 - 5y + 2z = 25 \quad (2')$$

and equation (3) − 3 × equation (1) gives:

$$0 - 5y - 5z = -10 \quad (3')$$

Leaving equations (1) and (2′) as they are gives:

$$x + y + z = 4 \quad (1)$$

$$0 - 5y + 2z = 25 \quad (2')$$

Equation (3′)−equation (2) gives: $\quad 0 + 0 - 7z = -35 \quad (3'')$

By appropriately manipulating the three original equations we have deliberately obtained zeros in the positions shown in equations (2′) and (3″).

Working backwards, from equation (3″), $z = \dfrac{-35}{-7} = \mathbf{5}$, from equation (2′), $-5y + 2(5) = 25$, from which, $\mathbf{y} = \dfrac{25-10}{-5} = \mathbf{-3}$ and from equation (1), $x + (-3) + 5 = 4$, from which, $\mathbf{x} = 4 + 3 - 5 = \mathbf{2}$
The above method is known as the **Gaussian elimination method**.

We conclude from the above example that if

$$a_{11}x + a_{12}y + a_{13}z = b_1 \quad (1)$$

$$a_{21}x + a_{22}y + a_{23}z = b_2 \quad (2)$$

$$a_{31}x + a_{32}y + a_{33}z = b_3 \quad (3)$$

the three-step **procedure** to solve simultaneous equations in three unknowns using the **Gaussian elimination method** is:

(i) Equation (2) $-\dfrac{a_{21}}{a_{11}}$ × equation (1) to form equation (2′) and equation (3) $-\dfrac{a_{31}}{a_{11}}$ × equation (1) to form equation (3′)

(ii) Equation (3′) $-\dfrac{a_{32}}{a_{22}}$ × equation (2′) to form equation (3″)

(iii) Determine z from equation (3″), then y from equation (2′) and finally, x from equation (1)

For example, a d.c. circuit comprises three closed loops. Applying Kirchhoff's laws to the closed loops gives the following equations for current flow in milliamperes:

$$2I_1 + 3I_2 - 4I_3 = 26 \quad (1)$$

$$I_1 - 5I_2 - 3I_3 = -87 \quad (2)$$

$$-7I_1 + 2I_2 + 6I_3 = 12 \quad (3)$$

Using the Gaussian elimination method to solve for I_1, I_2 and I_3:
Following the above procedure:

(i) $$2I_1 + 3I_2 - 4I_3 = 26 \quad (1)$$

Equation (2) $-\frac{1}{2}$ × equation (1) gives: $0 - 6.5I_2 - I_3 = -100$ (2′)

Equation (3) $-\frac{-7}{2}$ × equation (1) gives: $0 + 12.5I_2 - 8I_3 = 103$ (3′)

(ii) $$2I_1 + 3I_2 - 4I_3 = 26 \quad (1)$$

$$0 - 6.5I_2 - I_3 = -100 \quad (2')$$

Equation (3′) $-\dfrac{12.5}{-6.5}$ × equation (2′) gives:

$$0 + 0 - 9.923I_3 = -89.308 \quad (3'')$$

(iii) From equation (3″), $I_3 = \dfrac{-89.308}{-9.923} = \mathbf{9\ mA}$, from equation (2′), $-6.5I_2 - 9 = -100$, from which, $I_2 = \dfrac{-100 + 9}{-6.5} = \mathbf{14\ mA}$ and from equation (1), $2I_1 + 3(14) - 4(9) = 26$, from which, $I_1 = \dfrac{26 - 42 + 36}{2}$ $= \dfrac{20}{2} = \mathbf{10\ mA}$

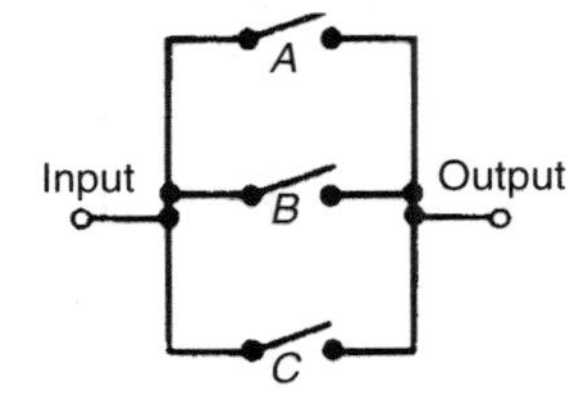

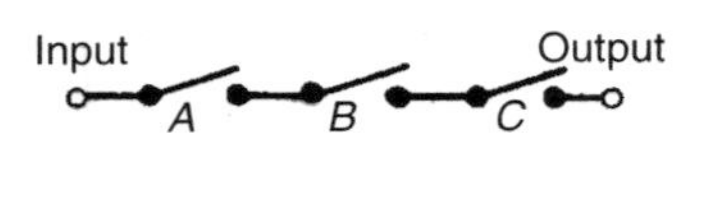

Input A	B	C	Output $Z = A+B+C$
0	0	0	0
0	0	1	1
0	1	0	1
0	1	1	1
1	0	0	1
1	0	1	1
1	1	0	1
1	1	1	1

(a) The or - function electrical circuit and truth table

Input A	B	C	Output $Z = A.B.C$
0	0	0	0
0	0	1	0
0	1	0	0
0	1	1	0
1	0	0	0
1	0	1	0
1	1	0	0
1	1	1	1

(b) The and - function electrical circuit and truth table

Figure 45.3

giving the equivalent of switches 5 to 6, 6 to 7 and 7 to 8 in series. Thus the output is given by: $\boldsymbol{Z = \overline{A}.(B.A + \overline{B}).\overline{B}}$

The truth table is as shown in Table 45.2. Columns 1 and 2 give all the possible combinations of switches A and B. Column 3 is the **and**-function applied to columns 1 and 2, giving $B.A$. Column 4 is $\overline{B}$, i.e. the opposite to column 2. Column 5 is the **or**-function applied to columns 3 and 4. Column 6 is $\overline{A}$, i.e. the opposite to column 1. The output is column 7 and is obtained by applying the **and**-function to columns 4, 5 and 6.

In another example, to derive the Boolean expression and construct a truth table for the switching circuit shown in Figure 45.6:

The parallel circuit 1 to 2 and 3 to 4 gives $(A + \overline{B})$ and this is equivalent to a single switching unit between 7 and 2. The parallel circuit 5 to 6 and 7 to 2 gives $C + (A + \overline{B})$ and this is equivalent to a single switching unit between 8 and 2. The series circuit 9 to 8 and 8 to 2 gives the output $\boldsymbol{Z = B.[C + (A + \overline{B})]}$

The truth table is shown in Table 45.3. Columns 1, 2 and 3 give all the possible combinations of A, B and C. Column 4 is $\overline{B}$ and is the opposite to column 2.

1 A	2 B	3 $A.B$	4 $\overline{A}.\overline{B}$	5 $Z = AB + \overline{A}.\overline{B}$
0	0	0	1	1
0	1	0	0	0
1	0	0	0	0
1	1	1	0	1

(a) Truth table for $Z = A.B + \overline{A}.\overline{B}$

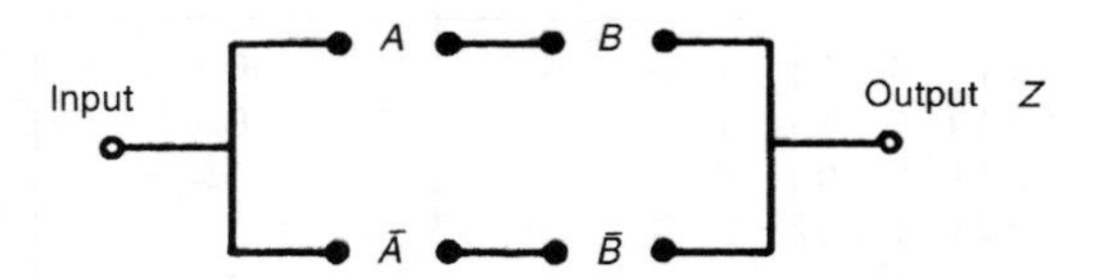

(b) Switching circuit for $Z = A.B + \overline{A}.\overline{B}$

Figure 45.4

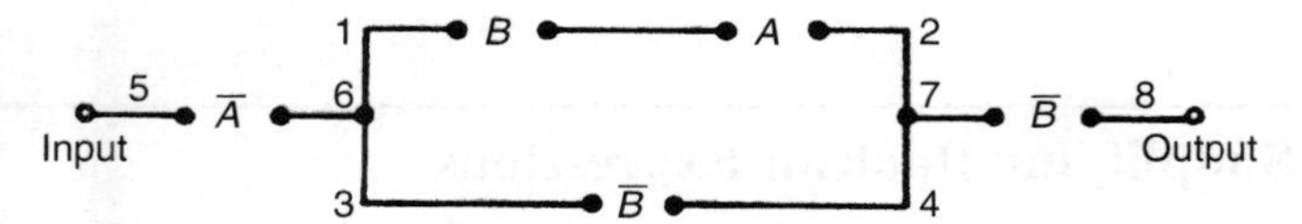

Figure 45.5

Table 45.2

1 A	2 B	3 $B.A$	4 $\overline{B}$	5 $B.A + \overline{B}$	6 $\overline{A}$	7 $Z = \overline{A}.(B.A + \overline{B}).\overline{B}$
0	0	0	1	1	1	1
0	1	0	0	0	1	0
1	0	0	1	1	0	0
1	1	1	0	1	0	0

Column 5 is the **or**-function applied to columns 1 and 4, giving $(A + \overline{B})$. Column 6 is the **or**-function applied to columns 3 and 5 giving $C + (A + \overline{B})$. The output is given in column 7 and is obtained by applying the **and**-function to columns 2 and 6, giving $Z = B.[C + (A + \overline{B})]$

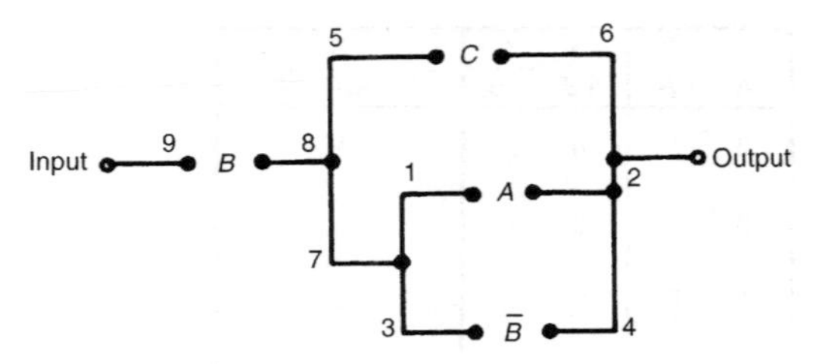

Figure 45.6

Table 45.3

1 A	2 B	3 C	4 $\overline{B}$	5 $A+\overline{B}$	6 $C+(A+\overline{B})$	7 $Z=B.[C+(A+\overline{B})]$
0	0	0	1	1	1	0
0	0	1	1	1	1	0
0	1	0	0	0	0	0
0	1	1	0	0	1	1
1	0	0	1	1	1	0
1	0	1	1	1	1	0
1	1	0	0	1	1	1
1	1	1	0	1	1	1

Simplifying Boolean Expressions

A Boolean expression may be used to describe a complex switching circuit or logic system. If the Boolean expression can be simplified, then the number of switches or logic elements can be reduced resulting in a saving in cost. Three principal ways of simplifying Boolean expressions are:
(a) by using the laws and rules of Boolean algebra,
(b) by applying de Morgan's laws, and
(c) by using Karnaugh maps.

Laws and rules of Boolean algebra

A summary of the principal laws and rules of Boolean algebra are given in Table 45.4.
For example, to simplify the Boolean expression: $\overline{P}.\overline{Q}+\overline{P}.Q+P.\overline{Q}$

With reference to Table 45.4: Reference

$(\overline{P}.\overline{Q})+\overline{P}.Q+P.\overline{Q}=\overline{P}.(\overline{Q}+Q)+P.\overline{Q}$ 5

$=\overline{P}.1+P.\overline{Q}$ 10

$=\boldsymbol{\overline{P}+P.\overline{Q}}$ 12

Table 45.4

Ref.	Name	Rule or law
1	Commutative laws	$A + B = B + A$
2		$A \cdot B = B \cdot A$
3	Associative laws	$(A + B) + C = A + (B + C)$
4		$(A \cdot B) \cdot C = A \cdot (B \cdot C)$
5	Distributive laws	$A \cdot (B + C) = A \cdot B + A \cdot C$
6		$A + (B \cdot C) = (A + B) \cdot (A + C)$
7	Sum rules	$A + 0 = A$
8		$A + 1 = 1$
9		$A + A = A$
10		$A + \overline{A} = 1$
11	Product rules	$A \cdot 0 = 0$
12		$A \cdot 1 = A$
13		$A \cdot A = A$
14		$A \cdot \overline{A} = 0$
15	Absorption rules	$A + A \cdot B = A$
16		$A \cdot (A + B) = A$
17		$A + \overline{A} \cdot B = A + B$

In another example, to simplify $(P + \overline{P}.Q).(Q + \overline{Q}.P)$

With reference to Table 45.4: Reference

$$
\begin{aligned}
(P + \overline{P}.Q).(Q + \overline{Q}.P) &= P.(Q + \overline{Q}.P) + \overline{P}.Q.(Q + \overline{Q}.P) && 5\\
&= P.Q + P.\overline{Q}.P + \overline{P}.Q.Q + \overline{P}.Q.\overline{Q}.P && 5\\
&= P.Q + P.\overline{Q} + \overline{P}.Q + \overline{P}.Q.\overline{Q}.P && 13\\
&= P.Q + P.\overline{Q} + \overline{P}.Q + 0 && 14\\
&= P.Q + P.\overline{Q} + \overline{P}.Q && 7\\
&= P.(Q + \overline{Q}) + \overline{P}.Q && 5\\
&= P.1 + \overline{P}.Q && 10\\
&= \boldsymbol{P + \overline{P}.Q} && 12
\end{aligned}
$$

In another example, to simplify $F.G.\overline{H} + F.G.H + \overline{F}.G.H$

With reference to Table 45.4: Reference

$$
\begin{aligned}
F.G.\overline{H} + F.G.H + \overline{F}.G.H &= F.G.(\overline{H} + H) + \overline{F}.G.H && 5\\
&= F.G.1 + \overline{F}.G.H && 10\\
&= F.G + \overline{F}.G.H && 12\\
&= \boldsymbol{G.(F + \overline{F}.H)} && 5
\end{aligned}
$$

De Morgan's laws

De Morgan's laws may be used to simplify **not**-functions having two or more elements. The laws state that:

$$\boxed{\overline{A+B} = \overline{A}.\overline{B}} \quad \textbf{and} \quad \boxed{\overline{A.B} = \overline{A}+\overline{B}}$$

and may be verified by using a truth table.

For example, to simplify the Boolean expression $\overline{\overline{A}.B} + \overline{\overline{A}+B}$ by using de Morgan's laws and the rules of Boolean algebra:

Applying de Morgan's law to the first term gives:

$$\overline{\overline{A}.B} = \overline{\overline{A}} + \overline{B} = A + \overline{B} \quad \text{since } \overline{\overline{A}} = A$$

Applying de Morgan's law to the second term gives:

$$\overline{\overline{A}+B} = \overline{\overline{A}}.\overline{B} = A.\overline{B}$$

Thus, $\overline{\overline{A}.B} + \overline{\overline{A}+B} = (A+\overline{B}) + A.\overline{B}$

Removing the bracket and reordering gives: $A + A.\overline{B} + \overline{B}$

But, by rule 15, Table 45.4, $A + A.B = A$. It follows that: $A + A.\overline{B} = A$. Thus:

$\boldsymbol{\overline{\overline{A}.B} + \overline{\overline{A}+B} = A + \overline{B}}$

In another example, to simplify the Boolean expression $\overline{(A.\overline{B}+C)}.\overline{(\overline{A}+B.\overline{C})}$ by using de Morgan's laws and the rules of Boolean algebra:

Applying de Morgan's laws to the first term gives:

$$\overline{(A.\overline{B}+C)} = \overline{A.\overline{B}}.\overline{C} = (\overline{A}+\overline{\overline{B}}).\overline{C} = (\overline{A}+B).\overline{C} = \overline{A}.\overline{C} + B.\overline{C}$$

Applying de Morgan's law to the second term gives:

$$(\overline{A} + \overline{B.\overline{C}}) = \overline{A} + (\overline{B} + \overline{\overline{C}}) = \overline{A} + (\overline{B} + C)$$

Thus $\overline{(A.\overline{B}+C)}.(\overline{A}+\overline{B\overline{C}}) = (\overline{A}.\overline{C} + B.\overline{C}).(\overline{A} + \overline{B} + C)$

$$= \overline{A}.\overline{A}.\overline{C} + \overline{A}.\overline{B}.\overline{C} + \overline{A}.\overline{C}.C + \overline{A}.B.\overline{C}$$

$$+ B.\overline{B}.\overline{C} + B.\overline{C}.C$$

But from Table 45.4, $\overline{A}.\overline{A} = \overline{A}$ and $\overline{C}.C = B.\overline{B} = 0$

Hence the Boolean expression becomes

$$\overline{A}.\overline{C} + \overline{A}.\overline{B}.\overline{C} + \overline{A}.B.\overline{C} = \overline{A}.\overline{C}(1 + \overline{B} + B) = \overline{A}.\overline{C}(1 + B) = \overline{A}.\overline{C}$$

Thus: $\boldsymbol{\overline{(A.\overline{B}+C)}.(\overline{A}+\overline{B.\overline{C}}) = \overline{A}.\overline{C}}$

Karnaugh maps

(i) Two-variable Karnaugh maps

A truth table for a two-variable expression is shown in Table 45.5(a), the '1' in the third row output showing that $Z = A.\overline{B}$. Each of the four possible Boolean expressions associated with a two-variable function can be depicted as shown in Table 45.5(b) in which one cell is allocated to each row of the truth table. A matrix similar to that shown in Table 45.5(b) can be used to depict $Z = A.\overline{B}$, by putting a 1 in the cell corresponding to $A.\overline{B}$ and 0's in the remaining cells. This method of depicting a Boolean expression is called a two-variable **Karnaugh map**, and is shown in Table 45.5(c).
To simplify a two-variable Boolean expression, the Boolean expression is depicted on a Karnaugh map, as outlined above. Any cells on the map having either a common vertical side or a common horizontal side are grouped together to form a **couple**. (This is a coupling together of cells, not just combining two together). The simplified Boolean expression for a couple is given by those variables common to all cells in the couple.

(ii) Three-variable Karnaugh maps

A truth table for a three-variable expression is shown in Table 45.6(a), the l's in the output column showing that: $Z = \overline{A}.\overline{B}.C + \overline{A}.B.C + A.B.\overline{C}$. Each of the eight possible Boolean expressions associated with a three-variable function can be depicted as shown in Table 45.6(b) in which one cell is allocated to each row of the truth table. A matrix similar to that shown in Table 45.6(b) can be used to depict: $Z = \overline{A}.\overline{B}.C + \overline{A}.B.C + A.B.\overline{C}$, by putting l's in the cells corresponding to the Boolean terms on the right of the Boolean equation and 0's in

Table 45.5

Inputs		*Output*	*Boolean*
A	*B*	*Z*	*expression*
0	0	0	$\overline{A} \cdot \overline{B}$
0	1	0	$\overline{A} \cdot B$
1	0	1	$A \cdot \overline{B}$
1	1	0	$A \cdot B$

(a)

B \ *A*	0 ($\overline{A}$)	1 (A)
0($\overline{B}$)	$\overline{A}.\overline{B}$	$A.\overline{B}$
1(B)	$\overline{A}.B$	$A.B$

(b)

B \ *A*	0	1
0	0	1
1	0	0

(c)

Table 45.6

Inputs			Output	Boolean
A	B	C	Z	expression
0	0	0	0	$\overline{A}\cdot\overline{B}\cdot\overline{C}$
0	0	1	1	$\overline{A}\cdot\overline{B}\cdot C$
0	1	0	0	$\overline{A}\cdot B\cdot\overline{C}$
0	1	1	1	$\overline{A}\cdot B\cdot C$
1	0	0	0	$A\cdot\overline{B}\cdot\overline{C}$
1	0	1	0	$A\cdot\overline{B}\cdot C$
1	1	0	1	$A\cdot B\cdot\overline{C}$
1	1	1	0	$A\cdot B\cdot C$

(a)

C \ A.B	00 ($\overline{A}.\overline{B}$)	01 ($\overline{A}.B$)	11 ($A.B$)	10 ($A.\overline{B}$)
0($\overline{C}$)	$\overline{A}.\overline{B}.\overline{C}$	$\overline{A}.B.\overline{C}$	$A.B.\overline{C}$	$A.\overline{B}.\overline{C}$
1(C)	$\overline{A}.\overline{B}.C$	$\overline{A}.B.C$	$A.B.C$	$A.\overline{B}.C$

(b)

C \ A.B	00	01	11	10
0	0	0	1	0
1	1	1	0	0

(c)

the remaining cells. This method of depicting a three-variable Boolean expression is called a three-variable Karnaugh map, and is shown in Table 45.6(c). To simplify a three-variable Boolean expression, the Boolean expression is depicted on a Karnaugh map as outlined above. Any cells on the map having common edges either vertically or horizontally are grouped together to form couples of four cells or two cells. During coupling the horizontal lines at the top and bottom of the cells are taken as a common edge, as are the vertical lines on the left and right of the cells. The simplified Boolean expression for a couple is given by those variables common to all cells in the couple.

(iii) Four-variable Karnaugh maps

A truth table for a four-variable expression is shown in Table 45.7(a), the l's in the output column showing that: $Z=\overline{A}.\overline{B}.C.\overline{D}+\overline{A}.B.C.\overline{D}+A.\overline{B}.C.\overline{D}+A.B.C.\overline{D}$ Each of the sixteen possible Boolean expressions associated with a four-variable function can be depicted as shown in Table 45.7(b), in which one cell is allocated to each row of the truth table. A matrix similar to that shown In Table 45.7(b) can be used to depict: $Z=\overline{A}.\overline{B}.C.\overline{D}+\overline{A}.B.C.\overline{D}+A.\overline{B}.C.\overline{D}+A.B.C.\overline{D}$ by putting l's in the cells corresponding to the Boolean terms on the right of the Boolean equation and 0's in the remaining cells. This method of depicting a four-variable expression is called a four-variable Karnaugh map, and is shown in Table 45.7(c).

To simplify a four-variable Boolean expression, the Boolean expression is depicted on a Karnaugh map as outlined above. Any cells on the map having common edges either vertically or horizontally are grouped together to form

Table 45.7

Inputs				Output	Boolean
A	B	C	D	Z	expression
0	0	0	0	0	$\overline{A}\cdot\overline{B}\cdot\overline{C}\cdot\overline{D}$
0	0	0	1	0	$\overline{A}\cdot\overline{B}\cdot\overline{C}\cdot D$
0	0	1	0	1	$\overline{A}\cdot\overline{B}\cdot C\cdot\overline{D}$
0	0	1	1	0	$\overline{A}\cdot\overline{B}\cdot C\cdot D$
0	1	0	0	0	$\overline{A}\cdot B\cdot\overline{C}\cdot\overline{D}$
0	1	0	1	0	$\overline{A}\cdot B\cdot\overline{C}\cdot D$
0	1	1	0	1	$\overline{A}\cdot B\cdot C\cdot\overline{D}$
0	1	1	1	0	$\overline{A}\cdot B\cdot C\cdot D$
1	0	0	0	0	$A\cdot\overline{B}\cdot\overline{C}\cdot\overline{D}$
1	0	0	1	0	$A\cdot\overline{B}\cdot\overline{C}\cdot D$
1	0	1	0	1	$A\cdot\overline{B}\cdot C\cdot\overline{D}$
1	0	1	1	0	$A\cdot\overline{B}\cdot C\cdot D$
1	1	0	0	0	$A\cdot B\cdot\overline{C}\cdot\overline{D}$
1	1	0	1	0	$A\cdot B\cdot\overline{C}\cdot D$
1	1	1	0	1	$A\cdot B\cdot C\cdot\overline{D}$
1	1	1	1	0	$A\cdot B\cdot C\cdot D$

(a)

C.D \ A.B	00 ($\overline{A}.\overline{B}$)	01 ($\overline{A}.B$)	11 ($A.B$)	10 ($A.\overline{B}$)
00 ($\overline{C}.\overline{D}$)	$\overline{A}.\overline{B}.\overline{C}.\overline{D}$	$\overline{A}.B.\overline{C}.\overline{D}$	$A.B.\overline{C}.\overline{D}$	$A.\overline{B}.\overline{C}.\overline{D}$
01 ($\overline{C}.D$)	$\overline{A}.\overline{B}.\overline{C}.D$	$\overline{A}.B.\overline{C}.D$	$A.B.\overline{C}.D$	$A.\overline{B}.\overline{C}.D$
11 ($C.D$)	$\overline{A}.\overline{B}.C.D$	$\overline{A}.B.C.D$	$A.B.C.D$	$A.\overline{B}.C.D$
10 ($C.\overline{D}$)	$\overline{A}.\overline{B}.C.\overline{D}$	$\overline{A}.B.C.\overline{D}$	$A.B.C.\overline{D}$	$A.\overline{B}.C.\overline{D}$

(b)

C.D \ A.B	0.0	0.1	1.1	1.0
0.0	0	0	0	0
0.1	0	0	0	0
1.1	0	0	0	0
1.0	1	1	1	1

(c)

couples of eight cells, four cells or two cells. During coupling, the horizontal lines at the top and bottom of the cells may be considered to be common edges, as are the vertical lines on the left and the right of the cells. The simplified Boolean expression for a couple is given by those variables common to all cells in the couple.

Summary of procedure when simplifying a Boolean expression using a Karnaugh map

(a) Draw a four, eight or sixteen-cell matrix, depending on whether there are two, three or four variables.
(b) Mark in the Boolean expression by putting 1's in the appropriate cells.

(c) Form couples of 8, 4 or 2 cells having common edges, forming the largest groups of cells possible. (Note that a cell containing a 1 may be used more than once when forming a couple. Also note that each cell containing a 1 must be used at least once.)
(d) The Boolean expression for a couple is given by the variables which are common to all cells in the couple.

For example, to simplify the expression: $\overline{P}.\overline{Q} + \overline{P}.Q$ using Karnaugh map techniques:
Using the above procedure:

(a) The two-variable matrix is drawn and is shown in Table 45.8.
(b) The term $\overline{P}.\overline{Q}$ is marked with a 1 in the top left-hand cell, corresponding to $P = 0$ and $Q = 0$; $\overline{P}.Q$ is marked with a 1 in the bottom left-hand cell corresponding to $P = 0$ and $Q = 1$.
(c) The two cells containing 1's have a common horizontal edge and thus a vertical couple, shown by the broken line, can be formed.
(d) The variable common to both cells in the couple is $P = 0$, i.e. $\overline{P}$ thus

$$\boldsymbol{\overline{P}.\overline{Q} + \overline{P}.Q = \overline{P}}$$

Table 45.8

Q \ P	0	1
0	1	0
1	1	0

In another example, to simplify $\overline{X}.Y.\overline{Z} + \overline{X}.\overline{Y}.Z + X.Y.\overline{Z} + X.\overline{Y}.Z$ using Karnaugh map techniques:
Using the above procedure:

(a) A three-variable matrix is drawn and is shown in Table 45.9.
(b) The 1's on the matrix correspond to the expression given, i.e. for $\overline{X}.Y.\overline{Z}$, $X = 0$, $Y = 1$ and $Z = 0$ and hence corresponds to the cell in the top row and second column, and so on.
(c) Two couples can be formed, shown by the broken lines. The couple in the bottom row may be formed since the vertical lines on the left and right of the cells are taken as a common edge.
(d) The variables common to the couple in the top row are $Y = 1$ and $Z = 0$, that is, $\boldsymbol{Y.\overline{Z}}$ and the variables common to the couple in the bottom row

Table 45.9

Z \ X.Y	0.0	0.1	1.1	1.0
0	0	1	1	0
1	1	0	0	1

Table 45.10

(a)

R \ P.Q	0.0	0.1	1.1	1.0
0	3 2	3 2	3 1	3 1
1	4 1	4 2	3 1	4 1

(b)

R \ P.Q	0.0	0.1	1.1	1.0
0	X	X		
1	X	X		X

are $Y = 0$, $Z = 1$, that is, $\boldsymbol{\overline{Y}.Z}$. Hence:

$$\boldsymbol{\overline{X}.Y.\overline{Z} + \overline{X}.\overline{Y}.Z + X.Y.\overline{Z} + X.\overline{Y}.Z = Y.\overline{Z} + \overline{Y}.Z}$$

In another example, to simplify $\overline{(P + \overline{Q}.R)} + \overline{(P.Q + \overline{R})}$ using a Karnaugh map technique:
The term $(P + \overline{Q}.R)$ corresponds to the cells marked 1 on the matrix in Table 45.10(a), hence $\overline{(P + \overline{Q}.R)}$ corresponds to the cells marked 2. Similarly, $(P.Q + \overline{R})$ corresponds to the cells marked 3 in Table 45.10(a), hence $\overline{(P.Q + \overline{R})}$ corresponds to the cells marked 4. The expression $\overline{(P + \overline{Q}.R)} + \overline{(P.Q + \overline{R})}$ corresponds to cells marked with either a 2 or with a 4 and is shown in Table 45.10(b) by X's. These cells may be coupled as shown by the broken lines. The variables common to the group of four cells is $P = 0$, i.e. $\boldsymbol{\overline{P}}$, and those common to the group of two cells are $Q = 0$, $R = 1$, i.e. $\boldsymbol{\overline{Q}.R}$

Thus: $\boldsymbol{\overline{(P + \overline{Q}.R)} + \overline{(P.Q + \overline{R})} = \overline{P} + \overline{Q}.R}$

---

46 Logic Circuits and Gates

Logic circuits

In practice, logic gates are used to perform the **and**, **or** and **not**-functions introduced in Chapter 45. Logic gates can be made from switches, magnetic devices or fluidic devices, but most logic gates in use are electronic devices. Various logic gates are available. For example, the Boolean expression $(A.B.C)$ can be produced using a three-input, **and**-gate and $(C + D)$ by using a two-input **or**-gate. The principal gates in common use are introduced below. The term 'gate' is used in the same sense as a normal gate, the open state being indicated by a binary '1' and the closed state by a binary '0'. A gate will only open when the requirements of the gate are met and, for example, there will only be a '1' output on a two-input **and**-gate when both the inputs to the gate are at a '1' state.

The and-gate

The different symbols used for a three-input, **and**-gate are shown in Figure 46.1(a) and the truth table is shown in Figure 46.1(b). This shows that there will only be a '1' output when A is 1 and B is 1 and C is 1, written as: $Z = A.B.C$

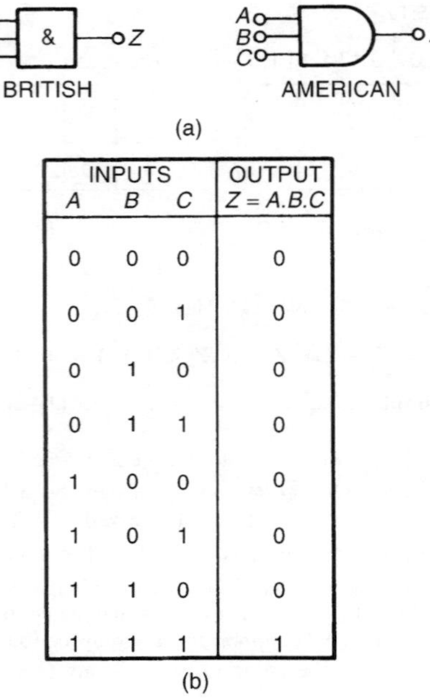

INPUTS			OUTPUT
A	B	C	$Z = A.B.C$
0	0	0	0
0	0	1	0
0	1	0	0
0	1	1	0
1	0	0	0
1	0	1	0
1	1	0	0
1	1	1	1

Figure 46.1

The or-gate

The different symbols used for a three-input **or-**gate are shown in Figure 46.2(a) and the truth table is shown in Figure 46.2(b). This shows that there will be a '1' output when A is 1, or B is 1, or C is 1, or any combination of A, B or C is 1, written as: $Z = A + B + C$

The invert-gate or not-gate

The different symbols used for an **invert**-gate are shown in Figure 46.3(a) and the truth table is shown in Figure 46.4(b). This shows that a '0' input gives a '1' output and vice versa, i.e. it is an 'opposite to' function. The invert of A is written $\overline{A}$ and is called 'not-A'

The nand-gate

The different symbols used for a **nand**-gate are shown in Figure 46.4(a) and the truth table is shown in Figure 46.4(b). This gate is equivalent to an **and**-gate and an **invert**-gate in series (not-and = nand) and the output is written as: $Z = \overline{A.B.C}$

The nor-gate

The different symbols used for a **nor**-gate are shown in Figure 46.5(a) and the truth table is shown in Figure 46.5(b). This gate is equivalent to an **or**-gate and an **invert**-gate in series, (not-or = nor), and the output is written as: $Z = \overline{A + B + C}$

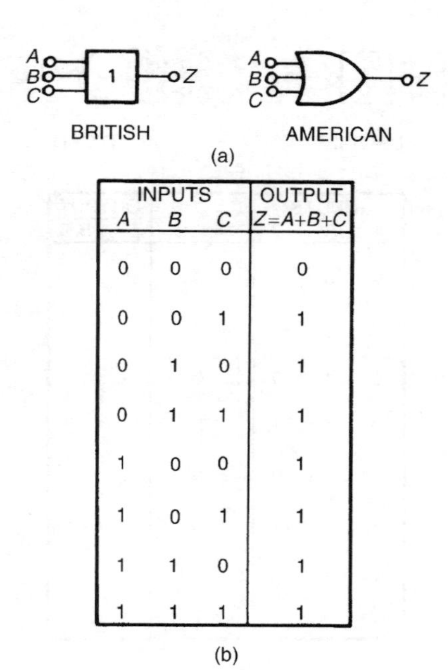

INPUTS			OUTPUT
A	B	C	$Z=A+B+C$
0	0	0	0
0	0	1	1
0	1	0	1
0	1	1	1
1	0	0	1
1	0	1	1
1	1	0	1
1	1	1	1

Figure 46.2

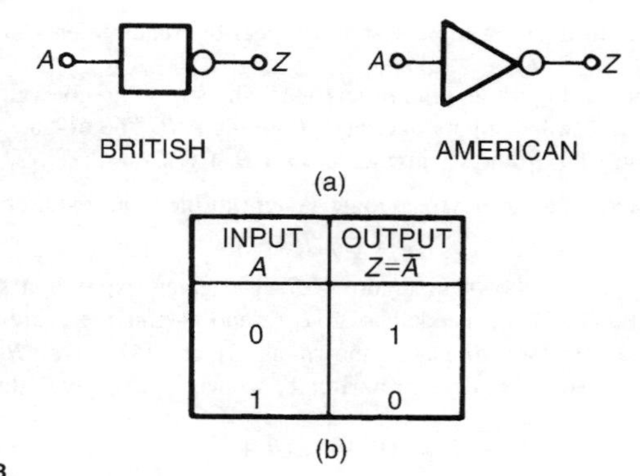

INPUT A	OUTPUT $Z=\overline{A}$
0	1
1	0

Figure 46.3

Combinational logic networks

In most logic circuits, more than one gate is needed to give the required output. Except for the **invert**-gate, logic gates generally have two, three or four inputs and are confined to one function only. Thus, for example, a two-input, **or**-gate or a four-input **and**-gate can be used when designing a logic circuit.

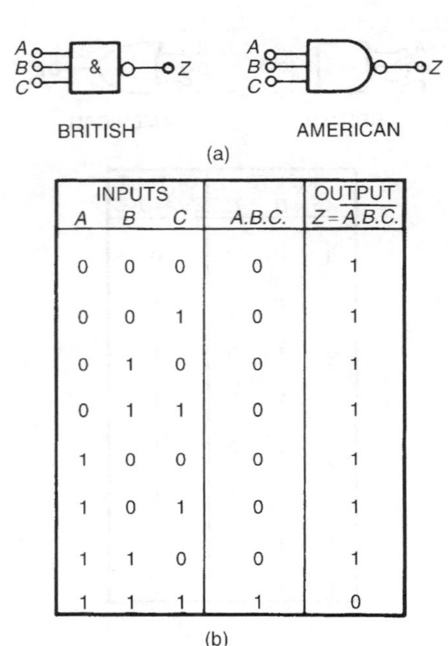

INPUTS				OUTPUT
A	B	C	$A.B.C.$	$Z=\overline{A.B.C.}$
0	0	0	0	1
0	0	1	0	1
0	1	0	0	1
0	1	1	0	1
1	0	0	0	1
1	0	1	0	1
1	1	0	0	1
1	1	1	1	0

Figure 46.4

For example, to devise a logic system to meet the requirements of:
$Z = A.\overline{B} + C$
With reference to Figure 46.6 an **invert**-gate, shown as (1), gives $\overline{B}$. The **and**-gate, shown as (2), has inputs of A and $\overline{B}$, giving $A.\overline{B}$. The **or**-gate, shown as (3), has inputs of $A.\overline{B}$ and C, giving: $\boldsymbol{Z = A.\overline{B} + C}$

In another example, to devise a logic system to meet the requirements of

$$(P + \overline{Q}).(\overline{R} + S)$$

The logic system is shown in Figure 46.7. The given expression shows that two **invert**-functions are needed to give $\overline{Q}$ and $\overline{R}$ and these are shown as gates (1) and (2). Two **or**-gates, shown as (3) and (4), give $(P + \overline{Q})$ and $(\overline{R} + S)$ respectively. Finally, an **and**-gate, shown as (5), gives the required output,

$$\boldsymbol{Z = (P + \overline{Q}).(\overline{R} + S)}$$

In another example, to devise a logic circuit to meet the requirements of the output given in Table 46.1, using as few gates as possible:
The '1' outputs in rows 6, 7 and 8 of Table 46.1 show that the Boolean expression is: $Z = A.\overline{B}.C + A.B.\overline{C} + A.B.C$
The logic circuit for this expression can be built using three, 3-input **and**-gates and one, 3-input **or**-gate, together with two **invert**-gates. However, the number of gates required can be reduced by using the techniques introduced in Chapter 45, resulting in the cost of the circuit being reduced. Any of the

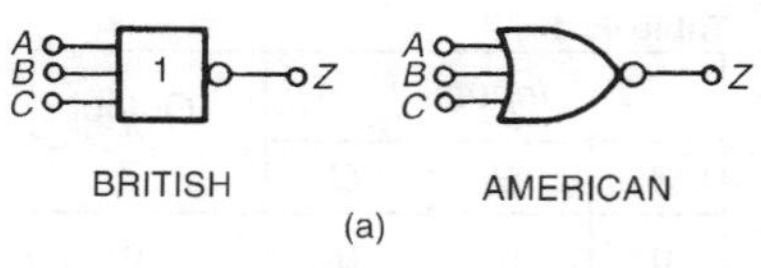

(a)

INPUTS				OUTPUT
A	B	C	$A+B+C$	$Z=\overline{A+B+C}$
0	0	0	0	1
0	0	1	1	0
0	1	0	1	0
0	1	1	1	0
1	0	0	1	0
1	0	1	1	0
1	1	0	1	0
1	1	1	1	0

(b)

Figure 46.5

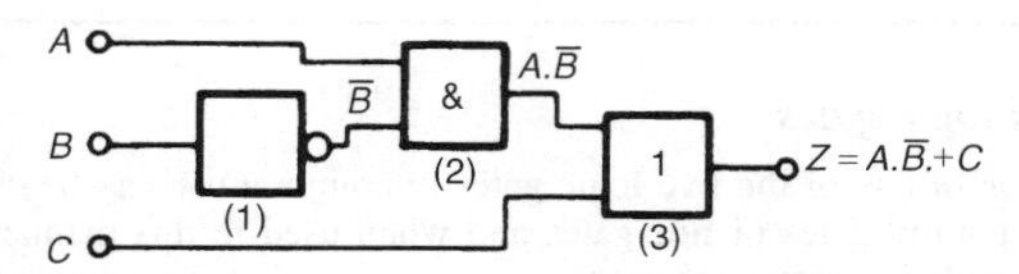

Figure 46.6

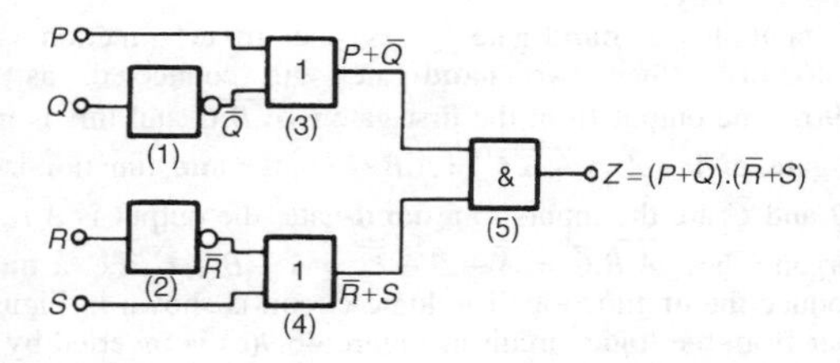

Figure 46.7

techniques can be used, and in this case, the rules of Boolean algebra (see Table 45.4) are used.

$$Z = A.\overline{B}.C + A.B.\overline{C} + A.B.C = A.[\overline{B}.C + B.\overline{C} + B.C]$$

$$= A.[\overline{B}.C + B(\overline{C} + C)] = A.[\overline{B}.C + B]$$

$$= A.[B + \overline{B}.C] = \mathbf{A.[B + C]}$$

Table 46.1

Inputs			*Output*
A	*B*	*C*	*Z*
0	0	0	0
0	0	1	0
0	1	0	0
0	1	1	0
1	0	0	0
1	0	1	1
1	1	0	1
1	1	1	1

The logic circuit to give this simplified expression is shown in Figure 46.8.

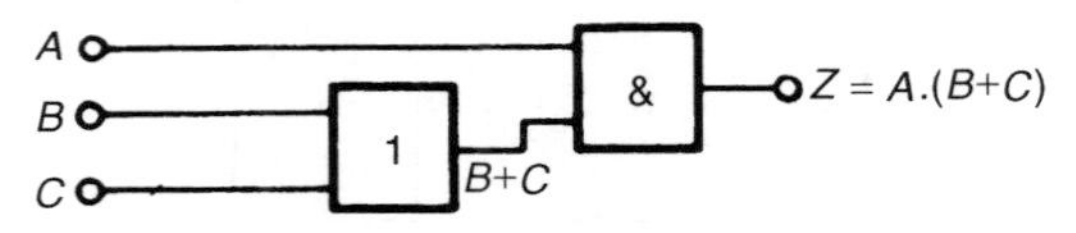

Figure 46.8

Universal logic gates

The function of any of the five logic gates in common use can be obtained by using either **nand**-gates or **nor**-gates and when used in this manner, the gate selected is called a **universal gate**.
For example, to show how **invert**, **and**, **or** and **nor**-functions can be produced using nand-gates only:
A single input to a **nand**-gate gives the **invert**-function, as shown in Figure 46.9(a). When two **nand**-gates are connected, as shown in Figure 46.9(b), the output from the first gate is $\overline{A.B.C}$ and this is inverted by the second gate, giving $Z = \overline{\overline{A.B.C}} = A.B.C$ i.e. the **and**-function is produced.
When $\overline{A}$, $\overline{B}$ and $\overline{C}$ are the inputs to a **nand**-gate, the output is $\overline{\overline{A}.\overline{B}.\overline{C}}$
By de Morgan's law, $\overline{\overline{A}.\overline{B}.\overline{C}} = \overline{\overline{A}} + \overline{\overline{B}} + \overline{\overline{C}} = A + B + C$, i.e. a **nand**-gate is used to produce the **or**-function. The logic circuit is shown in Figure 46.9(c).
If the output from the logic circuit in Figure 46.9(c) is inverted by adding an additional **nand**-gate, the output becomes the invert of an **or**-function, i.e. the **nor**-function, as shown in Figure 46.9(d).
In another example, to show how **invert**, **or**, **and** and **nand**-functions can be produced by using **nor**-gates only:
A single input to a **nor**-gate gives the **invert**-function, as shown in Figure 46.10(a). When two **nor**-gates are connected, as shown in Figure 46.10(b), the output from the first gate is $\overline{A + B + C}$ and this is inverted by the second gate, giving $Z = \overline{\overline{A + B + C}} = A + B + C$, i.e. the **or**-function

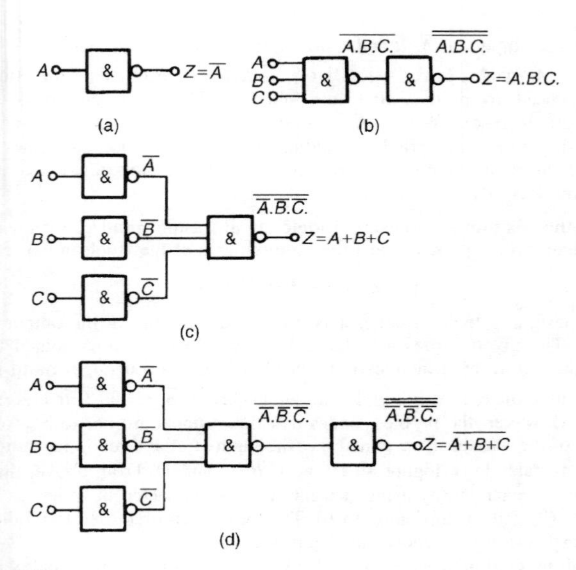

Figure 46.9

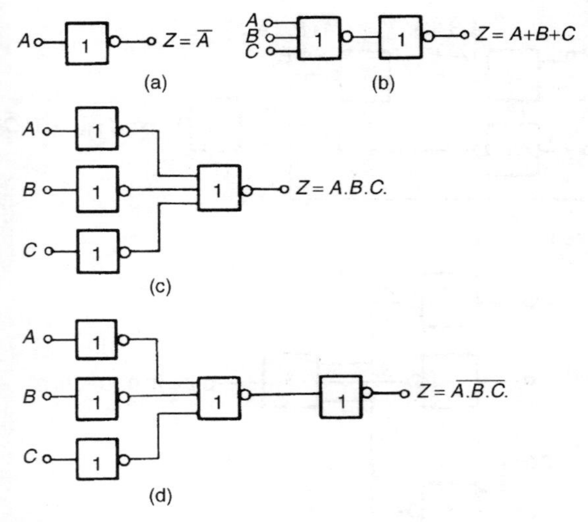

Figure 46.10

is produced. Inputs of $\overline{A}$, $\overline{B}$ and $\overline{C}$ to a **nor**-gate give an output of $\overline{\overline{A}+\overline{B}+\overline{C}}$ By de Morgan's law, $\overline{\overline{A}+\overline{B}+\overline{C}} = \overline{\overline{A}}.\overline{\overline{B}}.\overline{\overline{C}} = A.B.C$, i.e. the **nor**-gate can be used to produce the **and**-function. The logic circuit is shown in Figure 46.10(c). When the output of the logic circuit, shown in Figure 46.10(c), is inverted by adding an additional **nor**-gate, the output then becomes the invert of an **or**-function, i.e. the **nor**-function as shown in Figure 46.10(d).

In another example, to design a logic circuit, using **nand**-gates having not more than three inputs, to meet the requirements of the Boolean expression:

$$Z = \overline{A} + \overline{B} + C + \overline{D}$$

When designing logic circuits, it is often easier to start at the output of the circuit. The given expression shows there are four variables joined by **or**-functions. From the principles introduced above, if a four-input **nand**-gate is used to give the expression given, the inputs are $\overline{\overline{A}}$, $\overline{\overline{B}}$, $\overline{C}$ and $\overline{\overline{D}}$ that is A, B, $\overline{C}$ and D. However, the problem states that three-inputs are not to be exceeded so two of the variables are joined, i.e. the inputs to the three-input **nand**-gate, shown as gate (1) in Figure 46.10, is A, B, $\overline{C}$ and D. From above, the **and**-function is generated by using two **nand**-gates connected in series, as shown by gates (2) and (3) in Figure 46.10. The logic circuit required to produce the given expression is as shown in Figure 46.10.

In another example, an alarm indicator in a grinding mill complex should be activated if (a) the power supply to all mills is off and (b) the hopper feeding the mills is less than 10% full, and (c) if less than two of the three

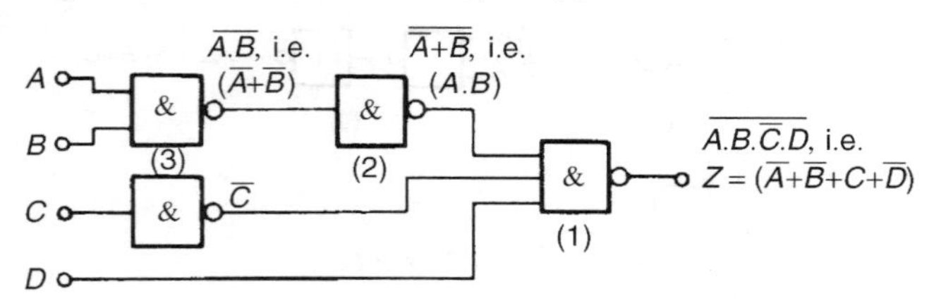

Figure 46.11

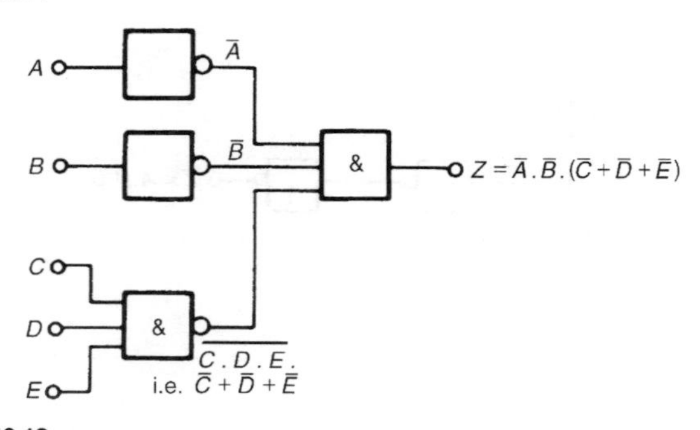

Figure 46.12

grinding mills are in action. To devise a logic system to meet these requirements:
Let variable A represent the power supply on to all the mills, then $\overline{A}$ represents the power supply off. Let B represent the hopper feeding the mills being more than 10% full, then $\overline{B}$ represents the hopper being less than 10% full. Let C, D and E represent the three mills respectively being in action, then $\overline{C}$, $\overline{D}$ and $\overline{E}$ represent the three mills respectively not being in action. The required expression to activate the alarm is: $Z = \overline{A}.\overline{B}.(\overline{C} + \overline{D} + \overline{E})$
There are three variables joined by **and**-functions in the output, indicating that a three-input **and**-gate is required, having inputs of $\overline{A}$, $\overline{B}$ and $(\overline{C} + \overline{D} + \overline{E})$. The term $(\overline{C} + \overline{D} + \overline{E})$ is produced by a three-input **nand**-gate. When variables C, D and E are the inputs to a **nand**-gate, the output is $\overline{C.D.E}$, which, by de Morgan's law is $\overline{C} + \overline{D} + \overline{E}$. Hence the required logic circuit is as shown in Figure 46.12.

Differential Calculus

47 Introduction to Differentiation

Introduction to calculus

Calculus is a branch of mathematics involving or leading to calculations dealing with continuously varying functions.
Calculus is a subject that falls into two parts:
(i) **differential calculus** (or **differentiation**) and
(ii) **integral calculus** (or **integration**).
Differentiation is used in calculations involving velocity and acceleration, rates of change and maximum and minimum values of curves.
Integration may be used to determine areas, volumes, mean and r.m.s. values, centroids and second moments of areas.

Functional notation

In an equation such as $y = 3x^2 + 2x - 5$, y is said to be a function of x and may be written as $y = f(x)$.
An equation written in the form $f(x) = 3x^2 + 2x - 5$ is termed **functional notation**.
The value of $f(x)$ when $x = 0$ is denoted by $f(0)$, and the value of $f(x)$ when $x = 2$ is denoted by $f(2)$ and so on.

For example, if $f(x) = 3x^2 + 2x - 5$, then

$$f(0) = 3(0)^2 + 2(0) - 5 = -5$$

$$f(2) = 3(2)^2 + 2(2) - 5 = 11$$

and $$f(-1) = 3(-1)^2 + 2(-1) - 5 = -4$$

The gradient of a curve

(a) If a tangent is drawn at a point P on a curve, then the gradient of this tangent is said to be the **gradient of the curve** at P. In Figure 47.1, the gradient of the curve at P is equal to the gradient of the tangent PQ.
(b) For the curve shown in Figure 47.2, let the points A and B have coordinates (x_1, y_1) and (x_2, y_2), respectively. In functional notation, $y_1 = f(x_1)$ and $y_2 = f(x_2)$ as shown.

The gradient of the chord AB $= \dfrac{BC}{AC} = \dfrac{BD - CD}{ED} = \dfrac{f(x_2) - f(x_1)}{(x_2 - x_1)}$

(c) For the curve $f(x) = x^2$ shown in Figure 47.3:
(i) the gradient of chord AB $= \dfrac{f(3) - f(1)}{3 - 1} = \dfrac{9 - 1}{2} = 4$

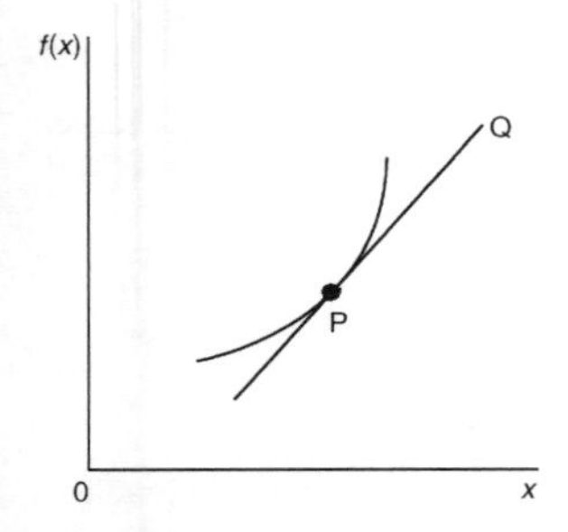

Figure 47.1

Figure 47.2

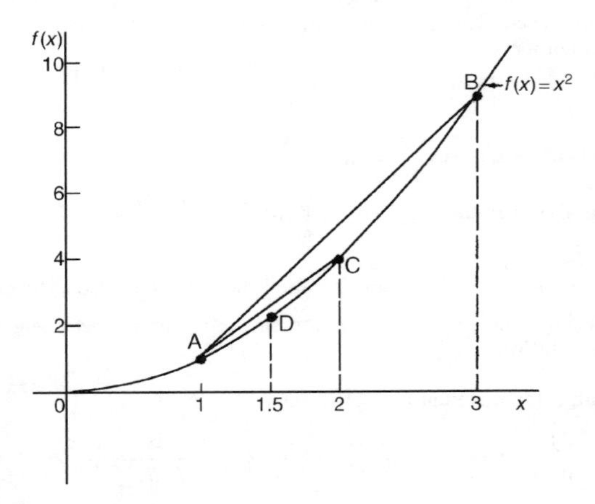

Figure 47.3

(ii) the gradient of chord AC $= \dfrac{f(2) - f(1)}{2 - 1} = \dfrac{4 - 1}{1} = 3$

(iii) the gradient of chord AD $= \dfrac{f(1.5) - f(1)}{1.5 - 1} = \dfrac{2.25 - 1}{0.5} = 2.5$

(iv) if E is the point on the curve $(1.1, f(1.1))$ then the gradient of chord AE $= \dfrac{f(1.1) - f(1)}{1.1 - 1} = \dfrac{1.21 - 1}{0.1} = 2.1$

(v) if F is the point on the curve $(1.01, f(1.01))$ then the gradient of chord AF $= \dfrac{f(1.01) - f(1)}{1.01 - 1} = \dfrac{1.0201 - 1}{0.01} = 2.01$

Thus as point B moves closer and closer to point A the gradient of the chord approaches nearer and nearer to the value 2. This is called the **limiting value** of the gradient of the chord AB and when B coincides with A the chord becomes the tangent to the curve.

Differentiation from first principles

(i) In Figure 47.4, A and B are two points very close together on a curve, δx (delta x) and δy (delta y) representing small increments in the x and y directions, respectively.

Gradient of chord AB $= \dfrac{\delta x}{\delta y}$, however $\delta y = f(x+\delta x) - f(x)$

Hence $\dfrac{\delta x}{\delta y} = \dfrac{f(x+\delta x) - f(x)}{\delta x}$

As δx approaches zero, $\dfrac{\delta x}{\delta y}$ approaches a limiting value and the gradient of the chord approaches the gradient of the tangent at A.

(ii) When determining the gradient of a tangent to a curve there are two notations used. The gradient of the curve at A in Figure 47.4 can either be written as:

$$\lim_{\delta x \to 0} \frac{\delta y}{\delta x} \quad \text{or} \quad \lim_{\delta x \to 0} \left\{ \frac{f(x+\delta x) - f(x)}{\delta x} \right\}$$

In **Leibniz notation,** $\boldsymbol{\dfrac{dy}{dx} = \lim_{\delta x \to 0} \dfrac{\delta y}{\delta x}}$

In **functional notation,** $\boldsymbol{f'(x) = \lim_{\delta x \to 0} \left\{ \dfrac{f(x+\delta x) - f(x)}{\delta x} \right\}}$

(iii) $\dfrac{dy}{dx}$ is the same as $f'(x)$ and is called the **differential coefficient** or the **derivative**. The process of finding the differential coefficient is called **differentiation**.

Summarising, the differential coefficient,

$$\frac{dy}{dx} = f'(x) = \lim_{\delta x \to 0} \frac{\delta y}{\delta x} = \lim_{\delta x \to 0} \left\{ \frac{f(x+\delta x) - f(x)}{\delta x} \right\}$$

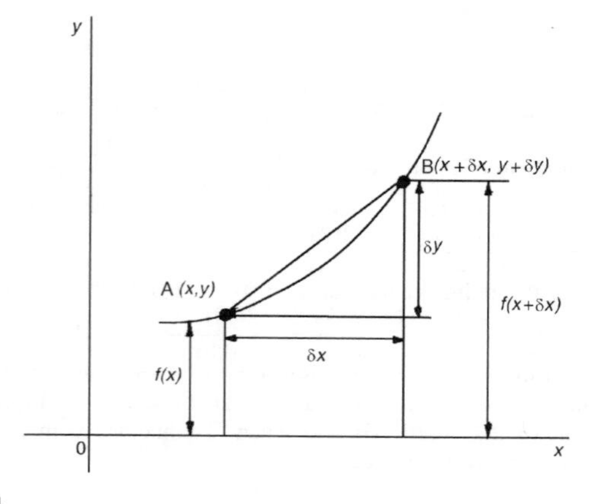

Figure 47.4

For example, differentiating from first principle $f(x) = x^2$ means 'to find $f'(x)$' by using the expression $f'(x) = \underset{\delta x \to 0}{\text{limit}} \left\{ \frac{f(x+\delta x) - f(x)}{\delta x} \right\}$

Substituting $(x + \delta x)$ for x gives

$$f(x+\delta x) = (x+\delta x)^2 = x^2 + 2x\delta x + \delta x^2,$$

hence
$$f'(x) = \underset{\delta x \to 0}{\text{limit}} \left\{ \frac{(x^2 + 2x\delta x + \delta x^2) - (x^2)}{\delta x} \right\}$$

$$= \underset{\delta x \to 0}{\text{limit}} \left\{ \frac{2x\delta x + \delta x^2}{\delta x} \right\} = \underset{\delta x \to 0}{\text{limit}} \{2x + \delta x\}$$

As $\delta x \to 0$, $[2x + \delta x] \to (2x + 0])$. Thus $\boldsymbol{f'(x) = 2x}$, i.e. the differential coefficient of x^2 is $2x$.
At, say, $x = 2$, the gradient of the curve, $f'(x) = 2(2) = \mathbf{4}$.

Differentiation of y = axⁿ by the general rule

From differentiation by first principles, a general rule for differentiating ax^n emerges where a and n are any constants. This rule is:

if $\qquad \boldsymbol{y = ax^n}$ **then** $\boldsymbol{\frac{dy}{dx} = anx^{n-1}}$

or, $\quad$ **if** $\boldsymbol{f(x) = ax^n}$ **then** $\boldsymbol{f'(x) = anx^{n-1}}$

When differentiating, results can be expressed in a number of ways. For example:

(i) if $y = 3x^2$ then $\frac{dy}{dx} = 6x$,
(ii) if $f(x) = 3x^2$ then $f'(x) = 6x$,
(iii) the differential coefficient of $3x^2$ is $6x$,
(iv) the derivative of $3x^2$ is $6x$, and
(v) $\frac{d}{dx}(3x^2) = 6x$

For example, using the general rule, differentiating the following with respect to x: (a) $y = 5x^7$ $\quad$ (b) $y = 3\sqrt{x}$ $\quad$ (c) $y = \frac{4}{x^2}$

(a) Comparing $y = 5x^7$ with $y = ax^n$ shows that $a = 5$ and $n = 7$. Using the general rule,

$$\frac{\mathbf{dy}}{\mathbf{dx}} = \text{anx}^{n-1} = (5)(7)x^{7-1} = \mathbf{35x^6}$$

(b) $y = 3\sqrt{x} = 3x^{\frac{1}{2}}$. Hence $a = 3$ and $n = \frac{1}{2}$

$$\frac{\mathbf{dy}}{\mathbf{dx}} = \text{anx}^{n-1} = (3)\frac{1}{2}x^{\frac{1}{2}-1} = \frac{3}{2}x^{-\frac{1}{2}} = \frac{3}{2x^{\frac{1}{2}}} = \frac{\mathbf{3}}{\mathbf{2\sqrt{x}}}$$

(c) $y = \frac{4}{x^2} = 4x^{-2}$. Hence $a = 4$ and $n = -2$

$$\frac{\mathbf{dy}}{\mathbf{dx}} = \text{anx}^{n-1} = (4)(-2)x^{-2-1} = -8x^{-3} = -\frac{\mathbf{8}}{\boldsymbol{x^3}}$$

Differentiation of sine and cosine functions

Figure 47.5(a) shows a graph of $y = \sin\theta$. The gradient is continually changing as the curve moves from O to A to B to C to D. The gradient, given by $\frac{dy}{d\theta}$, may be plotted in a corresponding position below $y = \sin\theta$, as shown in Figure 47.5(b).

(i) At 0, the gradient is positive and is at its steepest. Hence 0′ is a maximum positive value.

(ii) Between 0 and A the gradient is positive but is decreasing in value until at A the gradient is zero, shown as A′.

(iii) Between A and B the gradient is negative but is increasing in value until at B the gradient is at its steepest. Hence B′ is a maximum negative value.

(iv) If the gradient of $y = \sin\theta$ is further investigated between B and C and C and D then the resulting graph of $\frac{dy}{d\theta}$ is seen to be a cosine wave. Hence

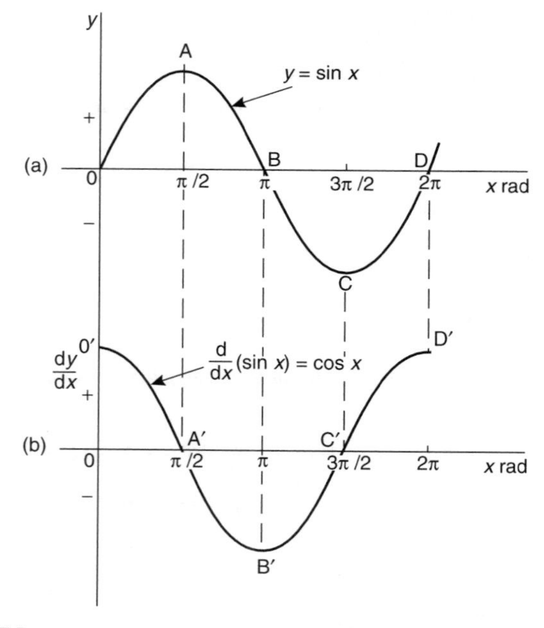

Figure 47.5

the rate of change of $\sin\theta$ is $\cos\theta$, i.e. **if $y = \sin\theta$ then $\frac{dy}{d\theta} = \cos\theta$**

It may also be shown that:

if $\quad y = \sin a\theta, \quad \frac{dy}{d\theta} = a\cos a\theta$ (where a is a constant)

and **if $y = \sin(a\theta + \alpha)$, $\frac{dy}{d\theta} = a\cos(a\theta + \alpha)$** (where a and α are constants).

If a similar exercise is followed for $y = \cos\theta$ then the graphs of Figure 47.6 result, showing $\frac{dy}{d\theta}$ to be a graph of $\sin\theta$, but displaced by π radians. If each point on the curve $y = \sin\theta$ (as shown in Figure 47.6(a)) were to be made negative, (i.e. $+\frac{\pi}{2}$ is made $-\frac{\pi}{2}$, $-\frac{3\pi}{2}$ is made $+\frac{3\pi}{2}$, and so on) then the graph shown in Figure 47.6(b) would result.

This latter graph therefore represents the curve of $-\sin\theta$. Thus,

if $\quad y = \cos\theta, \quad \frac{dy}{d\theta} = -\sin\theta.$

It may also be shown that:

if $\quad y = \cos a\theta, \quad \frac{dy}{d\theta} = -a\sin a\theta$ (where a is a constant)

and if $y = \cos(a\theta + \alpha)$, $\frac{dy}{d\theta} = -a\sin(a\theta + \alpha)$ (where a and α are constants).

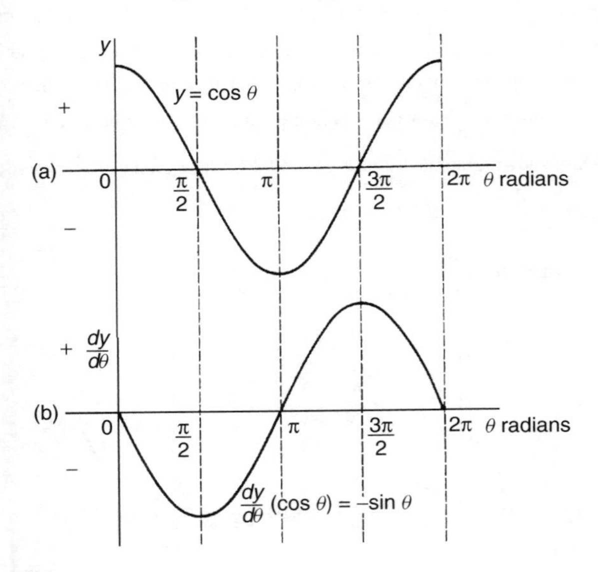

Figure 47.6

For example, if $y = 7\sin 2x - 3\cos 4x$ then
$\frac{dy}{dx} = (7)(2)\cos 2x - (3)(-4)\sin 4x = \mathbf{14\cos 2x + 12\sin 4x}$

In another example, if $f(\theta) = 5\sin(100\pi\theta - 0.40)$
$f'(\theta) = 5[100\pi\cos(100\pi\theta - 0.40)] = \mathbf{500\pi\cos(100\pi\theta - 0.40)}$

Differentiation of e^{ax} and $\ln ax$

A graph of $y = e^x$ is shown in Figure 47.7(a). The gradient of the curve at any point is given by $\frac{dy}{dx}$ and is continually changing. By drawing tangents to the curve at many points on the curve and measuring the gradient of the tangents, values of $\frac{dy}{dx}$ for corresponding values of x may be obtained. These values are shown graphically in Figure 47.7(b). The graph of $\frac{dy}{dx}$ against x is identical to the original graph of $y = e^x$. It follows that:

if $\quad \boldsymbol{y = e^x}$, then $\boldsymbol{\frac{dy}{dx} = e^x}$

It may also be shown that if $\boldsymbol{y = e^{ax}}$, then $\boldsymbol{\frac{dy}{dx} = ae^{ax}}$

For example, if $y = 2e^{6x}$, then $\frac{dy}{dx} = (2)(6e^{6x}) = \mathbf{12e^{6x}}$.

A graph of $y = \ln x$ is shown in Figure 47.8(a). The gradient of the curve at any point is given by $\frac{dy}{dx}$ and is continually changing. By drawing tangents to the curve at many points on the curve and measuring the gradient of the tangents, values of $\frac{dy}{dx}$ for corresponding values of x may be obtained. These values are shown graphically in Figure 47.8(b). The graph of $\frac{dy}{dx}$ against x is the graph of $\frac{dy}{dx} = \frac{1}{x}$. It follows that:

if $\quad \boldsymbol{y = \ln x}$, then $\boldsymbol{\frac{dy}{dx} = \frac{1}{x}}$

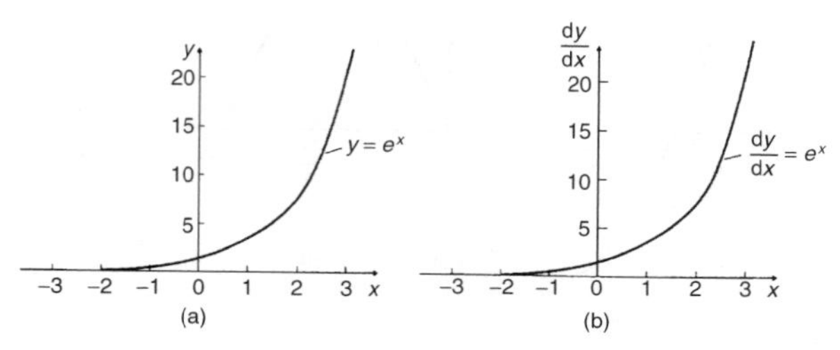

Figure 47.7

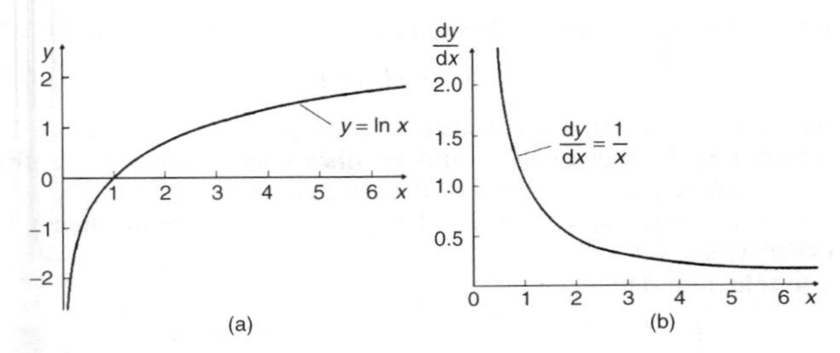

Figure 47.8

It may also be shown that

$$\text{if} \quad \boldsymbol{y = \ln ax}, \text{ then } \frac{\mathbf{dy}}{\mathbf{dx}} = \frac{\mathbf{1}}{\boldsymbol{x}}$$

(Note that in the latter expression 'a' does not appear in the $\frac{dy}{dx}$ term).

Thus if $y = 3\ln 4x$, then $\frac{dy}{dx} = (3)\left(\frac{1}{x}\right) = \boldsymbol{\frac{3}{x}}$

---

48 Methods of Differentiation

Differentiation of common functions

The **standard derivatives** summarised below were derived in Chapter 47 and are true for all real values of x.

y or $f(x)$	$\frac{dy}{dx}$ or $f'(x)$
ax^n	anx^{n-1}
$\sin ax$	$a\cos ax$
$\cos ax$	$-a\sin ax$
e^{ax}	ae^{ax}
$\ln ax$	$\frac{1}{x}$

For example, to differentiate $y = 6$:
$y = 6$ may be written as $y = 6x^0$, i.e. in the general rule $a = 6$ and $n = 0$. Hence $\frac{dy}{dx} = (6)(0)x^{0-1} = \mathbf{0}$.
In general, **the differential coefficient of a constant is always zero**.

In another example, to differentiate $y = 6x$: Since $y = 6x$, in the general rule $a = 6$ and $n = 1$. Hence $\frac{dy}{dx} = (6)(1)x^{1-1} = 6x^0 = \mathbf{6}$

In another example, to find the differential coefficients of $y = 3\sin 4x$: When $y = 3\sin 4x$ then $\frac{dy}{dx} = (3)(4\cos 4x) = \mathbf{12\cos 4x}$

In general, the differential coefficient of kx, where k is a constant, is k. The **differential coefficient of a sum or difference** is the sum or difference of the differential coefficients of the separate terms.

Thus, if $f(x) = p(x) + q(x) - r(x)$, (where f, p, q and r are functions), then $f'(x) = p'(x) + q'(x) - r'(x)$

For example, to differentiate

$$y = 5x^4 + 4x - \frac{1}{2x^2} + \frac{1}{\sqrt{x}} - 3 \text{ with respect to } x:$$

$$y = 5x^4 + 4x - \frac{1}{2x^2} + \frac{1}{\sqrt{x}} - 3 \text{ is rewritten as}$$

$$y = 5x^4 + 4x - \tfrac{1}{2}x^{-2} + x^{-1/2} - 3$$

Thus $$\frac{dy}{dx} = (5)(4)x^{4-1} + (4)(1)x^{1-1} - \tfrac{1}{2}(-2)x^{-2-1} + (1)\left(-\tfrac{1}{2}\right)x^{(-1/2)-1} - 0$$

$$= 20x^3 + 4 + x^{-3} - \tfrac{1}{2}x^{-3/2}$$

i.e. $$\frac{dy}{dx} = \mathbf{20x^3 + 4 - \frac{1}{x^3} - \frac{1}{2\sqrt{x^3}}}$$

In another example, to determine the derivative of $f(\theta) = \frac{2}{e^{3\theta}} + 6\ln 2\theta$:

$$f(\theta) = \frac{2}{e^{3\theta}} + 6\ln 2\theta = 2e^{-3\theta} + 6\ln 2\theta$$

Hence $$f'(\theta) = (2)(-3)e^{-3\theta} + 6\left(\frac{1}{\theta}\right) = -6e^{-3\theta} + \frac{6}{\theta} = \mathbf{\frac{-6}{e^{3\theta}} + \frac{6}{\theta}}$$

Differentiation of a product

When $y = uv$, and u and v are both functions of x,

then $$\boxed{\frac{dy}{dx} = u\frac{dv}{dx} + v\frac{du}{dx}}$$

This is known as the **product rule**.

For example, to find the differential coefficient of $y = 3x^2 \sin 2x$:

$3x^2 \sin 2x$ is a product of two terms $3x^2$ and $\sin 2x$

Let $u = 3x^2$ and $v = \sin 2x$

Using the product rule:

$$\frac{dy}{dx} = u \quad \frac{dv}{dx} \quad + \quad v \quad \frac{du}{dx}$$

$$\downarrow \qquad \downarrow \qquad\qquad \downarrow \quad \downarrow$$

gives: $$\frac{dy}{dx} = (3x^2)\ (2\cos 2x) + (\sin 2x)(6x)$$

i.e. $$\frac{dy}{dx} = 6x^2\cos 2x + 6x\sin 2x = \mathbf{6x\,(x\cos 2x + \sin 2x)}$$

Note that the differential coefficient of a product is **not** obtained by merely differentiating each term and multiplying the two answers together.

Differentiation of a quotient

When $y = \dfrac{u}{v}$, and u and v are both functions of x

then $$\boxed{\frac{\mathbf{dy}}{\mathbf{dx}} = \frac{v\dfrac{\mathbf{du}}{\mathbf{dx}} - u\dfrac{\mathbf{dv}}{\mathbf{dx}}}{v^2}}$$

This is known as the **quotient rule**.

For example, to find the differential coefficient of $y = \dfrac{4 \sin 5x}{5x^4}$:

$\dfrac{4 \sin 5x}{5x^4}$ is a quotient. Let $u = 4 \sin 5x$ and $v = 5x^4$.
(Note that v is **always** the denominator and u the numerator)

$$\frac{dy}{dx} = \frac{v\dfrac{du}{dx} - u\dfrac{dv}{dx}}{v^2} = \frac{(5x^4)(20 \cos 5x) - (4 \sin 5x)(20x^3)}{(5x^4)^2}$$

$$= \frac{100x^4 \cos 5x - 80x^3 \sin 5x}{25x^8} = \frac{20x^3[5x \cos 5x - 4 \sin 5x]}{25x^8}$$

i.e. $$\frac{\mathbf{dy}}{\mathbf{dx}} = \frac{\mathbf{4}}{\mathbf{5x^5}}(\mathbf{5x \cos 5x - 4 \sin 5x})$$

Note that the differential coefficient is **not** obtained by merely differentiating each term in turn and then dividing the numerator by the denominator.
In another example, to determine the differential coefficient of $y = \tan ax$:

$$y = \tan ax = \frac{\sin ax}{\cos ax}.$$

Differentiation of $\tan ax$ is thus treated as a quotient with $u = \sin ax$ and $v = \cos ax$

$$\frac{dy}{dx} = \frac{v\dfrac{du}{dx} - u\dfrac{dv}{dx}}{v^2}$$

$$= \frac{(\cos ax)(a \cos ax) - (\sin ax)(-a \sin ax)}{(\cos ax)^2}$$

$$= \frac{a \cos^2 ax + a \sin^2 ax}{(\cos ax)^2} = \frac{a(\cos^2 ax + \sin^2 ax)}{\cos^2 ax}$$

$$= \frac{a}{\cos^2 ax}, \quad \text{since } \cos^2 ax + \sin^2 ax = 1 \text{ (see Chapter 29)}$$

Hence $$\frac{\mathbf{dy}}{\mathbf{dx}} = \boldsymbol{a \sec^2 ax} \quad \text{since } \sec^2 ax = \frac{1}{\cos^2 ax} \text{ (see Chapter 25)}$$

Function of a function

It is often easier to make a substitution before differentiating.

If y is a function of x then $\boxed{\dfrac{\mathbf{d}y}{\mathbf{d}x} = \dfrac{\mathbf{d}y}{\mathbf{d}u} \times \dfrac{\mathbf{d}u}{\mathbf{d}x}}$

This is known as the **'function of a function'** rule (or sometimes the **chain rule**).

For example, if $y = (3x - 1)^9$ then, by making the substitution $u = (3x - 1)$, $y = u^9$, which is of the 'standard' form.

Hence $\dfrac{dy}{du} = 9u^8$ and $\dfrac{du}{dx} = 3$

Then $\dfrac{dy}{dx} = \dfrac{dy}{du} \times \dfrac{du}{dx} = (9u^8(3) = 27u^8$

Rewriting u as $(3x - 1)$ gives: $\mathbf{\dfrac{dy}{dx} = 27(3x - 1)^8}$. Since y is a function of u, and u is a function of x, then y is a function of a function of x.

In another example, to determine the differential coefficient of $y = \sqrt{3x^2 + 4x - 1}$:

$$y = \sqrt{3x^2 + 4x - 1} = (3x^2 + 4x - 1)^{1/2}$$

Let $u = 3x^2 + 4x - 1$ then $y = u^{1/2}$

Hence $\dfrac{du}{dx} = 6x + 4$ and $\dfrac{dy}{du} = \dfrac{1}{2}u^{-1/2} = \dfrac{1}{2\sqrt{u}}$

Using the function of a function rule,

$$\frac{dy}{dx} = \frac{dy}{du} \times \frac{du}{dx} = \left(\frac{1}{2\sqrt{u}}\right)(6x + 4) = \frac{3x + 2}{\sqrt{u}}$$

i.e. $$\mathbf{\frac{dy}{dx} = \frac{3x + 2}{\sqrt{(3x^2 + 4x - 1)}}}$$

Successive differentiation

When a function $y = f(x)$ is differentiated with respect to x the differential coefficient is written as $\dfrac{dy}{dx}$ or $f'(x)$. If the expression is differentiated again, the second differential coefficient is obtained and is written as $\dfrac{d^2y}{dx^2}$ (pronounced dee two y by dee x squared) or $f''(x)$ (pronounced f double-dash x). By successive differentiation further higher derivatives such as $\dfrac{d^3y}{dx^3}$ and $\dfrac{d^4y}{dx^4}$ may be obtained.

For example, if $y = 3x^4$,

$$\frac{dy}{dx} = 12x^3, \ \frac{d^2y}{dx^2} = 36x^2, \ \frac{d^3y}{dx^3} = 72x, \ \frac{d^4y}{dx^4} = 72$$

and $\quad \dfrac{d^5 y}{dx^5} = 0$

In another example, if $\quad f(x) = 2x^5 - 4x^3 + 3x - 5$

then $\quad f'(x) = 10x^4 - 12x^2 + 3$

and $\quad \boldsymbol{f''(x) = 40x^3 - 24x = 4x(10x^2 - 6)}$

Differentiation of hyperbolic functions

From Chapter 13, $\quad \dfrac{d}{dx}(\sinh x) = \dfrac{d}{dx}\left(\dfrac{e^x - e^{-x}}{2}\right) = \left[\dfrac{e^x - (-e^{-x})}{2}\right]$

$$= \left(\frac{e^x + e^{-x}}{2}\right) = \mathbf{\cosh x}$$

If $\boldsymbol{y = \sinh ax}$, where '$a$' is a constant, then $\boldsymbol{\dfrac{dy}{dx} = a \cosh ax}$

$$\frac{d}{dx}(\cosh x) = \frac{d}{dx}\left(\frac{e^x + e^{-x}}{2}\right) = \left[\frac{e^x + (-e^{-x})}{2}\right]$$

$$= \left(\frac{e^x - e^{-x}}{2}\right) = \mathbf{\sinh x}$$

If $\boldsymbol{y = \cosh ax}$, where '$a$' is a constant, then $\boldsymbol{\dfrac{dy}{dx} = a \sinh ax}$

Using the quotient rule of differentiation the derivatives of $\tanh x$, $\operatorname{sech} x$, $\operatorname{cosech} x$ and $\coth x$ may be determined using the above results. A summary is given below

y **or** $f(x)$	$\dfrac{dy}{dx}$ **or** $f'(x)$
$\sinh ax$	$a \cosh ax$
$\cosh ax$	$a \sinh ax$
$\tanh ax$	$a \operatorname{sech}^2 ax$
$\operatorname{sech} ax$	$-a \operatorname{sech} ax \tanh ax$
$\operatorname{cosech} ax$	$-a \operatorname{cosech} ax \coth ax$
$\coth ax$	$-a \operatorname{cosech}^2 ax$

For example, to differentiate the following with respect to x:

(a) $y = 4 \text{ sh } 2x - \dfrac{3}{7} \text{ ch } 3x \quad$ (b) $y = 5 \text{ th } \dfrac{x}{2} - 2 \coth 4x$

(a) $\dfrac{dy}{dx} = 4(2\cosh 2x) - \dfrac{3}{7}(3 \sinh 3x) = \mathbf{8 \cosh 2x - \dfrac{9}{7} \sinh 3x}$

(b) $\dfrac{dy}{dx} = 5\left(\dfrac{1}{2} \operatorname{sech}^2 \dfrac{x}{2}\right) - 2(-4 \operatorname{cosech}^2 4x) = \mathbf{\dfrac{5}{2} \operatorname{sech}^2 \dfrac{x}{2} + 8 \operatorname{cosech}^2 4x}$

In another example, to differentiate the following with respect to the variable:

(a) $y = 4 \sin 3t \text{ ch } 4t \quad$ (b) $y = \ln(\text{sh } 3\theta) - 4 \text{ ch }^2 3\theta$

(a) $y = 4 \sin 3t \text{ ch } 4t$ (i.e. a product)

$$\frac{dy}{dx} = (4 \sin 3t)(4 \text{ sh } 4t) + (\text{ch } 4t)(4)(3 \cos 3t)$$

$$= 16 \sin 3t \text{ sh } 4t + 12 \text{ ch } 4t \cos 3t$$

$$= \mathbf{4(4 \sin 3t \text{ sh } 4t + 3 \cos 3t \text{ ch } 4t)}$$

(b) $y = \ln(\text{sh } 3\theta) - 4 \text{ ch }^2 3\theta$ (i.e. a function of a function)

$$\frac{dy}{d\theta} = \left(\frac{1}{\text{sh } 3\theta}\right)(3 \text{ ch } 3\theta) - (4)(2 \text{ ch } 3\theta)(3 \text{ sh } 3\theta)$$

$$= 3 \coth 3\theta - 24 \text{ ch } 3\theta \text{ sh } 3\theta = \mathbf{3(\coth 3\theta - 8 \text{ ch } 3\theta \text{ sh } 3\theta)}$$

---

49 Some Applications of Differentiation

Rates of Change

If a quantity y depends on and varies with a quantity x then the rate of change of y with respect to x is $\frac{dy}{dx}$. Thus, for example, the rate of change of pressure p with height h is $\frac{dp}{dh}$.
A rate of change with respect to time is usually just called 'the rate of change', the 'with respect to time' being assumed. Thus, for example, a rate of change of current, i, is $\frac{di}{dt}$ and a rate of change of temperature, θ, is $\frac{d\theta}{dt}$, and so on.
For example, Newtons law of cooling is given by $\theta = \theta_0 e^{-kt}$, where the excess of temperature at zero time is θ_0°C and at time t seconds is θ°C. To determine the rate of change of temperature after 40 s, given that $\theta_0 = 16^\circ$C and $k = -0.03$:

The rate of change of temperature is $\frac{d\theta}{dt}$.

Since $\theta = \theta_0 e^{-kt}$ then $\frac{d\theta}{dt} = (\theta_0)(-k)e^{-kt} = -k\theta_0 e^{-kt}$

When $\theta_0 = 16, k = -0.03$ and $t = 40$

then $\frac{d\theta}{dt} = -(-0.03)(16)e^{-(-0.03)(40)} = 0.48e^{1.2} = \mathbf{1.594^\circ C/s}$

Velocity and Acceleration

If a body moves a distance x metres in a time t seconds then:

(i) **distance** $\boldsymbol{x = f(t)}$

(ii) **velocity** $\boldsymbol{v = f'(t)}$ **or** $\boldsymbol{\frac{dx}{dt}}$, which is the gradient of the distance/time graph

(iii) **acceleration** $\boldsymbol{a = \frac{dv}{dt} = f'' \text{ or } \frac{d^2x}{dt^2}}$, which is the gradient of the velocity/time graph.

For example, the distance x metres travelled by a vehicle in time t seconds after the brakes are applied is given by $x = 20t - \frac{5}{3}t^2$. To determine (a) the speed of the vehicle (in km/h) at the instant the brakes are applied, and (b) the distance the car travels before it stops:

(a) Distance, $x = 20t - \frac{5}{3}t^2$. Hence velocity $v = \frac{dx}{dt} = 20 - \frac{10}{3}t$. At the instant the brakes are applied, time = 0. Hence

$$\textbf{velocity} \quad v = 20 \text{ m/s} = \frac{20 \times 60 \times 60}{1000} \text{ km/h} = \mathbf{72\ km/h}.$$

(Note: changing from m/s to km/h merely involves multiplying by 3.6).

(b) When the car finally stops, the velocity is zero, i.e. $v = 20 - \frac{10}{3}t = 0$, from which, $20 = \frac{10}{3}t$, giving t = 6 s. Hence the distance travelled before the car stops is given by:

$$x = 20t - \tfrac{5}{3}t^2 = 20(6) - \tfrac{5}{3}(6)^2 = 120 - 60 = \mathbf{60\ m}$$

In another example, the angular displacement θ radians of a flywheel varies with time t seconds and follows the equation $\theta = 9t^2 - 2t^3$. To determine (a) the angular velocity and acceleration of the flywheel when time, $t = 1$ s, and (b) the time when the angular acceleration is zero:

(a) Angular displacement $\theta = 9t^2 - 2t^3$ rad

$$\text{Angular velocity } \omega = \frac{d\theta}{dt} = 18t - 6t^2 \text{ rad/s}$$

When time $t = 1$ s, $\omega = 18(1) - 6(1)^2 = \mathbf{12\ rad/s}$

$$\text{Angular acceleration } \alpha = \frac{d^2\theta}{dt^2} = 18 - 12t \text{ rad/s}$$

When time $t = 1$ s, $\boldsymbol{\alpha} = 18 - 12(1) = \mathbf{6\ rad/s^2}$

(b) When the angular acceleration is zero, $18 - 12t = 0$, from which, $18 = 12t$, giving time, $\mathbf{t = 1.5\ s}$

Turning points

In Figure 49.1, the gradient (or rate of change) of the curve changes from positive between O and P to negative between P and Q, and then positive again between Q and R. At point P, the gradient is zero and, as x increases, the gradient of the curve changes from positive just before P to negative just after. Such a point is called a **maximum point** and appears as the 'crest of a wave'. At point Q, the gradient is also zero and, as x increases, the gradient of the curve changes from negative just before Q to positive just after. Such a point is called a **minimum point**, and appears as the 'bottom of a valley'. Points such as P and Q are given the general name of **turning points**.

It is possible to have a turning point, the gradient on either side of which is the same. Such a point is given the special name of a **point of inflexion**, and examples are shown in Figure 49.2.

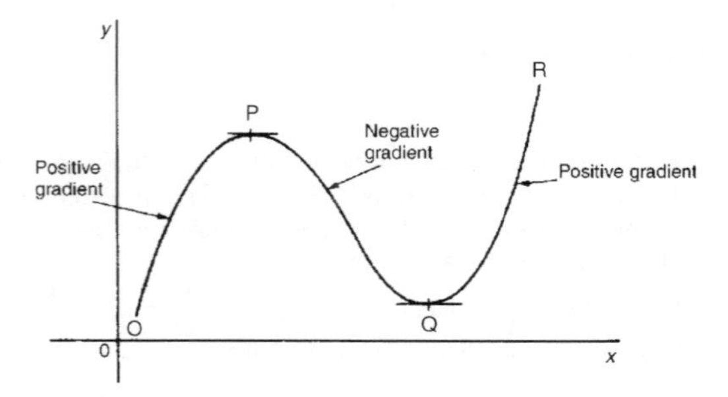

Figure 49.1

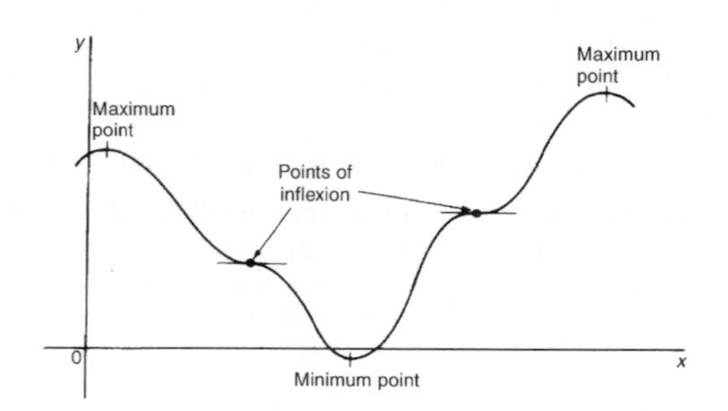

Figure 49.2

Maximum and minimum points and points of inflexion are given the general term of **stationary points**.

Procedure for finding and distinguishing between stationary points

(i) Given $y = f(x)$, determine $\frac{dy}{dx}$ (i.e. $f'(x)$)

(ii) Let $\frac{dy}{dx} = 0$ and solve for the values of x

(iii) Substitute the values of x into the original equation, $y = f(x)$, to find the corresponding y-ordinate values. This establishes the co-ordinates of the stationary points. To determine the nature of the stationary points:
Either

(iv) Find $\frac{d^2y}{dx^2}$ and substitute into it the values of x found in (ii).
If the result is: (a) positive — the point is a minimum one,
(b) negative — the point is a maximum one,
(c) zero — the point is a point of inflexion

or

(v) Determine the sign of the gradient of the curve just before and just after the stationary points. If the sign change for the gradient of the curve is:
(a) positive to negative — the point is a maximum one
(b) negative to positive — the point is a minimum one

(c) positive to positive or negative to negative — the point is a point of inflexion

For example, to find the maximum and minimum values of the curve $y = x^3 - 3x + 5$:

Since $y = x^3 - 3x + 5$ then $\frac{dy}{dx} = 3x^2 - 3$. For a maximum or minimum value $\frac{dy}{dx} = 0$. Hence $3x^2 - 3 = 0$, from which, $3x^2 = 3$ and $x = \pm 1$.

When $x = 1$, $y = (1)^3 - 3(1) + 5 = 3$

When $x = -1$, $y = (-1)^3 - 3(-1) + 5 = 7$.

Hence (1, 3) and (−1, 7) are the co-ordinates of the turning points.

Considering the point (1, 3):

If x is slightly less than 1, say 0.9, then $\frac{dy}{dx} = 3(0.9)^2 - 3$, which is negative.

If x is slightly more than 1, say 1.1, then $\frac{dy}{dx} = 3(1.1)^2 - 3$, which is positive.

Since the gradient changes from negative to positive, **the point (1, 3) is a minimum point**.

Considering the point (−1, 7):

If x is slightly less than -1, say -1.1, then $\frac{dy}{dx} = 3(-1.1)^2 - 3$, which is positive. If x is slightly more than -1, say -0.9, then $\frac{dy}{dx} = 3(-0.9)^2 - 3$, which is negative. Since the gradient changes from positive to negative, **the point (−1, 7) is a maximum point**.

Since $\frac{dy}{dx} = 3x^2 - 3$, then $\frac{d^2y}{dx^2} = 6x$. When $x = 1$, $\frac{d^2y}{dx^2}$ is positive, hence (1, 3) is a **minimum value**. When $x = -1$, $\frac{d^2y}{dx^2}$ is negative, hence (−1, 7) is a **maximum value**.

Thus the maximum value is 7 and the minimum value is 3.

It can be seen that the second differential method of determining the nature of the turning points is, in this case, quicker than investigating the gradient.

Practical problems involving maximum and minimum values

There are many **practical problems** involving maximum and minimum values which occur in science and engineering. Usually, am equation has to be determined from given data, and rearranged where necessary, so that it contains only one variable.

For example, to determine the area of the largest piece of rectangular ground that can be enclosed by 100 m of fencing, if part of an existing straight wall is used as one side:

Let the dimensions of the rectangle be x and y as shown in Figure 49.3, where PQ represents the straight wall.

From Figure 49.3, $\quad x + 2y = 100 \qquad (1)$

Area of rectangle, $\quad A = xy \qquad (2)$

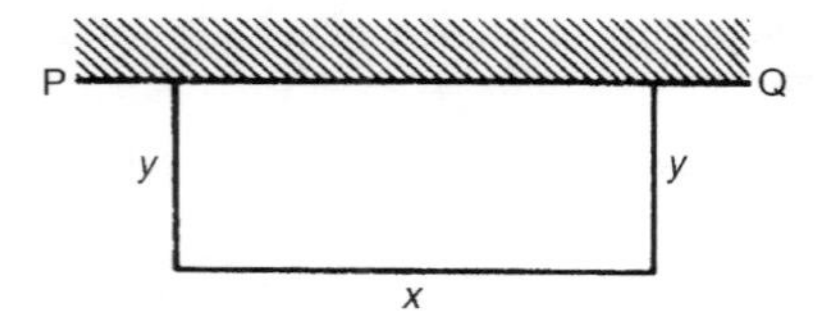

Figure 49.3

Since the maximum area is required, a formula for area A is needed in terms of one variable only. From equation (1), $x = 100 - 2y$

Hence area $\mathbf{A = xy = (100 - 2y)y = 100y - 2y^2}$

$$\frac{dA}{dy} = 100 - 4y = 0, \text{ for a turning point, from which, } y = 25 \text{ m.}$$

$$\frac{d^2A}{dy^2} = -4, \text{ which is negative, giving a maximum value.}$$

When $y = 25$ m, $x = 50$ m from equation (1).

Hence the **maximum possible area** $= xy = (50)(25) = \mathbf{1250\ m^2}$

In another example, an open rectangular box with square ends is fitted with an overlapping lid which covers the top and the front face. To determine the maximum volume of the box if 6 m^2 of metal are used in its construction:

A rectangular box having square ends of side x and length y is shown in Figure 49.4

Surface area of box, A, consists of two ends and five faces (since the lid also covers the front face).

Hence $\quad A = 2x^2 + 5xy = 6 \qquad (1)$

Since it is the maximum volume required, a formula for the volume in terms of one variable only is needed. Volume of box, $V = x^2y$

From equation (1), $\quad y = \dfrac{6 - 2x^2}{5x} = \dfrac{6}{5x} - \dfrac{2x}{5} \qquad (2)$

Hence volume $\quad V = x^2y = x^2\left(\dfrac{6}{5x} - \dfrac{2x}{5}\right) = \dfrac{6x}{5} - \dfrac{2x^3}{5}$

$$\frac{dV}{dx} = \frac{6}{5} - \frac{6x^2}{5} = 0 \text{ for a maximum or minimum value.}$$

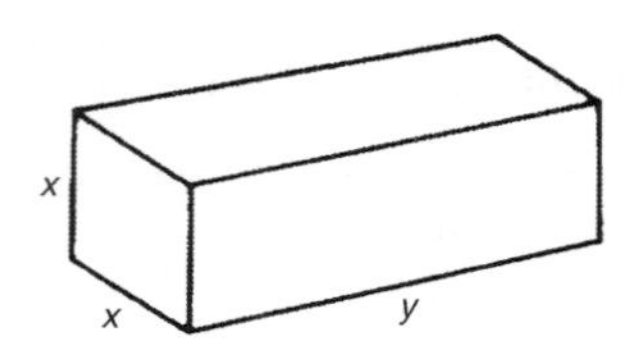

Figure 49.4

Hence $6 = 6x^2$, giving $x = 1$ m ($x = -1$ is not possible, and is thus neglected). $\frac{d^2V}{dx^2} = \frac{-12x}{5}$. When $x = 1$, $\frac{d^2V}{dx^2}$ is negative, giving a maximum value. From equation (2), when $x = 1$, $y = \frac{6}{5(1)} - \frac{2(1)}{5} = \frac{4}{5}$

Hence the maximum volume of the box is given by

$$V = x^2 y = (1)^2 \left(\frac{4}{5}\right) = \mathbf{\frac{4}{5}\ m^3}$$

Tangents and normals

Tangents

The equation of the tangent to a curve $y = f(x)$ at the point (x_1, y_1) is given by:

$$\boxed{\boldsymbol{y - y_1 = m(x - x_1)}}$$

where $m = \frac{dy}{dx}$ = gradient of the curve at (x_1, y_1).

For example, to find the equation of the tangent to the curve $y = x^2 - x - 2$ at the point $(1, -2)$:

Gradient, $m = \frac{dy}{dx} = 2x - 1$.

At the point $(1, -2)$, $x = 1$ and $m = 2(1) - 1 = 1$. Hence the equation of the tangent is: $y - y_1 = m(x - x_1)$

i.e. $\quad y - -2 = 1(x - 1)$

i.e. $\quad y + 2 = x - 1$

or $\quad \boldsymbol{y = x - 3}$

The graph of $y = x^2 - x - 2$ is shown in Figure 49.5. The line AB is the tangent to the curve at the point C, i.e. $(1, -2)$, and the equation of this line is $y = x - 3$.

Normals

The normal at any point on a curve is the line that passes through the point and is at right angles to the tangent. Hence, in Figure 49.5, the line CD is the normal.

It may be shown that if two lines are at right angles then the product of their gradients is -1. Thus if m is the gradient of the tangent, then the gradient of the normal is $-\frac{1}{m}$

Hence the equation of the normal at the point (x_1, y_1) is given by:

$$\boxed{\boldsymbol{y - y_1 = -\frac{1}{m}(x - x_1)}}$$

For example, to find the equation of the normal to the curve $y = x^2 - x - 2$ at the point $(1, -2)$:

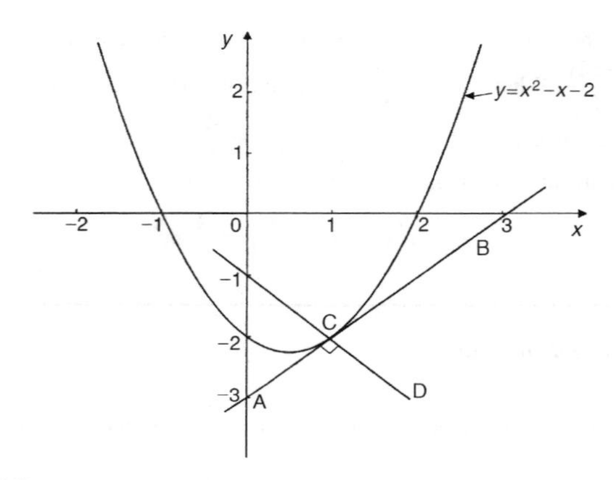

Figure 49.5

$m = 1$ from above, hence the equation of the normal is

$$y - (-2) = -\tfrac{1}{1}(x - 1)$$

i.e. $\quad y + 2 = -x + 1 \quad$ or $\quad \boldsymbol{y = -x - 1}$

Thus the line CD in Figure 49.5 has the equation $y = -x - 1$

Small changes

If y is a function of x, i.e. $y = f(x)$, and the approximate change in y corresponding to a small change δx in x is required, then:

$$\frac{\delta y}{\delta x} \approx \frac{\mathrm{d}y}{\mathrm{d}x}$$

and $\quad \boldsymbol{\delta y \approx \frac{dy}{dx}.\delta x}$ or $\boldsymbol{\delta y \approx f'(x).\delta x}$

For example, the time of swing T of a pendulum is given by $T = k\sqrt{l}$, where k is a constant. To determine the percentage change in the time of swing if the length of the pendulum l changes from 32.1 cm to 32.0 cm:

If $T = k\sqrt{l} = kl^{1/2}$, then $\dfrac{\mathrm{d}T}{\mathrm{d}l} = k\left(\dfrac{1}{2}l^{-1/2}\right) = \dfrac{k}{2\sqrt{l}}$

Approximate change in T, $\delta t \approx \dfrac{\mathrm{d}T}{\mathrm{d}l}\delta l \approx \left(\dfrac{k}{2\sqrt{l}}\right)\delta l$

$\approx \left(\dfrac{k}{2\sqrt{l}}\right)(-0.1)$ (negative since l decreases)

$$\text{Percentage error} = \left(\frac{\text{approximate change in T}}{\text{original value of T}}\right)100\%$$

$$= \frac{\left(\frac{k}{2\sqrt{l}}\right)(-0.1)}{k\sqrt{l}} \times 100\% = \left(\frac{-0.1}{2l}\right)100\%$$

$$= \left(\frac{-0.1}{2(32.1)}\right)100\% = \mathbf{-0.156\%}$$

Hence the change in the time of swing is a decrease of 0.156%

---

50 Differentiation of Parametric Equations

Introduction

Certain mathematical functions can be expressed more simply by expressing, say, x and y separately in terms of a third variable. For example, $y = r\sin\theta$, $x = r\cos\theta$. Then, any value given to θ will produce a pair of values for x and y, which may be plotted to provide a curve of $y = f(x)$.
The third variable, θ, is called a **parameter** and the two expressions for y and x are called **parametric equations**.
The above example of $y = r\sin\theta$ and $x = r\cos\theta$ are the parametric equations for a circle. The equation of any point on a circle, centre at the origin and of radius r is given by: $x^2 + y^2 = r^2$.
To show that $y = r\sin\theta$ and $x = r\cos\theta$ are suitable parametric equations for such a circle:

$$\text{left hand side of equation} = x^2 + y^2$$
$$= (r\cos\theta)^2 + (r\sin\theta)^2$$
$$= r^2\cos^2\theta + r^2\sin^2\theta$$
$$= r^2(\cos^2\theta + \sin^2\theta)$$
$$= r^2 = \text{right hand side (since } \cos^2\theta + \sin^2\theta = 1)$$

Some common parametric equations

The following are some of the more common parametric equations, and Figure 50.1 shows typical shapes of these curves.

(a) Ellipse $x = a\cos\theta,\ y = b\sin\theta$
(b) Parabola $x = \text{a}t^2,\ y = 2\text{a}t$
(c) Hyperbola $x = a\sec\theta,\ y = b\tan\theta$
(d) Rectangular hyperbola $x = \text{c}t,\ y = \dfrac{c}{t}$
(e) Cardioid $x = a(2\cos\theta - \cos 2\theta),\ y = a(2\sin\theta - \sin 2\theta)$
(f) Astroid $x = a\cos^3\theta,\ y = a\sin^3\theta$
(g) Cycloid $x = a(\theta - \sin\theta),\ y = a(1 - \cos\theta)$

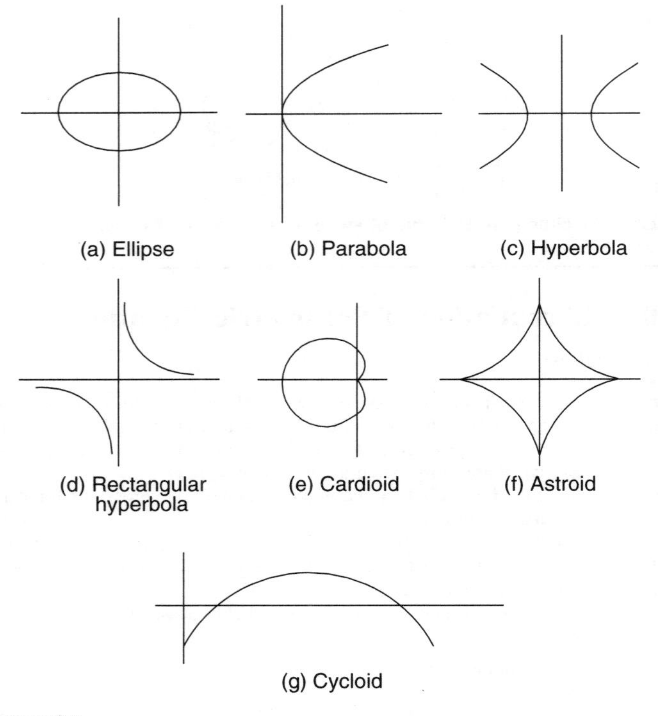

Figure 50.1

Differentiation in parameters

When x and y are given in terms of a parameter, say, θ, then by the function of a function rule of differentiation:

$$\frac{\mathrm{d}y}{\mathrm{d}x} = \frac{\mathrm{d}y}{\mathrm{d}\theta} \times \frac{\mathrm{d}\theta}{\mathrm{d}x}$$

It may be shown that this can be written as:

$$\boxed{\frac{\mathrm{d}y}{\mathrm{d}x} = \frac{\dfrac{\mathrm{d}y}{\mathrm{d}\theta}}{\dfrac{\mathrm{d}x}{\mathrm{d}\theta}}} \qquad (1)$$

For the second differential, $\frac{d^2y}{dx^2} = \frac{d}{dx}\left(\frac{dy}{dx}\right) = \frac{d}{d\theta}\left(\frac{dy}{dx}\right)\frac{d\theta}{dx}$

or $$\boxed{\frac{\mathbf{d^2y}}{\mathbf{dx^2}} = \frac{\frac{\mathbf{d}}{\mathbf{d\theta}}\left(\frac{\mathbf{dy}}{\mathbf{dx}}\right)}{\frac{\mathbf{dx}}{\mathbf{d\theta}}}} \qquad (2)$$

For example, given $x = 5\theta - 1$ and $y = 2\theta(\theta - 1)$, to determine $\frac{dy}{dx}$ in terms of θ:

$$x = 5\theta - 1, \text{ hence } \frac{dx}{d\theta} = 5$$

$$y = 2\theta(\theta - 1) = 2\theta^2 - 2\theta, \text{ hence } \frac{dy}{d\theta} = 4\theta - 2 = 2(2\theta - 1)$$

From equation (1), $\frac{\mathbf{dy}}{\mathbf{dx}} = \frac{\frac{dy}{d\theta}}{\frac{dx}{d\theta}} = \frac{2(2\theta - 1)}{5}$ or $\mathbf{\frac{2}{5}(2\theta - 1)}$

In another example, when determining the surface tension of a liquid, the radius of curvature ρ, of part of the surface is given by:

$$\rho = \frac{\sqrt{\left[1 + \left(\frac{dy}{dx}\right)^2\right]^3}}{\frac{d^2y}{dx^2}}$$

To find the radius of curvature of the part of the surface having the parametric equations $x = 3t^2$, $y = 6t$ at the point $t = 2$:

$x = 3t^2$, hence $\frac{dx}{dt} = 6t$ and $y = 6t$, hence $\frac{dy}{dt} = 6$

From equation (1), $$\frac{dy}{dx} = \frac{\frac{dy}{dt}}{\frac{dx}{dt}} = \frac{6}{6t} = \frac{1}{t}$$

From equation (2), $$\frac{d^2y}{dt^2} = \frac{\frac{d}{dt}\left(\frac{dy}{dx}\right)}{\frac{dx}{dt}} = \frac{\frac{d}{dt}\left(\frac{1}{t}\right)}{6t} = \frac{\frac{d}{dt}(t^{-1})}{6t}$$

$$= \frac{-t^{-2}}{6t} = \frac{-\frac{1}{t^2}}{6t} = \frac{-1}{6t^3}$$

Hence radius of curvature,

$$\rho = \frac{\sqrt{\left[1 + \left(\frac{dy}{dx}\right)^2\right]^3}}{\frac{d^2y}{dx^2}} = \frac{\sqrt{\left[1 + \left(\frac{1}{t}\right)^2\right]^3}}{\frac{-1}{6t^3}}$$

When $t = 2$,
$$\rho = \frac{\sqrt{\left[1 + \left(\frac{1}{2}\right)^2\right]^3}}{-\frac{1}{6(2)^3}} = \frac{\sqrt{(1.25)^3}}{-\frac{1}{48}}$$

$$= -48\sqrt{1.25^3} = -48 = \mathbf{-67.08}$$

---

51 Differentiation of Implicit Functions

Implicit functions

When an equation can be written in the form $y = f(x)$ it is said to be an **explicit function** of x. Examples of explicit functions include $y = 2x^3 - 3x + 4$, $y = 2x \ln x$ and $y = \frac{3e^x}{\cos x}$. In these examples y may be differentiated with respect to x by using standard derivatives, the product rule and the quotient rule of differentiation respectively.

Sometimes with equations involving, say, y and x, it is impossible to make y the subject of the formula. The equation is then called an **implicit function** and examples of such functions include $y^3 + 2x^2 = y^2 - x$ and $\sin y = x^2 + 2xy$

Differentiating implicit functions

It is possible to **differentiate an implicit function** by using the **function of a function rule**, which may be stated as

$$\frac{du}{dx} = \frac{du}{dy} \times \frac{dy}{dx}$$

Thus, to differentiate y^3 with respect to x, the substitution $u = y^3$ is made, from which, $\frac{du}{dy} = 3y^2$. Hence, $\frac{d}{dx}(y^3) = (3y^2) \times \frac{dy}{dx}$, by the function of a function rule.

A simple rule for differentiating an implicit function is summarised as:

$$\boxed{\frac{\mathbf{d}}{\mathbf{d}x}[f(y)] = \frac{\mathbf{d}}{\mathbf{d}y}[f(y)] \times \frac{\mathbf{d}y}{\mathbf{d}x}} \qquad (1)$$

For example, to differentiate $u = \sin 3t$ with respect to x:

$$\frac{du}{dx} = \frac{du}{dt} \times \frac{dt}{dx} = \frac{d}{dt}(\sin 3t) \times \frac{dt}{dx} = \mathbf{3\cos 3t\,\frac{dt}{dx}}$$

In another example, to differentiate $u = 4\ln 5y$ with respect to t:

$$\frac{du}{dt} = \frac{du}{dy} \times \frac{dy}{dt} = \frac{d}{dy}(4\ln 5y) \times \frac{dy}{dt} = \left(\frac{4}{y}\right)\frac{dy}{dt}$$

Differentiating implicit functions containing products and quotients

The product and quotient rules of differentiation must be applied when differentiating functions containing products and quotients of two variables.

For example, $\frac{d}{dx}(x^2y) = (x^2)\frac{d}{dx}(y) + (y)\frac{d}{dx}(x^2)$, by the product rule

$$= (x^2)\left(1\frac{dy}{dx}\right) + y(2x), \text{ by using equation (1)}$$

$$= x^2\frac{dy}{dx} + 2xy$$

In another example,

$$\frac{d}{dx}\left(\frac{3y}{2x}\right) = \frac{(2x)\frac{d}{dx}(3y) - (3y)\frac{d}{dx}(2x)}{(2x)^2} = \frac{(2x)\left(3\frac{dy}{dx}\right) - (3y)(2)}{4x^2}$$

$$= \frac{6x\frac{dy}{dx} - 6y}{4x^2} = \frac{3}{2x^2}\left(x\frac{dy}{dx} - y\right)$$

Further implicit differentiation

An implicit function such as $3x^2 + y^2 - 5x + y = 2$, may be differentiated term by term with respect to x. This gives:

$$\frac{d}{dx}(3x^2) + \frac{d}{dx}(y^2) - \frac{d}{dx}(5x) + \frac{d}{dx}(y) = \frac{d}{dx}(2)$$

i.e. $6x + 2y\frac{dy}{dx} - 5 + 1\frac{dy}{dx} = 0$, using equation (1) and standard derivatives.

An expression for the derivative $\frac{dy}{dx}$ in terms of x and y may be obtained by rearranging this latter equation. Thus:

$$(2y + 1)\frac{dy}{dx} = 5 - 6x$$

from which, $$\frac{dy}{dx} = \frac{5 - 6x}{2y + 1}$$

---

52 Logarithmic Differentiation

Introduction to logarithmic differentiation

With certain functions containing more complicated products and quotients, differentiation is often made easier if the logarithm of the function is taken before differentiating. This technique, called **'logarithmic differentiation'** is achieved with a knowledge of (i) the laws of logarithms, (ii) the differential coefficients of logarithmic functions, and (iii) the differentiation of implicit functions.

Laws of logarithms

Three laws of logarithms may be expressed as:

(i) $\log(A \times B) = \log A + \log B$ (ii) $\log\left(\frac{A}{B}\right) = \log A - \log B$

(iii) $\log A^n = n \log A$

In calculus, Napierian logarithms (i.e. logarithms to a base of 'e') are invariably used. Thus for two functions $f(x)$ and $g(x)$ the laws of logarithms may be expressed as:

(i) $\ln[f(x).g(x)] = \ln f(x) + \ln g(x)$ (ii) $\ln\left(\frac{f(x)}{g(x)}\right) = \ln f(x) - \ln g(x)$

(iii) $\ln[f(x)]^n = n \ln f(x)$

Taking Napierian logarithms of both sides of the equation $y = \frac{f(x).g(x)}{h(x)}$

gives: $\ln y = \ln\left(\frac{f(x).g(x)}{h(x)}\right)$

which may be simplified using the above laws of logarithms, giving:

$$\ln y = \ln f(x) + \ln g(x) - \ln h(x)$$

This latter form of the equation is often easier to differentiate.

Differentiation of logarithmic functions

The differential coefficient of the logarithmic function $\ln x$ is given by:

$$\frac{\mathbf{d}}{\mathbf{d}x}(\ln x) = \frac{1}{x}$$

More generally, it may be shown that:

$$\frac{\mathbf{d}}{\mathbf{d}x}[\ln f(x)] = \frac{f'(x)}{f(x)} \qquad (1)$$

For example, if $y = \ln(3x^2 + 2x - 1)$ then $\frac{dy}{dx} = \frac{6x + 2}{3x^2 + 2x - 1}$

In another example, if $y = \ln(\sin 3x)$, then $\dfrac{dy}{dx} = \dfrac{3\cos 3x}{\sin 3x} = 3\cot 3x$.
As explained in Chapter 51, by using the function of a function rule:

$$\frac{d}{dx}(\ln y) = \left(\frac{1}{y}\right)\frac{dy}{dx} \qquad (2)$$

For example, differentiation of $y = \dfrac{(1+x)^2\sqrt{(x-1)}}{x\sqrt{(x+2)}}$ may be achieved by using the product and quotient rules of differentiation; however the working would be rather complicated. With logarithmic differentiation the following procedure is adopted:

(i) Take Napierian logarithms of both sides of the equation. Thus

$$\ln y = \ln\left\{\frac{(1+x)^2\sqrt{(x-1)}}{x\sqrt{(x+2)}}\right\} = \ln\left\{\frac{(1+x)^2(x-1)^{1/2}}{x(x+2)^{1/2}}\right\}$$

(ii) Apply the laws of logarithms.
Thus $\ln y = \ln(1+x)^2 + \ln(x-1)^{1/2} - \ln x - \ln(x+2)^{1/2}$, by laws (i) and (ii)
i.e. $\ln y = 2\ln(1+x) + \frac{1}{2}\ln(x-1) - \ln x - \frac{1}{2}\ln(x+2)$, by law (iii)

(iii) Differentiate each term in turn with respect to x using equations (1) and (2)

$$\text{Thus } \frac{1}{y}\frac{dy}{dx} = \frac{2}{(1+x)} + \frac{\frac{1}{2}}{(x-1)} - \frac{1}{x} - \frac{\frac{1}{2}}{(x+2)}$$

(iv) Rearrange the equation to make $\dfrac{dy}{dx}$ the subject.

$$\text{Thus } \frac{dy}{dx} = y\left\{\frac{2}{(1+x)} + \frac{1}{2(x-1)} - \frac{1}{x} - \frac{1}{2(x+2)}\right\}$$

(v) Substitute for y in terms of x

$$\text{Thus } \frac{dy}{dx} = \frac{(1+x)^2\sqrt{(x-1)}}{x\sqrt{(x+2)}}\left\{\frac{2}{(1+x)} + \frac{1}{2(x-1)} - \frac{1}{x} - \frac{1}{2(x+2)}\right\}$$

Differentiation of $[f(x)]^x$

Whenever an expression to be differentiated contains a term raised to a power which is itself a function of the variable, then logarithmic differentiation must be used. For example, the differentiation of expressions such as x^x, $(x+2)^x$, $\sqrt[x]{(x-1)}$ and x^{3x+2} can only be achieved using logarithmic differentiation.

For example, to determine $\dfrac{dy}{dx}$ given $y = x^x$:
Taking Napierian logarithms of both sides of $y = x^x$ gives:
$\ln y = \ln x^x = x\ln x$, by law (iii)
Differentiating both sides with respect to x gives:

$$\frac{1}{y}\frac{dy}{dx} = (x)\left(\frac{1}{x}\right) + (\ln x)(1), \text{ using the product rule}$$

i.e. $\frac{1}{y}\frac{dy}{dx} = 1 + \ln x$, from which, $\frac{dy}{dx} = y(1 + \ln x)$

i.e. $\mathbf{\frac{dy}{dx} = x^x(1 + \ln x)}$

---

53 Differentiation of Inverse Trigonometric and Hyperbolic Functions

Inverse functions

If $y = 3x - 2$, then by transposition, $x = \frac{y+2}{3}$. The function $x = \frac{y+2}{3}$ is called the inverse **function** of $y = 3x - 2$.

Inverse trigonometric functions are denoted by prefixing the function with 'arc'. For example, if $y = \sin x$, then $x = \arcsin y$. Similarly, if $y = \cos x$, then $x = \arccos y$, and so on. Alternatively, if $y = \sin x$, then $x = \sin^{-1} y$. A sketch of each of the inverse trigonometric functions is shown in Figure 53.1.

Inverse hyperbolic functions are denoted by prefixing the function with 'ar'. For example, if $y = \sinh x$, then $x = \text{arsinh}\, y$. Similarly, if $y = \text{sech}\, x$, then $x = \text{arsech}\, y$, and so on. Alternatively, if $y = \sinh x$, then $x = \sinh^{-1} y$. A sketch of each of the inverse hyperbolic functions is shown in Figure 53.2.

Differentiation of inverse trigonometric functions

The differential coefficients of inverse trigonometric functions are summarised in Table 53.1.

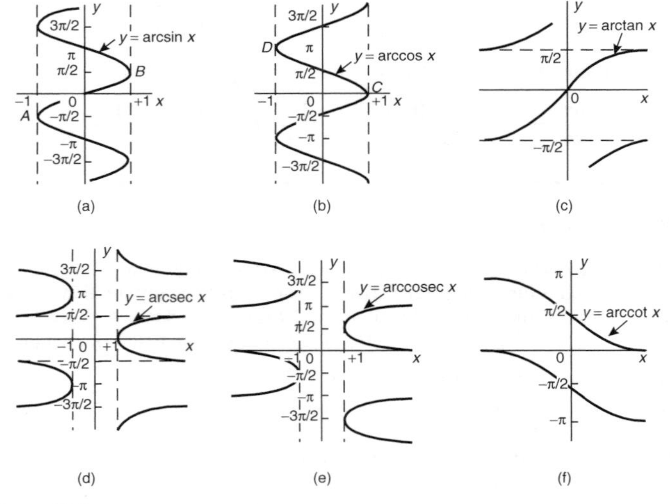

Figure 53.1

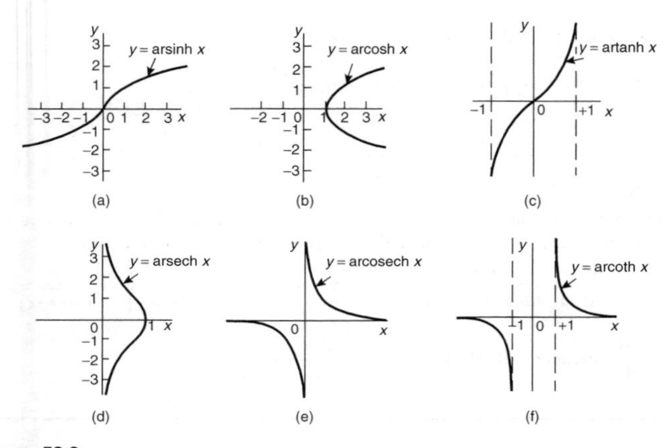

Figure 53.2

Table 53.1 Differential coefficients of inverse trigonometric functions

***y* or *f*(*x*)**	$\frac{dy}{dx}$ **or *f′*(*x*)**
(i) $\arcsin \frac{x}{a}$	$\frac{1}{\sqrt{a^2 - x^2}}$
$\arcsin f(x)$	$\frac{f'(x)}{\sqrt{1 - [f(x)]^2}}$
(ii) $\arccos \frac{x}{a}$	$\frac{-1}{\sqrt{a^2 - x^2}}$
$\arccos f(x)$	$\frac{-f'(x)}{\sqrt{1 - [f(x)]^2}}$
(iii) $\arctan \frac{x}{a}$	$\frac{a}{a^2 + x^2}$
$\arctan f(x)$	$\frac{f'(x)}{1 + [f(x)]^2}$
(iv) $\text{arcsec} \frac{x}{a}$	$\frac{a}{x\sqrt{x^2 - a^2}}$
$\text{arcsec} f(x)$	$\frac{f'(x)}{f(x)\sqrt{[f(x)]^2 - 1}}$
(v) $\text{arccosec} \frac{x}{a}$	$\frac{-a}{x\sqrt{x^2 - a^2}}$
$\text{arccosec} f(x)$	$\frac{-f'(x)}{f(x)\sqrt{[f(x)]^2 - 1}}$
(vi) $\text{arccot} \frac{x}{a}$	$\frac{-a}{a^2 + x^2}$
$\text{arccot} f(x)$	$\frac{-f'(x)}{1 + [f(x)]^2}$

For example, to find $\frac{dy}{dx}$ given $y = \arcsin 5x^2$.

From Table 53.1(i), if $y = \arcsin f(x)$ then $\frac{dy}{dx} = \frac{f'(x)}{\sqrt{1-[f(x)]^2}}$.

Hence, if $y = \arcsin 5x^2$ then $f(x) = 5x^2$ and $f'(x) = 10x$.

Thus $\frac{dy}{dx} = \frac{10x}{\sqrt{1-(5x^2)^2}} = \mathbf{\frac{10x}{\sqrt{1-25x^4}}}$.

In another example, to find the differential coefficient of $y = \ln(\arccos 3x)$:
Let $u = \arccos 3x$ then $y = \ln u$. By the function of a function rule,

$$\frac{dy}{dx} = \frac{dy}{du}\cdot\frac{du}{dx} = \frac{1}{u}\times\frac{d}{dx}(\arccos 3x) = \frac{1}{\arccos 3x}\left\{\frac{-3}{\sqrt{1-(3x)^2}}\right\}$$

i.e. $$\mathbf{\frac{d}{dx}[\ln(\arccos 3x)] = \frac{-3}{\sqrt{1-9x^2}\,\arccos 3x}}$$

Logarithmic forms of the inverse hyperbolic functions

Inverse hyperbolic functions may be evaluated most conveniently when expressed in a **logarithmic form**.

For example, if $y = \text{arcsinh}\,\frac{x}{a}$ then $\frac{x}{a} = \sinh y$.

From Chapter 13, $e^y = \cosh y + \sinh y$ and $\cosh^2 y - \sinh^2 y = 1$, from which, $\cosh y = \sqrt{1+\sinh^2 y}$ which is positive since $\cosh y$ is always positive (see Figure 13.2, page 57).

Hence $$e^y = \sqrt{1+\sinh^2 y} + \sinh y$$

$$= \sqrt{\left[1+\left(\frac{x}{a}\right)^2\right]} + \frac{x}{a} = \sqrt{\left(\frac{a^2+x^2}{a^2}\right)} + \frac{x}{a}$$

$$= \frac{\sqrt{a^2+x^2}}{a} + \frac{x}{a} \quad \text{or} \quad \frac{x+\sqrt{a^2+x^2}}{a}$$

Taking Napierian logarithms of both sides gives:

$$y = \ln\left\{\frac{x+\sqrt{a^2+x^2}}{a}\right\}$$

Hence $$\mathbf{\text{arcsinh}\frac{x}{a} = \ln\left\{\frac{x+\sqrt{a^2+x^2}}{a}\right\}} \qquad (1)$$

For example, to evaluate arsinh $\frac{3}{4}$, let $x = 3$ and $a = 4$ in equation (1).

Then $$\text{arcsinh}\,\frac{3}{4} = \ln\left\{\frac{3+\sqrt{4^2+3^2}}{4}\right\} = \ln\left(\frac{3+5}{4}\right)$$

$$= \ln 2 = 0.6931$$

By similar reasoning to the above it may be shown that:

$$\mathbf{arccosh}\frac{x}{a} = \ln\left\{\frac{x+\sqrt{x^2-a^2}}{a}\right\} \text{ and } \mathbf{arctanh}\frac{x}{a} = \frac{1}{2}\ln\left(\frac{a+x}{a-x}\right)$$

In another example, to evaluate $\cosh^{-1} 1.4$, correct to 3 decimal places:
From above, $\cosh^{-1}\frac{x}{a} = \ln\left\{\frac{x \pm \sqrt{x^2-a^2}}{a}\right\}$ and

$\cosh^{-1} 1.4 = \cosh^{-1}\frac{14}{10} = \cosh^{-1}\frac{7}{5}$

In the equation for $\cosh^{-1}\frac{x}{a}$, let $x = 7$ and $a = 5$ then $\cosh^{-1}\frac{7}{5} =$
$\ln\left\{\frac{7+\sqrt{7^2-5^2}}{5}\right\} = \ln 2.3798 = \mathbf{0.867}$, correct to 3 decimal places

---

Differentiation of inverse hyperbolic functions

The differential coefficients of inverse hyperbolic functions are summarised in Table 53.2

For example, to find the differential coefficient of $y = \text{arcsinh}\, 2x$:
From Table 53.2(i),

$$\frac{d}{dx}[\text{arcsinh}\, f(x)] = \frac{f'(x)}{\sqrt{[f(x)]^2+1}}$$

Hence $$\frac{d}{dx}(\text{arcsinh}\, 2x) = \frac{2}{\sqrt{[(2x)^2+1]}} = \mathbf{\frac{2}{\sqrt{[4x^2+1]}}}$$

In another example, to determine $\frac{d}{dx}[\cosh^{-1}\sqrt{(x^2+1)}]$

If $$y = \cosh^{-1} f(x),\ \frac{dy}{dx} = \frac{f'(x)}{\sqrt{\{[f(x)]^2-1\}}}$$

If $$y = \cosh^{-1}\sqrt{(x^2+1)},\ \text{then } f(x) = \sqrt{(x^2+1)}$$

and $$f'(x) = \frac{1}{2}(x+1)^{-1/2}(2x) = \frac{x}{\sqrt{(x^2+1)}}$$

Hence $$\frac{d}{dx}[\cosh^{-1}\sqrt{(x^2+1)}] = \frac{\frac{x}{\sqrt{(x^2+1)}}}{\sqrt{\{[\sqrt{(x^2+1)}]^2-1\}}} = \frac{\frac{x}{\sqrt{(x^2+1)}}}{\sqrt{(x^2+1-1)}}$$

$$= \frac{\frac{x}{\sqrt{(x^2+1)}}}{x} = \mathbf{\frac{1}{\sqrt{(x^2+1)}}}$$

Table 53.2 Differential coefficients of inverse hyperbolic functions

***y* or *f*(*x*)**	$\frac{dy}{dx}$ **or *f*′(*x*)**
(i) $\text{arcsinh}\,\frac{x}{a}$	$\frac{1}{\sqrt{x^2+a^2}}$
$\text{arcsinh}\,f(x)$	$\frac{f'(x)}{\sqrt{[f(x)]^2+1}}$
(ii) $\text{arccosh}\,\frac{x}{a}$	$\frac{1}{\sqrt{x^2-a^2}}$
$\text{arccosh}\,f(x)$	$\frac{f'(x)}{\sqrt{[f(x)]^2-1}}$
(iii) $\text{arctanh}\,\frac{x}{a}$	$\frac{a}{a^2-x^2}$
$\text{arctanh}\,f(x)$	$\frac{f'(x)}{1-[f(x)]^2}$
(iv) $\text{arcsech}\,\frac{x}{a}$	$\frac{-a}{x\sqrt{a^2-x^2}}$
$\text{arcsech}\,f(x)$	$\frac{-f'(x)}{f(x)\sqrt{1-[f(x)]^2}}$
(v) $\text{arccosech}\,\frac{x}{a}$	$\frac{-a}{x\sqrt{x^2+a^2}}$
$\text{arccosech}\,f(x)$	$\frac{-f'(x)}{f(x)\sqrt{[f(x)]^2+1}}$
(vi) $\text{arccoth}\,\frac{x}{a}$	$\frac{a}{a^2-x^2}$
$\text{arccoth}\,f(x)$	$\frac{f'(x)}{1-[f(x)]^2}$

54 Partial Differentiation

Introduction to partial derivatives

In engineering, it sometimes happens that the variation of one quantity depends on changes taking place in two, or more, other quantities. For example, the volume V of a cylinder is given by $V = \pi r^2 h$. The volume will change if either radius r or height h is changed. The formula for volume may be stated mathematically as $V = f(r, h)$ which means 'V is some function of r and h'. Some other practical examples include:

(i) time of oscillation, $t = 2\pi\sqrt{\frac{l}{g}}$ i.e. $t = f(l, g)$

(ii) torque $T = I\alpha$, i.e. $T = f(I, \alpha)$

(iii) pressure of an ideal gas $p = \frac{mRT}{V}$ i.e. $p = f(T, V)$

(iv) resonant frequency $f_r = \frac{1}{2\pi\sqrt{LC}}$ i.e. $f_r = f(L, C)$, and so on.

When differentiating a function having two variables, one variable is kept constant and the differential coefficient of the other variable is found with respect to that variable. The differential coefficient obtained is called a **partial derivative** of the function.

First order partial derivatives

A 'curly dee', ∂, is used to denote a differential coefficient in an expression containing more than one variable.

Hence if $V = \pi r^2 h$ then $\frac{\partial V}{\partial r}$ means 'the partial derivative of V with respect to r, with h remaining constant'. Thus $\frac{\partial V}{\partial r} = (\pi h)\frac{d}{dr}(r^2) = (\pi h)(2r) = 2\pi rh$.

Similarly, $\frac{\partial V}{\partial h}$ means 'the partial derivative of V with respect to h, with r remaining constant'. Thus $\frac{\partial V}{\partial h} = (\pi r^2)\frac{d}{dh}(h) = (\pi r^2(1)) = \pi r^2$.

$\frac{\partial V}{\partial r}$ and $\frac{\partial V}{\partial h}$ are examples of **first order partial derivatives**, since $n = 1$ when written in the form $\frac{\partial^n V}{\partial r^n}$.

First order partial derivatives are used when finding the total differential, rates of change and errors for functions of two or more variables (see Chapter 55), and when finding maxima, minima and saddle points for functions of two variables (see Chapter 56).

For example, if $Z = 5x^4 + 2x^3y^2 - 3y$

then $$\frac{\partial Z}{\partial x} = \frac{d}{dx}(5x^4) + (2y^2)\frac{d}{dx}(x^3) - (3y)\frac{d}{dx}(l)$$
$$= 20x^3 + (2y^2)(3x^2) - (3y)(0) = \mathbf{20x^3 + 6x^2y^2}$$

and $$\frac{\partial Z}{\partial y} = (5x^4)\frac{d}{dy}(1) + (2x^3)\frac{d}{dy}(y^2) - 3\frac{d}{dy}(y)$$
$$= 0 + (2x^3)(2y) - 3 = \mathbf{4x^3y - 3}$$

Second order partial derivatives

As with ordinary differentiation where a differential coefficient may be differentiated again, a partial derivative may be differentiated partially again, to give higher order partial derivatives.

If $V = \pi r^2 h$, then $\dfrac{\partial V}{\partial r} = (\pi h)\dfrac{d}{dr}(r^2) = (\pi h)(2r) = 2\pi rh$

and $\dfrac{\partial V}{\partial h} = (\pi r^2)\dfrac{d}{dh}(h) = (\pi r^2)(1) = \pi r^2$

from the previous section

(i) Differentiating $\dfrac{\partial V}{\partial r}$ with respect to r, keeping h constant, gives $\dfrac{\partial}{\partial r}\left(\dfrac{\partial V}{\partial r}\right)$, which is written as $\dfrac{\partial^2 V}{\partial r^2}$

Thus if $V = \pi r^2 h$ then $\dfrac{\partial^2 V}{\partial r^2} = \dfrac{\partial}{\partial r}(2\pi rh) = \mathbf{2\pi h}$

(ii) Differentiating $\dfrac{\partial V}{\partial h}$ with respect to h, keeping r constant, gives $\dfrac{\partial}{\partial h}\left(\dfrac{\partial V}{\partial h}\right)$, which is written as $\dfrac{\partial^2 V}{\partial h^2}$

Thus $\dfrac{\partial^2 V}{\partial h^2} = \dfrac{\partial}{\partial h}(\pi r^2) = \mathbf{0}$

(iii) Differentiating $\dfrac{\partial V}{\partial h}$ with respect to r, keeping h constant, gives $\dfrac{\partial}{\partial r}\left(\dfrac{\partial V}{\partial h}\right)$, which is written as $\dfrac{\partial^2 V}{\partial r \partial h}$

Thus $\dfrac{\partial^2 V}{\partial r \partial h} = \dfrac{\partial}{\partial r}\left(\dfrac{\partial V}{\partial h}\right) = \dfrac{\partial}{\partial r}(\pi r^2) = \mathbf{2\pi r}$

(iv) Differentiating $\dfrac{\partial V}{\partial r}$ with respect to h, keeping r constant, gives $\dfrac{\partial}{\partial h}\left(\dfrac{\partial V}{\partial r}\right)$, which is written as $\dfrac{\partial^2 V}{\partial h \partial r}$.

Thus $\dfrac{\partial^2 V}{\partial h \partial r} = \dfrac{\partial}{\partial h}\left(\dfrac{\partial V}{\partial r}\right) = \dfrac{\partial}{\partial h}(2\pi rh) = \mathbf{2\pi r}$

(v) $\dfrac{\partial^2 V}{\partial r^2}$, $\dfrac{\partial^2 V}{\partial h^2}$, $\dfrac{\partial^2 V}{\partial r \partial h}$ and $\dfrac{\partial^2 V}{\partial h \partial r}$ are examples of **second order partial derivatives**.

(vi) It is seen from (iii) and (iv) that $\dfrac{\partial^2 V}{\partial r \partial h} = \dfrac{\partial^2 V}{\partial h \partial r}$ and such a result is always true for continuous functions (i.e. a graph of the function has no sudden jumps or breaks).

Second order partial derivatives are used in the solution of partial differential equations, in waveguide theory, in such areas of thermodynamics covering entropy and the continuity theorem, and when finding maxima, minima and saddle points for functions of two variables (see Chapter 56).

For example, to find

(a) $\frac{\partial^2 Z}{\partial x^2}$ (b) $\frac{\partial^2 Z}{\partial y^2}$ (c) $\frac{\partial^2 Z}{\partial x \partial y}$ (d) $\frac{\partial^2 Z}{\partial y \partial x}$ given $Z = 4x^2y^3 - 2x^3 + 7y^2$:

(a) $\frac{\partial Z}{\partial x} = 8xy^3 - 6x^2$

$$\frac{\partial^2 Z}{\partial x^2} = \frac{\partial}{\partial x}\left(\frac{\partial Z}{\partial x}\right) = \frac{\partial}{\partial x}(8xy^3 - 6x^2) = \mathbf{8y^3 - 12x}$$

(b) $\frac{\partial Z}{\partial y} = 12x^2y^2 + 14y$

$$\frac{\partial^2 Z}{\partial y^2} = \frac{\partial}{\partial y}\left(\frac{\partial Z}{\partial y}\right) = \frac{\partial}{\partial y}(12x^2y^2 + 14y) = \mathbf{24x^2y + 14}$$

(c) $\frac{\partial^2 Z}{\partial x \partial y} = \frac{\partial}{\partial x}\left(\frac{\partial Z}{\partial y}\right) = \frac{\partial}{\partial x}(12x^2y^2 + 14y) = \mathbf{24xy^2}$

(d) $\frac{\partial^2 Z}{\partial y \partial x} = \frac{\partial}{\partial y}\left(\frac{\partial Z}{\partial x}\right) = \frac{\partial}{\partial y}(8xy^3 - 6x^2) = \mathbf{24xy^2}$

---

55 Total Differential, Rates of Change and Small Changes

Total differential

In Chapter 54, partial differentiation is introduced for the case where only one variable changes at a time, the other variables being kept constant. In practice, variables may all be changing at the same time.

If $Z = f(u, v, w, \ldots)$, then the **total differential**, **dZ**, is given by the sum of the separate partial differentials of Z

i.e. $$\mathbf{dZ = \frac{\partial Z}{\partial u} du + \frac{\partial Z}{\partial v} dv + \frac{\partial Z}{\partial w} dw + \ldots} \quad (1)$$

For example, if $Z = f(u, v, w)$ and $Z = 3u^2 - 2v + 4w^3v^2$ the total differential $dZ = \frac{\partial Z}{\partial u} du + \frac{\partial Z}{\partial v} dv + \frac{\partial Z}{\partial w} dw$

$$\frac{\partial Z}{\partial u} = 6u \quad \text{(i.e. } v \text{ and } w \text{ are kept constant)}$$

$$\frac{\partial Z}{\partial v} = -2 + 8w^3v \quad \text{(i.e. } u \text{ and } w \text{ are kept constant)}$$

$$\frac{\partial Z}{\partial w} = 12w^2v^2 \quad \text{(i.e. } u \text{ and } v \text{ are kept constant)}$$

Hence $\mathbf{dZ = 6u\, du + (8vw^3 - 2)\, dv + (12v^2w^2)\, dw}$

Rates of change

Sometimes it is necessary to solve problems in which different quantities have different rates of change. From equation (1), the rate of change of Z, $\frac{dZ}{dt}$, is given by:

$$\frac{dZ}{dt} = \frac{\partial Z}{\partial u}\frac{du}{dt} + \frac{\partial Z}{\partial v}\frac{dv}{dt} + \frac{\partial Z}{\partial w}\frac{dw}{dt} + \ldots \tag{2}$$

For example, if the height of a right circular cone is increasing at 3 mm/s and its radius is decreasing at 2 mm/s, then the rate at which the volume is changing (in cm^3/s) when the height is 3.2 cm and the radius is 1.5 cm, is determined as follows:

Volume of a right circular cone, $V = \frac{1}{3}\pi r^2 h$

Using equation (2), the rate of change of volume,

$$\frac{dV}{dt} = \frac{\partial V}{\partial r}\frac{dr}{dt} + \frac{\partial V}{\partial h}\frac{dh}{dt}$$

$$\frac{\partial V}{\partial r} = \frac{2}{3}\pi rh \quad \text{and} \quad \frac{\partial V}{\partial h} = \frac{1}{3}\pi r^2$$

Since the height is increasing at 3 mm/s, i.e. 0.3 cm/s, then $\frac{dh}{dt} = +0.3$ and since the radius is decreasing at 2 mm/s, i.e. 0.2 cm/s, then $\frac{dr}{dt} = -0.2$

Hence $$\frac{dV}{dt} = \left(\frac{2}{3}\pi rh\right)(-0.2) + \left(\frac{1}{3}\pi r^2\right)(+0.3) = \frac{-0.4}{3}\pi rh + 0.1\pi r^2$$

However, $h = 3.2$ cm and $r = 1.5$ cm.

Hence $$\frac{dV}{dt} = \frac{-0.4}{3}\pi(1.5)(3.2) + (0.1)\pi(1.5)^2$$

$$= -2.011 + 0.707 = -1.304 \text{ cm}^3/\text{s}$$

Thus the rate of change of volume is 1.30 cm³/s decreasing

Small changes

It is often useful to find an approximate value for the change (or error) of a quantity caused by small changes (or errors) in the variables associated with the quantity. If $Z = f(u, v, w, \ldots)$ and δu, δv, δw, ... denote **small changes** in u, v, w, ... respectively, then the corresponding approximate change δZ in Z is obtained from equation (1) by replacing the differentials by the small changes

Thus $$\delta Z \approx \frac{\partial Z}{\partial u}\delta u + \frac{\partial Z}{\partial v}\delta v + \frac{\partial Z}{\partial w}\delta w + \ldots \tag{3}$$

For example, if the modulus of rigidity $G = (R^4\theta)/L$, where R is the radius, θ the angle of twist and L the length, the approximate percentage error in G

when R is increased by 2%, θ is reduced by 5% and L is increased by 4% is determined as follows:

From equation (3), $\delta G \approx \frac{\partial G}{\partial R}\delta R + \frac{\partial G}{\partial \theta}\delta\theta + \frac{\partial G}{\partial L}\delta L$

Since $G = \frac{R^4\theta}{L}$, $\frac{\partial G}{\partial R} = \frac{4R^3\theta}{L}$, $\frac{\partial G}{\partial \theta} = \frac{R^4}{L}$

and $\frac{\partial G}{\partial L} = \frac{-R^4\theta}{L^2}$

Since R is increased by 2%, $\delta R = \frac{2}{100}R = 0.02\ R$. Similarly, $\delta\theta = -0.05\theta$ and $\delta L = 0.04\ L$

$$\text{Hence}\quad \delta G \approx \left(\frac{4R^3\theta}{L}\right)(0.02R) + \left(\frac{R^4}{L}\right)(-0.05\theta) + \left(-\frac{R^4\theta}{L^2}\right)(0.04L)$$

$$\approx \frac{R^4\theta}{L}[0.08 - 0.05 - 0.04] \approx -0.01\frac{R^4\theta}{L}$$

i.e. $\delta G \approx -\frac{1}{100}G$

Hence the approximate percentage error in G is a 1% decrease.

---

56 Maxima, Minima and Saddle Points of Functions of two Variables

Functions of two independent variables

If a relation between two real variables, x and y, is such that when x is given, y is determined, then y is said to be a function of x and is denoted by $y = f(x)$; x is called the independent variable and y the dependent variable.

If $y = f(u, v)$, then y is a function of two independent variables u and v. For example, if, say, $y = f(u, v) = 3u^2 - 2v$ then when $u = 2$ and $v = 1$, $y = 3(2)^2 - 2(1) = 10$. This may be written as $f(2, 1) = 10$. Similarly, if $u = 1$ and $v = 4$, $f(1, 4) = -5$.

Consider a function of two variables x and y defined by $z = f(x, y) = 3x^2 - 2y$. If $(x, y) = (0, 0)$, then $f(0, 0) = 0$ and if $(x, y) = (2, 1)$, then $f(2, 1) = 10$. Each pair of numbers, (x, y), may be represented by a point P in the (x, y) plane of a rectangular Cartesian co-ordinate system as shown in Figure 56.1. The corresponding value of $z = f(x, y)$ may be represented by a line PP' drawn parallel to the z-axis. Thus, if, for example, $z = 3x^2 - 2y$, as above, and P is the co-ordinate (2, 3) then the length of PP' is $3(2)^2 - 2(3) = 6$. Figure 56.2 shows that when a large number of (x, y) co-ordinates are taken for a function $f(x, y)$, and then $f(x, y)$ calculated for each, a large number of lines such as PP' can be constructed, and in the limit when all points in the (x, y) plane are considered, a surface is seen to result as shown in

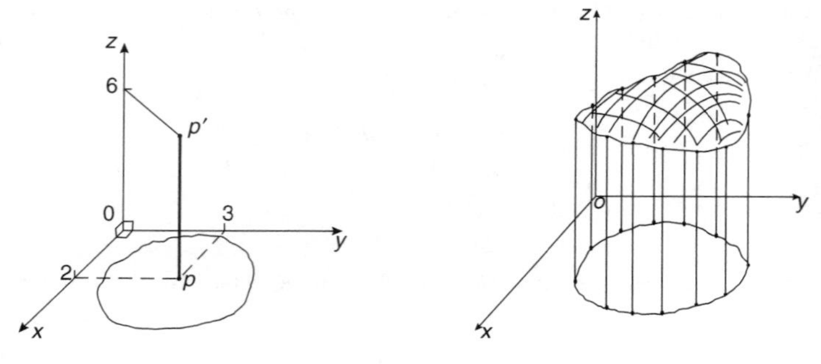

Figure 56.1

Figure 56.2

Figure 56.2. Thus the function $z = f(x, y)$ represents a surface, and not a curve.

Maxima, minima and saddle points

Partial differentiation is used when determining stationary points for functions of two variables. A function $f(x, y)$ is said to be a maximum at a point (x, y) if the value of the function there is greater than at all points in the immediate vicinity, and is a minimum if less than at all points in the immediate vicinity. Figure 56.3 shows geometrically a maximum value of a function of two variables and it is seen that the surface $z = f(x, y)$ is higher at $(x, y) = (a, b)$ than at any point in the immediate vicinity.
Figure 56.4 shows a minimum value of a function of two variables and it is seen that the surface $z = f(x, y)$ is lower at $(x, y) = (p, q)$ than at any point in the immediate vicinity.
If $z = f(x, y)$ and a maximum occurs at (a, b), the curve lying in the two planes $x = a$ and $y = b$ must also have a maximum point (a, b) as shown in Figure 56.5. Consequently, the tangents (shown as t_1 and t_2) to the curves at (a, b) must be parallel to Ox and Oy respectively. This requires that $\frac{\partial z}{\partial x} = 0$

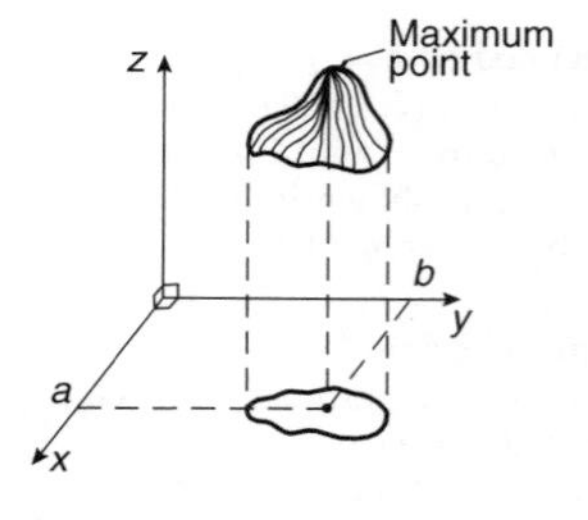

Figure 56.3

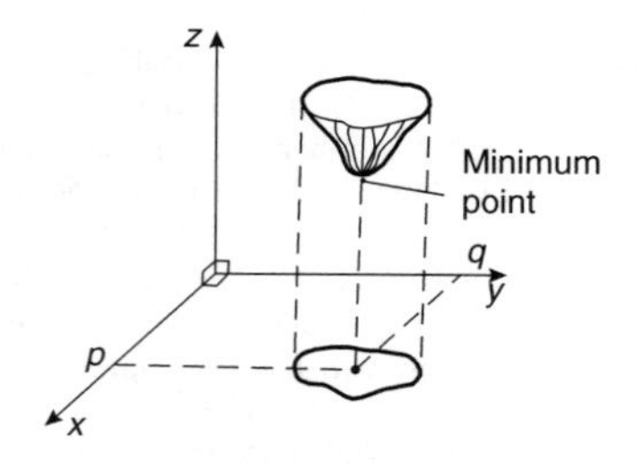

Figure 56.4

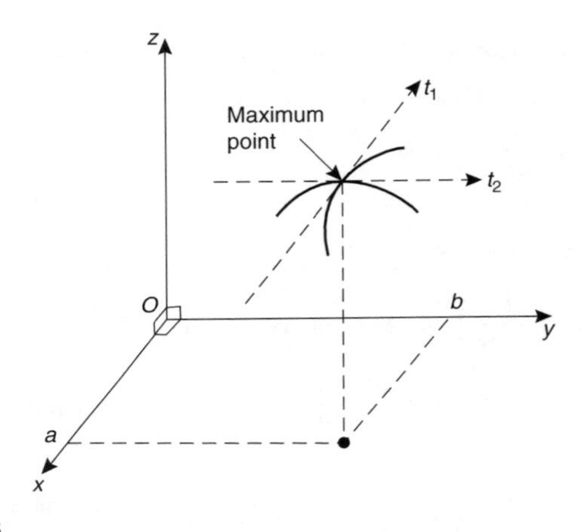

Figure 56.5

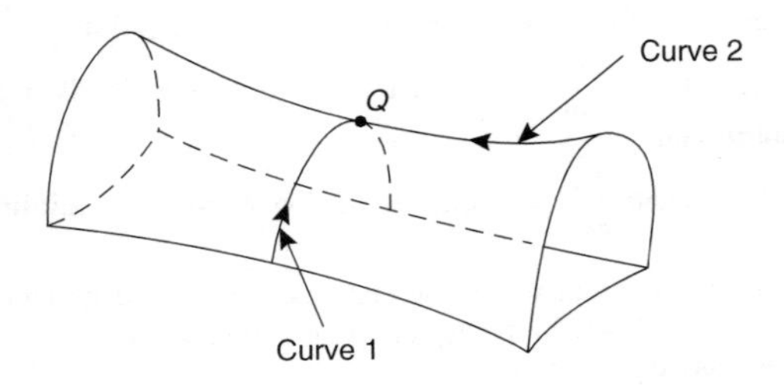

Figure 56.6

and $\frac{\partial z}{\partial y} = 0$ at all maximum and minimum values, and the solution of these equations gives the stationary (or critical) points of z.

With functions of two variables there are three types of stationary points possible, these being a maximum point, a minimum point, and a **saddle point**. A saddle point Q is shown in Figure 56.6 and is such that a point Q is a maximum for curve 1 and a minimum for curve 2.

Procedure to determine maxima, minima and saddle points for functions of two variables

Given $z = f(x, y)$:

(i) determine $\frac{\partial z}{\partial x}$ and $\frac{\partial z}{\partial y}$

(ii) for stationary points, $\frac{\partial z}{\partial x} = 0$ and $\frac{\partial z}{\partial y} = 0$,

(iii) solve the simultaneous equations $\frac{\partial z}{\partial x} = 0$ and $\frac{\partial z}{\partial y} = 0$ for x and y, which gives the co-ordinates of the stationary points,

(iv) determine $\frac{\partial^2 z}{\partial x^2}$, $\frac{\partial^2 z}{\partial y^2}$ and $\frac{\partial^2 z}{\partial x \partial y}$

(v) for each of the co-ordinates of the stationary points, substitute values of x and y into $\frac{\partial^2 z}{\partial x^2}$, $\frac{\partial^2 z}{\partial y^2}$ and $\frac{\partial^2 z}{\partial x \partial y}$ and evaluate each,

(vi) evaluate $\left(\frac{\partial^2 z}{\partial x \partial y}\right)^2$ for each stationary point,

(vii) substitute the values of $\frac{\partial^2 z}{\partial x^2}$, $\frac{\partial^2 z}{\partial y^2}$ and $\frac{\partial^2 z}{\partial x \partial y}$ into the equation

$$\Delta = \left(\frac{\partial^2 z}{\partial x \partial y}\right)^2 - \left(\frac{\partial^2 z}{\partial x^2}\right)\left(\frac{\partial^2 z}{\partial y^2}\right) \text{ and evaluate,}$$

(viii) (a) if $\boldsymbol{\Delta > 0}$ then the stationary point is a **saddle point**

(b) if $\boldsymbol{\Delta < 0}$ and $\boldsymbol{\frac{\partial^2 z}{\partial x^2} < 0}$, then the stationary point is a **maximum point**, and

(c) if $\boldsymbol{\Delta < 0}$ and $\boldsymbol{\frac{\partial^2 z}{\partial x^2} > 0}$, then the stationary point is a **minimum point**

For example, the co-ordinates of the stationary point and its nature for the function $z = (x-1)^2 + (y-2)^2$ is determined as follows:
Following the above procedure:

(i) $\frac{\partial z}{\partial x} = 2(x-1)$ and $\frac{\partial z}{\partial y} = 2(y-2)$

(ii) $2(x-1) = 0$ (1)

$2(y-2) = 0$ (2)

(iii) From equations (1) and (2), $x = 1$ and $y = 2$, thus the only stationary point exists at (1, 2)

(iv) Since $\frac{\partial z}{\partial x} = 2(x-1) = 2x - 2$, $\frac{\partial^2 z}{\partial x^2} = 2$

and since $\frac{\partial z}{\partial y} = 2(y-2) = 2y - 4$, $\frac{\partial^2 z}{\partial y^2} = 2$

and $\frac{\partial^2 z}{\partial x \partial y} = \frac{\partial}{\partial x}\left(\frac{\partial z}{\partial y}\right) = \frac{\partial}{\partial x}(2y - 4) = 0$

(v) $\frac{\partial^2 z}{\partial x^2} = \frac{\partial^2 z}{\partial y^2} = 2$ and $\frac{\partial^2 z}{\partial x \partial y} = 0$

(vi) $\left(\frac{\partial^2 z}{\partial x \partial y}\right)^2 = 0$

(vii) $\Delta = (0)^2 - (2)(2) = -4$

(viii) Since $\Delta < 0$ and $\dfrac{\partial^2 z}{\partial x^2} > 0$, **the stationary point (1, 2) is a minimum**.

The surface $z = (x - 1)^2 + (y - 2)^2$ is shown in three dimensions in Figure 56.7. Looking down towards the x-y plane from above, it is possible to produce a **contour map**. A contour is a line on a map that gives places having the same vertical height above a datum line (usually the mean sea-level on a geographical map). A contour map for $z = (x - 1)^2 + (y - 2)^2$ is shown in Figure 56.8. The values of z are shown on the map and these give an indication of the rise and fall to a stationary point.

In another example, an open rectangular container is to have a volume of 62.5 m^3. The least surface area of material required is determined as follows: Let the dimensions of the container be x, y and z as shown in Figure 56.9.

Volume $\qquad V = xyz = 62.5 \qquad (1)$

Surface area, $\qquad S = xy + 2yz + 2xz \qquad (2)$

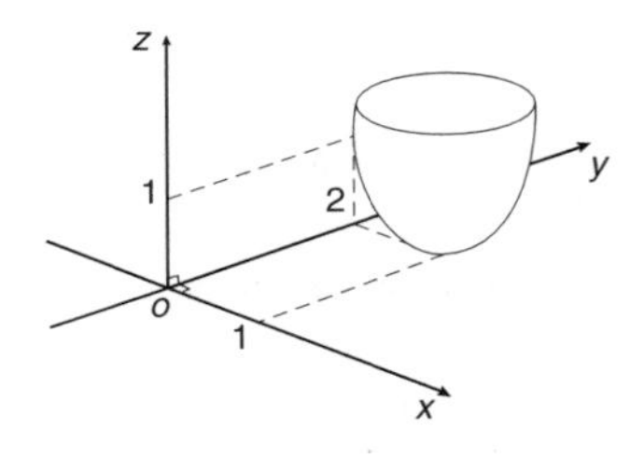

Figure 56.7

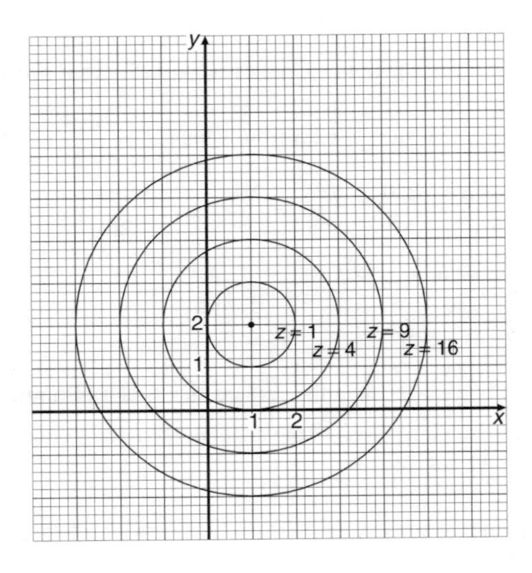

Figure 56.8

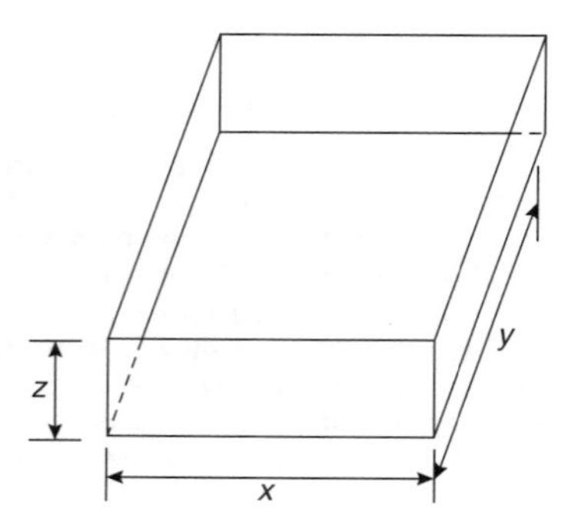

Figure 56.9

From equation (1), $z = \dfrac{62.5}{xy}$

Substituting in equation (2) gives: $S = xy + 2y\left(\dfrac{62.5}{xy}\right) + 2x\left(\dfrac{62.5}{xy}\right)$

i.e. $S = xy + \dfrac{125}{x} + \dfrac{125}{y}$ which is a function of two variables

$$\frac{\partial S}{\partial x} = y - \frac{125}{x^2} = 0 \text{ for a stationary point, hence } x^2y = 125 \qquad (3)$$

$$\frac{\partial S}{\partial y} = x - \frac{125}{y^2} = 0 \text{ for a stationary point, hence } xy^2 = 125 \qquad (4)$$

Dividing equation (3) by (4) gives: $\dfrac{x^2y}{xy^2} = 1$, i.e. $\dfrac{x}{y} = 1$, i.e. $x = y$

Substituting $y = x$ in equation (3) gives $x^3 = 125$, from which, $x = 5$ m. Hence $y = 5$ m also.

From equation (1), $(5)(5)(z) = 62.5$ from which, $z = \dfrac{62.5}{25} = 2.5$ m

$$\frac{\partial^2 S}{\partial x^2} = \frac{250}{x^3}, \quad \frac{\partial^2 S}{\partial y^2} = \frac{250}{y^3} \quad \text{and} \quad \frac{\partial^2 S}{\partial x \partial y} = 1$$

When $x = y = 5$, $\dfrac{\partial^2 S}{\partial x^2} = 2$, $\dfrac{\partial^2 S}{\partial y^2} = 2$ and $\dfrac{\partial^2 S}{\partial x \partial y} = 1$

$\Delta = (1)^2 - (2)(2) = -3$

Since $\Delta < 0$ and $\dfrac{\partial^2 S}{\partial x^2} > 0$, then the surface area S is a **minimum**

Hence the minimum dimensions of the container to have a volume of 62.5 m^3 are **5 m by 5 m by 2.5 m**

From equation (2),

minimum surface area, $S = (5)(5) + 2(5)(2.5) + 2(5)(2.5) = \mathbf{75\ m^2}$

Integral Calculus

57 Introduction to Integration

The Process of integration

The process of integration reverses the process of differentiation. In differentiation, if $f(x) = 2x^2$ then $f'(x) = 4x$. Thus the integral of $4x$ is $2x^2$, i.e. integration is the process of moving from $f'(x)$ to $f(x)$. By similar reasoning, the integral of $2t$ is t^2.
Integration is a process of summation or adding parts together and an elongated S, shown as $\int$, is used to replace the words 'the integral of'. Hence, from above, $\int 4x = 2x^2$ and $\int 2t$ is t^2.
In differentiation, the differential coefficient $\frac{dy}{dx}$ indicates that a function of x is being differentiated with respect to x, the dx indicating that it is 'with respect to x'. In integration the variable of integration is shown by adding d(the variable) after the function to be integrated.
Thus $\int 4x\,dx$ means 'the integral of $4x$ with respect to x', and $\int 2t\,dt$ means 'the integral of $2t$ with respect to t'
As stated above, the differential coefficient of $2x^2$ is $4x$, hence $\int 4x\,dx = 2x^2$. However, the differential coefficient of $2x^2 + 7$ is also $4x$. Hence $\int 4x\,dx$ is also equal to $2x^2 + 7$. To allow for the possible presence of a constant, whenever the process of integration is performed, a constant 'c' is added to the result.

Thus $$\int 4x\,dx = 2x^2 + c \text{ and } \int 2t\,dt = t^2 + c$$

'c' is called the **arbitrary constant of integration**.

The general solution of integrals of the form ax^n

The general solution of integrals of the form $\int ax^n\,dx$, where a and n are constants is given by: $\int \boldsymbol{ax^n}\,\mathbf{dx} = \dfrac{\boldsymbol{ax^{n+1}}}{\boldsymbol{n+1}} + \boldsymbol{c}$
This rule is true when n is fractional, zero, or a positive or negative integer, with the exception of $n = -1$.

For example, $$\int 3x^4\,dx = \frac{3x^{4+1}}{4+1} + c = \mathbf{\frac{3}{5}x^5 + c}$$

In another example,

$$\int \frac{2}{x^2}\,dx = \int 2x^{-2}\,dx = \frac{2x^{-2+1}}{-2+1} + c = \frac{2x^{-1}}{-1} + c = \mathbf{\frac{-2}{x} + c}$$

In another example,

$$\int \sqrt{x}\,dx = \int x^{1/2}\,dx = \frac{x^{\frac{1}{2}+1}}{\frac{1}{2}+1} + c = \frac{x^{\frac{3}{2}}}{\frac{3}{2}} + c = \tfrac{2}{3}\sqrt{x^3} + c$$

Each of these three results may be checked by differentiation.
The integral of a constant k is $kx + c$.

For example, $\int 8\,dx = 8x + c$

When a sum of several terms is integrated the result is the sum of the integrals of the separate terms.
For example,

$$\int (3x + 2x^2 - 5)\,dx = \int 3x\,dx + \int 2x^2 dx - \int 5\,dx = \frac{3x^2}{2} + \frac{2x^3}{3} - 5x + c$$

Standard integrals

Since integration is the reverse process of differentiation the **standard integrals** listed in Table 57.1 may be deduced and readily checked by differentiation.

Table 57.1 Standard integrals

(i)	$\int ax^n\,dx$	$= \frac{ax^{n+1}}{n+1} + c$ (except when $n = -1$)
(ii)	$\int \cos ax\,dx$	$= \frac{1}{a}\sin ax + c$
(iii)	$\int \sin ax\,dx$	$= -\frac{1}{a}\cos ax + c$
(iv)	$\int \sec^2 ax\,dx$	$= \frac{1}{a}\tan ax + c$
(v)	$\int \text{cosec}^2\, ax\,dx$	$= -\frac{1}{a}\cot ax + c$
(vi)	$\int \text{cosec}\, ax \cot ax\,dx$	$= -\frac{1}{a}\text{cosec}\, ax + c$
(vii)	$\int \sec ax \tan ax\,dx$	$= \frac{1}{a}\sec ax + c$
(viii)	$\int e^{ax}\,dx$	$= \frac{1}{a}e^{ax} + c$
(ix)	$\int \frac{1}{x}\,dx$	$= \ln x + c$

For example,

$$\int \frac{2x^3 - 3x}{4x}\,dx = \int \frac{2x^3}{4x} - \frac{3x}{4x}\,dx = \int \frac{x^2}{2} - \frac{3}{4}\,dx$$

$$= \left(\frac{1}{2}\right)\frac{x^{2+1}}{2+1} - \frac{3}{4}x + c = \left(\frac{1}{2}\right)\frac{x^3}{3} - \frac{3}{4}x + c = \frac{1}{6}x^3 - \frac{3}{4}x + c$$

In another example,

$$\int \frac{-5}{9\sqrt[4]{t^3}}\,dt = \int \frac{-5}{9t^{3/4}}\,dt = \int \left(-\frac{5}{9}\right) t^{-\frac{3}{4}}\,dt = \left(-\frac{5}{9}\right)\frac{t^{-\frac{3}{4}+1}}{-\frac{3}{4}+1} + c$$

$$= \left(-\frac{5}{9}\right)\frac{t^{\frac{1}{4}}}{\frac{1}{4}} + c = \left(-\frac{5}{9}\right)\left(\frac{4}{1}\right) t^{\frac{1}{4}} + c = \mathbf{-\frac{20}{9}\sqrt[4]{t} + c}$$

In another example,

$$\int (4\cos 3x - 5\sin 2x)\,dx = (4)\left(\frac{1}{3}\right)\sin 3x - (5)\left(-\frac{1}{2}\right)\cos 2x$$

$$= \mathbf{\frac{4}{3}\sin 3x + \frac{5}{2}\cos 2x + c}$$

In another example,

$$\int (7\sec^2 4t + 3\,\mathrm{cosec}^2\, 2t)\,dt = (7)\left(\frac{1}{4}\right)\tan 4t + (3)\left(-\frac{1}{2}\right)\cot 2t + c$$

$$= \mathbf{\frac{7}{4}\tan 4t - \frac{3}{2}\cot 2t + c}$$

In another example, $\int \frac{2}{3e^{4t}}\,dt = \int \frac{2}{3}e^{-4t}\,dt = \left(\frac{2}{3}\right)\left(-\frac{1}{4}\right)e^{-4t} + c$

$$= -\frac{1}{6}e^{-4t} + c = \mathbf{-\frac{1}{6e^{4t}} + c}$$

In another example, $\int \frac{3}{5x}\,dx = \int \left(\frac{3}{5}\right)\left(\frac{1}{x}\right)dx = \mathbf{\frac{3}{5}\ln x + c}$

Definite Integrals

Integrals containing an arbitrary constant c in their results are called **indefinite integrals** since their precise value cannot be determined without further information. **Definite integrals** are those in which limits are applied.
If an expression is written as $[x]_a^b$, 'b' is called the upper limit and 'a' the lower limit.
The operation of applying the limits is defined as $[x]_a^b = (b) - (a)$
The increase in the value of the integral x^2 as x increases from 1 to 3 is written as $\int_1^3 x^2\,dx$
Applying the limits gives:

$$\int_1^3 x^2\,dx = \left[\frac{x^3}{3} + c\right]_1^3 = \left(\frac{3^3}{3} + c\right) - \left(\frac{1^3}{3} + c\right)$$

$$= (9 + c) - \left(\frac{1}{3} + c\right) = \mathbf{8\frac{2}{3}}$$

Note that the 'c' term always cancels out when limits are applied and it need not be shown with definite integrals.
For example,

$$\int_0^{\pi/2} 3\sin 2x\,dx = \left[(3)\left(-\frac{1}{2}\right)\cos 2x\right]_0^{\pi/2} = \left[-\frac{3}{2}\cos 2x\right]_0^{\pi/2}$$

$$= \left\{-\frac{3}{2}\cos 2\left(\frac{\pi}{2}\right)\right\} - \left\{-\frac{3}{2}\cos 2(0)\right\}$$

$$= \left\{-\frac{3}{2}\cos\pi\right\} - \left\{-\frac{3}{2}\cos 0\right\}$$

$$= \left\{-\frac{3}{2}(-1)\right\} - \left\{-\frac{3}{2}(1)\right\} = \frac{3}{2} + \frac{3}{2} = \mathbf{3}$$

In another example,

$$\int_1^2 4\cos 3t\,dt = \left[(4)\left(\frac{1}{3}\right)\sin 3t\right]_1^2 = \left[\frac{4}{3}\sin 3t\right]_1^2 = \left\{\frac{4}{3}\sin 6\right\} - \left\{\frac{4}{3}\sin 3\right\}$$

Note that limits of trigonometric functions are always expressed in radians, thus, for example, $\sin 6$ means the sine of 6 radians $= -0.279415\ldots$

Hence $$\int_1^2 4\cos 3t\,dt = \left\{\frac{4}{3}(-0.279415\ldots)\right\} - \left\{\frac{4}{3}(0.141120\ldots)\right\}$$

$$= (-0.37255) - (0.18816) = \mathbf{-0.5607}$$

In another example, $$\int_1^2 4e^{2x}\,dx = \left[\frac{4}{2}e^{2x}\right]_1^2 = 2[e^{2x}]_1^2 = 2[e^4 - e^2]$$

$$= 2[54.5982 - 7.3891] = \mathbf{94.42}$$

In another example,

$$\int_1^4 \frac{3}{4u}\,du = \left[\frac{3}{4}\ln u\right]_1^4 = \frac{3}{4}[\ln 4 - \ln 1] = \frac{3}{4}[1.3863 - 0] = \mathbf{1.040}$$

---

58 Integration Using Algebraic Substitutions

Introduction

Functions that require integrating are not always in the 'standard form' shown in Chapter 57. However, it is often possible to change a function into a form that can be integrated by using either:

(i) an algebraic substitution (see below),

(ii) trigonometric and hyperbolic substitutions (see Chapters 59 and 61),

(iii) partial fractions (see Chapter 60),

(iv) integration by parts (see Chapter 62), or
(v) reduction formulae (see Chapter 63).

Algebraic substitutions

With **algebraic substitutions**, the substitution usually made is to let u be equal to $f(x)$ such that $f(u)\,du$ is a standard integral. It is found that integrals of the forms: $k\int[f(x)]^n f'(x)\,dx$ and $k\int \frac{f'(x)}{[f(x)]^n}\,dx$, (where k and n are constants) can both be integrated by substituting u for $f(x)$.

For example, to determine $\int \cos(3x+7)\,dx$:
$\int \cos(3x+7)\,dx$ is not a standard integral of the form shown in Table 57.1, page 306, thus an algebraic substitution is made.
Let $u = 3x+7$ then $\frac{du}{dx} = 3$ and rearranging gives $dx = \frac{du}{3}$

Hence $$\int \cos(3x+7)\,dx = \int (\cos u)\frac{du}{3} = \int \frac{1}{3}\cos u\,du,$$

which is a standard integral

$$= \frac{1}{3}\sin u + c$$

Rewriting u as $(3x+7)$ gives: $\int \cos(3x+7)\,dx = \mathbf{\frac{1}{3}\sin(3x+7)+c}$, which may be checked by differentiating it.

In another example, to find $\int (2x-5)^7\,dx$:
$(2x-5)$ may be multiplied by itself 7 times and then each term of the result integrated. However, this would be a lengthy process, and thus an algebraic substitution is made.

Let $u = (2x-5)$ then $\frac{du}{dx} = 2$ and $dx = \frac{du}{2}$

Hence $$\int (2x-5)^7\,dx = \int u^7\frac{du}{2} = \frac{1}{2}\int u^7\,du = \frac{1}{2}\left(\frac{u^8}{8}\right) + c = \frac{1}{16}u^8 + c$$

Rewriting u as $(2x-5)$ gives: $$\int \mathbf{(2x-5)^7\,dx = \frac{1}{16}(2x-5)^8 + c}$$

In another example, to find $\int \frac{4}{(5x-3)}\,dx$:

Let $u = (5x-3)$ then $\frac{du}{dx} = 5$ and $dx = \frac{du}{5}$

Hence $$\int \frac{4}{(5x-3)}\,dx = \int \frac{4}{u}\frac{du}{5} = \frac{4}{5}\int \frac{1}{u}\,du$$

$$= \frac{4}{5}\ln u + c = \mathbf{\frac{4}{5}\ln(5x-3)+c}$$

In another example, to evaluate $\int_0^{\pi/6} 24\sin^5\theta\cos\theta\,d\theta$:
Let $u = \sin\theta$ then $\dfrac{du}{d\theta} = \cos\theta$ and $d\theta = \dfrac{du}{\cos\theta}$

Hence $\int 24\sin^5\theta\cos\theta\,d\theta = \int 24u^5\cos\theta\dfrac{du}{\cos\theta} = 24\int u^5\,du$, by cancelling

$$= 24\frac{u^6}{6} + c = 4u^6 + c = 4(\sin\theta)^6 + c = 4\sin^6\theta + c$$

Thus $$\int_0^{\pi/6} 24\sin^5\theta\cos\theta\,d\theta = \left[4\sin^6\theta\right]_0^{\pi/6} = 4\left[\left(\sin\frac{\pi}{6}\right)^6 - (\sin 0)^6\right]$$

$$= 4\left[\left(\frac{1}{2}\right)^6 - 0\right] = \mathbf{\frac{1}{16}} \text{ or } \mathbf{0.0625}$$

Change of limits

When evaluating definite integrals involving substitutions it is sometimes more convenient to **change the limits** of the integral.
For example, to evaluate $\int_1^3 5x\sqrt{2x^2+7}\,dx$, taking positive values of square roots only:
Let $u = 2x^2 + 7$, then $\dfrac{du}{dx} = 4x$ and $dx = \dfrac{du}{4x}$
When $x = 3$, $u = 2(3)^2 + 7 = 25$ and when $x = 1$, $u = 2(1)^2 + 7 = 9$

Hence, $$\int_{x=1}^{x=3} 5x\sqrt{2x^2+7}\,dx = \int_{u=9}^{u=25} 5x\sqrt{u}\frac{du}{4x}$$

$$= \frac{5}{4}\int_9^{25}\sqrt{u}\,du = \frac{5}{4}\int_9^{25} u^{1/2}\,du$$

Thus the limits have been changed, and it is unnecessary to change the integral back in terms of x.

Thus, $$\int_{x=1}^{x=3} 5x\sqrt{2x^2+7}\,dx = \frac{5}{4}\left[\frac{u^{3/2}}{3/2}\right]_9^{25} = \frac{5}{6}\left[\sqrt{u^3}\right]_9^{25}$$

$$= \frac{5}{6}[\sqrt{25^3} - \sqrt{9^3}] = \frac{5}{6}(125 - 27) = \mathbf{81\frac{2}{3}}$$

---

59 Integration Using Trigonometric and Hyperbolic Substitutions

Table 59.1 gives a summary of **trigonometric and hyperbolic substitutions**.

Table 59.1 Integrals using trigonometric substitutions

$f(x)$	$\int f(x)\,dx$	Method
1. $\cos^2 x$	$\frac{1}{2}\left(x + \frac{\sin 2x}{2}\right) + c$	Use $\cos 2x = 2\cos^2 x - 1$
2. $\sin^2 x$	$\frac{1}{2}\left(x - \frac{\sin 2x}{2}\right) + c$	Use $\cos 2x = 1 - 2\sin^2 x$
3. $\tan^2 x$	$\tan x - x + c$	Use $1 + \tan^2 x = \sec^2 x$
4. $\cot^2 x$	$-\cot x - x + c$	Use $\cot^2 x + 1 = \text{cosec}^2 x$
5. $\cos^m x \sin^n x$	(a) If either m or n is odd (but not both), use $\cos^2 x + \sin^2 x = 1$ (b) If both m and n are even, use either $\cos 2x = 2\cos^2 x - 1$ or $\cos 2x = 1 - 2\sin^2 x$	
6. $\sin A \cos B$		Use $\frac{1}{2}[\sin(A+B) + \sin(A-B)]$
7. $\cos A \sin B$		Use $\frac{1}{2}[\sin(A+B) - \sin(A-B)]$
8. $\cos A \cos B$		Use $\frac{1}{2}[\cos(A+B) + \cos(A-B)]$
9. $\sin A \sin B$		Use $-\frac{1}{2}[\cos(A+B) - \cos(A-B)]$
10. $\frac{1}{\sqrt{(a^2 - x^2)}}$	$\sin^{-1}\frac{x}{a} + c$	Use $x = a\sin\theta$ substitution
11. $\sqrt{a^2 - x^2}$	$\frac{a^2}{2}\sin^{-1}\frac{x}{a} + \frac{x}{2}\sqrt{a^2 - x^2} + c$	
12. $\frac{1}{a^2 + x^2}$	$\frac{1}{a}\tan^{-1}\frac{x}{a} + c$	Use $x = a\tan\theta$ substitution
13. $\frac{1}{\sqrt{(x^2 + a^2)}}$	$\text{arsinh}\,\frac{x}{a} + c$ or $\ln\left\{\frac{x + \sqrt{(x^2 + a^2)}}{a}\right\} + c$	Use $x = a\sinh\theta$ substitution
14. $\sqrt{(x^2 + a^2)}$	$\frac{a^2}{2}\text{arsinh}\,\frac{x}{a} + \frac{x}{2}\sqrt{(x^2 + a^2)} + c$	
15. $\frac{1}{\sqrt{(x^2 - a^2)}}$	$\text{arcosh}\,\frac{x}{a} + c$ or $\ln\left\{\frac{x + \sqrt{(x^2 - a^2)}}{a}\right\} + c$	Use $x = a\cosh\theta$ substitution
16. $\sqrt{(x^2 - a^2)}$	$\frac{x}{2}\sqrt{(x^2 - a^2)} - \frac{a^2}{2}\text{arcosh}\,\frac{x}{a} + c$	

For example, to evaluate $\int_0^{\pi/4} 2\cos^2 4t\,dt$:
Since $\cos 2t = 2\cos^2 t - 1$ (from Chapter 29), then $\cos^2 t = \frac{1}{2}(1 + \cos 2t)$ and $\cos^2 4t = \frac{1}{2}(1 + \cos 8t)$.

$$\text{Hence } \int_0^{\pi/4} 2\cos^2 4t\,dt = 2\int_0^{\pi/4} \frac{1}{2}(1 + \cos 8t)\,dt = \left[t + \frac{\sin 8t}{8}\right]_0^{\pi/4}$$

$$= \left[\frac{\pi}{4} + \frac{\sin 8\left(\frac{\pi}{4}\right)}{8}\right] - \left[0 + \frac{\sin 0}{8}\right] = \boldsymbol{\frac{\pi}{4}} \text{ or } \mathbf{0.7854}$$

In another example, to find $3\int \tan^2 4x\,dx$:
Since $1+\tan^2 x = \sec^2 x$, then $\tan^2 x = \sec^2 x - 1$ and $\tan^2 4x = \sec^2 4x - 1$
Hence $3\int \tan^2 4x\,dx = 3\int (\sec^2 4x - 1)\,dx = \mathbf{3\left(\dfrac{\tan 4x}{4} - x\right) + c}$

In another example, to determine $\int \sin^5\theta\,d\theta$:

Since $\cos^2\theta + \sin^2\theta = 1$ then $\sin^2\theta = (1-\cos^2\theta)$

$$\text{Hence}\quad \int \sin^5\theta\,d\theta = \int \sin\theta(\sin^2\theta)^2\,d\theta = \int \sin\theta(1-\cos^2\theta)^2\,d\theta$$

$$= \int \sin\theta(1 - 2\cos^2\theta + \cos^4\theta)\,d\theta$$

$$= \int (\sin\theta - 2\sin\theta\cos^2\theta + \sin\theta\cos^4\theta)\,d\theta$$

$$= \mathbf{-\cos\theta + \frac{2\cos^3\theta}{3} - \frac{\cos^5\theta}{5} + c}$$

[Whenever a power of a cosine is multiplied by a sine of power 1, or vice-versa, the integral may be determined by inspection as shown.

In general, $\int \cos^n\theta\sin\theta\,d\theta = \dfrac{-\cos^{n+1}\theta}{(n+1)} + c$ and

$\int \sin^n\theta\cos\theta\,d\theta = \dfrac{\sin^{n+1}\theta}{(n+1)} + c$

Alternatively, an algebraic substitution may be used as shown in Chapter 58.]
In another example, to find $\int \sin^2 t\cos^4 t\,dt$:

$$\int \sin^2 t\cos^4 t\,dt = \int \sin^2 t(\cos^2 t)^2\,dt$$

$$= \int \left(\frac{1-\cos 2t}{2}\right)\left(\frac{1+\cos 2t}{2}\right)^2 dt$$

$$= \frac{1}{8}\int (1-\cos 2t)(1 + 2\cos 2t + \cos^2 2t)\,dt$$

$$= \frac{1}{8}\int (1 + 2\cos 2t + \cos^2 2t - \cos 2t - 2\cos^2 2t - \cos^3 2t)\,dt$$

$$= \frac{1}{8}\int (1 + \cos 2t - \cos^2 2t - \cos^3 2t)\,dt$$

$$= \frac{1}{8}\int \left[1 + \cos 2t - \left(\frac{1+\cos 4t}{2}\right) - \cos 2t(1 - \sin^2 2t)\right] dt$$

$$= \frac{1}{8}\int \left(\frac{1}{2} - \frac{\cos 4t}{2} + \cos 2t\sin^2 2t\right) dt$$

$$= \mathbf{\frac{1}{8}\left(\frac{t}{2} - \frac{\sin 4t}{8} + \frac{\sin^3 2t}{6}\right) + c}$$

In another example, to determine $\int \sin 3t \cos 2t \, dt$:

$$\int \sin 3t \cos 2t \, dt = \int \frac{1}{2}[\sin(3t + 2t) + \sin(3t - 2t)] \, dt, \text{ from 6 of Table 59.1,}$$

$$= \frac{1}{2}\int (\sin 5t + \sin t) \, dt = \mathbf{\frac{1}{2}\left(\frac{-\cos 5t}{5} - \cos t\right) + c}$$

In another example, to evaluate $\int_0^3 \frac{1}{\sqrt{(9 - x^2)}} \, dx$:

From 10 of Table 59.1,

$$\int_0^3 \frac{1}{\sqrt{(9 - x^2)}} \, dx = \left[\sin^{-1}\frac{x}{3}\right]_0^3, \text{ since } a = 3$$

$$= (\sin^{-1} 1 - \sin^{-1} 0) = \mathbf{\frac{\pi}{2}} \text{ or } \mathbf{1.5708}$$

In another example, to evaluate $\int_0^4 \sqrt{16 - x^2} \, dx$:
From 11 of Table 59.1,

$$\int_0^4 \sqrt{16 - x^2} \, dx = \left[\frac{16}{2}\arcsin\frac{x}{4} + \frac{x}{2}\sqrt{(16 - x^2)}\right]_0^4$$

$$= [8 \sin^{-1} 1 + 2\sqrt{0}] - [8 \sin^{-1} 0 + 0]$$

$$= 8 \sin^{-1} 1 = 8\left(\frac{\pi}{2}\right) = \mathbf{4\pi} \text{ or } \mathbf{12.57}$$

In another example, to evaluate $\int_0^2 \frac{1}{(4 + x^2)} \, dx$:

From 12 of Table 59.1,

$$\int_0^2 \frac{1}{(4 + x^2)} \, dx = \frac{1}{2}\left[\tan^{-1}\frac{x}{2}\right]_0^2 \text{ since } a = 2$$

$$= \frac{1}{2}(\tan^{-1} 1 - \tan^{-1} 0) = \frac{1}{2}\left(\frac{\pi}{4} - 0\right) = \mathbf{\frac{\pi}{8}} \text{ or } \mathbf{0.3927}$$

In another example, to evaluate $\int_0^2 \frac{1}{\sqrt{(x^2 + 4)}} \, dx$, correct to 4 decimal places:

$$\int_0^2 \frac{1}{\sqrt{(x^2 + 4)}} \, dx = \left[\operatorname{arsinh}\frac{x}{2}\right]_0^2 \text{ or } \left[\ln\left\{\frac{x + \sqrt{(x^2 + 4)}}{2}\right\}\right]_0^2$$

from 13 of Table 59.1, where $a = 2$

Using the logarithmic form,

$$\int_0^2 \frac{1}{\sqrt{(x^2 + 4)}} \, dx = \left[\ln\left(\frac{2 + \sqrt{8}}{2}\right) - \ln\left(\frac{0 + \sqrt{4}}{2}\right)\right] = \ln 2.4142 - \ln 1$$

$$= \mathbf{0.8814}, \text{ correct to 4 decimal places}$$

In another example, to determine $\int \frac{2x-3}{\sqrt{(x^2-9)}}\,dx$:

$$\int \frac{2x-3}{\sqrt{(x^2-9)}}\,dx = \int \frac{2x}{\sqrt{(x^2-9)}}\,dx - \int \frac{3}{\sqrt{(x^2-9)}}\,dx$$

The first integral is determined using the algebraic substitution $u = (x^2 - 9)$, and the second integral is of the form $\int \frac{1}{\sqrt{(x^2-a^2)}}\,dx$ (see 15 of Table 59.1).

Hence

$$\int \frac{2x}{\sqrt{(x^2-9)}}\,dx - \int \frac{3}{\sqrt{(x^2-9)}}\,dx = \mathbf{2\sqrt{(x^2-9)} - 3\,\text{arcosh}\,\frac{x}{3} + c}$$

60 Integration Using Partial Fractions

Introduction

The process of expressing a fraction in terms of simpler fractions — called **partial fractions** — is discussed in Chapter 14, with the forms of partial fractions used being summarised in Table 14.1, page 61.
Certain functions have to be resolved into partial fractions before they can be integrated.

Linear factors

For example, to determine $\int \frac{11-3x}{x^2+2x-3}\,dx$:

As shown on page 61: $\frac{11-3x}{x^2+2x-3} \equiv \frac{2}{(x-1)} - \frac{5}{(x+3)}$

Hence $\int \frac{11-3x}{x^2+2x-3}\,dx = \int \left\{\frac{2}{(x-1)} - \frac{5}{(x+3)}\right\}dx$

$$= \mathbf{2\ln(x-1) - 5\ln(x+3) + c}$$

by algebraic substitutions (see Chapter 58) or $\mathbf{\ln\left\{\frac{(x-1)^2}{(x+3)^5}\right\} + c}$ by the laws of logarithms

In another example, to evaluate $\int_2^3 \frac{x^3-2x^2-4x-4}{x^2+x-2}\,dx$, correct to 4 significant figures:
By dividing out and resolving into partial fractions, it was shown on page 62:
$\frac{x^3-2x^2-4x-4}{x^2+x-2} \equiv x - 3 + \frac{4}{(x+2)} - \frac{3}{(x-1)}$

Hence

$$\int_2^3 \frac{x^3 - 2x^2 - 4x - 4}{x^2 + x - 2}\,dx \equiv \int_2^3 \left\{x - 3 + \frac{4}{(x+2)} - \frac{3}{(x-1)}\right\} dx$$

$$= \left[\frac{x^2}{2} - 3x + 4\ln(x+2) - 3\ln(x-1)\right]_2^3$$

$$= \left(\frac{9}{2} - 9 + 4\ln 5 - 3\ln 2\right)$$

$$-(2 - 6 + 4\ln 4 - 3\ln 1) = \mathbf{-1.687},$$

correct to 4 significant figures

Repeated linear factors

For example, to find $\int \frac{5x^2 - 2x - 19}{(x+3)(x-1)^2}\,dx$:

It was shown on page 63:

$$\frac{5x^2 - 2x - 19}{(x+3)(x-1)^2} \equiv \frac{2}{(x+3)} + \frac{3}{(x-1)} - \frac{4}{(x-1)^2}$$

Hence

$$\int \frac{5x^2 - 2x - 19}{(x+3)(x-1)^2}\,dx \equiv \int \left\{\frac{2}{(x+3)} + \frac{3}{(x-1)} - \frac{4}{(x-1)^2}\right\} dx$$

$$= \mathbf{2\ln(x+3) + 3\ln(x-1) + \frac{4}{(x-1)} + c}$$

or $\mathbf{\ln(x+3)^2(x-1)^3 + \frac{4}{(x-1)} + c}$

Quadratic factors

For example, to find $\int \frac{3 + 6x + 4x^2 - 2x^3}{x^2(x^2+3)}\,dx$:

It was shown on page 63: $\frac{3 + 6x + 4x^2 - 2x^2}{x^2(x^2+3)} \equiv \frac{2}{x} + \frac{1}{x^2} + \frac{3 - 4x}{(x^2+3)}$

Thus

$$\int \frac{3 + 6x + 4x^2 - 2x^3}{x^2(x^2+3)}\,dx \equiv \int \left(\frac{2}{x} + \frac{1}{x^2} + \frac{3 - 4x}{(x^2+3)}\right) dx$$

$$= \int \left\{\frac{2}{x} + \frac{1}{x^2} + \frac{3}{(x^2+3)} - \frac{4x}{(x^2+3)}\right\} dx$$

$\int \frac{3}{(x^2+3)}\,dx = 3\int \frac{1}{x^2+(\sqrt{3})^2}\,dx = \frac{3}{\sqrt{3}}\tan^{-1}\frac{x}{\sqrt{3}}$, from 12, Table 59.1, page 311.

$\int \frac{4x}{x^2+3}\,dx$ is determined using the algebraic substitution $u = (x^2+3)$

Hence, $\int \left\{\frac{2}{x} + \frac{1}{x^2} + \frac{3}{(x^2+3)} - \frac{4x}{(x^2+3)}\right\} dx$

$$= 2\ln x - \frac{1}{x} + \frac{3}{\sqrt{3}}\tan^{-1}\frac{x}{\sqrt{3}} - 2\ln(x^2+3) + c$$

$$= \mathbf{\ln\left(\frac{x}{x^2+3}\right)^2 - \frac{1}{x} + \sqrt{3}\tan^{-1}\frac{x}{\sqrt{3}} + c}$$

---

61 The $t = \tan\frac{\theta}{2}$ Substitution

Integrals of the form $\int \frac{1}{a\cos\theta + b\sin\theta + c}\,d\theta$, where a, b and c are constants, may be determined by using the substitution $t = \tan\frac{\theta}{2}$. The reason is explained below.

If angle A in the right-angled triangle ABC shown in Figure 61.1 is made equal to $\frac{\theta}{2}$ then, since tangent $= \frac{\text{opposite}}{\text{adjacent}}$, if BC $= t$ and AB $= 1$, then $\tan\frac{\theta}{2} = t$.

By Pythagoras' theorem, AC $= \sqrt{1+t^2}$

Therefore $\sin\frac{\theta}{2} = \frac{t}{\sqrt{1+t^2}}$ and $\cos\frac{\theta}{2} = \frac{1}{\sqrt{1+t^2}}$

Since $\sin 2x = 2\sin x\cos x$ (from double angle formulae, Chapter 31), then $\sin\theta = 2\sin\frac{\theta}{2}\cos\frac{\theta}{2} = 2\left(\frac{t}{\sqrt{1+t^2}}\right)\left(\frac{1}{\sqrt{1+t^2}}\right)$

i.e. $$\boxed{\sin\theta = \frac{2t}{(1+t^2)}} \qquad (1)$$

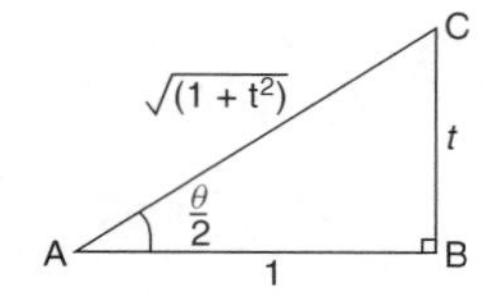

Figure 61.1

Since $\cos 2x = \cos^2 x - \sin^2 x$ (from double angle formulae), then

$$\cos\theta = \cos^2\frac{\theta}{2} - \sin^2\frac{\theta}{2} = \left(\frac{1}{\sqrt{1+t^2}}\right)^2 - \left(\frac{t}{\sqrt{1+t^2}}\right)^2$$

i.e. $$\boldsymbol{\cos\theta = \frac{1-t^2}{1+t^2}} \quad (2)$$

Also, since $t = \tan\frac{\theta}{2}$, $\frac{dt}{d\theta} = \frac{1}{2}\sec^2\frac{\theta}{2} = \frac{1}{2}\left(1+\tan^2\frac{\theta}{2}\right)$ from trigonometric identities, i.e. $\frac{dt}{d\theta} = \frac{1}{2}(1+t^2)$

from which, $$\boldsymbol{d\theta = \frac{2\,dt}{1+t^2}} \quad (3)$$

Equations (1), (2) and (3) are used to determine integrals of the form $\int \frac{1}{a\cos\theta + b\sin\theta + c}\,d\theta$ where a, b or c may be zero.

For example, to determine $\int \frac{d\theta}{\sin\theta}$:

If $t = \tan\frac{\theta}{2}$ then $\sin\theta = \frac{2t}{1+t^2}$ and $d\theta = \frac{2\,dt}{1+t^2}$ from equations (1) and (3).

Thus $$\int \frac{d\theta}{\sin\theta} = \int \frac{1}{\frac{2t}{1+t^2}}\left(\frac{2\,dt}{1+t^2}\right) = \int \frac{1}{t}\,dt = \ln t + c$$

Hence $$\boldsymbol{\int \frac{d\theta}{\sin\theta} = \ln\left(\tan\frac{\theta}{2}\right) + c}$$

In another example, to determine $\int \frac{d\theta}{5+4\cos\theta}$:

If $t = \tan\frac{\theta}{2}$ then $\cos\theta = \frac{1-t^2}{1+t^2}$ and $d\theta = \frac{2\,dt}{1+t^2}$ from equations (2) and (3).

Thus $$\int \frac{d\theta}{5+4\cos\theta} = \int \frac{1}{5+4\left(\frac{1-t^2}{1+t^2}\right)}\left(\frac{2\,dt}{1+t^2}\right)$$

$$= \int \frac{1}{\frac{5(1+t^2)+4(1-t^2)}{1+t^2}}\left(\frac{2\,dt}{(1+t^2)}\right)$$

$$= 2\int \frac{dt}{t^2+9} = 2\int \frac{dt}{t^2+3^2}$$

$$= 2\left(\frac{1}{3}\tan^{-1}\frac{t}{3}\right) + c$$

Hence $$\boldsymbol{\int \frac{d\theta}{5+4\cos\theta} = \frac{2}{3}\tan^{-1}\left(\frac{1}{3}\tan\frac{\theta}{2}\right) + c}$$

In another example, to determine $\int \frac{\mathrm{d}x}{\sin x + \cos x}$:

If $t = \tan\frac{x}{2}$ then $\sin x = \frac{2t}{1+t^2}$, $\cos x = \frac{1-t^2}{1+t^2}$ and $\mathrm{d}x = \frac{2\,dt}{1+t^2}$ from equations (1), (2) and (3).

Thus
$$\int \frac{\mathrm{d}x}{\sin x + \cos x} = \int \frac{\dfrac{2\,\mathrm{d}t}{1+t^2}}{\left(\dfrac{2t}{1+t^2}\right) + \left(\dfrac{1-t^2}{1+t^2}\right)} = \int \frac{\dfrac{2\,\mathrm{d}t}{1+t^2}}{\dfrac{2t+1-t^2}{1+t^2}}$$

$$= \int \frac{2\,\mathrm{d}t}{1+2t-t^2} = \int \frac{-2\,\mathrm{d}t}{t^2-2t-1}$$

$$= \int \frac{-2\,\mathrm{d}t}{(t-1)^2-2} = \int \frac{2\,\mathrm{d}t}{(\sqrt{2})^2-(t-1)^2}$$

$$= 2\left[\frac{1}{2\sqrt{2}} \ln\left\{\frac{\sqrt{2}+(t-1)}{\sqrt{2}-(t-1)}\right\}\right] + c$$

(by using partial fractions)

i.e.
$$\int \frac{\mathrm{d}x}{\sin x + \cos x} = \boldsymbol{\frac{1}{\sqrt{2}} \ln\left\{\frac{\sqrt{2}-1+\tan\frac{x}{2}}{\sqrt{2}+1-\tan\frac{x}{2}}\right\} + c}$$

---

62 Integration by Parts

From the product rule of differentiation: $\frac{\mathrm{d}}{\mathrm{d}x}(uv) = v\frac{\mathrm{d}u}{\mathrm{d}x} + u\frac{\mathrm{d}v}{\mathrm{d}x}$, where u and v are both functions of x.

Rearranging gives: $u\frac{\mathrm{d}v}{\mathrm{d}x} = \frac{\mathrm{d}}{\mathrm{d}x}(uv) - v\frac{\mathrm{d}u}{\mathrm{d}x}$

Integrating both sides with respect to x gives:

$$\int u\frac{\mathrm{d}v}{\mathrm{d}x}\,\mathrm{d}x = \int \frac{\mathrm{d}}{\mathrm{d}x}(uv)\,\mathrm{d}x - \int v\frac{\mathrm{d}u}{\mathrm{d}x}\,\mathrm{d}x$$

i.e.
$$\boxed{\boldsymbol{u\frac{\mathrm{d}v}{\mathrm{d}x}\,\mathrm{d}x = \int uv - \int v\frac{\mathrm{d}u}{\mathrm{d}x}\,\mathrm{d}x}} \text{ or } \boxed{\boldsymbol{\int u\,\mathrm{d}v = uv - \int v\,\mathrm{d}u}}$$

This is known as the **integration by parts formula** and provides a method of integrating such products of simple functions as $\int xe^x\,\mathrm{d}x$, $\int t\sin t\,\mathrm{d}t$, $\int \mathrm{e}^\theta \cos\theta\,\mathrm{d}\theta$ and $\int x\ln x\,\mathrm{d}x$.

Given a product of two terms to integrate the initial choice is: 'which part to make equal to du' and 'which part to make equal to dv'. The choice must be such that the 'u part' becomes a constant after successive differentiation and the 'dv part' can be integrated from standard integrals. Invariable, the following rule holds: 'If a product to be integrated contains an algebraic term (such as x, t^2 or 3θ) then this term is chosen as the u part. The one exception to this rule is when a '$\ln x$' term is involved; in this case $\ln x$ is chosen as the 'u part'

For example, to determine $\int x \cos x \, dx$:

From the integration by parts formula, $\int u \, dv = uv - \int v \, du$

Let $u = x$, from which $\frac{du}{dx} = 1$, i.e. $du = dx$, and let $dv = \cos x \, dx$, from which $v = \int \cos x \, dx = \sin x$

Expressions for u, du, v and dv are now substituted into the 'by parts' formula as shown below.

$$\int \boxed{u} \, \boxed{dv} = \boxed{u} \, \boxed{v} - \int \boxed{v} \, \boxed{du}$$

$$\int \boxed{x} \, \boxed{\cos x \, dx} = \boxed{(x)} \, \boxed{(\sin x)} - \int \boxed{(\sin x)} \, \boxed{(dx)}$$

i.e. $\int x \cos x \, dx = x \sin x - (-\cos x) + c = \mathbf{x \sin x + \cos x + c}$

[This result may be checked by differentiating the right hand side,

i.e. $\frac{d}{dx}(x \sin x + \cos x + c)$

$$= [(x)(\cos x) + (\sin x)(1)] - \sin x + 0 \text{ using the product rule}$$

$$= x \cos x, \text{ which is the function being integrated]}$$

In another example, to find $\int 3te^{2t} \, dt$:

Let $u = 3t$, from which, $\frac{du}{dt} = 3$, i.e. $du = 3 \, dt$, and let $dv = e^{2t} \, dt$, from which, $v = \int e^{2t} \, dt = \frac{1}{2}e^{2t}$

Substituting into $\int u \, dv = uv - \int v \, du$ gives:

$$\int 3te^{2t} \, dt = (3t)\left(\frac{1}{2}e^{2t}\right) - \int \left(\frac{1}{2}e^{2t}\right)(3 \, dt) = \frac{3}{2}te^{2t} - \frac{3}{2}\int e^{2t} \, dt$$

$$= \frac{3}{2}te^{2t} - \frac{3}{2}\left(\frac{e^{2t}}{2}\right) + c$$

Hence $\int \mathbf{3te^{2t} \, dt} = \mathbf{\frac{3}{2}e^{2t}\left(t - \frac{1}{2}\right) + c}$, which may be checked by differentiating

In another example, to determine $\int x^2 \sin x \, dx$:

Let $u = x^2$, from which, $\frac{du}{dx} = 2x$, i.e. $du = 2x \, dx$, and let $dv = \sin x \, dx$, from which, $v = \int \sin x \, dx = -\cos x$

Substituting into $\int u \, dv = uv - \int v \, du$ gives :

$$\int x^2 \sin x \, dx = (x^2)(-\cos x) - \int (-\cos x)(2x \, dx)$$

$$= -x^2 \cos x + 2\left[\int x \cos x \, dx\right]$$

The integral, $\int x \cos x \, dx$, is not a 'standard integral' and it can only be determined by using the integration by parts formula again.
From the first example, $\int x \cos x \, dx = x \sin x + \cos x$

$$\text{Hence} \quad \int x^2 \sin x \, dx = -x^2 \cos x + 2\{x \sin x + \cos x\} + c$$

$$= -x^2 \cos x + 2x \sin x + 2 \cos x + c$$

$$= \mathbf{(2 - x^2) \cos x + 2x \sin x + c}$$

In general, if the algebraic term of a product is of power n, then the integration by parts formula is applied n times.

In another example, to find $\int x \ln x \, dx$:
The logarithmic function is chosen as the 'u part'

Thus when $u = \ln x$, then $\dfrac{du}{dx} = \dfrac{1}{x}$, i.e. $du = \dfrac{dx}{x}$

Letting $dv = x \, dx$ gives $v = \int x \, dx = \dfrac{x^2}{2}$

Substituting into $\int u \, dv = uv - \int v \, du$ gives:

$$\int x \ln x \, dx = (\ln x)\left(\frac{x^2}{2}\right) - \int \left(\frac{x^2}{2}\right)\frac{dx}{x}$$

$$= \frac{x^2}{2} \ln x - \frac{1}{2}\int x \, dx$$

$$= \frac{x^2}{2} \ln x - \frac{1}{2}\left(\frac{x^2}{2}\right) + c$$

$$\text{Hence} \int x \ln x \, dx = \mathbf{\frac{x^2}{2}\left(\ln x - \frac{1}{2}\right) + c} \text{ or } \mathbf{\frac{x^2}{4}(2 \ln x - 1) + c}$$

---

63 Reduction Formulae

Introduction

When using integration by parts in Chapter 62, an integral such as $\int x^2 e^x \, dx$ requires integration by parts twice. Similarly, $\int x^3 e^x \, dx$ requires integration by parts three times. Thus, integrals such as $\int x^5 e^x \, dx$, $\int x^6 \cos x \, dx$ and $\int x^8 \sin 2x \, dx$ for example, would take a long time to determine using integration by parts. **Reduction formulae** provide a quicker method for determining such integrals.

Using reduction formulae for integrals of the form $\int x^n e^x\,dx$

To determine $\int x^n e^x\,dx$ using integration by parts, let $u = x^n$ from which, $\frac{du}{dx} = nx^{n-1}$ and $du = nx^{n-1}\,dx$, and $dv = e^x\,dx$ from which, $v = \int e^x\,dx = e^x$

Thus, $\int x^n e^x\,dx = x^n e^x - \int e^x\,nx^{n-1}\,dx$ using the integration by parts formula

$$= x^n e^x - n\int x^{n-1} e^x\,dx$$

The integral on the far right is seen to be of the same form as the integral on the left-hand side, except that n has been replaced by $n - 1$.

Thus, if we let $\int x^n e^x\,dx = I_n$, then $\int x^{n-1} e^x\,dx = I_{n-1}$

Hence $\int x^n e^x\,dx = x^n e^x - n\int x^{n-1} e^x\,dx$

can be written as:

$$\boxed{\boldsymbol{I_n = x^n e^x - nI_{n-1}}} \qquad (1)$$

Equation (1) is an example of a reduction formula since it expresses an integral in n in terms of the same integral in $n - 1$

For example, to determine $\int x^3 e^x\,dx$ using a reduction formula:
From equation (1), $I_n = x^n e^x - nI_{n-1}$

Hence $\int x^3 e^x\,dx = I_3 = x^3 e^x - 3\,I_2, \qquad I_2 = x^2 e^x - 2I_1,$

$$I_1 = x^1 e^x - 1\,I_0$$

and $I_0 = \int x^0 e^x\,dx = \int e^x\,dx = e^x$

Thus

$$\int x^3 e^x\,dx = x^3 e^x - 3\left[x^2 e^x - 2I_1\right]$$
$$= x^3 e^x - 3\left[x^2 e^x - 2(xe^x - I_0)\right]$$
$$= x^3 e^x - 3\left[x^2 e^x - 2(xe^x - e^x)\right]$$
$$= x^3 e^x - 3x^2 e^x + 6(xe^x - e^x)$$
$$= x^3 e^x - 3x^2 e^x + 6xe^x - 6e^x$$

i.e.

$$\boldsymbol{\int x^3 e^x\,dx = e^x(x^3 - 3x^2 + 6x - 6) + c}$$

Using reduction formulae for integrals of the form $\int x^n \cos x\, dx$

Let $I_n = \int x^n \cos x\, dx$ then, using integration by parts:

if $\quad u = x^n$ then $\dfrac{du}{dx} = nx^{n-1}$ and if

$$dv = \cos x\, dx \text{ then } v = \int \cos x\, dx = \sin x$$

Hence $\quad I_n = x^n \sin x - \int (\sin x) n x^{n-1}\, dx$

$$= x^n \sin x - n \int x^{n-1} \sin x\, dx$$

Using integration by parts again, this time with $u = x^{n-1}$:
$\dfrac{du}{dx} = (n-1)x^{n-2}$, and $dv = \sin x\, dx$, from which, $v = \int \sin x\, dx = -\cos x$

Hence $\quad I_n = x^n \sin x - n \left[x^{n-1}(-\cos x) - \int (-\cos x)(n-1)x^{n-2}\, dx \right]$

$$= x^n \sin x + n \left[x^{n-1} \cos x - n(n-1) \int x^{n-2} \cos x\, dx \right]$$

i.e. $$\boxed{I_n = x^n \sin x + nx^{n-1} \cos x - n(n-1) I_{n-2}} \qquad (2)$$

For example , to determine $\int x^2 \cos x\, dx$ using a reduction formula:

Using the reduction formula of equation (2):

$$\int x^2 \cos x\, dx = I_2 = x^2 \sin x + 2x^1 \cos x - 2(1) I_0$$

and $$I_0 = \int x^0 \cos x\, dx = \int \cos x\, dx = \sin x$$

Hence $$\int \boldsymbol{x^2 \cos x\, dx = x^2 \sin x + 2x \cos x - 2 \sin x + c}$$

Using reduction formulae for integrals of the form $\int x^n \sin x\, dx$

Let $I_n = \int x^n \sin x\, dx$. Using integration by parts, if $u = x^n$ then $\dfrac{du}{dx} = nx^{n-1}$ and if $dv = \sin x\, dx$ then $v = \int \sin x\, dx = \cos x$

Hence $\quad \int x^n \sin x\, dx = I_n = x^n(-\cos x) - \int (-\cos x) n x^{n-1}\, dx$

$$= -x^n \cos x + n \int x^{n-1} \cos x\, dx$$

Using integration by parts again, with $u = x^{n-1}$, from which, $\frac{du}{dx} = (n-1)x^{n-2}$, and $dv = \cos x$, from which, $v = \int \cos x\,dx = \sin x$

Hence $\quad I_n = -x^n \cos x + n\left[x^{n-1}(\sin x) - \int (\sin x)(n-1)x^{n-2}\,dx\right]$

$$= -x^n \cos x + nx^{n-1}(\sin x) - n(n-1)\int x^{n-2}\sin x\,dx$$

i.e. $$\boxed{\boldsymbol{I_n = -x^n \cos x + nx^{n-1}(\sin x) - n(n-1)I_{n-2}}} \qquad (3)$$

For example, to determine $\int x^3 \sin x\,dx$ using a reduction formula:

Using equation (3),

$$\int x^3 \sin x\,dx = I_3 = -x^3 \cos x + 3x^2 \sin x - 3(2)I_1$$

and $$I_1 = -x^1 \cos x + 1x^0 \sin x = -x\cos x + \sin x$$

Hence $$\int x^3 \sin x\,dx = -x^3 \cos x + 3x^2 \sin x - 6[-x\cos x + \sin x]$$

$$= \boldsymbol{-x^3 \cos x + 3x^2 \sin x + 6x \cos x - 6\sin x + c}$$

Using reduction formulae for integrals of the form $\int \sin^n x\,dx$

Let $I_n = \int \sin^n x\,dx \equiv \int \sin^{n-1} x \sin x\,dx$ from laws of indices.
Using integration by parts, let $u = \sin^{n-1} x$, from which, $\frac{du}{dx} = (n-1)\sin^{n-2} x \cos x$ and $du = (n-1)\sin^{n-2} x \cos x\,dx$, and let $dv = \sin x\,dx$, from which, $v = \int \sin x\,dx = -\cos x$

Hence $\quad I_n = \int \sin^{n-1} x \sin x\,dx = (\sin^{n-1} x)(-\cos x)$

$$- \int (-\cos x)(n-1)\sin^{n-2} x \cos x\,dx$$

$$= -\sin^{n-1} x \cos x + (n-1)\int \cos^2 x \sin^{n-2} x\,dx$$

$$= -\sin^{n-1} x \cos x + (n-1)\int (1 - \sin^2 x)\sin^{n-2} x\,dx$$

$$= -\sin^{n-1} x \cos x + (n-1)\left\{\int \sin^{n-2} x\,dx - \int \sin^n x\,dx\right\}$$

i.e. $$I_n = -\sin^{n-1} x \cos x + (n-1)I_{n-2} - (n-1)I_n$$

i.e. $$I_n + (n-1)I_n = -\sin^{n-1} x \cos x + (n-1)I_{n-2}$$

and $$nI_n = -\sin^{n-1} x \cos x + (n-1)I_{n-2}$$

from which, $$\int \sin^n x\,dx = \boxed{\boldsymbol{I_n = -\frac{1}{n}\sin^{n-1} x \cos x + \frac{n-1}{n} I_{n-2}}} \qquad (4)$$

For example, to determine $\int \sin^4 x\,dx$ using a reduction formula:
Using equation (4), $\int \sin^4 x\,dx = I_4 = -\frac{1}{4}\sin^3 x\cos x + \frac{3}{4}I_2$

$$I_2 = -\frac{1}{2}\sin^1 x\cos x + \frac{1}{2}I_0 \quad \text{and} \quad I_0 = \int \sin^0 x\,dx = \int 1\,dx = x$$

Hence
$$\int \sin^4 x\,dx = I_4 = -\frac{1}{4}\sin^3 x\cos x + \frac{3}{4}\left[-\frac{1}{2}\sin x\cos x + \frac{1}{2}(x)\right]$$

$$= -\frac{1}{4}\sin^3 x\cos x - \frac{3}{8}\sin x\cos x + \frac{3}{8}x + c$$

Using reduction formulae for integrals of the form $\int \cos^n x\,dx$

Let $I_n = \int \cos^n x\,dx \equiv \int \cos^{n-1} x\cos x\,dx$ from laws of indices
Using integration by parts, let $u = \cos^{n-1} x$ from which,
$\frac{du}{dx} = (n-1)\cos^{n-2} x(-\sin x)$ and $du = (n-1)\cos^{n-2} x(-\sin x)dx$, and let
$dv = \cos x\,dx$ from which, $v = \int \cos x\,dx = \sin x$. Then

$$I_n = (\cos^{n-1} x)(\sin x) - \int (\sin x)(n-1)\cos^{n-2} x(-\sin x)\,dx$$

$$= (\cos^{n-1} x)(\sin x) + (n-1)\int \sin^2 x\cos^{n-2} x\,dx$$

$$= (\cos^{n-1} x)(\sin x) + (n-1)\int (1-\cos^2 x)\cos^{n-2} x\,dx$$

$$= (\cos^{n-1} x)(\sin x) + (n-1)\left\{\int \cos^{n-2} x\,dx - \int \cos^n x\,dx\right\}$$

i.e. $\quad I_n = (\cos^{n-1} x)(\sin x) + (n-1)I_{n-2} - (n-1)I_n$

i.e. $\quad I_n + (n-1)I_n = (\cos^{n-1} x)(\sin x) + (n-1)I_{n-2}$

i.e. $\quad nI_n = (\cos^{n-1} x)(\sin x) + (n-1)I_{n-2}$

Thus
$$\boxed{\mathbf{I_n = \frac{1}{n}\cos^{n-1} x\sin x + \frac{n-1}{n}I_{n-2}}} \qquad (5)$$

For example, to determine a reduction formula for $\int_0^{\pi/2} \cos^n x\,dx$ and hence evaluate $\int_0^{\pi/2} \cos^5 x\,dx$:
From equation (5),

$$\int \cos^n x\,dx = \frac{1}{n}\cos^{n-1} x\sin x + \frac{n-1}{n}I_{n-2}$$

and hence $$\int_0^{\pi/2} \cos^n x\,dx = \left[\frac{1}{n}\cos^{n-1}x\sin x\right]_0^{\pi/2} + \frac{n-1}{n}I_{n-2}$$

$$= [0-0] + \frac{n-1}{n}I_{n-2}$$

i.e. $$\boldsymbol{\int_0^{\pi/2} \cos^n x\,dx = I_n = \frac{n-1}{n}I_{n-2}} \qquad (6)$$

(This result is usually known as **Wallis's formula**)

Thus, from equation (6), $\int_0^{\pi/2} \cos^5 x\,dx = \frac{4}{5}I_3$

$$I_3 = \frac{2}{3}I_1 \text{ and } I_1 = \int_0^{\pi/2} \cos^1 x\,dx = [\sin x]_0^{\pi/2} = (1-0) = 1$$

Hence $$\int_0^{\pi/2} \cos^5 x\,dx = \frac{4}{5}I_3 = \frac{4}{5}\left[\frac{2}{3}I_1\right] = \frac{4}{5}\left[\frac{2}{3}(1)\right] = \mathbf{\frac{8}{15}}$$

Further reduction formulae

For example, to determine a reduction formula for $\int \tan^n x\,dx$ and hence find $\int \tan^7 x\,dx$:

Let $$I_n = \int \tan^n x\,dx \equiv \int \tan^{n-2}x\tan^2 x\,dx \text{ by the laws of indices}$$

$$= \int \tan^{n-2}x(\sec^2 x - 1)\,dx \text{ since } 1+\tan^2 x = \sec^2 x$$

$$= \int \tan^{n-2}x\sec^2 x\,dx - \int \tan^{n-2}x\,dx$$

$$= \int \tan^{n-2}x\sec^2 x\,dx - I_{n-2}$$

i.e. $$\boldsymbol{I_n = \frac{\tan^{n-1}x}{n-1} - I_{n-2}}$$

When $n = 7$, $$I_7 = \int \tan^7 x\,dx = \frac{\tan^6 x}{6} - I_5$$

$$I_5 = \frac{\tan^4 x}{4} - I_3 \quad \text{and} \quad I_3 = \frac{\tan^2 x}{2} - I_1$$

$$I_1 = \int \tan x\,dx = \ln(\sec x) \text{ using } \tan x = \frac{\sin x}{\cos x} \text{ and letting } u = \cos x$$

Thus $$\int \tan^7 x\,dx = \frac{\tan^6 x}{6} - \left[\frac{\tan^4 x}{4} - \left(\frac{\tan^2 x}{2} - (\ln(\sec x))\right)\right]$$

Hence $$\boldsymbol{\int \tan^7 x\,dx = \frac{1}{6}\tan^6 x - \frac{1}{4}\tan^4 x + \frac{1}{2}\tan^2 x - \ln(\sec x) + c}$$

---

64 Numerical Integration

Introduction

Even with advanced methods of integration there are many mathematical functions which cannot be integrated by analytical methods and thus approximate methods have then to be used. Approximate methods of definite integrals may be determined by what is termed **numerical integration**.

It may be shown that determining the value of a definite integral is, in fact, finding the area between a curve, the horizontal axis and the specified ordinates. Three methods of finding approximate areas under curves are the trapezoidal rule, the mid-ordinate rule and Simpson's rule, and these rules are used as a basis for numerical integration.

The trapezoidal rule

Let a required definite integral be denoted by $\int_a^b y\,dx$ and be represented by the area under the graph of $y = f(x)$ between the limits $x = a$ and $x = b$ as shown in Figure 64.1.

Let the range of integration be divided into n equal intervals each of width d, such that $nd = b - a$, i.e. $d = \frac{b-a}{n}$

The ordinates are labelled $y_1, y_2, y_3, \ldots\ldots y_{n+1}$ as shown.

The trapezoidal rule states:

$$\int_a^b y\,dx \approx \left(\begin{array}{c}\textbf{width of}\\ \textbf{interval}\end{array}\right)\left\{\frac{1}{2}\left(\begin{array}{c}\textbf{first + last}\\ \textbf{ordinate}\end{array}\right)+\left(\begin{array}{c}\textbf{sum of}\\ \textbf{remaining}\\ \textbf{ordinates}\end{array}\right)\right\} \quad (1)$$

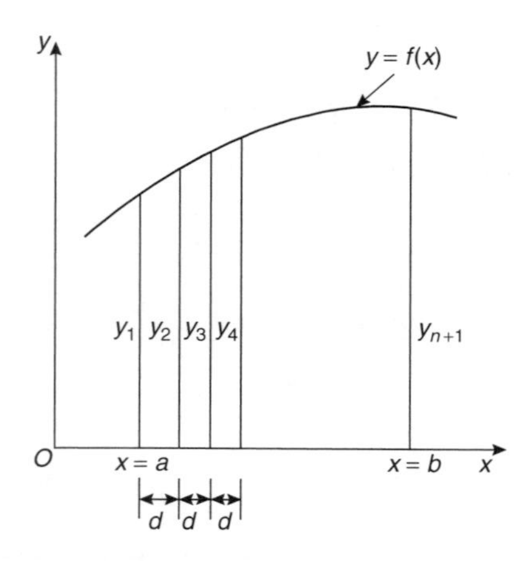

Figure 64.1

For example, using the trapezoidal rule with 8 intervals to evaluate $\int_1^3 \frac{2}{\sqrt{x}}\,dx$, correct to 3 decimal places:

With 8 intervals, the width of each is $\frac{3-1}{8}$ i.e. 0.25 giving ordinates at 1.00, 1.25, 1.50, 1.75, 2.00, 2.25, 2.50, 2.75 and 3.00. Corresponding values of $\frac{2}{\sqrt{x}}$ are shown in the table below.

x	1.00	1.25	1.50	1.75	2.00
$\frac{2}{\sqrt{x}}$	2.0000	1.7889	1.6330	1.5119	1.4142

x	2.25	2.50	2.75	3.00
$\frac{2}{\sqrt{x}}$	1.3333	1.2649	1.2060	1.1547

From equation (1):

$$\int_1^3 \frac{2}{\sqrt{x}}\,dx \approx (0.25)\left\{\begin{array}{r}\frac{1}{2}(2.000+1.1547)+1.7889+1.6330\\ +1.5119+1.4142+1.3333\\ +1.2649+1.2060\end{array}\right\}$$

$$= \mathbf{2.932}, \text{ correct to 3 decimal places}$$

The greater the number of intervals chosen (i.e. the smaller the interval width) the more accurate will be the value of the definite integral. The exact value is found when the number of intervals is infinite, which is, of course, what the process of integration is based upon. Using integration:

$$\int_1^3 \frac{2}{\sqrt{x}}\,dx = \int_1^3 2x^{-(1/2)}\,dx = \left[\frac{2x^{(-1/2)+1}}{-\frac{1}{2}+1}\right]_1^3 = [4x^{1/2}]_1^3 = 4[\sqrt{x}]_1^3$$

$$= 4[\sqrt{3}-\sqrt{1}] = \mathbf{2.928}, \text{ correct to 3 decimal places}$$

The mid-ordinate Rule

Let a required definite integral be denoted again by $\int_a^b y\,dx$ and represented by the area under the graph of $y = f(x)$ between the limits $x = a$ and $x = b$, as shown in Figure 64.2.

With the mid-ordinate rule each interval of width d is assumed to be replaced by a rectangle of height equal to the ordinate at the middle point of each interval, shown as $y_1, y_2, y_3, \ldots y_n$ in Figure 64.2.

The mid-ordinate rule states:

$$\int_a^b y\,dx \approx \textbf{(width of interval)(sum of mid-ordinates)} \qquad (2)$$

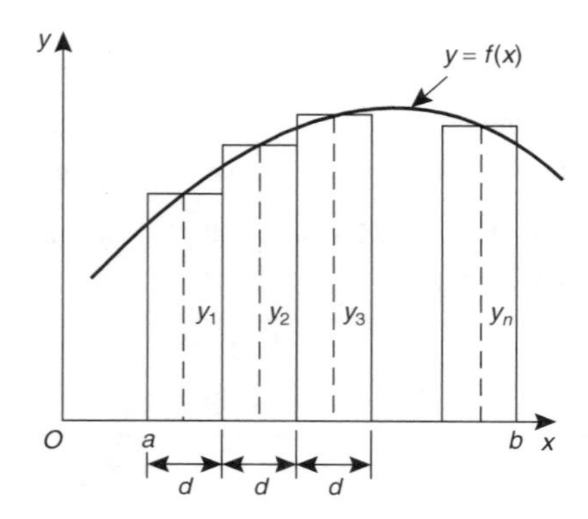

Figure 64.2

For example, using the mid-ordinate rule with 8 intervals, to evaluate $\int_1^3 \frac{2}{\sqrt{x}}\, dx$, correct to 3 decimal places:

With 8 intervals, each will have a width of 0.25 and the ordinates will occur at 1.00, 1.25, 1.50, 1.75, ... and thus mid-ordinates at 1.125, 1.375, 1.625......

Corresponding values of $\frac{2}{\sqrt{x}}$ are shown in the following table.

x	1.125	1.375	1.625	1.875	2.125	2.375	2.625	2.875
$\frac{2}{\sqrt{x}}$	1.8856	1.7056	1.5689	1.4606	1.3720	1.2978	1.2344	1.1795

From equation (2):

$$\int_1^3 \frac{2}{\sqrt{x}}\, dx \approx (0.25)[1.8856 + 1.7056 + 1.5689 + 1.4606$$

$$+1.3720 + 1.2978 + 1.2344 + 1.1795]$$

$$= \mathbf{2.926}, \text{ correct to 3 decimal places}$$

As previously, the greater the number of intervals the nearer the result is to the true value (of 2.928, correct to 3 decimal places).

Simpson's rule

The approximation made with the trapezoidal rule is to join the top of two successive ordinates by a straight line, i.e. by using a linear approximation of the form $a + bx$. With Simpson's rule, the approximation made is to join the tops of three successive ordinates by a parabola, i.e. by using a quadratic approximation of the form $a + bx + cx^2$

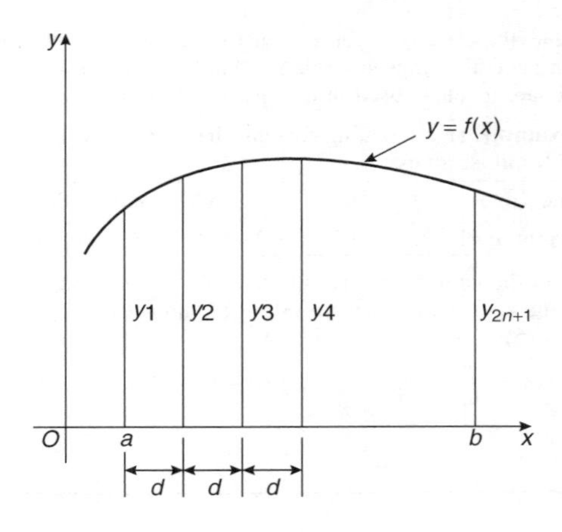

Figure 64.3

Let a definite integral be denoted by $\int_a^b y\,dx$ and represented by the area under the graph of $y = f(x)$ between the limits $x = a$ and $x = b$, as shown in Figure 64.3. The range of integration, $b - a$, is divided into an **even** number of intervals, say $2n$, each of width d.
Simpson's rule states:

$$\int_a^b y\,dx \approx \frac{1}{3}\begin{pmatrix}\text{width of}\\ \text{interval}\end{pmatrix}\left\{\begin{pmatrix}\text{first + last}\\ \text{ordinate}\end{pmatrix} + 4\begin{pmatrix}\text{sum of even}\\ \text{ordinates}\end{pmatrix} + 2\begin{pmatrix}\text{sum of remaining}\\ \text{odd ordinates}\end{pmatrix}\right\} \quad (5)$$

Note that Simpson's rule can only be applied when an even number of intervals is chosen, i.e. an odd number of ordinates.

For example, using Simpson's rule with 8 intervals, to evaluate $\displaystyle\int_1^3 \frac{2}{\sqrt{x}}\,dx$, correct to 3 decimal places:

With 8 intervals, each will have a width of $\dfrac{3-1}{8}$, i.e. 0.25 and the ordinates occur at 1.00, 1.25, 1.50, 1.75, ..., 3.0. The values of the ordinates are as shown in the table on page 327.
Thus, from equation (5):

$$\int_1^3 \frac{2}{\sqrt{x}}\,dx \approx \frac{1}{3}(0.25)[(2.0000 + 1.1547) + 4(1.7889 + 1.5119 + 1.3333 + 1.2060) + 2(1.6330 + 1.4142 + 1.2649)]$$

$$= \frac{1}{3}(0.25)[3.1547 + 23.3604 + 8.6242]$$

$$= \mathbf{2.928}, \text{ correct to 3 decimal places}$$

It is noted that the latter answer is exactly the same as that obtained by integration. In general, Simpson's rule is regarded as the most accurate of the three approximate methods used in numerical integration.

In another example, an alternating current i has the following values at equal intervals of 2.0 milliseconds.

Time (ms)	0	2.0	4.0	6.0	8.0	10.0	12.0
Current $i(A)$	0	3.5	8.2	10.0	7.3	2.0	0

Charge, q, in millicoulombs, is given by $q = \int_0^{12.0} i\,dt$. Using Simpson's rule to determine the approximate charge in the 12 millisecond period:
From equation (5):

$$\text{Charge, } q = \int_0^{12.0} i\,dt \approx \frac{1}{3}(2.0)[(0+0)+4(3.5+10.0+2.0) +2(8.2+7.3)] = \mathbf{62\ mC}$$

65 Areas Under and Between Curves

Area under a curve

The area shown shaded in Figure 65.1 may be determined using approximate methods (such as the trapezoidal rule, the mid-ordinate rule or Simpson's rule) or, more precisely, by using integration.
Let A be the area shown shaded in Figure 65.1 and let this area be divided into a number of strips each of width δx. One such strip is shown and let the area of this strip be δA.
Then: $\quad \delta A \approx y\delta x \qquad (1)$
The accuracy of statement (1) increases when the width of each strip is reduced, i.e. area A is divided into a greater number of strips.
Area A is equal to the sum of all the strips from $x = a$ to $x = b$,

i.e.
$$A = \lim_{\delta x \to 0} \sum_{x=a}^{x=b} y\delta x \qquad (2)$$

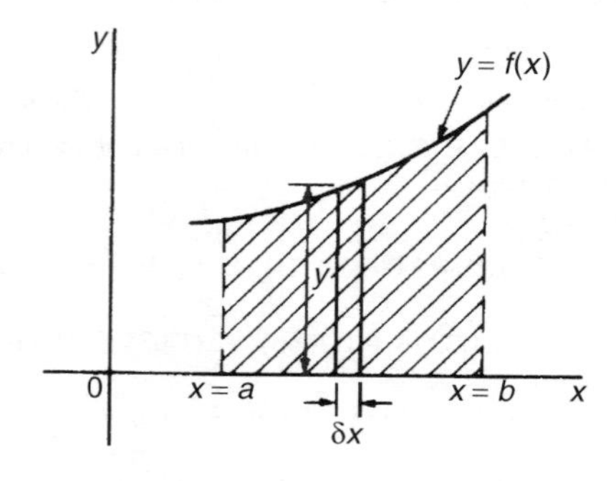

Figure 65.1

From statement (1), $\dfrac{\delta A}{\delta x} \approx y$ (3)

In the limit, as δx approaches zero, $\dfrac{\delta A}{\delta x}$ becomes the differential coefficient $\dfrac{dA}{dx}$.

Hence $\lim_{\delta x \to 0}\left(\dfrac{\delta A}{\delta x}\right) = \dfrac{dA}{dx} = y$, from statement (3).

By integration, $\displaystyle\int \frac{dA}{dx}\,dx = \int y\,dx$ i.e. $A = \displaystyle\int y\,dx$

The ordinates $x = a$ and $x = b$ limit the area and such ordinate values are shown as limits. Hence

$$A = \int_a^b y\,dx \tag{4}$$

Equating statements (2) and (4) gives:

$$\text{Area } A = \lim_{\delta x \to 0} \sum_{x=a}^{x=b} y\delta x = \int_a^b y\,dx = \int_a^b f(x)\,dx$$

If the area between a curve $x = f(y)$, the y-axis and ordinates $y = p$ and $y = q$ is required then area $= \int_p^q x\,dy$

Thus determining the area under a curve by integration merely involves evaluating a definite integral.

There are several instances in engineering and science where the area beneath a curve needs to be accurately determined. For example, the areas between limits of a velocity/time graph gives distance travelled, force/distance graph gives work done, voltage/current graph gives power, and so on.

Should a curve drop below the x-axis, then $y(= f(x))$ becomes negative and $f(x)\,dx$ is negative. When determining such areas by integration, a negative sign is placed before the integral. For the curve shown in Figure 65.2, the total shaded area is given by (area E + area F + area G).

By integration,

total shaded area $= \int_a^b f(x)\,dx - \int_b^c f(x)\,dx + \int_c^d f(x)\,dx$.

(Note that this is **not** the same as $\int_a^d f(x)\,dx$).

It is usually necessary to sketch a curve in order to check whether it crosses the x-axis.

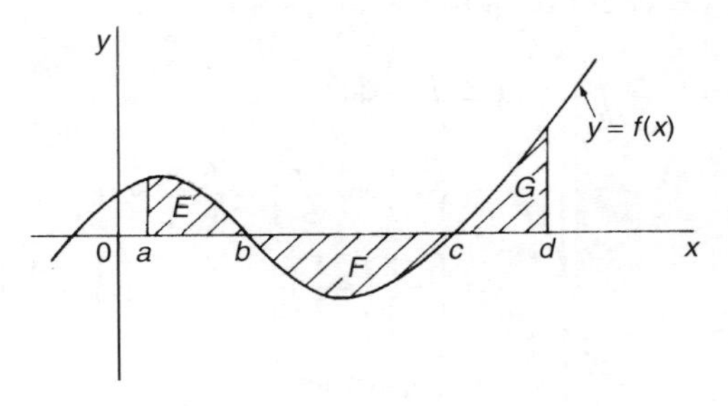

Figure 65.2

For example, the velocity v of a body t seconds after a certain instant is $(2t^2 + 5)$m/s. To find by integration how far it moves in the interval from $t = 0$ to $t = 4$ s:
Since $2t^2 + 5$ is a quadratic expression, the curve $v = 2t^2 + 5$ is a parabola cutting the v-axis at $v = 5$, as shown in Figure 65.3.
The distance travelled is given by the area under the v/t curve, shown shaded in Figure 65.3.

By integration, shaded area $= \int_0^4 v\,dt = \int_0^4 (2t^2 + 5)\,dt = \left[\frac{2t^3}{3} + 5t\right]_0^4$

i.e. **distance travelled = 62.67 m**

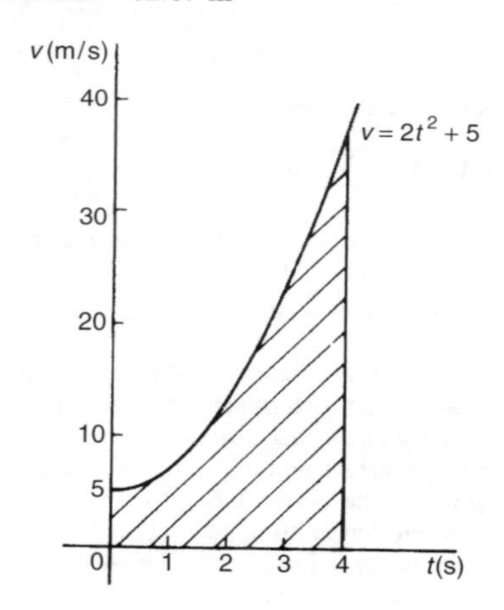

Figure 65.3

In another example, to find the area enclosed by the curve $y = \sin 2x$, the x-axis and the ordinates $x = 0$ and $x = \dfrac{\pi}{3}$:
A sketch of $y = \sin 2x$ is shown in Figure 65.4.
(Note that $y = \sin 2x$ has a period of $\dfrac{2\pi}{2}$, i.e. π radians)

$$\text{Shaded area} = \int_0^{\pi/3} y\,dx = \int_0^{\pi/3} \sin 2x\,dx = \left[-\frac{1}{2}\cos 2x\right]_0^{\pi/3}$$

$$= \left\{-\frac{1}{2}\cos\frac{2\pi}{3}\right\} - \left\{-\frac{1}{2}\cos 0\right\}$$

$$= \left\{-\frac{1}{2}\left(-\frac{1}{2}\right)\right\} - \left\{-\frac{1}{2}(1)\right\} = \frac{1}{4} + \frac{1}{2}$$

$$= \mathbf{\frac{3}{4}\,\text{square units}}$$

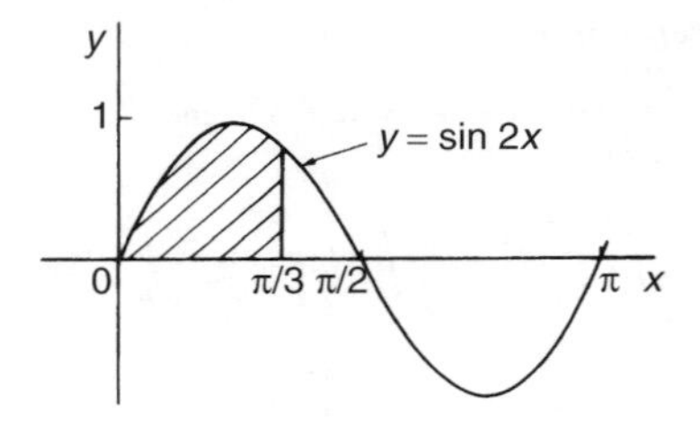

Figure 65.4

In another example, to determine the area between the curve $y = x^3 - 2x^2 - 8x$ and the x-axis:

$$y = x^3 - 2x^2 - 8x = x(x^2 - 2x - 8) = x(x + 2)(x - 4)$$

When $\quad y = 0, x = 0$ or $(x + 2) = 0$ or $(x - 4) = 0$,
i.e. when $y = 0$, $x = 0$ or -2 or 4, which means that the curve crosses the x-axis at 0, -2, and 4. Since the curve is a continuous function, only one other co-ordinate value needs to be calculated before a sketch of the curve can be produced. When $x = 1$, $y = -9$, showing that the part of the curve between $x = 0$ and $x = 4$ is negative. A sketch of $y = x^3 - 2x^2 - 8x$ is shown in Figure 65.5. (Another method of sketching Figure 65.5 would have been to draw up a table of values).

$$\text{Shaded area} = \int_{-2}^{0} (x^3 - 2x^2 - 8x)\,\mathrm{d}x - \int_{0}^{4} (x^3 - 2x^2 - 8x)\,\mathrm{d}x$$

$$= \left[\frac{x^4}{4} - \frac{2x^3}{3} - \frac{8x^2}{2}\right]_{-2}^{0} - \left[\frac{x^4}{4} - \frac{2x^3}{3} - \frac{8x^2}{2}\right]_{0}^{4}$$

$$= \left(6\frac{2}{3}\right) - \left(-42\frac{2}{3}\right) = \mathbf{49\frac{1}{3}\ square\ units}$$

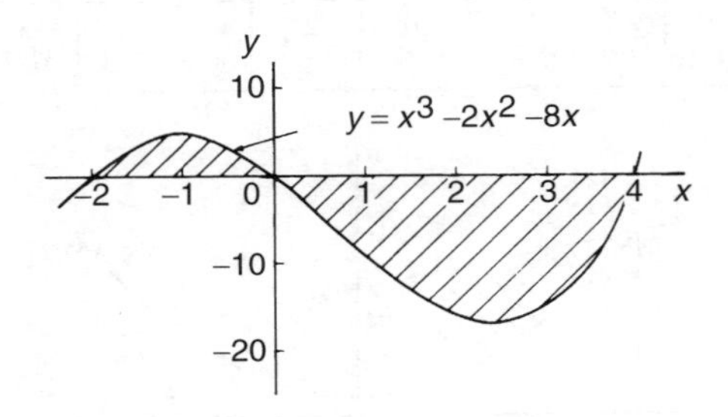

Figure 65.5

The area between curves

The area enclosed between curves $y = f_1(x)$ and $y = f_2(x)$, shown shaded in Figure 65.6, is given by:

$$\text{shaded area} = \int_a^b f_2(x)\,dx - \int_a^b f_1(x)\,dx = \int_a^b [f_2(x) - f_1(x)]\,dx$$

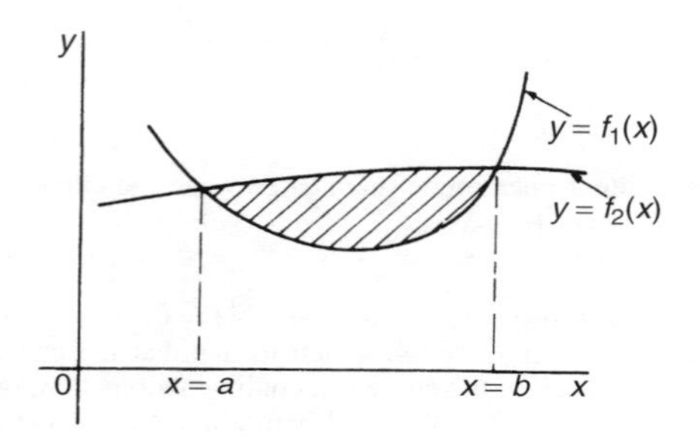

Figure 65.6

For example, to determine the area enclosed between the curves $y = x^2 + 1$ and $y = 7 - x$:

At the points of intersection the curves are equal. Thus, equating the y values of each curve gives:

$$x^2 + 1 = 7 - x$$

from which, $\quad x^2 + x - 6 = 0$

Factorising gives: $\quad (x - 2)(x + 3) = 0$

from which $\quad x = 2$ and $x = -3$

By firstly determining the points of intersection the range of x-values has been found. Tables of values are produced as shown below.

x	−3	−2	−1	0	1	2
$y = x^2 + 1$	10	5	2	1	2	5

x	−3	0	2
$y = 7 - x$	10	7	5

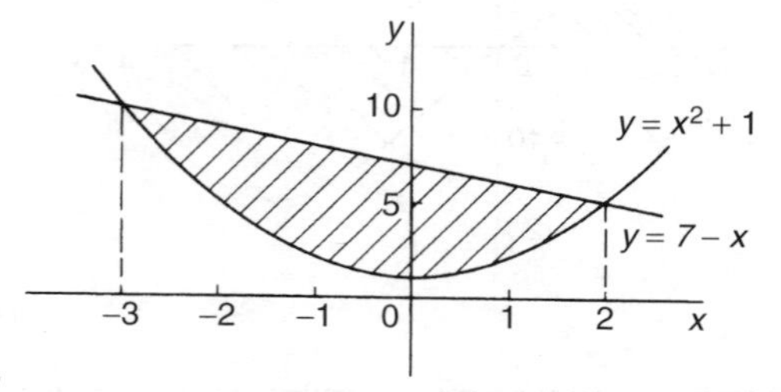

Figure 65.7

A sketch of the two curves is shown in Figure 65.7.

$$\text{Shaded area} = \int_{-3}^{2} (7 - x)\,dx - \int_{-3}^{2} (x^2 + 1)\,dx$$

$$= \int_{-3}^{2} [(7 - x) - (x^2 + 1)]\,dx$$

$$= \int_{-3}^{2} (6 - x - x^2)\,dx = \left[6x - \frac{x^2}{2} - \frac{x^3}{3}\right]_{-3}^{2}$$

$$= \left(12 - 2 - \frac{8}{3}\right) - \left(-18 - \frac{9}{2} + 9\right)$$

$$= \left(7\frac{1}{3}\right) - \left(-13\frac{1}{2}\right) = \mathbf{20\frac{5}{6}\ sq.\ units}$$

In another example, to determine by integration the area bounded by the three straight lines $y = 4 - x$, $y = 3x$ and $3y = x$:
Each of the straight lines are shown sketched in Figure 65.8.

$$\text{Shaded area} = \int_{0}^{1} \left(3x - \frac{x}{3}\right) dx + \int_{1}^{3} \left[(4 - x) - \frac{x}{3}\right] dx$$

$$= \left[\frac{3x^2}{2} - \frac{x^2}{6}\right]_0^1 + \left[4x - \frac{x^2}{2} - \frac{x^2}{6}\right]_1^3$$

$$= \left[\left(\frac{3}{2} - \frac{1}{6}\right) - (0)\right] + \left[\left(12 - \frac{9}{2} - \frac{9}{6}\right) - \left(4 - \frac{1}{2} - \frac{1}{6}\right)\right]$$

$$= \left(1\frac{1}{3}\right) + \left(6 - 3\frac{1}{3}\right) = \mathbf{4\ square\ units}$$

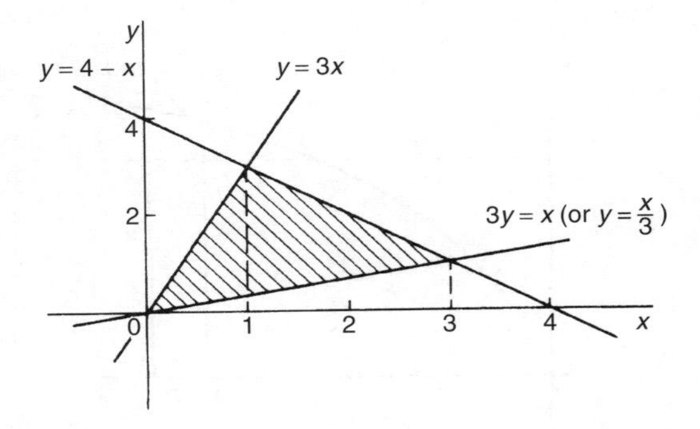

Figure 65.8

66 Mean and Root Mean Square Values

Mean or average values

The mean or average value of the curve shown in Figure 66.1, between $x = a$ and $x = b$, is given by:

$$\textbf{mean or average value,}\quad \bar{y} = \frac{\textbf{area under curve}}{\textbf{length of base}}$$

When the area under a curve may be obtained by integration then:

mean or average value, $\quad \bar{y} = \dfrac{\int_a^b y\,dx}{b-a}$

i.e. $$\boxed{\bar{y} = \frac{1}{b-a}\int_a^b f(x)\,dx}$$

For a periodic function, such as a sine wave, the mean value is assumed to be 'the mean value over half a cycle', since the mean value over a complete cycle is zero.

For example, to determine, using integration, the mean value of $y = 5x^2$ between $x = 1$ and $x = 4$:

Mean value, $$\bar{y} = \frac{1}{4-1}\int_1^4 y\,dx = \frac{1}{3}\int_1^4 5x^2\,dx$$

$$= \frac{1}{3}\left[\frac{5x^3}{3}\right]_1^4 = \frac{5}{9}[x^3]_1^4 = \frac{5}{9}(64-1) = \mathbf{35}$$

In another example, a sinusoidal voltage is given by $v = 100\sin\omega t$ volts. To determine the mean value of the voltage over half a cycle using integration: Half a cycle means the limits are 0 to π radians.

Mean value, $$\bar{v} = \frac{1}{\pi - 0}\int_0^\pi v\,d(\omega t) = \frac{1}{\pi}\int_0^\pi 100\sin\omega t\,d(\omega t)$$

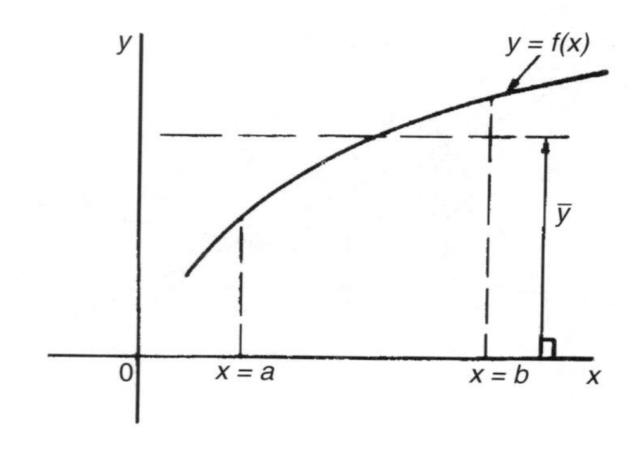

Figure 66.1

$$= \frac{100}{\pi}[-\cos \omega t]_0^\pi = \frac{100}{\pi}[(-\cos \pi) - (-\cos 0)]$$

$$= \frac{100}{\pi}[(+1) - (-1)] = \frac{200}{\pi} = \mathbf{63.66\ volts}$$

[Note that for a sine wave, **mean value** $= \boldsymbol{\frac{2}{\pi}} \times$ **maximum value**

In this case, mean value $= \frac{2}{\pi} \times 100 = 63.66$ V]

Root mean square values

The **root mean square value** of a quantity is 'the square root of the mean value of the squared values of the quantity' taken over an interval. With reference to Figure 66.1, the r.m.s. value of $y = f(x)$ over the range $x = a$ to $x = b$ is given by:

$$\textbf{r.m.s. value} = \sqrt{\left\{\frac{1}{b-a}\int_a^b y^2\,dx\right\}}$$

One of the principal applications of r.m.s. values is with alternating currents and voltages. The r.m.s. value of an alternating current is defined as 'that current which will give the same heating effect as the equivalent direct current'.

For example, to determine the r.m.s. value of $y = 2x^2$ between $x = 1$ and $x = 4$:

$$\text{R.m.s. value} = \sqrt{\left\{\frac{1}{4-1}\int_1^4 y^2\,dx\right\}} = \sqrt{\left\{\frac{1}{3}\int_1^4 (2x^2)^2\,dx\right\}}$$

$$= \sqrt{\left\{\frac{1}{3}\int_1^4 4x^4\,dx\right\}} = \sqrt{\left\{\frac{4}{3}\left[\frac{x^5}{5}\right]_1^4\right\}}$$

$$= \sqrt{\left\{\frac{4}{15}(1024-1)\right\}} = \sqrt{272.8} = \mathbf{16.5}$$

In another example, a sinusoidal voltage has a maximum value of 100 V. To calculate its r.m.s. value:

A sinusoidal voltage v having a maximum value of 10 V may be written as $v = 10\sin\theta$. Over the range $\theta = 0$ to $\theta = \pi$,

$$\text{r.m.s. value} = \sqrt{\left\{\frac{1}{\pi-0}\int_0^\pi v^2\,d\theta\right\}} = \sqrt{\left\{\frac{1}{\pi}\int_0^\pi (100\sin\theta)^2\,d\theta\right\}}$$

$$= \sqrt{\left\{\frac{10000}{\pi}\int_0^\pi \sin^2\theta\,d\theta\right\}} \quad \text{which is not a 'standard' integral}$$

It is shown in Chapter 31 that $\cos 2A = 1 - 2\sin^2 A$ and this formula is used whenever $\sin^2 A$ needs to be integrated.
Rearranging $\cos 2A = 1 - 2\sin^2 A$ gives $\sin^2 A = \frac{1}{2}(1 - \cos 2A)$

Hence $\sqrt{\left\{\dfrac{10\,000}{\pi}\displaystyle\int_0^{\pi} \sin^2\theta \, d\theta\right\}}$

$$= \sqrt{\left\{\frac{10\,000}{\pi}\int_0^{\pi} \frac{1}{2}(1 - \cos 2\theta)\, d\theta\right\}}$$

$$= \sqrt{\left\{\frac{10\,000}{\pi}\,\frac{1}{2}\left[\theta - \frac{\sin 2\theta}{2}\right]_0^{\pi}\right\}}$$

$$= \sqrt{\left\{\frac{10\,000}{\pi}\,\frac{1}{2}\left[\left(\pi - \frac{\sin 2\pi}{2}\right) - \left(0 - \frac{\sin 0}{2}\right)\right]\right\}}$$

$$= \sqrt{\left\{\frac{10\,000}{\pi}\,\frac{1}{2}[\pi]\right\}} = \sqrt{\left\{\frac{10000}{2}\right\}}$$

$$= \frac{100}{\sqrt{2}} = \mathbf{70.71\ volts}$$

[Note that for a sine wave, **r.m.s. value** $= \dfrac{\mathbf{1}}{\sqrt{\mathbf{2}}}$ **× maximum value**.

In this case, r.m.s. value $= \dfrac{1}{\sqrt{2}} \times 100 = 70.71$ V]

67 Volumes of Solids of Revolution

Introduction

If the area under the curve $y = f(x)$, (shown in Figure 67.1(a)), between $x = a$ and $x = b$ is rotated 360° about the x-axis, then a volume known as a **solid of revolution** is produced as shown in Figure 67.1(b).
The volume of such a solid may be determined precisely using integration.
Let the area shown in Figure 67.1(a) be divided into a number of strips each of width δx. One such strip is shown shaded.
When the area is rotated 360°about the x-axis, each strip produces a solid of revolution approximating to a circular disc of radius y and thickness δx.

Volume of disc = (circular cross-sectional area) (thickness) = $(\pi y^2)(\delta x)$

Total volume, V, between ordinates $x = a$ and $x = b$ is given by:

$$\textbf{Volume, } V = \lim_{\delta x \to 0} \sum_{x=a}^{x=b} \pi y^2 \delta x = \int_a^b \pi y^2 \, dx$$

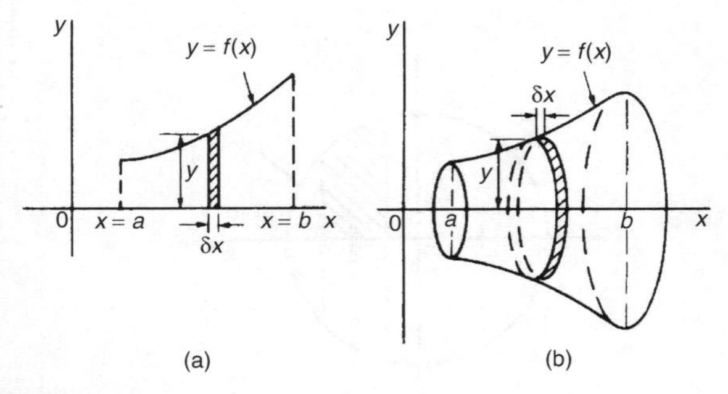

Figure 67.1

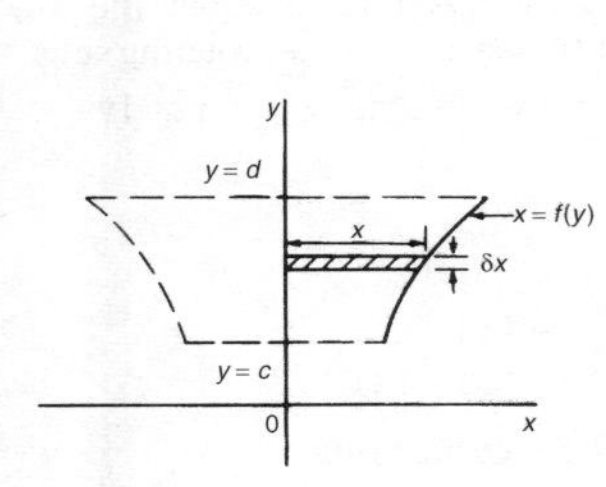

Figure 67.2

Figure 67.3

If a curve $x = f(y)$ is rotated about the y-axis 360° between the limits $y = c$ and $y = d$, as shown in Figure 67.2, then the volume generated is given by:

$$\textbf{Volume, } V = \lim_{\delta y \to 0} \sum_{y=c}^{y=d} \pi x^2 \delta y = \int_a^b \pi x^2 \, dy$$

For example, the curve $y = x^2 + 4$ is rotated one revolution about the x-axis between the limits $x = 1$ and $x = 4$. To determine the volume of the solid of revolution produced:

Revolving the shaded area shown in Figure 67.3 about the x-axis 360° produces a solid of revolution given by:

$$\text{Volume } = \int_1^4 \pi y^2 \, dx = \int_1^4 \pi (x^2 + 4)^2 \, dx = \int_1^4 \pi (x^4 + 8x^2 + 16) \, dx$$

$$= \pi \left[\frac{x^5}{5} + \frac{8x^3}{3} + 16x \right]_1^4$$

$$= \pi[(204.8 + 170.67 + 64) - (0.2 + 2.67 + 16)]$$

$$= \mathbf{420.6\pi \text{ cubic units}}$$

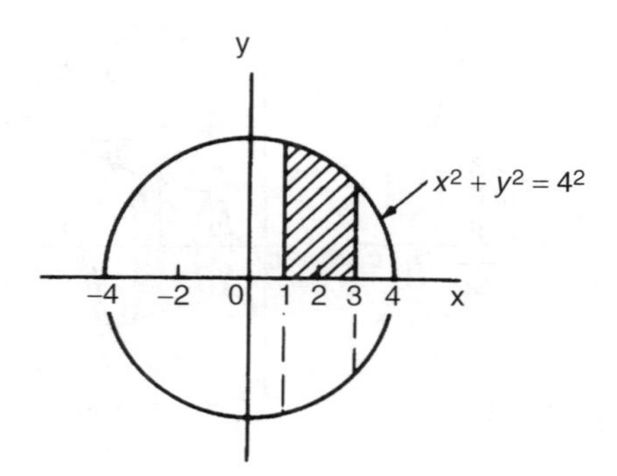

Figure 67.4

The volume produced when the curve $y = x^2 + 4$ is rotated about the y-axis between $y = 5$ (when $x = 1$) and $y = 20$ (when $x = 4$), i.e. rotating area ABCD of Figure 67.3 about the y-axis is given by: volume $= \int_5^{20} \pi x^2 \, dy$

Since $\quad y = x^2 + 4, \;$ then $x^2 = y - 4$

Hence $\quad$ volume $= \displaystyle\int_5^{20} \pi(y - 4)\, dy = \pi \left[\frac{y^2}{2} - 4y \right]_5^{20}$

$$= \pi[(120) - (-7.5)] = \mathbf{127.5\pi \text{ cubic units}}$$

In another example, to calculate the volume of a frustum of a sphere of radius 4 cm that lies between two parallel planes at 1 cm and 3 cm from the centre and on the same side of it:
The volume of a frustum of a sphere may be determined by integration by rotating the curve $x^2 + y^2 = 4^2$ (i.e. a circle, centre 0, radius 4) one revolution about the x-axis, between the limits $x = 1$ and $x = 3$ (i.e. rotating the shaded area of Figure 67.4).

$$\text{Volume of frustum} = \int_1^3 \pi y^2 \, dx = \int_1^3 \pi(4^2 - x^2)\, dx = \pi \left[16x - \frac{x^3}{3} \right]_1^3$$

$$= \pi \left[(39) - \left(15\frac{2}{3} \right) \right] = \mathbf{23\frac{1}{3}\pi \text{ cubic units}}$$

68 Centroids of Simple Shapes

Centroids

A **lamina** is a thin flat sheet having uniform thickness. The **centre of gravity** of a lamina is the point where it balances perfectly, i.e. the lamina's **centre of mass**.

When dealing with an area (i.e. a lamina of negligible thickness and mass) the term **centre of area** or **centroid** is used for the point where the centre of gravity of a lamina of that shape would lie.

The first moment of area

The **first moment of area** is defined as the product of the area and the perpendicular distance of its centroid from a given axis in the plane of the area. In Figure 68.1, the first moment of area A about axis XX is given by $(A\ y)$ cubic units.

Centroid of area between a curve and the x-axis

Figure 68.2 shows an area PQRS bounded by the curve $y = f(x)$, the x-axis and ordinates $x = a$ and $x = b$. Let this area be divided into a large number of strips, each of width δx. A typical strip is shown shaded drawn at point (x, y) on $f(x)$.

The area of the strip is approximately rectangular and is given by $y\delta x$.

The centroid, C, has coordinates $\left(x, \ \dfrac{y}{2}\right)$

First moment of area of shaded strip about axis OY $= (y\,\delta x)(x) = xy\,\delta x$

Total first moment of area PQRS about axis OY

$$= \lim_{\delta x \to 0} \sum_{x=a}^{x=b} xy\,\delta x = \int_a^b xy\,\mathrm{d}x$$

First moment of area of shaded strip about axis OX

$$= (y\,\delta x)\left(\frac{y}{2}\right) = \frac{1}{2}y^2 x$$

Total first moment of area PQRS about axis OX

$$= \lim_{\delta x \to 0} \sum_{x=a}^{x=b} \frac{1}{2}y^2\,\delta x = \frac{1}{2}\int_a^b y^2\,\mathrm{d}x$$

Area of PQRS, $A = \int_a^b y\,\mathrm{d}x$ (from Chapter 65)

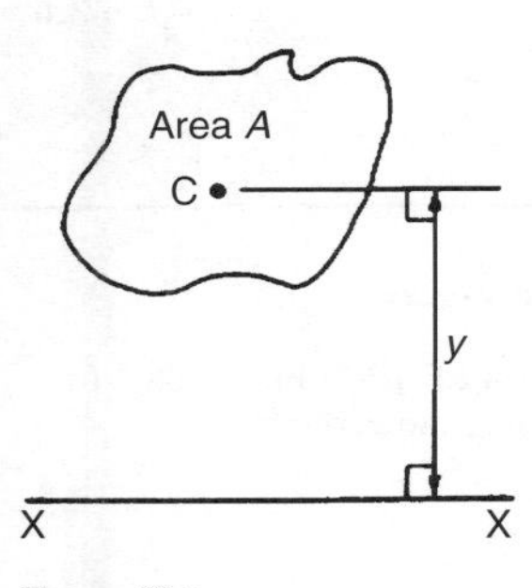

Figure 68.1

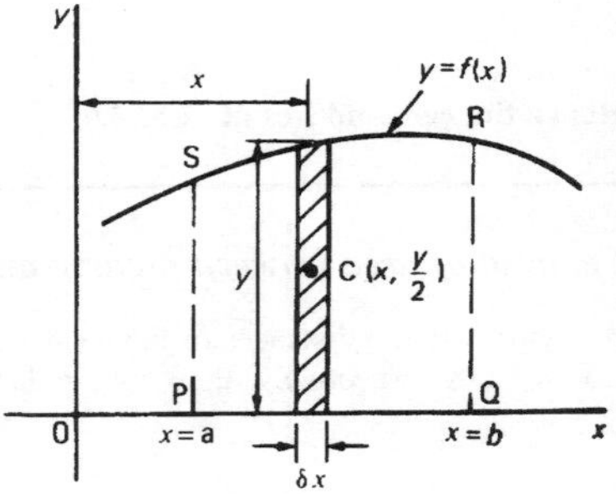

Figure 68.2

Let $\overline{x}$ and $\overline{y}$ be the distances of the centroid of area A about OY and OX respectively then:

$(\overline{x})(A) =$ total first moment of area A about axis OY $= \int_a^b xy\,\mathrm{d}x$

from which, $$\boldsymbol{\overline{x} = \frac{\int_a^b xy\,\mathrm{d}x}{\int_a^b y\,\mathrm{d}x}}$$

and $(\overline{y})(A) =$ total first moment of area A about axis OX $= \dfrac{1}{2}\int_a^b y^2\,\mathrm{d}x$

from which, $$\boldsymbol{\overline{y} = \frac{\frac{1}{2}\int_a^b y^2\,\mathrm{d}x}{\int_a^b y\,\mathrm{d}x}}$$

For example, to find the position of the centroid of the area bounded by the curve $y = 3x^2$, the x-axis and the ordinates $x = 0$ and $x = 2$:

If $(\overline{x}, \overline{y})$ are the co-ordinates of the centroid of the given area then:

$$\overline{x} = \frac{\int_0^2 xy\,\mathrm{d}x}{\int_0^2 y\,\mathrm{d}x} = \frac{\int_0^2 x(3x^2)\,\mathrm{d}x}{\int_0^2 3x^2\,\mathrm{d}x} = \frac{\int_0^2 3x^3\,\mathrm{d}x}{\int_0^2 3x^2\,\mathrm{d}x} = \frac{\left[\frac{3x^4}{4}\right]_0^2}{[x^3]_0^2} = \frac{12}{8} = \mathbf{1.5}$$

$$\overline{y} = \frac{\frac{1}{2}\int_0^2 y^2\,\mathrm{d}x}{\int_0^2 y\,\mathrm{d}x} = \frac{\frac{1}{2}\int_0^2 (3x^2)^2\,\mathrm{d}x}{8} = \frac{\frac{1}{2}\int_0^2 9x^4\,\mathrm{d}x}{8} = \frac{\frac{9}{2}\left[\frac{x^5}{5}\right]_0^2}{8} = \frac{\frac{9}{2}\left(\frac{32}{5}\right)}{8}$$

$$= \frac{18}{5} = \mathbf{3.6}$$

Hence the centroid lies at (1.5, 3.6)

Centroid of area between a curve and the y-axis

If x and y are the distances of the centroid of area EFGH in Figure 68.3 from OY and OX respectively, then, by similar reasoning to earlier:

$$(\overline{x})\ (\text{total area}) = \underset{\delta y \to 0}{\text{limit}} \sum_{y=c}^{y=d} x\delta y\left(\frac{x}{2}\right) = \frac{1}{2}\int_c^d x^2\,\mathrm{d}y$$

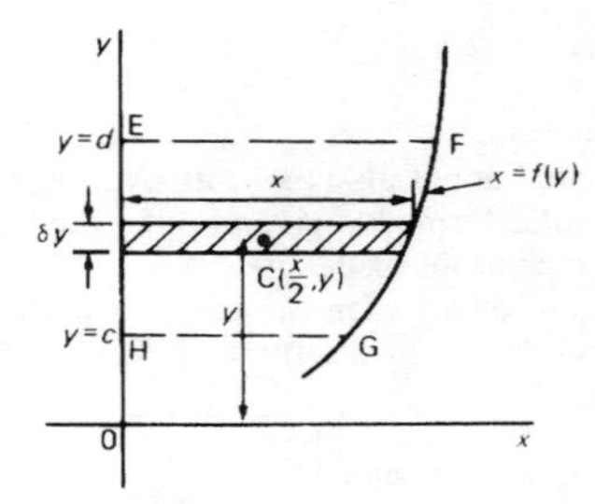

Figure 68.3

from which, $$\boxed{\overline{x} = \frac{\frac{1}{2}\int_c^d x^2\,dy}{\int_c^d x\,dy}}$$

and $$(\overline{y})\ (\text{total area}) = \underset{\delta y \to 0}{\text{limit}} \sum_{y=c}^{y=d} (x\delta y)y = \int_c^d xy\,dy$$

from which, $$\boxed{\overline{y} = \frac{\int_c^d xy\,dy}{\int_c^d x\,dy}}$$

For example, to locate the centroid of the area enclosed by the curve $y = 2x^2$, the y-axis and ordinates $y = 1$ and $y = 4$, correct to 3 decimal places:

$$\overline{x} = \frac{\frac{1}{2}\int_1^4 x^2\,dy}{\int_1^4 x\,dy} = \frac{\frac{1}{2}\int_1^4 \frac{y}{2}\,dy}{\int_1^4 \sqrt{\frac{y}{2}}\,dy} = \frac{\frac{1}{2}\left[\frac{y^2}{4}\right]_1^4}{\left[\frac{2y^{3/2}}{3\sqrt{2}}\right]_1^4} = \frac{\frac{15}{8}}{\frac{14}{3\sqrt{2}}} = 0.568$$

$$\overline{y} = \frac{\int_1^4 xy\,dy}{\int_1^4 x\,dy} = \frac{\int_1^4 \sqrt{\frac{y}{2}}(y)\,dy}{\frac{14}{3\sqrt{2}}} = \frac{\int_1^4 \frac{y^{3/2}}{\sqrt{2}}\,dy}{\frac{14}{3\sqrt{2}}} = \frac{\frac{1}{\sqrt{2}}\left[\frac{y^{5/2}}{\frac{5}{2}}\right]_1^4}{\frac{14}{3\sqrt{2}}}$$

$$= \frac{\frac{2}{5\sqrt{2}}(31)}{\frac{14}{3\sqrt{2}}} = 2.657$$

Hence the position of the centroid is at (0.568, 2.657)

Theorem of Pappus

A theorem of Pappus states:
'If a plane area is rotated about an axis in its own plane but not intersecting it, the volume of the solid formed is given by the product of the area and the distance moved by the centroid of the area'.
With reference to Figure 68.4, when the curve $y = f(x)$ is rotated one revolution about the x-axis between the limits $x = a$ and $x = b$, the volume V generated is given by:

volume $V = (A)(2\pi\overline{y})$, from which, $\boxed{\overline{y} = \dfrac{V}{2\pi A}}$

For example, to determine the position of the centroid of a semicircle of radius r by using the theorem of Pappus:

A semicircle is shown in Figure 68.5 with its diameter lying on the x-axis and its centre at the origin. Area of semicircle $= \dfrac{\pi r^2}{2}$. When the area is rotated about the x-axis one revolution a sphere is generated of volume $\frac{4}{3}\pi r^3$
Let centroid C be at a distance $\overline{y}$ from the origin as shown in Figure 68.5.
From the theorem of Pappus, volume generated = area × distance moved through by centroid

i.e. $$\frac{4}{3}\pi r^3 = \left(\frac{\pi r^2}{2}\right)(2\pi\overline{y}) \quad \text{Hence} \quad \overline{y} = \frac{\frac{4}{3}\pi r^3}{\pi^2 r^2} = \frac{4r}{3\pi}$$

$$\left[\text{By integration, } \overline{y} = \frac{\frac{1}{2}\int_{-r}^{r} y^2\,dx}{\text{area}} = \frac{\frac{1}{2}\int_{-r}^{r}(r^2 - x^2)\,dx}{\frac{\pi r^2}{2}}\right.$$

$$\left. = \frac{\frac{1}{2}\left[r^2x - \frac{x^3}{3}\right]_{-r}^{r}}{\frac{\pi r^2}{2}} = \frac{\frac{1}{2}\left[\left(r^3 - \frac{r^3}{3}\right) - \left(-r^3 + \frac{r^3}{3}\right)\right]}{\frac{\pi r^2}{2}} = \frac{4r}{3\pi}\right]$$

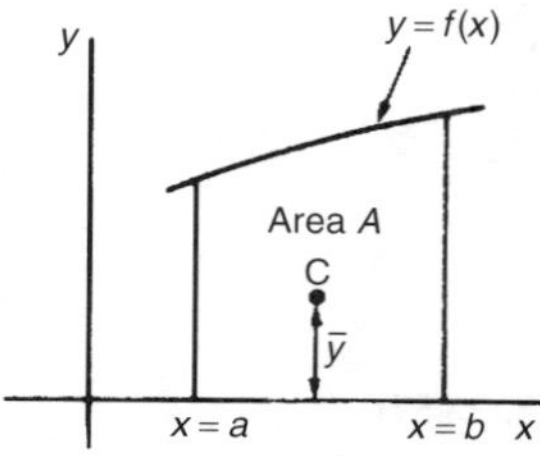

Figure 68.4

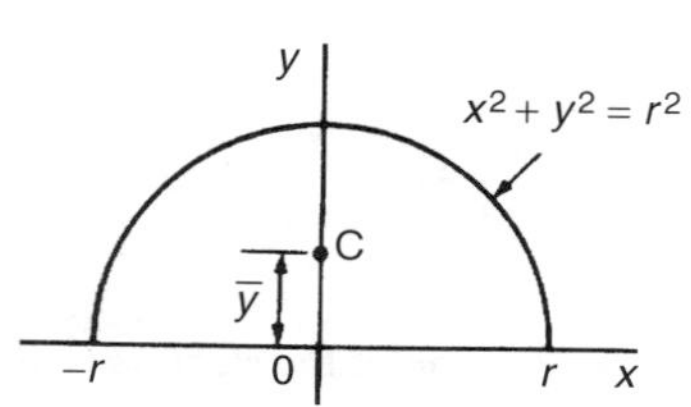

Figure 68.5

Hence the centroid of a semicircle lies on the axis of symmetry, distance $\frac{4r}{3\pi}$ (or 0.424 r) from its diameter.

In another example: (a) To calculate the area bounded by the curve $y = 2x^2$, the x-axis and ordinates $x = 0$ and $x = 3$, (b) if the area in part (a) is revolved (i) about the x-axis and (ii) about the y-axis, to find the volumes of the solids produced, and (c) to locate the position of the centroid using (i) integration, and (ii) the theorem of Pappus:

(a) The required area is shown shaded in Figure 68.6.

$$\text{Area} = \int_0^3 y\,dx = \int_0^3 2x^2\,dx = \left[\frac{2x^3}{3}\right]_0^3 = \mathbf{18\ square\ units}$$

(b) (i) When the shaded area of Figure 68.6 is revolved 360° about the x-axis, the volume generated

$$= \int_0^3 \pi y^2\,dx = \int_0^3 \pi(2x^2)^2\,dx = \int_0^3 4\pi x^4\,dx$$

$$= 4\pi\left[\frac{x^5}{5}\right]_0^3 = 4\pi\left(\frac{243}{5}\right) = \mathbf{194.4\pi\ cubic\ units}$$

(ii) When the shaded area of Figure 68.6 is revolved 360° about the y-axis, the volume generated = (volume generated by $x = 3$)

−(volume generated by $y = 2x^2$)

$$= \int_0^{18} \pi(3)^2\,dy - \int_0^{18} \pi\left(\frac{y}{2}\right)dy$$

$$= \pi\int_0^{18}\left(9 - \frac{y}{2}\right)dy = \pi\left[9y - \frac{y^2}{4}\right]_0^{18}$$

$= \mathbf{81\pi\ cubic\ units}$

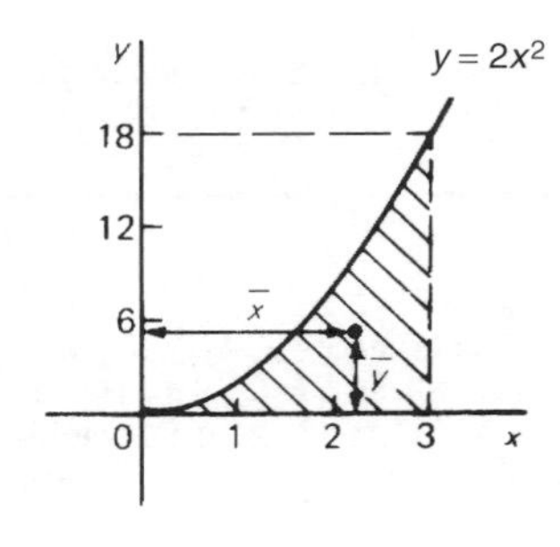

Figure 68.6

(c) If the co-ordinates of the centroid of the shaded area in Figure 68.6 are $(\overline{x}, \overline{y})$ then:

(i) by integration,

$$\overline{x} = \frac{\int_0^3 xy\,dx}{\int_0^3 y\,dx} = \frac{\int_0^3 x(2x^2)\,dx}{18} = \frac{\int_0^3 2x^3\,dx}{18} = \frac{\left[\frac{2x^4}{4}\right]_0^3}{18} = \frac{81}{36} = \mathbf{2.25}$$

$$\overline{y} = \frac{\frac{1}{2}\int_0^3 y^2\,dx}{\int_0^3 y\,dx} = \frac{\frac{1}{2}\int_0^3 (2x^2)^2\,dx}{18} = \frac{\frac{1}{2}\int_0^3 4x^4\,dx}{18} = \frac{\frac{1}{2}\left[\frac{4x^5}{5}\right]_0^3}{18} = \mathbf{5.4}$$

(ii) using the theorem of Pappus:

Volume generated when shaded area is revolved about OY $= (\text{area})(2\pi\overline{x})$

i.e. $81\pi = (18)(2\pi\overline{x})$, from which, $\overline{x} = \dfrac{81\pi}{36\pi} = \mathbf{2.25}$

Volume generated when shaded area is revolved about OX $= (\text{area})(2\pi\overline{y})$

i.e. $194.4\pi = (18)(2\pi\overline{y})$, from which, $\overline{y} = \dfrac{194.4\pi}{36\pi} = \mathbf{5.4}$

Hence the centroid of the shaded area in Figure 68.6 is at (2.25, 5.4)

---

69 Second Moments of Area of Regular Sections

Moments of area

The **first moment of area** about a fixed axis of a lamina of area A, perpendicular distance y from the centroid of the lamina is defined as Ay cubic units. The **second moment of area** of the same lamina as above is given by Ay^2, i.e. the perpendicular distance from the centroid of the area to the fixed axis is squared.

Second moments of areas are usually denoted by I and have units of mm^4, cm^4, and so on.

Radius of gyration

Several areas, $a_1, a_2, a_3, \ldots$ at distances $y_1, y_2, y_3, \ldots$ from a fixed axis, may be replaced by a single area A, where $A = a_1 + a_2 + a_3 + \ldots$ at distance k from the axis, such that $Ak^2 = \sum ay^2$. k is called the **radius of gyration** of area A about the given axis. Since $Ak^2 = \sum ay^2 = I$ then the radius of gyration, $\boldsymbol{k = \sqrt{\frac{I}{A}}}$

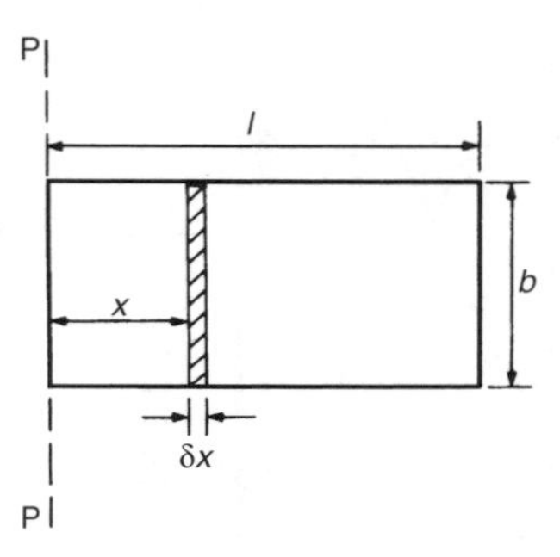

Figure 69.1

The second moment of area is a quantity much used in the theory of bending of beams, in the torsion of shafts, and in calculations involving water planes and centres of pressure.

The procedure to determine the second moment of area of regular sections about a given axis is (i) to find the second moment of area of a typical element and (ii) to sum all such second moments of area by integrating between appropriate limits. For example, the second moment of area of the rectangle shown in Figure 69.1 about axis PP is found by initially considering an elemental strip of width δx, parallel to and distance x from axis PP. Area of shaded strip $= b\delta x$. Second moment of area of the shaded strip about PP $= (x^2)(b\delta x)$.

The second moment of area of the whole rectangle about PP is obtained by summing all such strips between $x = 0$ and $x = l$, i.e. $\sum_{x=0}^{x=l} x^2 b\delta x$

It is a fundamental theorem of integration that $\lim_{\delta x \to 0} \sum_{x=0}^{x=l} x^2 b\delta x = \int_0^l x^2 b\,\mathrm{d}x$

Thus the second moment of area of the rectangle about PP

$$= b\int_0^l x^2\,\mathrm{d}x = b\left[\frac{x^3}{3}\right]_0^l = \boldsymbol{\frac{bl^3}{3}}$$

Since the total area of the rectangle, $A = lb$, then $I_{pp} = (lb)\left(\frac{l^2}{3}\right) = \boldsymbol{\frac{Al^2}{3}}$

$I_{pp} = Ak_{pp}^2$ thus $k_{pp}^2 = \frac{l^2}{3}$

i.e. the radius of gyration about axis PP, $\boldsymbol{k_{pp}} = \sqrt{\frac{l^2}{3}} = \boldsymbol{\frac{l}{\sqrt{3}}}$

Parallel axis theorem

In Figure 69.2, axis GG passes through the centroid C of area A. Axes DD and GG are in the same plane, are parallel to each other and distance d apart.

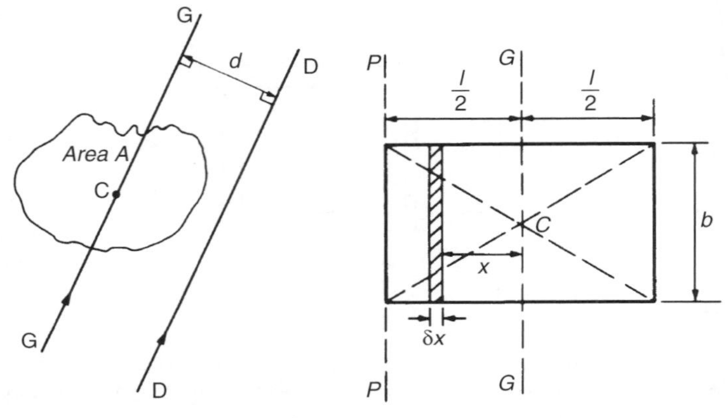

Figure 69.2 **Figure 69.3**

The parallel axis theorem states:

$$I_{DD} = I_{GG} + Ad^2$$

Using the parallel axis theorem the second moment of area of a rectangle about an axis through the centroid may be determined. In the rectangle shown in Figure 69.3, $I_{pp} = \dfrac{bl^3}{3}$ (from above)

From the parallel axis theorem $I_{pp} = I_{GG} + (bl)\left(\dfrac{l}{2}\right)^2$

i.e. $\dfrac{bl^3}{3} = I_{GG} + \dfrac{bl^3}{4}$ from which, $\boldsymbol{I_{GG}} = \dfrac{bl^3}{3} - \dfrac{bl^3}{4} = \boldsymbol{\dfrac{bl^3}{12}}$

Perpendicular axis theorem

In Figure 69.4, axes OX, OY and OZ are mutually perpendicular. If OX and OY lie in the plane of area A then the perpendicular axis theorem states:

$$I_{OZ} = I_{OX} + I_{OY}$$

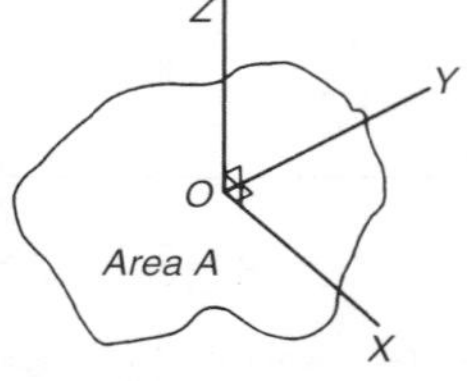

Figure 69.4

Figure 69.5

A summary of derived standard results for the second moment of area and radius of gyration of regular sections are listed in Table 69.1.

For example, to determine the second moment of area and the radius of gyration about axes AA, BB and CC for the rectangle shown in Figure 69.5:
From Table 69.1, the second moment of area about axis AA,

$$\boldsymbol{I_{AA}} = \frac{bl^3}{3} = \frac{(4.0)(12.0)^3}{3} = \mathbf{2304\ cm^4}$$

Radius of gyration, $$\mathbf{k_{AA}} = \frac{l}{\sqrt{3}} = \frac{12.0}{\sqrt{3}} = \mathbf{6.93\ cm}$$

Similarly, $$\boldsymbol{I_{BB}} = \frac{lb^3}{3} = \frac{(12.0)(4.0)^3}{3} = \mathbf{256\ cm^4}$$

and $$\boldsymbol{k_{BB}} = \frac{b}{\sqrt{3}} = \frac{4.0}{\sqrt{3}} = \mathbf{2.31\ cm}$$

Table 69.1 Summary of standard results of the second moments of areas of regular sections

Shape	*Position of axis*	*Second moment of area, I*	*Radius of gyration, k*
Rectangle	(1) Coinciding with b	$\frac{bl^3}{3}$	$\frac{l}{\sqrt{3}}$
length l	(2) Coinciding with l	$\frac{lb^3}{3}$	$\frac{b}{\sqrt{3}}$
breadth b	(3) Through centroid, parallel to b	$\frac{bl^3}{12}$	$\frac{l}{\sqrt{12}}$
	(4) Through centroid, parallel to l	$\frac{lb^3}{12}$	$\frac{b}{\sqrt{12}}$
Triangle	(1) Coinciding with b	$\frac{bh^3}{12}$	$\frac{h}{\sqrt{6}}$
Perpendicular	(2) Through centroid, parallel to base	$\frac{bh^3}{36}$	$\frac{h}{\sqrt{18}}$
height h, base b	(3) Through vertex, parallel to base	$\frac{bh^3}{4}$	$\frac{h}{\sqrt{2}}$
Circle radius r	(1) Through centre, perpendicular to plane (i.e. polar axis)	$\frac{\pi r^4}{2}$	$\frac{r}{\sqrt{2}}$
	(2) Coinciding with diameter	$\frac{\pi r^4}{4}$	$\frac{r}{2}$
	(3) About a tangent	$\frac{5\pi r^4}{4}$	$\frac{\sqrt{5}}{2}$ r
Semicircle radius r	Coinciding with diameter	$\frac{\pi r^4}{8}$	$\frac{r}{2}$

The second moment of area about the centroid of a rectangle is $\frac{bl^3}{12}$ when the axis through the centroid is parallel with the breadth b. In this case, the axis CC is parallel with the length l

Hence $\quad I_{CC} = \frac{lb^3}{12} = \frac{(12.0)(4.0)^3}{12} = \mathbf{64\ cm^4}$

and $\quad k_{CC} = \frac{b}{\sqrt{12}} = \frac{4.0}{\sqrt{12}} = \mathbf{1.15\ cm}$

In another example, to find the second moment of area and the radius of gyration about axis PP for the rectangle shown in Figure 69.6:

$$I_{GG} = \frac{lb^3}{12} \text{ where } l = 40.0 \text{ mm and } b = 15.0 \text{ mm}$$

Hence $\quad I_{GG} = \frac{(40.0)(15.0)^3}{12} = 11250\ \text{mm}^4$

From the parallel axis theorem, $I_{PP} = I_{GG} + Ad^2$, where $A = 40.0 \times 15.0 = 600\ \text{mm}^2$ and $d = 25.0 + 7.5 = 32.5$ mm, the perpendicular distance between GG and PP. Hence $\boldsymbol{I_{PP}} = 11\,250 + (600)(32.5)^2 = \mathbf{645\,000\ mm^4}$

$$I_{PP} = Ak_{PP}^2,$$

from which, $\quad \mathbf{k_{PP}} = \sqrt{\frac{I_{PP}}{\text{area}}} = \sqrt{\left(\frac{645\,000}{600}\right)} = \mathbf{32.79\ mm}$

In another example, to determine the second moment of area and radius of gyration about axis QQ of the triangle BCD shown in Figure 69.7:

Using the parallel axis theorem: $I_{QQ} = I_{GG} + Ad^2$, where I_{GG} is the second moment of area about the centroid of the triangle,

i.e. $\quad \frac{bh^3}{36} = \frac{(8.0)(12.0)^3}{36} = 384\ \text{cm}^4,$

A is the area of the triangle $= \frac{1}{2}bh = \frac{1}{2}(8.0)(12.0) = 48\ \text{cm}^2$

and d is the distance between axes GG and QQ $= 6.0 + \frac{1}{3}(12.0) = 10$ cm

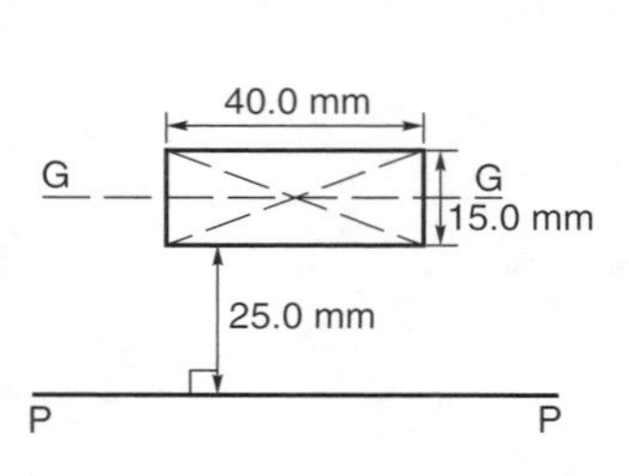

Figure 69.6

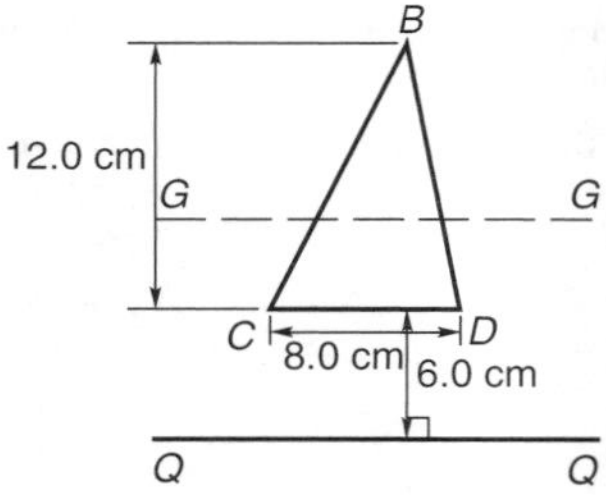

Figure 69.7

Hence the second moment of area about axis QQ,

$$I_{QQ} = 384 + (48)(10)^2 = \mathbf{5184\ cm^4}$$

Radius of gyration, $$k_{QQ} = \sqrt{\frac{I_{QQ}}{\text{area}}} = \sqrt{\left(\frac{5184}{48}\right)} = \mathbf{10.4\ cm}$$

In another example, to determine the polar second moment of area of the propeller shaft cross-section shown in Figure 69.8:

The polar second moment of area of a circle $= \dfrac{\pi r^4}{2}$

The polar second moment of area of the shaded area is given by the polar second moment of area of the 7.0 cm diameter circle minus the polar second moment of area of the 6.0 cm diameter circle.

Hence the polar second moment of area of the cross-section shown

$$= \frac{\pi}{2}\left(\frac{7.0}{2}\right)^4 - \frac{\pi}{2}\left(\frac{6.0}{2}\right)^4 = 235.7 - 127.2 = \mathbf{108.5\ cm^4}$$

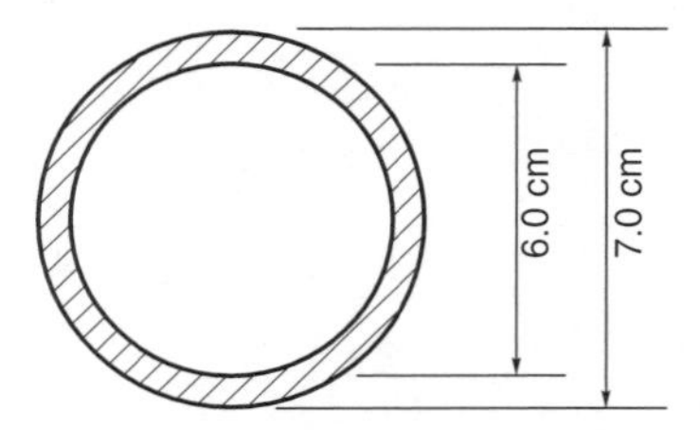

Figure 69.8

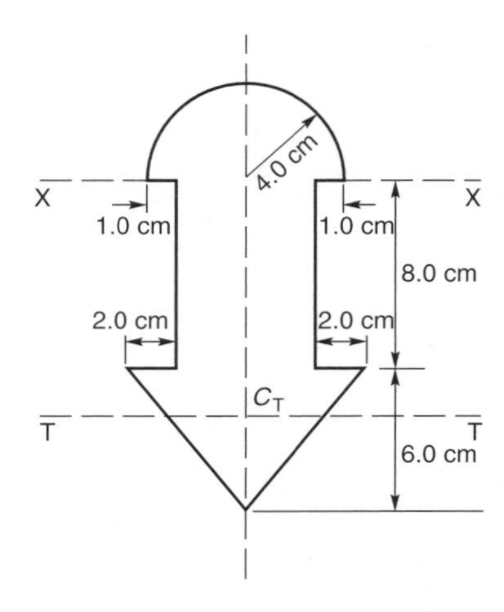

Figure 69.9

In another example, to determine correct to 3 significant figures, the second moment of area about axis XX for the composite area shown in Figure 69.9:

For the semicircle, $I_{XX} = \dfrac{\pi r^4}{8} = \dfrac{\pi(4.0)^4}{8} = 100.5 \text{ cm}^4$

For the rectangle, $I_{XX} = \dfrac{bl^3}{3} = \dfrac{(6.0)(8.0)^3}{3} = 1024 \text{ cm}^4$

For the triangle, about axis TT through centroid C_T,

$$I_{TT} = \frac{bh^3}{36} = \frac{(10)(6.0)^3}{36} = 60 \text{ cm}^4$$

By the parallel axis theorem, the second moment of area of the triangle about axis XX $= 60 + \left[\frac{1}{2}(10)(6.0)\right]\left[8.0 + \frac{1}{3}(6.0)\right]^2 = 3060 \text{ cm}^4$

Total second moment of area about XX $= 100.5 + 1024 + 3060 = 4184.5$

$= \mathbf{4180 \text{ cm}^4}$, correct to 3 significant figures

---

Differential Equations

70 Solution of First Order Differential Equations by Separation of Variables

Family of curves

Integrating both sides of the derivative $\frac{dy}{dx} = 3$ with respect to x gives $y = \int 3\,dx$, i.e. $y = 3x + c$, where c is an arbitrary constant. $y = 3x + c$ represents a **family of curves**, each of the curves in the family depending on the value of c. Examples include $y = 3x + 8$, $y = 3x + 3$, $y = 3x$ and $y = 3x - 10$ and these are shown in Figure 70.1. Each are straight lines of gradient 3. A particular curve of a family may be determined when a point on the curve is specified. Thus, if $y = 3x + c$ passes through the point (1, 2) then $2 = 3(1) + c$, from which, $c = -1$. The equation of the curve passing through (1, 2) is therefore $y = 3x - 1$.

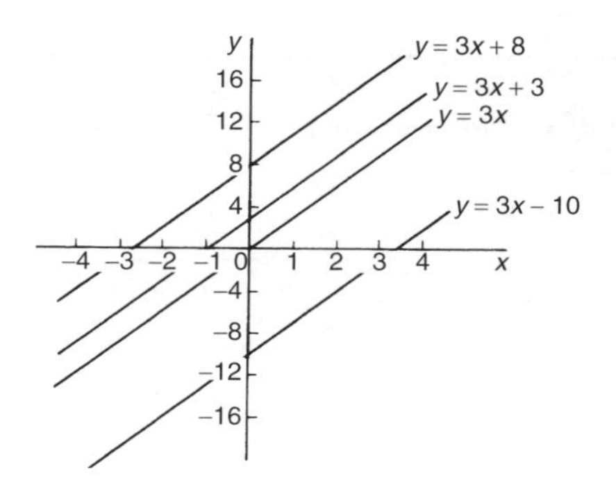

Figure 70.1

Differential equations

A **differential equation** is one that contains differential coefficients. Examples include (i) $\frac{dy}{dx} = 7x$ and (ii) $\frac{d^2y}{dx^2} + 5\frac{dy}{dx} + 2y = 0$

Differential equations are classified according to the highest derivative that occurs in them. Thus example (i) above is a **first order differential equation**, and example (ii) is a **second order differential equation**.

The **degree** of a differential equation is that of the highest power of the highest differential which the equation contains after simplification.

Thus $\left(\frac{d^2x}{dt^2}\right)^3 + 2\left(\frac{dx}{dt}\right)^5 = 7$ is a second order differential equation of degree three.

Starting with a differential equation it is possible, by integration and by being given sufficient data to determine unknown constants, to obtain the original function. This process is called **'solving the differential equation'**.

A solution to a differential equation that contains one or more arbitrary constants of integration is called the **general solution** of the differential equation. When additional information is given so that constants may be calculated the **particular solution** of the differential equation is obtained. The additional information is called **boundary conditions**. It was shown above that $y = 3x + c$ is the general solution of the differential equation $\frac{dy}{dx} = 3$. Given the boundary conditions $x = 1$ and $y = 2$, produces the particular solution of $y = 3x - 1$.

Equations which can be written in the form $\frac{dy}{dx} = f(x)$, $\frac{dy}{dx} = f(y)$ and $\frac{dy}{dx} = f(x).f(y)$ can all be solved by integration. In each case it is possible to separate the y's to one side of the equation and the x's to the other. Solving such equations is therefore known as solution by **separation of variables.**

The solution of equations of the form $\frac{dy}{dx} = f(x)$

A differential equation of the form $\frac{dy}{dx} = f(x)$ is solved by direct integration,

i.e. $$\boxed{y = \int f(x)\,dx}$$

For example, to find the particular solution of the differential equation $5\frac{dy}{dx} + 2x = 3$, given the boundary conditions $y = 1\frac{2}{5}$ when $x = 2$:

Since $5\frac{dy}{dx} + 2x = 3$ then $\frac{dy}{dx} = \frac{3 - 2x}{5} = \frac{3}{5} - \frac{2x}{5}$

Hence $$y = \int \left(\frac{3}{5} - \frac{2x}{5}\right) dx$$

i.e. $y = \frac{3x}{5} - \frac{x^2}{5} + c$, which is the general solution.

Substituting the boundary conditions $y = 1\frac{2}{5}$ and $x = 2$ to evaluate c gives:

$$1\tfrac{2}{5} = \tfrac{6}{5} - \tfrac{4}{5} + c, \text{ from which, } c = 1.$$

Hence the particular solution is $\boldsymbol{y = \frac{3x}{5} - \frac{x^2}{5} + 1}$

The solution of equations of the form $\dfrac{dy}{dx} = f(y)$

A differential equation of the form $\dfrac{dy}{dx} = f(y)$ is initially rearranged to give $dx = \dfrac{dy}{f(y)}$ and then the solution is obtained by direct integration,

i.e. $$\boxed{\int dx = \int \frac{dy}{f(y)}}$$

For example, to determine the particular solution of $(y^2 - 1)\dfrac{dy}{dx} = 3y$ given that $y = 1$ when $x = 2\frac{1}{6}$:

Rearranging gives: $$dx = \left(\frac{y^2 - 1}{3y}\right)dy = \left(\frac{y}{3} - \frac{1}{3y}\right)dy$$

Integrating gives: $$\int dx = \int \left(\frac{y}{3} - \frac{1}{3y}\right)dy$$

i.e. $$x = \frac{y^2}{6} - \frac{1}{3}\ln y + c,$$ which is the general solution

When $y = 1$, $x = 2\dfrac{1}{6}$, thus $2\dfrac{1}{6} = \dfrac{1}{6} - \dfrac{1}{3}\ln 1 + c$, from which, $c = 2$

Hence the particular solution is: $\boldsymbol{x = \dfrac{y^2}{6} - \dfrac{1}{3}\ln y + 2}$

The solution of equations of the form $\dfrac{dy}{dx} = f(x).f(y)$

A differential equation of the form $\dfrac{dy}{dx} = f(x).f(y)$, where $f(x)$ is a function of x only and $f(y)$ is a function of y only, may be rearranged as $\dfrac{dy}{f(y)} = f(x)\,dx$, and then the solution is obtained by direct integration, i.e.

$$\boxed{\int \frac{dy}{f(y)} = \int f(x)\,dx}$$

For example, to solve the equation $4xy\dfrac{dy}{dx} = y^2 - 1$:

Separating the variables gives: $$\left(\frac{4y}{y^2 - 1}\right)dy = \frac{1}{x}\,dx$$

Integrating both sides gives: $$\int \left(\frac{4y}{y^2 - 1}\right)dy = \int \left(\frac{1}{x}\right)dx$$

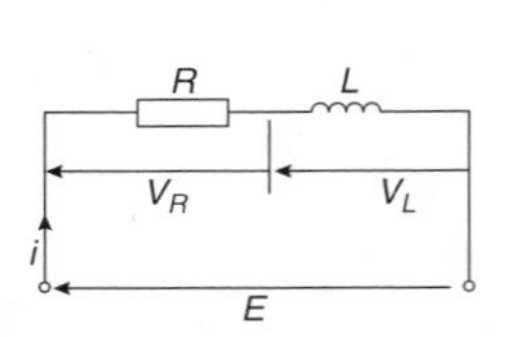

Figure 70.2

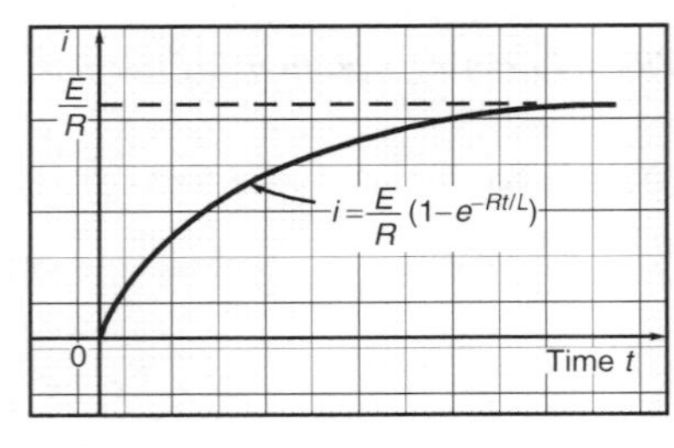

Figure 70.3

Using the substitution $u = y^2 - 1$, the general solution is:

$$\mathbf{2\ln(y^2 - 1) = \ln x + c}$$

In another example, the current i in an electric circuit containing resistance R and inductance L in series with a constant voltage source E is given by the differential equation $E - L\left(\frac{di}{dt}\right) = Ri$. Solving the equation to find i in terms of time t, given that when $t = 0$, $i = 0$:

In the R-L series circuit shown in Figure 70.2, the supply p.d., E, is given by

$$E = V_R + V_L$$

$V_R = iR$ and $V_L = L\frac{di}{dt}$

Hence $\quad E = iR + L\frac{di}{dt} \quad$ from which $\quad E - L\frac{di}{dt} = Ri$

Most electrical circuits can be reduced to a differential equation.

Rearranging $E - L\frac{di}{dt} = Ri$ gives $\qquad \frac{di}{dt} = \frac{E - Ri}{L}$

and separating the variables gives: $\qquad \frac{di}{E - Ri} = \frac{dt}{L}$

Integrating both sides gives: $\qquad \int \frac{di}{E - Ri} = \int \frac{dt}{L}$

Hence the general solution is: $\qquad -\frac{1}{R}\ln(E - Ri) = \frac{t}{L} + c$

(by making a substitution $u = E - Ri$, see Chapter 58)

When $t = 0$, $i = 0$, thus $-\frac{1}{R}\ln E = c$

Thus the particular solution is: $\qquad -\frac{1}{R}\ln(E - Ri) = \frac{t}{L} - \frac{1}{R}\ln E$

Transposing gives: $\qquad -\frac{1}{R}\ln(E - Ri) + \frac{1}{R}\ln E = \frac{t}{L}$

$$\frac{1}{R}[\ln E - \ln(E - Ri)] = \frac{t}{L}$$

$$\ln\left(\frac{E}{E - Ri}\right) = \frac{Rt}{L} \text{ from which } \frac{E}{E - Ri} = e^{Rt/L}$$

Hence $\dfrac{E - Ri}{E} = e^{-Rt/L}$ and $E - Ri = Ee^{-Rt/L}$ and $Ri = E - Ee^{-Rt/L}$

Hence current, $\boldsymbol{i = \frac{E}{R}(1 - e^{-Rt/L})}$, which represents the law of growth of current in an inductive circuit as shown in Figure 70.3.

71 Homogeneous First Order Differential Equations

Introduction

Certain first order differential equations are not of the 'variable-separable' type but can be made separable by changing the variable.

An equation of the form $P\dfrac{dy}{dx} = Q$, where P and Q are functions of both x and y of the same degree throughout, is said to be **homogeneous** in y and x. For example, $f(x, y) = x^2 + 3xy + y^2$ is a homogeneous function since each of the three terms are of degree 2. Similarly, $f(x, y) = \dfrac{x - 3y}{2x + y}$ is homogeneous in x and y since each of the four terms are of degree 1. However, $f(x, y) = \dfrac{x^2 - y}{2x^2 + y^2}$ is not homogeneous since the term in y in the numerator is of degree 1 and the other three terms are of degree 2.

Procedure to solve differential equations of the form $P\dfrac{dy}{dx} = Q$

(i) Rearrange $P\dfrac{dy}{dx} = Q$ into the form $\dfrac{dy}{dx} = \dfrac{Q}{P}$

(ii) Make the substitution $y = vx$ (where v is a function of x), from which,
$$\frac{dy}{dx} = v(1) + x\frac{dv}{dx} \text{ by the product rule.}$$

(iii) Substitute for both y and $\dfrac{dy}{dx}$ in the equation $\dfrac{dy}{dx} = \dfrac{Q}{P}$. Simplify, by cancelling, and an equation results in which the variables are separable.

(iv) Separate the variables and solve using the method shown in Chapter 70.

(v) Substitute $v = \dfrac{y}{x}$ to solve in terms of the original variables.

For example, to determine the particular solution of the equation $x\dfrac{dy}{dx} = \dfrac{x^2 + y^2}{y}$, given the boundary conditions that $x = 1$ when $y = 4$:

Using the above procedure:

(i) Rearranging $x\frac{dy}{dx} = \frac{x^2 + y^2}{y}$ gives $\frac{dy}{dx} = \frac{x^2 + y^2}{xy}$ which is homogeneous in x and y since each of the three terms on the right hand side are of the same degree (i.e. degree 2).

(ii) Let $y = vx$ then $\frac{dy}{dx} = v(1) + x\frac{dv}{dx}$

(iii) Substituting for y and $\frac{dy}{dx}$ in the equation $\frac{dy}{dx} = \frac{x^2 + y^2}{xy}$ gives:

$$v + x\frac{dv}{dx} = \frac{x^2 + (vx)^2}{x(vx)} = \frac{x^2 + v^2x^2}{vx^2} = \frac{1 + v^2}{v}$$

(iv) Separating the variables give:

$$x\frac{dv}{dx} = \frac{1 + v^2}{v} - v = \frac{1 + v^2 - v^2}{v} = \frac{1}{v}$$

Hence, $\quad v\,dv = \frac{1}{x}\,dx$

Integrating both sides gives: $\int v\,dv = \int \frac{1}{x}\,dx$ i.e. $\frac{v^2}{2} = \ln x + c$

(v) Replacing v by $\frac{y}{x}$ gives: $\frac{y^2}{2x^2} = \ln x + c$, which is the general solution.

When $x = 1$, $y = 4$, thus: $\frac{16}{2} = \ln 1 + c$, from which, $c = 8$

Hence, **the particular solution is:** $\boldsymbol{\frac{y^2}{2x^2} = \ln x + 8}$ or $\boldsymbol{y^2 = 2x^2(\ln x + 8)}$

---

72 Linear First Order Differential Equations

Introduction

An equation of the form $\frac{dy}{dx} + Py = Q$, where P and Q are functions of x only is called a **linear differential equation** since y and its derivatives are of the first degree.

The solution of $\frac{dy}{dx} + Py = Q$ is obtained by multiplying throughout by what is termed an **integrating factor**.

Multiplying $\frac{dy}{dx} + Py = Q$ by say R, a function of x only, gives:

$$R\frac{dy}{dx} + RPy = RQ \qquad (1)$$

The differential coefficient of a product Ry is obtained using the product rule, i.e. $\frac{d}{dx}(Ry) = R\frac{dy}{dx} + y\frac{dR}{dx}$, which is the same as the left hand side of equation (1), when R is chosen such that $RP = \frac{dR}{dx}$. If $\frac{dR}{dx} = RP$, then separating the variables gives $\frac{dR}{R} = P\,dx$.

Integrating both sides gives:

$$\int \frac{dR}{R} = \int P\,dx \quad \text{i.e.} \quad \ln R = \int P\,dx + c$$

from which, $R = e^{\int P\,dx + c} = e^{\int P\,dx}e^{c}$

i.e. $R = Ae^{\int P\,dx}$, where $A = e^{c} = a$ constant

Substituting $R = Ae^{\int P\,dx}$ in equation (1) gives:

$$Ae^{\int P\,dx}\left(\frac{dy}{dx}\right) + Ae^{\int P\,dx}Py = Ae^{\int P\,dx}Q$$

i.e. $$e^{\int P\,dx}\left(\frac{dy}{dx}\right) + e^{\int P\,dx}Py = e^{\int P\,dx}Q \qquad (2)$$

The left hand side of equation (2) is $\frac{d}{dx}(ye^{\int P\,dx})$ which may be checked by differentiating $ye^{\int P\,dx}$ with respect to x, using the product rule.

From equation (2), $\frac{d}{dx}(ye^{\int P\,dx}) = e^{\int P\,dx}Q$

Integrating both sides gives: $$\boxed{ye^{\int P\,dx} = \int e^{\int P\,dx}Q\,dx} \qquad (3)$$

$e^{\int P\,dx}$ is the **integrating factor**.

Procedure to solve differential equations of the form $\frac{dy}{dx} + Py = Q$

(i) Rearrange the differential equation into the form $\frac{dy}{dx} + Py = Q$, where P and Q are functions of x

(ii) Determine $\int P\,dx$

(iii) Determine the integrating factor $e^{\int P\,dx}$

(iv) Substitute $e^{\int P\,dx}$ into equation (3)

(v) Integrate the right hand side of equation (3) to give the general solution of the differential equation. Given boundary conditions, the particular solution may be determined.

For example, to solve the differential equation $\frac{1}{x}\frac{dy}{dx} + 4y = 2$, given the boundary conditions $x = 0$ when $y = 4$:

Using the above procedure:

(i) Rearranging gives $\frac{dy}{dx} + 4xy = 2x$, which is of the form $\frac{dy}{dx} + Py = Q$, where $P = 4x$ and $Q = 2x$

(ii) $\int P\,dx = \int 4x\,dx = 2x^2$

(iii) Integrating factor, $e^{\int P\,dx} = e^{2x^2}$

(iv) Substituting into equation (3) gives: $ye^{2x^2} = \int e^{2x^2}(2x)\,dx$

(v) Hence the general solution is: $ye^{2x^2} = \frac{1}{2}e^{2x^2} + c$, by using the substitution $u = 2x^2$

When $x = 0$, $y = 4$, thus $4e^0 = \frac{1}{2}e^0 + c$, from which, $c = \frac{7}{2}$

Hence the particular solution is: $ye^{2x^2} = \frac{1}{2}e^{2x^2} + \frac{7}{2}$

i.e. $\boldsymbol{y = \frac{1}{2} + \frac{7}{2}e^{-2x^2}}$ or $\boldsymbol{y = \frac{1}{2}(1 + 7e^{-2x^2})}$

---

73 Second Order Differential Equations of the Form $a\frac{d^2y}{dx^2} + b\frac{dy}{dx} + cy = 0$

Introduction

An equation of the form a $\frac{d^2y}{dx^2} + b\frac{dy}{dx} + cy = 0$, where a, b and c are constants, is called a **linear second order differential equation with constant coefficients.** When the right-hand side of the differential equation is zero, it is referred to as a **homogeneous differential equation.** When the right-hand side is not equal to zero (as in Chapter 74) it is referred to as a **non-homogeneous differential equation**.

There are numerous engineering examples of second order differential equations. Two examples are:

(i) $L\frac{d^2q}{dt^2} + R\frac{dq}{dt} + \frac{1}{C}q = 0$, representing an equation for charge q in an electrical circuit containing resistance R, inductance L and capacitance C in series.

(ii) $m\frac{d^2s}{dt^2} + a\frac{ds}{dt} + ks = 0$, defining a mechanical system, where s is the distance from a fixed point after t seconds, m is a mass, a the damping factor and k the spring stiffness.

If D represents $\frac{d}{dx}$ and D^2 represents $\frac{d^2}{dx^2}$ then the above equation may be stated as $(aD^2 + bD + c)y = 0$. This equation is said to be in '***D*-operator**' form.

If $y = Ae^{mx}$ then $\frac{dy}{dx} = Ame^{mx}$ and $\frac{d^2y}{dx^2} = Am^2e^{mx}$

Substituting these values into $a\frac{d^2y}{dx^2} + b\frac{dy}{dx} + cy = 0$ gives:

$$a(Am^2e^{mx}) + b(Ame^{mx}) + c(Ae^{mx}) = 0$$

i.e. $$Ae^{mx}(am^2 + bm + c) = 0$$

Thus $y = Ae^{mx}$ is a solution of the given equation provided that

$$(am^2 + bm + c) = 0.$$

$am^2 + bm + c = 0$ is called the **auxiliary equation,** and since the equation is a quadratic, m may be obtained either by factorising or by using the quadratic formula. Since, in the auxiliary equation, a, b and c are real values, then the equation may have either

(i) two different real roots (when $b^2 > 4ac$)
or (ii) two equal real roots (when $b^2 = 4ac$)
or (iii) two complex roots (when $b^2 < 4ac$)

Procedure to solve differential equations of the form $a\dfrac{d^2y}{dx^2} + b\dfrac{dy}{dx} + cy = 0$

(a) Rewrite the differential equation $a\dfrac{d^2y}{dx^2} + b\dfrac{dy}{dx} + cy = 0$ as $(aD^2 + bD + c)y = 0$

(b) Substitute m for D and solve the auxiliary equation $am^2 + bm + c = 0$ for m

(c) If the roots of the auxiliary equation are:

(i) **real and different,** say $m = \alpha$ and $m = \beta$, then the general solution is

$$\boldsymbol{y = Ae^{\alpha x} + Be^{\beta x}}$$

(ii) **real and equal**, say $m = \alpha$ twice, then the general solution is

$$\boldsymbol{y = (Ax + B)e^{\alpha x}}$$

(iii) **complex,** say $m = \alpha \pm j\beta$, then the general solution is

$$\boldsymbol{y = e^{\alpha x}\{A\cos\beta x + B\sin\beta x\}}$$

(d) Given boundary conditions, constants A and B, may be determined and the **particular solution** of the differential equation obtained. The particular solution obtained with differential equations may be verified by substituting expressions for y, $\dfrac{dy}{dx}$ and $\dfrac{d^2y}{dx^2}$ into the original equation.

For example, to solve $2\dfrac{d^2y}{dx^2} + 5\dfrac{dy}{dx} - 3y = 0$, given that when $x = 0$, $y = 4$ and $\dfrac{dy}{dx} = 9$:

Using the above procedure:

(a) $2\dfrac{d^2y}{dx^2} + 5\dfrac{dy}{dx} - 3y = 0$ in D-operator form is $(2D^2 + 5D - 3)y = 0$, where $D \equiv \dfrac{d}{dx}$

(b) Substituting m for D gives the auxiliary equation

$$2m^2 + 5m - 3 = 0$$

Factorising gives: $(2m - 1)(m + 3) = 0$,

from which, $m = \frac{1}{2}$ or $m = -3$

(c) Since the roots are real and different the **general solution is** $\boldsymbol{y = Ae^{\frac{1}{2}x} + Be^{-3x}}$

(d) When $x = 0$, $y = 4$, hence $4 = A + B$ (1)

Since $y = Ae^{\frac{1}{2}x} + Be^{-3x}$ then $\dfrac{dy}{dx} = \frac{1}{2}Ae^{\frac{1}{2}x} - 3Be^{-3x}$

When $x = 0$, $\dfrac{dy}{dx} = 9$ thus $9 = \frac{1}{2}A - 3B$ (2)

Solving the simultaneous equations (1) and (2) gives $A = 6$ and $B = -2$

Hence the particular solution is $\boldsymbol{y = 6e^{\frac{1}{2}x} - 2e^{-3x}}$

In another example, to solve $9\dfrac{d^2y}{dt^2} - 24\dfrac{dy}{dt} + 16y = 0$ given that when $t = 0$, $y = \dfrac{dy}{dt} = 3$:

(a) $9\dfrac{d^2y}{dt^2} - 24\dfrac{dy}{dt} + 16y = 0$ in D-operator form is $(9D^2 - 24D + 16)y = 0$ where $D \equiv \dfrac{d}{dt}$

(b) Substituting m for D gives the auxiliary equation

$$9m^2 - 24m + 16 = 0$$

Factorising gives: $(3m - 4)(3m - 4) = 0$, i.e. $m = \frac{4}{3}$ twice.

(c) Since the roots are real and equal, **the general solution is** $\boldsymbol{y = (At + B)e^{\frac{4}{3}t}}$

(d) When $t = 0$, $y = 3$ hence $3 = (0 + B)e^0$, i.e. $B = 3$

Since $y = (At + B)e^{\frac{4}{3}t}$ then $\dfrac{dy}{dt} = (At + B)(\frac{4}{3}e^{\frac{4}{3}t}) + Ae^{\frac{4}{3}t}$,

by the product rule.

When $t = 0$, $\dfrac{dy}{dt} = 3$ thus $3 = (0 + B)\frac{4}{3}e^0 + Ae^0$

i.e. $3 = \frac{4}{3}B + A$ from which, $A = -1$, since $B = 3$

Hence the particular solution is $\boldsymbol{y = (-t + 3)e^{\frac{4}{3}t}}$ or $\boldsymbol{y = (3 - t)e^{\frac{4}{3}t}}$

In another example, to solve $\dfrac{d^2y}{dx^2} + 6\dfrac{dy}{dx} + 13y = 0$, given that when $x = 0$, $y = 3$ and $\dfrac{dy}{dx} = 7$:

(a) $\dfrac{d^2y}{dx^2} + 6\dfrac{dy}{dx} + 13y = 0$ in D-operator form is $(D^2 + 6D + 13)y = 0$, where $D \equiv \dfrac{d}{dx}$

(b) Substituting m for D gives the auxiliary equation $m^2 + 6m + 13 = 0$
Using the quadratic formula:

$$m = \frac{-6 \pm \sqrt{[(6)^2 - 4(1)(13)]}}{2(1)} = \frac{-6 \pm \sqrt{(-16)}}{2}$$

i.e. $$m = \frac{-6 \pm j4}{2} = -3 \pm j2$$

(c) Since the roots are complex, **the general solution is** $\boldsymbol{y = e^{-3x}(A\cos 2x + B\sin 2x)}$

(d) When $x = 0$, $y = 3$, hence $3 = e^0(A\cos 0 + B\sin 0)$, i.e. $A = 3$.

Since $y = e^{-3x}(A\cos 2x + B\sin 2x)$

then
$$\frac{dy}{dx} = e^{-3x}(-2A\sin 2x + 2B\cos 2x) - 3e^{-3x}(A\cos 2x + B\sin 2x), \text{ by the product rule,}$$
$$= e^{-3x}[(2B - 3A)\cos 2x - (2A + 3B)\sin 2x]$$

When $x = 0$, $\frac{dy}{dx} = 7$,

hence $7 = e^0[(2B - 3A)\cos 0 - (2A + 3B)\sin 0]$

i.e. $7 = 2B - 3A$, from which, $B = 8$, since $A = 3$

Hence the particular solution is $\boldsymbol{y = e^{-3x}(3\cos 2x + 8\sin 2x)}$

---

74 Second Order Differential Equations of the Form $a\frac{d^2y}{dx^2} + b\frac{dy}{dx} + cy = f(x)$

Complementary function and particular integral

If in the differential equation

$$a\frac{d^2y}{dx^2} + b\frac{dy}{dx} + cy = f(x) \qquad (1)$$

the substitution $y = u + v$ is made then:

$$a\frac{d^2(u+v)}{dx^2} + b\frac{d(u+v)}{dx} + c(u+v) = f(x)$$

Rearranging gives:

$$\left(a\frac{d^2u}{dx^2} + b\frac{du}{dx} + cu\right) + \left(a\frac{d^2v}{dx^2} + b\frac{dv}{dx} + cv\right) = f(x)$$

If we let $$a\frac{d^2v}{dx^2} + b\frac{dv}{dx} + cv = f(x) \qquad (2)$$

then $$a\frac{d^2u}{dx^2} + b\frac{du}{dx} + cu = 0 \qquad (3)$$

The general solution, u, of equation (3) will contain two unknown constants, as required for the general solution of equation (1). The method of solution of equation (3) is shown in Chapter 73. The function u is called the **complementary function (C.F.)**

If the particular solution, v, of equation (2) can be determined without containing any unknown constants then $y = u + v$ will give the general solution of equation (1). The function v is called the **particular integral (P.I.)**. Hence the general solution of equation (1) is given by:

$$\mathbf{y = C.F. + P.I.}$$

Procedure to solve differential equations of the form $a\frac{d^2y}{dx^2} + b\frac{dy}{dx} + cy = f(x)$

(i) Rewrite the given differential equation as $(aD^2 + bD + c)y = f(x)$

(ii) Substitute m for D, and solve the auxiliary equation $am^2 + bm + c = 0$ for m

(iii) Obtain the complementary function, u, which is achieved using the same procedure as in Chapter 74, page 361.

(iv) To determine the particular integral, v, firstly assume a particular integral which is suggested by $f(x)$, but which contains undetermined coefficients. Table 74.1 gives some suggested substitutions for different functions $f(x)$.

(v) Substitute the suggested P.I. into the differential equation $(aD^2 + bD + c)v = f(x)$ and equate relevant coefficients to find the constants introduced.

(vi) The general solution is given by $y = \text{C.F.} + \text{P.I.}$, i.e. $y = u + v$

(vii) Given boundary conditions, arbitrary constants in the C.F. may be determined and the particular solution of the differential equation obtained.

For example, to solve $2\frac{d^2y}{dx^2} - 11\frac{dy}{dx} + 12y = 3x - 2$:

(i) $2\frac{d^2y}{dx^2} - 11\frac{dy}{dx} + 12y = 3x - 2$ in D-operator form is $(2D^2 - 11D + 12)y = 3x - 2$

(ii) Substituting m for D gives the auxiliary equation

$$2m^2 - 11m + 12 = 0$$

Factorising gives: $(2m - 3)(m - 4) = 0$,

from which, $m = \frac{3}{2}$ or $m = 4$

(iii) Since the roots are real and different, the C.F., $\boldsymbol{u = Ae^{\frac{3}{2}x} + Be^{4x}}$

(iv) Since $f(x) = 3x - 2$ is a polynomial, let the P.I., $v = ax + b$ (see Table 74.1(b))

Table 74.1 Form of particular integral for different functions

Type	*Straightforward cases* *Try as particular integral:*	*'Snag' cases* *Try as particular integral:*
(a) $f(x)$ = a constant	$v = k$	$v = kx$ (used when C.F. contains a constant)
(b) $f(x)$ = polynomial (i.e. $f(x) = L + Mx + Nx^2 + ..$ where any of the coefficients may be zero)	$v = a + bx + cx^2 + ..$	
(c) $f(x)$ = an exponential function (i.e. $f(x) = Ae^{ax}$)	$v = ke^{ax}$	(i) $v = kxe^{ax}$ (used when e^{ax} appears in the C.F.) (ii) $v = kx^2e^{ax}$ (used when e^{ax} **and** xe^{ax} both appear in the C.F.) etc.
(d) $f(x)$ = a sine or cosine function (i.e. $f(x) = a\sin px + b\cos px$ where a or b may be zero)	$v = A\sin px + B\cos px$	$v = x(A\sin px + B\cos px)$ (used when $\sin px$ and/or $\cos px$ appears in the C.F.)
(e) $f(x)$ = a sum e.g. (i) $f(x) = 4x^2 - 3\sin 2x$ (ii) $f(x) = 2 - x + e^{3x}$	(i) $v = ax^2 + bx + c + d\sin 2x + e\cos 2x$ (ii) $v = ax + b + ce^{3x}$	
(f) $f(x)$ = a product e.g. $f(x) = 2e^x \cos 2x$	$v = e^x(A\sin 2x + B\cos 2x)$	

(v) Substituting $v = ax + b$ into $(2D^2 - 11D + 12)v = 3x - 2$ gives:

$$(2D^2 - 11D + 12)(ax + b) = 3x - 2,$$

i.e. $$2D^2(ax + b) - 11D(ax + b) + 12(ax + b) = 3x - 2$$

i.e. $$0 - 11a + 12ax + 12b = 3x - 2$$

Equating the coefficients of x gives: $12a = 3$, from which, $a = \frac{1}{4}$
Equating the constant terms gives: $-11a + 12b = -2$

i.e. $$-11\left(\frac{1}{4}\right) + 12b = -2$$

from which, $$12b = -2 + \frac{11}{4} = \frac{3}{4} \quad \text{i.e.} \quad b = \frac{1}{16}$$

Hence the P.I., $v = ax + b = \mathbf{\frac{1}{4}x + \frac{1}{16}}$

(vi) The general solution is given by $y = u + v$, i.e.

$$\mathbf{y = Ae^{\frac{3}{2}x} + Be^{4x} + \frac{1}{4}x + \frac{1}{16}}$$

In another example, to solve $\frac{d^2y}{dx^2} - 2\frac{dy}{dx} + y = 3e^{4x}$ given that when $x = 0$, $y = -\frac{2}{3}$ and $\frac{dy}{dx} = 4\frac{1}{3}$:

(i) $\frac{d^2y}{dx^2} - 2\frac{dy}{dx} + y = 3e^{4x}$ in D-operator form is $(D^2 - 2D + 1)y = 3e^{4x}$

(ii) Substituting m for D gives the auxiliary equation

$$m^2 - 2m + 1 = 0$$

Factorising gives: $(m - 1)(m - 1) = 0$,

from which, $m = 1$ twice

(iii) Since the roots are real and equal the C.F., $\boldsymbol{u = (Ax + B)e^x}$

(iv) Let the particular integral, $v = ke^{4x}$ (see Table 74.1(c))

(v) Substituting $v = ke^{4x}$ into $(D^2 - 2D + 1)v = 3e^{4x}$ gives:

$$(D^2 - 2D + 1)ke^{4x} = 3e^{4x}$$

i.e. $D^2(ke^{4x}) - 2D(ke^{4x}) + 1(ke^{4x}) = 3e^{4x}$

i.e. $16ke^{4x} - 8ke^{4x} + ke^{4x} = 3e^{4x}$

Hence $9ke^{4x} = 3e^{4x}$, from which, $k = \frac{1}{3}$

Hence the P.I., $\boldsymbol{v = ke^{4x} = \frac{1}{3}e^{4x}}$

(vi) The general solution is given by $y = u + v$,

i.e. $\boldsymbol{y = (Ax + B)e^x + \frac{1}{3}e^{4x}}$

(vii) When $x = 0$, $y = -\frac{2}{3}$ thus $-\frac{2}{3} = (0 + B)e^0 + \frac{1}{3}e^0$, from which, $B = -1$

$$\frac{dy}{dx} = (Ax + B)e^x + e^x(A) + \frac{4}{3}e^{4x}$$

When $x = 0$, $\frac{dy}{dx} = 4\frac{1}{3}$, thus $\frac{13}{3} = B + A + \frac{4}{3}$

from which, $A = 4$, since $B = -1$

Hence the particular solution is: $\boldsymbol{y = (4x - 1)e^x + \frac{1}{3}e^{4x}}$

In another example, to solve $2\frac{d^2y}{dx^2} + 3\frac{dy}{dx} - 5y = 6\sin 2x$:

(i) $2\frac{d^2y}{dx^2} + 3\frac{dy}{dx} - 5y = 6\sin 2x$ in D-operator form is $(2D^2 + 3D - 5)y = 6\sin 2x$

(ii) The auxiliary equation is $2m^2 + 3m - 5 = 0$, from which,

$$(m - 1)(2m + 5) = 0, \text{ i.e. } m = 1 \text{ or } m = -\tfrac{5}{2}$$

(iii) Since the roots are real and different the C.F., $\boldsymbol{u = Ae^x + Be^{-\frac{5}{2}x}}$

(iv) Let the P.I., $v = A\sin 2x + B\cos 2x$ (see Table 74.1(d))

(v) Substituting $v = A\sin 2x + B\cos 2x$ into $(2D^2 + 3D - 5)v = 6\sin 2x$ gives: $(2D^2 + 3D - 5)(A\sin 2x + B\cos 2x) = 6\sin 2x$

$$D(A\sin 2x + B\cos 2x) = 2A\cos 2x - 2B\sin 2x$$

$$D^2(A\sin 2x + B\cos 2x) = D(2A\cos 2x - 2B\sin 2x) = -4A\sin 2x - 4B\cos 2x$$

Hence $(2D^2 + 3D - 5)(A\sin 2x + B\cos 2x) = -8A\sin 2x - 8B\cos 2x$
$+ 6A\cos 2x - 6B\sin 2x - 5A\sin 2x - 5B\cos 2x = 6\sin 2x$

Equating coefficient of $\sin 2x$ gives:

$$-13A - 6B = 6 \quad (1)$$

Equating coefficients of $\cos 2x$ gives:

$$6A - 13B = 0 \quad (2)$$

$6 \times (1)$ gives: $-78A - 36B = 36$ (3)

$13 \times (2)$ gives: $78A - 169B = 0$ (4)

$(3) + (4)$ gives: $-205B = 36$

from which, $B = \dfrac{-36}{205}$

Substituting $B = \dfrac{-36}{205}$ into equation (1) or (2) gives $A = \dfrac{-78}{205}$

Hence the P.I., $\boldsymbol{v = \dfrac{-78}{205}\sin 2x - \dfrac{36}{205}\cos 2x}$

(vi) The general solution, $y = u + v$,

i.e. $\boldsymbol{y = Ae^{x} + Be^{-\frac{5}{2}x} - \dfrac{2}{205}(39\sin 2x + 18\cos 2x)}$

75 Numerical Methods for First Order Differential Equations

Introduction

Not all first order differential equations may be solved using the methods used in Chapters 70 to 72. A number of other analytical methods of solving differential equations exist; however the differential equations that can be solved by such analytical methods is fairly restricted.

Where a differential equation and known boundary conditions are given, an approximate solution may be obtained by applying a **numerical method**. There are a number of such numerical methods available and the simplest of these is called **Euler's method**.

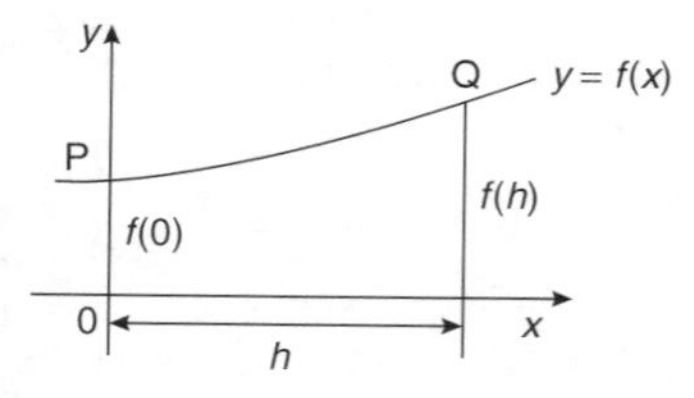

Figure 75.1

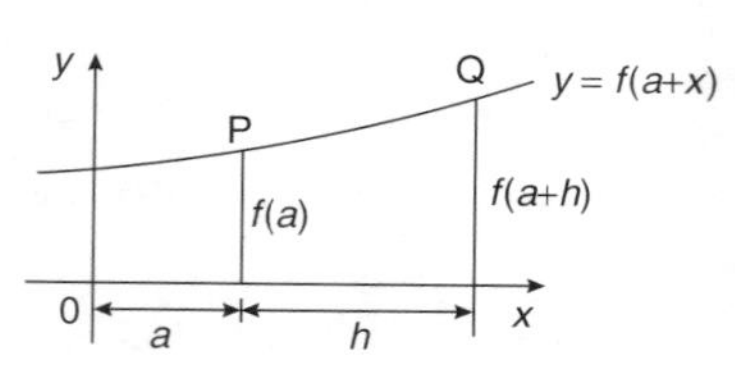

Figure 75.2

Euler's method

From Chapter 17, Maclaurin's series may be stated as:

$$f(x) = f(0) + x\,f'(0) + \frac{x^2}{2!}f''(0) + \ldots$$

Hence at some point $f(h)$ in Figure 75.1:

$$f(h) = f(0) + h\,f'(0) + \frac{h^2}{2!}f''(0) + \ldots$$

If the y-axis and origin are moved a units to the left, as shown in Figure 75.2, the equation of the same curve relative to the new axis becomes $y = f(a + x)$ and the function value at P is $f(a)$.
At point Q in Figure 75.2:

$$\boldsymbol{f(a+h) = f(a) + h\,f'(a) + \frac{h^2}{2!}f''(a) + \ldots} \qquad (1)$$

which is a statement called **Taylor's series**.
If h is the interval between two new ordinates y_0 and y_1, as shown in Figure 75.3, and if $f(a) = y_0$ and $y_1 = f(a + h)$, then Euler's method states:

$$f(a + h) = f(a) + h\,f'(a)$$

i.e. $$\boldsymbol{y_1 = y_0 + h(y')_0} \qquad (2)$$

The approximation used with Euler's method is to take only the first two terms of Taylor's series shown in equation (1).
Hence if y_0, h and $(y')_0$ are known, y_1, which is an approximate value for the function at Q in Figure 75.3, can be calculated.

For example, to obtain a numerical solution of the differential equation $\frac{dy}{dx} = 3(1 + x) - y$ given the initial conditions that $x = 1$ when $y = 4$, for the range $x = 1.0$ to $x = 2.0$ with intervals of 0.2 is determined as follows: $\frac{dy}{dx} = y' = 3(1 + x) - y$

With $x_0 = 1$ and $y_0 = 4$, $\boldsymbol{(y')_0} = 3(1 + 1) - 4 = \mathbf{2}$

By Euler's method: $y_1 = y_0 + h(y')_0$, from equation(2)

Hence $\boldsymbol{y_1} = 4 + (0.2)(2) = \mathbf{4.4}$, since $h = 0.2$

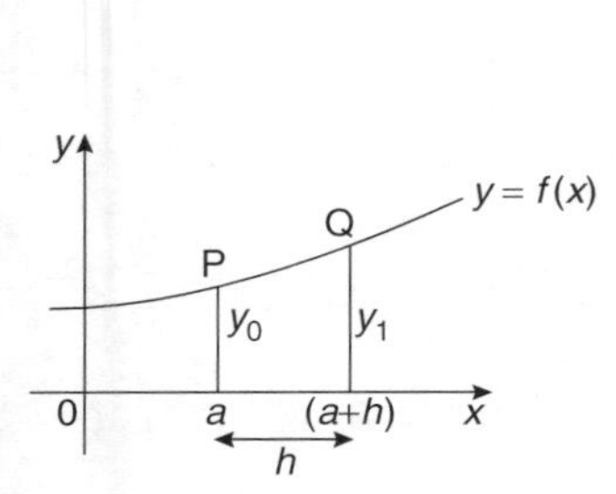

Figure 75.3

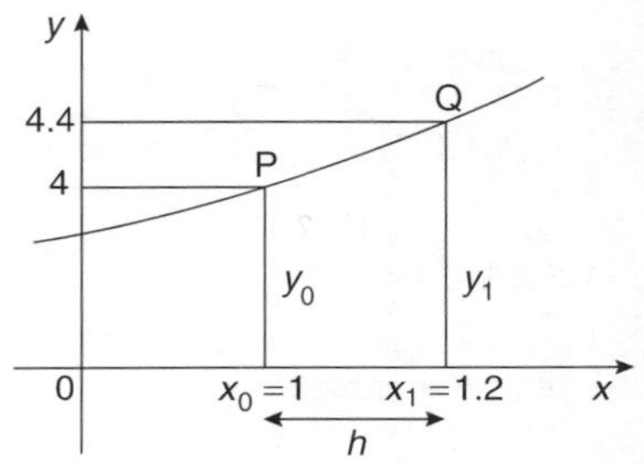

Figure 75.4

At point Q in Figure 75.4, $x_1 = 1.2$, $y_1 = 4.4$

and $\quad (y')_1 = 3(1 + x_1) - y_1$

i.e. $\quad \mathbf{(y')_1} = 3(1 + 1.2) - 4.4 = \mathbf{2.2}$

If the values of x, y and y' found for point Q are regarded as new starting values of x_0, y_0 and $(y')_0$, the above process can be repeated and values found for the point R shown in Figure 75.5.

Thus at point R, $\quad \mathbf{y_1} = y_0 + h(y')_0 \quad$ from equation (2)

$$= 4.4 + (0.2)(2.2) = \mathbf{4.84}$$

When $x_1 = 1.4$ and $y_1 = 4.84$, $\mathbf{(y')_1} = 3(1 + 1.4) - 4.84 = \mathbf{2.36}$
This step by step Euler's method can be continued and it is easiest to list the results in a table, as shown in Table 75.1. The results for lines 1 to 3 have been produced above.
For line 4, where $x_0 = 1.6$:

$$\mathbf{y_1} = y_0 + h(y')_0 = 4.84 + (0.2)(2.36) = \mathbf{5.312}$$

and $\quad \mathbf{(y')_0} = 3(1 + 1.6) - 5.312 = \mathbf{2.488}$

For line 5, where $x_0 = 1.8$:

$$\mathbf{y_1} = y_0 + h(y')_0 = 5.312 + (0.2)(2.488) = \mathbf{5.8096}$$

and $\quad \mathbf{(y')_0} = 3(1 + 1.8) - 5.8096 = \mathbf{2.5904}$

For line 6, where $x_0 = 2.0$:

$$\mathbf{y_1} = y_0 + h(y')_0 = 5.8096 + (0.2)(2.5904) = \mathbf{6.32768}$$

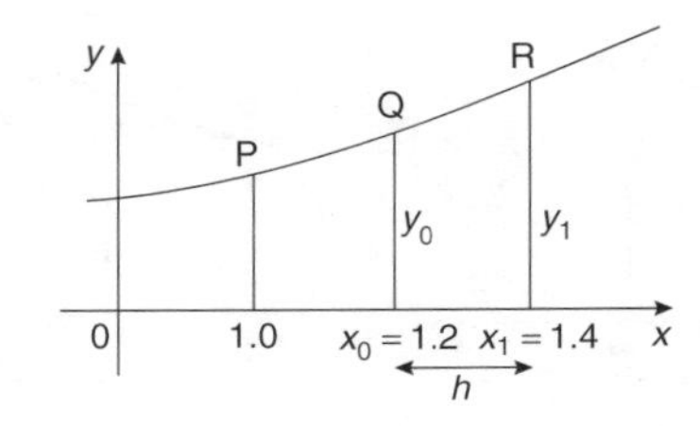

Figure 75.5

Table 75.1

	x_0	y_0	$(y')_0$
1.	1	4	2
2.	1.2	4.4	2.2
3.	1.4	4.84	2.36
4.	1.6	5.312	2.488
5.	1.8	5.8096	2.5904
6.	2.0	6.32768	

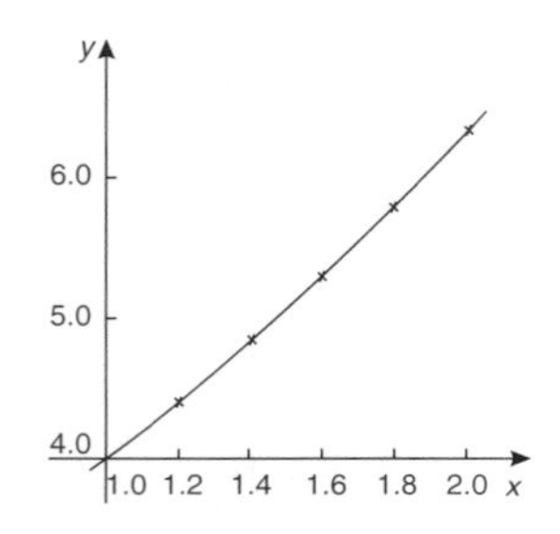

Figure 75.6

(As the range is 1.0 to 2.0 there is no need to calculate $(y')_0$ in line 6) The particular solution is given by the value of y against x.

A graph of the solution of $\frac{dy}{dx} = 3(1 + x) - y$ with initial conditions $x = 1$ and $y = 4$ is shown in Figure 75.6.

In practice it is probably best to plot the graph as each calculation is made, which checks that there is a smooth progression and that no calculation errors have occurred.

An improved Euler method

In the above Euler's method, the gradient $(y')_0$ at $P_{(x_0, y_0)}$ in Figure 75.7 across the whole interval h is used to obtain an approximate value of y_1 at point Q. QR in Figure 75.7 is the resulting error in the result.

In an improved Euler method, called the **Euler-Cauchy method**, the gradient at $P_{(x_0, y_0)}$ across half the interval is used and then continues with a line whose gradient approximates to the gradient of the curve at x_1, shown in Figure 75.8 Let y_{P_1} be the predicted value at point R using Euler's method, i.e. length RZ, where

$$\boldsymbol{y_{P_1} = y_0 + h(y')_0} \tag{3}$$

The error shown as QT in Figure 75.8 is now less than the error QR used in the basic Euler method and the calculated results will be of greater accuracy.

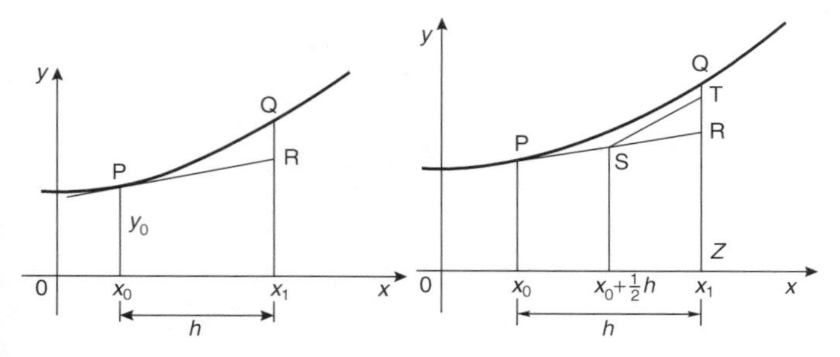

Figure 75.7 **Figure 75.8**

The corrected value, y_{C_1} in the improved Euler method is given by:

$$y_{C_1} = y_0 + \tfrac{1}{2}h[(y')_0 + f(x_1, y_{P_1})] \quad (4)$$

For example, applying the Euler-Cauchy method to solve the differential equation $\frac{dy}{dx} = y - x$ in the range 0(0.1)0.5, given the initial conditions that at $x = 0$, $y = 2$, is achieved as follows:

$$\frac{dy}{dx} = y' = y - x$$

Since the initial conditions are $x_0 = 0$ and $y_0 = 2$ then $(y')_0 = 2 - 0 = 2$
Interval $h = 0.1$, hence $x_1 = x_0 + h = 0 + 0.1 = 0.1$

From equation (3), $y_{P_1} = y_0 + h(y')_0 = 2 + (0.1)(2) = 2.2$

From equation (4), $y_{C_1} = y_0 + \frac{1}{2}h[(y')_0 + f(x_1, y_{P_1})]$

$= y_0 + \frac{1}{2}h[(y')_0 + (y_{P_1} - x_1)]$ in this case

$= 2 + \frac{1}{2}(0.1)[2 + (2.2 - 0.1)] = \mathbf{2.205}$

$(y')_1 = y_{C_1} - x_1 = 2.205 - 0.1 = 2.105$

If we produce a table of values, as in Euler's method, we have so far determined lines 1 and 2 of Table 75.2.

Table 75.2

	x	y	y'
1.	0	2	2
2.	0.1	2.205	2.105
3.	0.2	2.421025	2.221025
4.	0.3	2.649232625	2.349232625
5.	0.4	2.89090205	2.49090205
6.	0.5	3.147446765	

The results in line 2 are now taken as x_0, y_0 and $(y')_0$ for the next interval and the process is repeated.

For line 3, $x_1 = 0.2$

$y_{P_1} = y_0 + h(y')_0 = 2.205 + (0.1)(2.105) = 2.4155$

$y_{C_1} = y_0 + \frac{1}{2}h[(y')_0 + f(x_1, y_{P_1})]$

$= 2.205 + \frac{1}{2}(0.1)[2.105 + (2.4155 - 0.2)] = \mathbf{2.421025}$

$(y')_0 = y_{C_1} - x_1 = 2.421025 - 0.2 = 2.221025$

and so on.

$\frac{dy}{dx} = y - x$ may be solved analytically by the integrating factor method of Chapter 72, with the solution $y = x + 1 + e^x$. Substituting values of x of 0,

Table 75.3

	x	*Euler method* y	*Euler-Cauchy method* y	*Exact value* $y = x + 1 + e^x$
1.	0	2	2	2
2.	0.1	2.2	2.205	2.205170918
2.	0.2	2.41	2.421025	2.421402758
3.	0.3	2.631	2.649232625	2.649858808
4.	0.4	2.8641	2.89090205	2.891824698
5.	0.5	3.11051	3.147446765	3.148721271

Table 75.4

x	*Error in Euler method*	*Error in Euler-Cauchy method*
0	0	0
0.1	0.234%	0.00775%
0.2	0.472%	0.0156%
0.3	0.712%	0.0236%
0.4	0.959%	0.0319%
0.5	1.214%	0.0405%

0.1, 0.2, .. give the exact values shown in Table 75.3. Also shown in the Table are the values that would result from using the Euler method.
The percentage error for each method for each value of x is shown in Table 75.4. For example when $x = 0.3$,

$$\% \text{ error with Euler method} = \left(\frac{\text{actual} - \text{estimated}}{\text{actual}}\right) \times 100\%$$

$$= \left(\frac{2.649858808 - 2.631}{2.649858808}\right) \times 100\% = \mathbf{0.712\%}$$

% error with Euler-Cauchy method

$$= \left(\frac{2.649858808 - 2.649232625}{2.649858808}\right) \times 100\% = \mathbf{0.0236\%}$$

This calculation and the others listed in Table 75.4 show the Euler-Cauchy method to be more accurate than the Euler method.

Statistics and Probability

76 Presentation of Statistical Data

Some statistical terminology

Data are obtained largely by two methods:
(a) by counting — for example, the number of stamps sold by a post office in equal periods of time, and
(b) by measurement — for example, the heights of a group of people.
When data are obtained by counting and only whole numbers are possible, the data are called **discrete**. Measured data can have any value within certain limits and are called **continuous**.
A **set** is a group of data and an individual value within the set is called a **member** of the set. Thus, if the masses of five people are measured correct to the nearest 0.1 kilogram and are found to be 53.1 kg, 59.4 kg, 62.1 kg, 77.8 kg and 64.4 kg, then the set of masses in kilograms for these five people is:

$$\{53.1,\ 59.4,\ 62.1,\ 77.8,\ 64.4\}$$

and one of the members of the set is 59.4
A set containing all the members is called a **population**. Some members selected at random from a population are called a **sample**. Thus all car registration numbers form a population, but the registration numbers of, say, 20 cars taken at random throughout the country are a sample drawn from that population.
The number of times that the value of a member occurs in a set is called the **frequency** of that member. Thus in the set: {2, 3, 4, 5, 4, 2, 4, 7, 9}, member 4 has a frequency of three, member 2 has a frequency of 2 and the other members have a frequency of one.
The **relative frequency** with which any member of a set occurs is given by the ratio: $\dfrac{\text{frquency of member}}{\text{total frequency of all members}}$
For the set: {2, 3, 5, 4, 7, 5, 6, 2, 8}, the relative frequency of member 5 is $\frac{2}{9}$.
Often, relative frequency is expressed as a percentage and the **percentage relative frequency** is: (relative frequency $\times$ 100)%

Presentation of ungrouped data

Ungrouped data can be presented diagrammatically in several ways and these include:
(a) **pictograms**, in which pictorial symbols are used to represent quantities,
(b) **horizontal bar charts**, having data represented by equally spaced horizontal rectangles, and
(c) **vertical bar charts**, in which data are represented by equally spaced vertical rectangles.

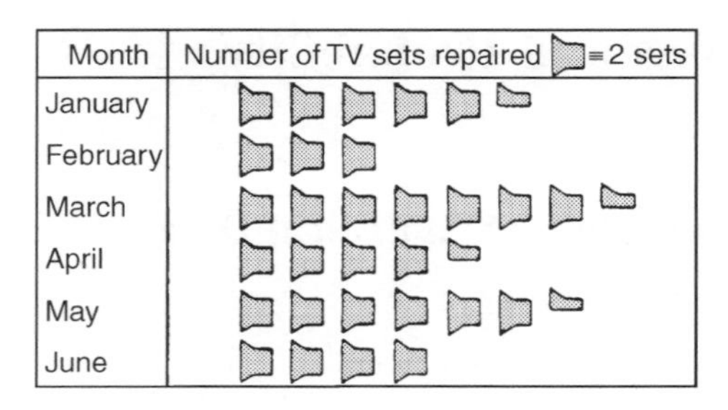

Figure 76.1

For example, the number of television sets repaired in a workshop by a technician in six, one-month periods is as shown below.

Month	January	February	March	April	May	June
Number repaired	11	6	15	9	13	8

This data may be represented as a pictogram as shown in Figure 76.1 where each symbol represents two television sets repaired. Thus, in January, $5\frac{1}{2}$ symbols are used to represent the 11 sets repaired, in February, 3 symbols are used to represent the 6 sets repaired, and so on.

In another example, The distance in miles traveled by four salesmen in a week are as shown below.

Salesmen	P	Q	R	S
Distance traveled (miles)	413	264	597	143

To represent these data diagrammatically by a horizontal bar chart, equally spaced horizontal rectangles of any width, but whose length is proportional to the distance traveled, are used. Thus, the length of the rectangle for salesman P is proportional to 413 miles, and so on. The horizontal bar chart depicting these data is shown in Figure 76.2.

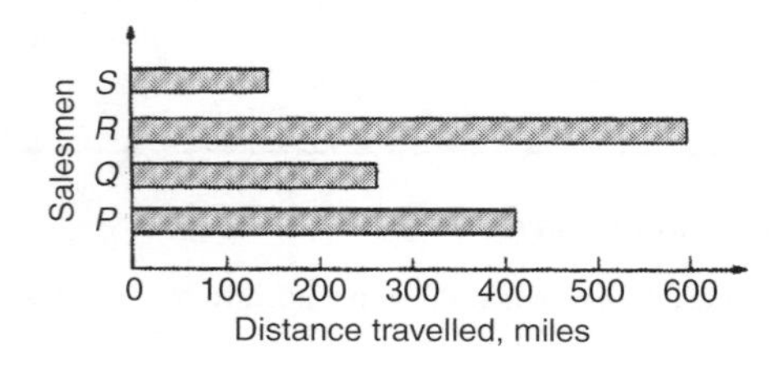

Figure 76.2

In another example, the number of issues of tools or materials from a store in a factory is observed for seven, one-hour periods in a day, and the results of the survey are as follows:

Period	1	2	3	4	5	6	7
Number of issues	34	17	9	5	27	13	6

In a vertical bar chart, equally spaced vertical rectangles of any width, but whose height is proportional to the quantity being represented, are used. Thus the height of the rectangle for period 1 is proportional to 34 units, and so on. The vertical bar chart depicting these data is shown in Figure 76.3.

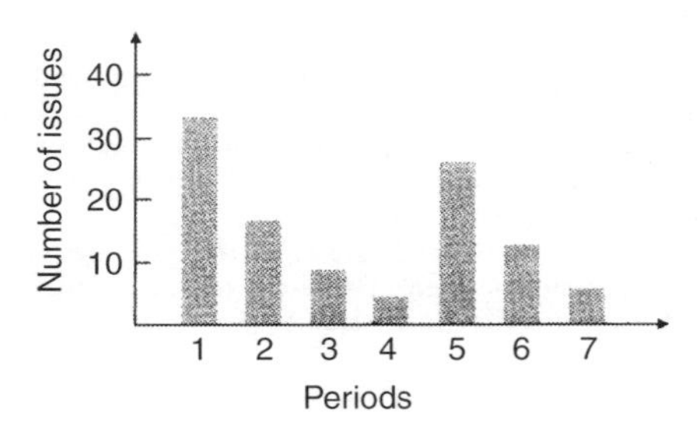

Figure 76.3

Percentage component bar chart

Trends in ungrouped data over equal periods of time can be presented diagrammatically by a **percentage component bar chart**. In such a chart, equally spaced rectangles of any width, but whose height corresponds to 100%, are constructed. The rectangles are then subdivided into values corresponding to the percentage relative frequencies of the members.

For example, the numbers of various types of dwellings sold by a company annually over a three-year period are as shown below.

	Year 1	Year 2	Year3
4-roomed bungalows	24	17	7
5-roomed bungalows	38	71	118
4-roomed houses	44	50	53
5-roomed houses	64	82	147
6-roomed houses	30	30	25

To draw percentage component bar charts to present these data, a table of percentage relative frequency values, correct to the nearest 1%, is the first requirement. Since, percentage relative frequency = $\dfrac{\text{frequency of member} \times 100}{\text{total frequency}}$ then for 4-roomed bungalows in year 1:

$$\text{percentage relative frequency} = \frac{24 \times 100}{24 + 38 + 44 + 64 + 30} = 12\%$$

The percentage relative frequencies of the other types of dwellings for each of the three years are similarly calculated and the results are as shown in the table below.

	Year 1	Year 2	Year 3
4-roomed bungalows	12%	7%	2%
5-roomed bungalows	19%	28%	34%
4-roomed houses	22%	20%	15%
5-roomed houses	32%	33%	42%
6-roomed houses	15%	12%	7%

The percentage component bar chart is produced by constructing three equally spaced rectangles of any width, corresponding to the three years. The heights of the rectangles correspond to 100% relative frequency, and are subdivided into the values in the table of percentages shown above. A key is used (different types of shading or different colour schemes) to indicate corresponding percentage values in the rows of the table of percentages. The percentage component bar chart is shown in Figure 76.4.

A **pie diagram** is used to show diagrammatically the parts making up the whole. In a pie diagram, the area of a circle represents the whole, and the areas of the sectors of the circle are made proportional to the parts that make up the whole.

For example, the retail price of a product costing £2 is made up as follows: materials 10p, labour 20p, research and development 40p, overheads 70p, profit 60p.

To present these data on a pie diagram, a circle of any radius is drawn, and the area of the circle represents the whole, which in this case is £2. The circle is subdivided into sectors so that the areas of the sectors are proportional to the parts, i.e. the parts that make up the total retail price. For the area of a sector to be proportional to a part, the angle at the centre of the circle must be proportional to that part. The whole, £2 or 200p, corresponds to 360°.

Therefore, 10p corresponds to $360 \times \frac{10}{200}$ degrees, i.e. 18°

$$20\text{p corresponds to } 360 \times \frac{20}{200} \text{ degrees, i.e. } 36^\circ$$

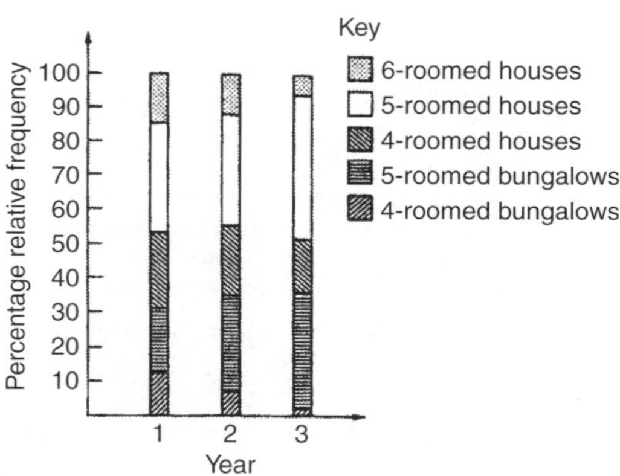

Figure 76.4

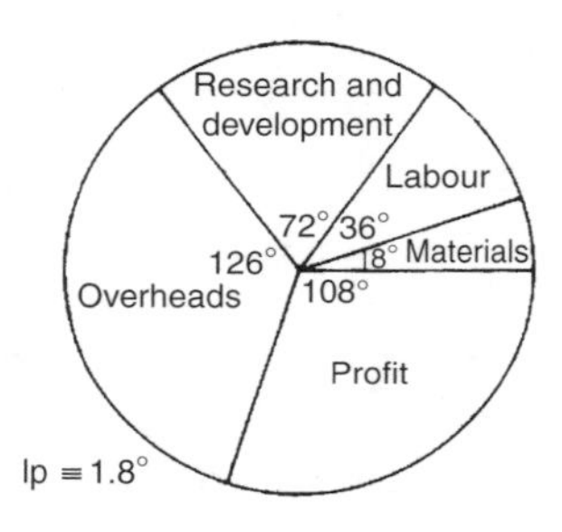

Figure 76.5

and so on, giving the angles at the centre of the circle for the parts of the retail price as: 18°, 36°, 72°, 126° and 108°, respectively.
The pie diagram is shown in Figure 76.5.

Presentation of grouped data

When the number of members in a set is small, say ten or less, the data can be represented diagrammatically without further analysis, by means of pictograms, bar charts, percentage components bar charts or pie diagrams.
For sets having more than ten members, those members having similar values are grouped together in **classes** to form a **frequency distribution**. To assist in accurately counting members in the various classes, a **tally diagram** is used. A frequency distribution is merely a table showing classes and their corresponding frequencies.
The new set of values obtained by forming a frequency distribution is called **grouped data**.
The terms used in connection with grouped data are shown in Figure 76.6(a). The size or range of a class is given by the **upper class boundary value**

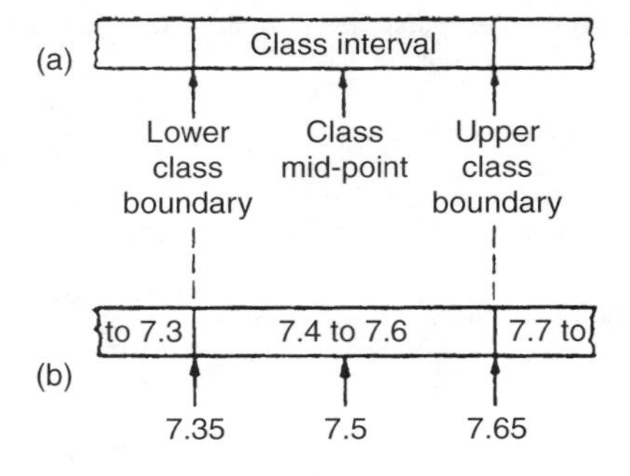

Figure 76.6

minus the **lower class boundary value**, and in Figure 76.6 is 7.65 −7.35, i.e. 0.30. The **class interval** for the class shown in Figure 76.6(b) is 7.4 to 7.6 and the class mid-point value is given by

$$\frac{(\text{upper class boundary value}) + (\text{lower class boundary value})}{2}$$

and in Figure 76.6 is $\frac{7.65 + 7.35}{2}$, i.e. 7.5

One of the principal ways of presenting grouped data diagrammatically is by using a **histogram**, in which the **areas** of vertical, adjacent rectangles are made proportional to frequencies of the classes. When class intervals are equal, the heights of the rectangles of a histogram are equal to the frequencies of the classes. For histograms having unequal class intervals, the area must be proportional to the frequency. Hence, if the class interval of class A is twice the class interval of class B, then for equal frequencies, the height of the rectangle representing A is half that of B.

Another method of presenting grouped data diagrammatically is by using a **frequency polygon**, which is the graph produced by plotting frequency against class mid-point values and joining the co-ordinates with straight lines.

A **cumulative frequency distribution** is a table showing the cumulative frequency for each value of upper class boundary. The cumulative frequency for a particular value of upper class boundary is obtained by adding the frequency of the class to the sum of the previous frequencies.

The curve obtained by joining the co-ordinates of cumulative frequency (vertically) against upper class boundary (horizontally) is called an **ogive** or a **cumulative frequency distribution curve**.

For example, the masses of 50 ingots, in kilograms, are measured correct to the nearest 0.1 kg and the results are as shown below.

8.0	8.6	8.2	7.5	8.0	9.1	8.5	7.6	8.2	7.8
8.3	7.1	8.1	8.3	8.7	7.8	8.7	8.5	8.4	8.5
7.7	8.4	7.9	8.8	7.2	8.1	7.8	8.2	7.7	7.5
8.1	7.4	8.8	8.0	8.4	8.5	8.1	7.3	9.0	8.6
7.4	8.2	8.4	7.7	8.3	8.2	7.9	8.5	7.9	8.0

The **range** of the data is the member having the largest value minus the member having the smallest value. Inspection of the set of data shows that:

$$\text{range} = 9.1 - 7.1 = 2.0$$

The size of each class is given approximately by $\frac{\text{range}}{\text{number of classes}}$

If about seven classes are required, the size of each class is 2.0/7, that is approximately 0.3, and thus the **class limits** are selected as 7.1 to 7.3, 7.4 to 7.6, 7.7 to 7.9, and so on.

The **class mid-point** for the 7.1 to 7.3 class is $\frac{7.35 + 7.05}{2}$, i.e. 7.2, for the 7.4 to 7.6 class is $\frac{7.65 + 7.35}{2}$, i.e. 7.5, and so on.

To assist with accurately determining the number in each class, a **tally diagram** is produced as shown in Table 76.1. This is obtained by listing the classes in

Table 76.1

Class	*Tally*
7.1 to 7.3	111
7.4 to 7.6	~~1111~~
7.7 to 7.9	~~1111~~ 1111
8.0 to 8.2	~~1111~~ ~~1111~~ 1111
8.3 to 8.5	~~1111~~ ~~1111~~ 1
8.6 to 8.8	~~1111~~ 1
8.9 to 9.1	11

Table 76.2

Class	*Class mid-point*	*Frequency*
7.1 to 7.3	7.2	3
7.4 to 7.6	7.5	5
7.7 to 7.9	7.8	9
8.0 to 8.2	8.1	14
8.3 to 8.5	8.4	11
8.6 to 8.8	8.7	6
8.9 to 9.1	9.0	2

the left-hand column and then inspecting each of the 50 members of the set of data in turn and allocating it to the appropriate class by putting a '1' in the appropriate row. Each fifth '1' allocated to a particular row is marked as an oblique line to help with final counting.

A **frequency distribution** for the data is shown in Table 76.2 and lists classes and their corresponding frequencies. Class mid-points are also shown in this table, since they are used when constructing the frequency polygon and histogram.

A **frequency polygon** is shown in Figure 76.7, the co-ordinates corresponding to the class mid-point/frequency values, given in Table 76.2. The co-ordinates are joined by straight lines and the polygon is 'anchored-down' at each end by joining to the next class mid-point value and zero frequency.

A **histogram** is shown in Figure 76.8, the width of a rectangle corresponding to (upper class boundary value – lower class boundary value) and height corresponding to the class frequency. The easiest way to draw a histogram is to mark class mid-point values on the horizontal scale and to draw the rectangles symmetrically about the appropriate class mid-point values and touching one another. A histogram for the data given in Table 76.2 is shown in Figure 76.8.

A **cumulative frequency distribution** is a table giving values of cumulative frequency for the values of upper class boundaries, and is shown in Table 76.3. Columns 1 and 2 show the classes and their frequencies. Column 3 lists the upper class boundary values for the classes given in column 1. Column 4 gives the cumulative frequency values for all frequencies less than the upper class boundary values given in column 3. Thus, for example, for the 7.7 to 7.9 class shown in row 3, the cumulative frequency value is the sum of all frequencies having values of less than 7.95, i.e. $3 + 5 + 9 = 17$, and so on.

The **ogive** for the cumulative frequency distribution given in Table 76.3 is shown in Figure 76.9. The co-ordinates corresponding to each upper class boundary/cumulative frequency value are plotted and the co-ordinates are

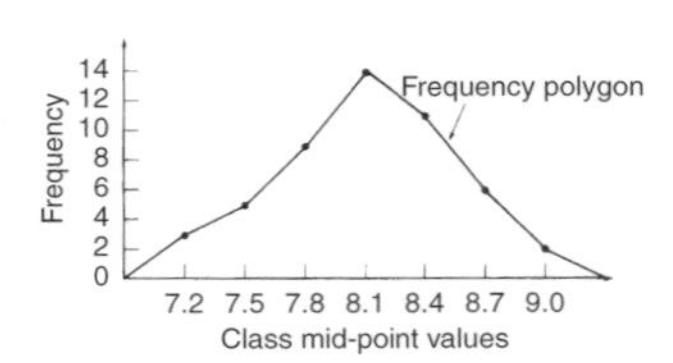

Figure 76.7

Frequency
14
12
10
8
6
4
2
0
Histogram
7.2 7.35 7.5 7.65 7.8 7.95 8.1 8.25 8.4 8.55 8.7 8.85 9.0 9.15
Class mid-point values

Figure 76.8

Table 76.3

1 Class	2 Frequency	3 Upper class boundary	4 Cumulative frequency
		Less than	
7.1-7.3	3	7.35	3
7.4-7.6	5	7.65	8
7.7-7.9	9	7.95	17
8.0-8.2	14	8.25	31
8.3-8.5	11	8.55	42
8.6-8.8	6	8.85	48
8.9-9.1	2	9.15	50

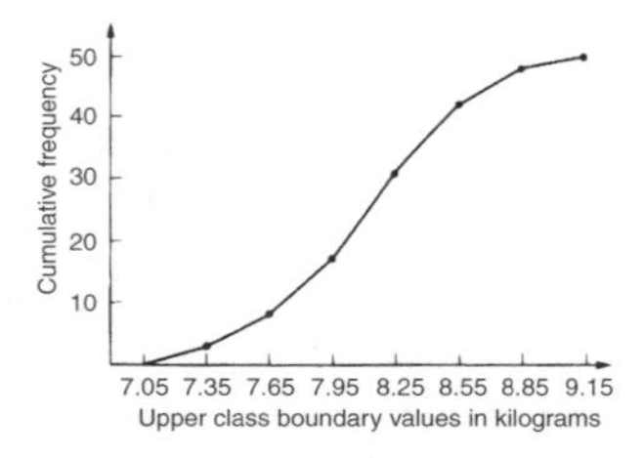

Figure 76.9

joined by straight lines (–not the best curve drawn through the co-ordinates as in experimental work). The ogive is 'anchored' at its start by adding the co-ordinate (7.05, 0).

77 Measures of Central Tendency and Dispersion

Measures of central tendency

A single value, which is representative of a set of values, may be used to give an indication of the general size of the members in a set, the word '**average**' often being used to indicate the single value.

The statistical term used for 'average' is the arithmetic mean or just the **mean**. Other measures of central tendency may be used and these include the **median** and the **modal** values.

Mean, median and mode for discrete data

Mean

The **arithmetic mean value** is found by adding together the values of the members of a set and dividing by the number of members in the set. Thus, the mean of the set of numbers: {4, 5, 6, 9} is:

$$\frac{4+5+6+9}{4}, \text{ i.e. } 6$$

In general, the mean of the set: $\{x_1, x_2, x_3, \ldots x_n\}$ is

$$\overline{x} = \frac{x_1 + x_2 + x_3 + \cdots + x_n}{n}, \text{ written as } \frac{\sum x}{n}$$

where $\sum$ is the Greek letter 'sigma' and means 'the sum of', and $\overline{x}$ (called x-bar) is used to signify a mean value.

Median

The **median value** often gives a better indication of the general size of a set containing extreme values. The set: {7, 5, 74, 10} has a mean value of 24, which is not really representative of any of the values of the members of the set. The median value is obtained by:

(a) **ranking** the set in ascending order of magnitude, and
(b) selecting the value of the **middle member** for sets containing an odd number of members, or finding the value of the mean of the two middle members for sets containing an even number of members.

For example, the set: {7, 5, 74, 10} is ranked as {5, 7, 10, 74}, and since it contains an even number of members (four in this case), the mean of 7 and 10 is taken, giving a median value of 8.5.

In another example, the set: {3, 81, 15, 7, 14} is ranked as {3, 7, 14, 15, 81} and the median value is the value of the middle member, i.e. 14.

Mode

The **modal value**, or **mode**, is the most commonly occurring value in a set. If two values occur with the same frequency, the set is 'bi-modal'.
For example, the set: {5, 6, 8, 2, 5, 4, 6, 5, 3} has a modal value of 5, since the member having a value of 5 occurs three times.

Mean, median and mode for grouped data

The mean value for a set of grouped data is found by determining the sum of the (frequency × class mid-point values) and dividing by the sum of the frequencies,

$$\text{i.e. mean value } \overline{x} = \frac{f_1x_1 + f_2x_2 + \ldots f_nx_n}{f_1 + f_2 + \cdots + f_n} = \frac{\sum(fx)}{\sum f}$$

where f is the frequency of the class having a mid-point value of x, and so on.

For example, the frequency distribution for the value of resistance in ohms of 48 resistors is:

20.5–20.9	3,	21.0–21.4	10,	21.5–21.9	11,
22.0–22.4	13,	22.5–22.9	9,	23.0–23.4	2

The class mid-point/frequency values are:

20.7 3, 21.2 10, 21.7 11, 22.2 13, 22.7 9 and 23.2 2

For grouped data, the mean value is given by: $\bar{x} = \dfrac{\sum(fx)}{\sum f}$

where f is the class frequency and x is the class mid-point value. Hence mean value,

$$\bar{x} = \frac{(3 \times 20.7) + (10 \times 21.2) + (11 \times 21.7) + (13 \times 22.2) + (9 \times 22.7) + (2 \times 23.2)}{48} = \frac{1052.1}{48} = 21.919..$$

i.e. **the mean value is 21.9 ohms**, correct to 3 significant figures.

Histogram

The mean, median and modal values for grouped data may be determined from a **histogram**. In a histogram, frequency values are represented vertically and variable values horizontally. The mean value is given by the value of the variable corresponding to a vertical line drawn through the centroid of the histogram. The median value is obtained by selecting a variable value such that the area of the histogram to the left of a vertical line drawn through the selected variable value is equal to the area of the histogram on the right of the line. The modal value is the variable value obtained by dividing the width of the highest rectangle in the histogram in proportion to the heights of the adjacent rectangles.

For example, the time taken in minutes to assemble a device is measured 50 times and the results are as shown below:.

14.5–15.5	5,	16.5–17.5	8,	18.5–19.5	16,
20.5–21.5	12,	22.5–23.5	6,	24.5–25.5	3

The mean, median and modal values of the distribution may be determined from a histogram depicting the data:

The histogram is shown in Figure 77.1. The mean value lies at the centroid of the histogram. With reference to any arbitrary axis, say YY shown at a time of 14 minutes, the position of the horizontal value of the centroid can be obtained from the relationship $AM = \sum(am)$, where A is the area of the histogram, M is the horizontal distance of the centroid from the axis YY, a is the area of a rectangle of the histogram and m is the distance of the centroid of the rectangle from YY. The areas of the individual rectangles are shown circled on the histogram giving a total area of 100 square units. The positions, m, of the centroids of the individual rectangles are $1, 3, 5, \ldots$ units from YY. Thus

$$100M = (10 \times 1) + (16 \times 3) + (32 \times 5) + (24 \times 7) + (12 \times 9) + (6 \times 11)$$

i.e. $M = \dfrac{560}{100} = 5.6$ units from YY

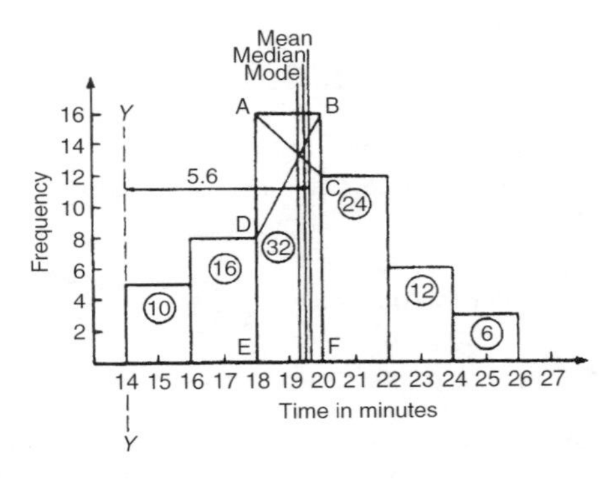

Figure 77.1

Thus the position of the **mean** with reference to the time scale is $14 + 5.6$, i.e. **19.6 minutes**.

The median is the value of time corresponding to a vertical line dividing the total area of the histogram into two equal parts. The total area is 100 square units, hence the vertical line must be drawn to give 50 units of area on each side. To achieve this with reference to Figure 77.1, rectangle ABFE must be split so that $50 - (10 + 16)$ units of area lie on one side and $50 - (24 + 12 + 6)$ units of area lie on the other. This shows that the area of ABFE is split so that 24 units of area lie to the left of the line and 8 units of area lie to the right, i.e. the vertical line must pass through 19.5 minutes. Thus the **median value** of the distribution is **19.5 minutes**.

The mode is obtained by dividing the line AB, which is the height of the highest rectangle, proportionally to the heights of the adjacent rectangles. With reference to Figure 77.1, this is done by joining AC and BD and drawing a vertical line through the point of intersection of these two lines. This gives the **mode** of the distribution and is **19.3 minutes**.

Standard deviation with discrete data

The standard deviation of a set of data gives an indication of the amount of dispersion, or the scatter, of members of the set from the measure of central tendency. Its value is the root-mean-square value of the members of the set and for discrete data is obtained as follows:

(a) determine the measure of central tendency, usually the mean value, (occasionally the median or modal values are specified),

(b) calculate the deviation of each member of the set from the mean, giving

$$(x_1 - \overline{x}), \quad (x_2 - \overline{x}), \quad (x_3 - \overline{x}), \ldots,$$

(c) determine the squares of these deviations, i.e.

$$(x_1 - \overline{x})^2, \quad (x_2 - \overline{x})^2, \quad (x_3 - \overline{x})^2, \ldots,$$

(d) find the sum of the squares of the deviations, that is

$$(x_1 - \overline{x})^2 + (x_2 - \overline{x})^2 + (x_3 - \overline{x})^2, \ldots,$$

(e) divide by the number of members in the set, n, giving

$$\frac{(x_1 - \overline{x})^2 + (x_2 - x)^2 + (x^3 - \overline{x})^2 + \ldots}{n}$$

(f) determine the square root of (e)

The standard deviation is indicated by σ (the Greek letter small 'sigma') and is written mathematically as:

$$\textbf{standard deviation, } \sigma = \sqrt{\left\{\frac{\sum(x - \overline{x})^2}{n}\right\}}$$

where x is a member of the set, $\overline{x}$ is the mean value of the set and n is the number of members in the set. The value of standard deviation gives an indication of the distance of the members of a set from the mean value. The set: $\{1, 4, 7, 10, 13\}$ has a mean value of 7 and a standard deviation of about 4.2. The set $\{5, 6, 7, 8, 9\}$ also has a mean value of 7, but the standard deviation is about 1.4. This shows that the members of the second set are mainly much closer to the mean value than the members of the first set.

For example, to determine the standard deviation from the mean of the set of numbers: $\{5, 6, 8, 4, 10, 3\}$, correct to 4 significant figures:

The arithmetic mean, $\overline{x} = \dfrac{\sum x}{n} = \dfrac{5 + 6 + 8 + 4 + 10 + 3}{6} = 6$

Standard deviation, $\sigma = \sqrt{\left\{\dfrac{\sum(x - \overline{x})^2}{n}\right\}}$

The $(x - \overline{x})^2$ values are: $(5-6)^2$, $(6-6)^2$, $(8-6)^2$, $(4-6)^2$, $(10-6)^2$ and $(3-6)^2$
The sum of the $(x - \overline{x})^2$ values,

i.e. $\sum(x - \overline{x})^2 = 1 + 0 + 4 + 4 + 16 + 9 = 34$

and $\dfrac{\sum(x - \overline{x})^2}{n} = \dfrac{34}{6} = 5.\dot{6}$ since there are 6 members in the set.

Hence, **standard deviation,** $\sigma = \sqrt{\left\{\dfrac{\sum(x - \overline{x})^2}{n}\right\}} = \sqrt{5.\dot{6}} = \mathbf{2.380}$,

correct to 4 significant figures

Standard deviation with grouped data

For **grouped data, standard deviation** $\sigma = \sqrt{\left\{\dfrac{\sum\{f(x - \overline{x})^2\}}{\sum f}\right\}}$

where f is the class frequency value, x is the class mid-point value and $\overline{x}$ is the mean value of the grouped data.

For example, the frequency distribution for the values of resistance in ohms of 48 resistors is:

20.5–20.9	3,	21.0–21.4	10,	21.5–21.9	11,
22.0–22.4	13,	22.5–22.9	9,	23.0–23.4	2

To find the standard deviation:
From earlier, the distribution mean value, $\overline{x} = 21.92$, correct to 4 significant figures.
The 'x-values' are the class mid-point values, i.e. 20.7, 21.2, 21.7, ...
Thus the $(x - \overline{x}^2)$ values are

$$(20.7\text{–}21.92)^2,\ (21.2\text{–}21.92)^2,\ (21.7\text{–}21.92)^2, \ldots,$$

and the $f(x - \overline{x})^2$ values are

$$3(20.7\text{–}21.92)^2,\ 10(21.2\text{–}21.92)^2,\ 11(21.7\text{–}21.92)^2, \ldots.$$

The $\sum f(x - \overline{x})^2$ values are

$$4.4652 + 5.1840 + 0.5324 + 1.0192 + 5.4756 + 3.2768 = 19.9532$$

$$\frac{\sum\{f(x-\overline{x})^2\}}{\sum f} = \frac{19.9532}{48} = 0.41569$$

and **standard deviation,** $\sigma = \sqrt{\left\{\dfrac{\sum\{f(x-\overline{x})^2\}}{\sum f}\right\}} = \sqrt{0.41569}$

$= \mathbf{0.645}$, correct to 3 significant figures

Quartiles, deciles and percentiles

Other measures of dispersion which are sometimes used are the quartile, decile and percentile values. The **quartile values** of a set of discrete data are obtained by selecting the values of members which divide the set into four equal parts. Thus for the set: {2, 3, 4, 5, 5, 7, 9, 11, 13, 14, 17} there are 11 members and the values of the members dividing the set into four equal parts are 4, 7, and 13. These values are signified by Q_1, Q_2 and Q_3 and called the first, second and third quartile values, respectively. It can be seen that the second quartile value, Q_2, is the value of the middle member and hence is the median value of the set.
For grouped data the ogive may be used to determine the quartile values. In this case, points are selected on the vertical cumulative frequency values of the ogive, such that they divide the total value of cumulative frequency into four equal parts. Horizontal lines are drawn from these values to cut the ogive. The values of the variable corresponding to these cutting points on the ogive give the quartile values.
For example, the frequency distribution given below refers to the overtime worked by a group of craftsmen during each of 48 working weeks in a year.

25–29	5,	30–34	4,	35–39	7,	40–44	11,
45–49	12,	50–54	8,	55–59	1		

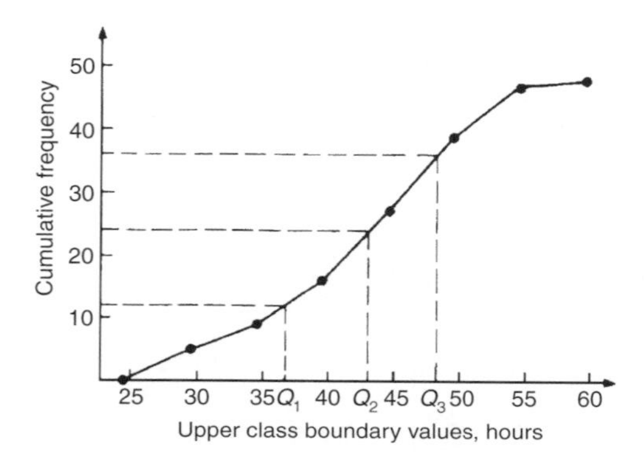

Figure 77.2

The cumulative frequency distribution (i.e. upper class boundary/cumulative frequency values) is:

29.5 5, 34.5 9, 39.5 16, 44.5 27, 49.5 39, 54.5 47, 59.5 48

The ogive is formed by plotting these values on a graph, as shown in Figure 77.2. The total frequency is divided into four equal parts, each having a range of 48/4, i.e. 12. This gives cumulative frequency values of 0 to 12 corresponding to the first quartile, 12 to 24 corresponding to the second quartile, 24 to 36 corresponding to the third quartile and 36 to 48 corresponding to the fourth quartile of the distribution, i.e. the distribution is divided into four equal parts. The quartile values are those of the variable corresponding to cumulative frequency values of 12, 24 and 36, marked Q_1, Q_2 and Q_3 in Figure 77.2. These values, correct to the nearest hour, are **37 hours, 43 hours and 48 hours**, respectively. The Q_2 value is also equal to the median value of the distribution. One measure of the dispersion of a distribution is called the **semi-interquartile range** and is given by $\frac{Q_2 - Q_1}{2}$, and is $\frac{48 - 37}{2}$ in this case, i.e. **$5\frac{1}{2}$ hours**.

When a set contains a large number of members, the set can be split into ten parts, each containing an equal number of members. These ten parts are then called **deciles** . For sets containing a very large number of members, the set may be split into one hundred parts, each containing an equal number of members. One of these parts is called a **percentile**.

78 Probability

Introduction to probability

The **probability** of something happening is the likelihood or chance of it happening. Values of probability lie between 0 and 1, where 0 represents an absolute impossibility and 1 represents an absolute certainty. The probability of an event happening usually lies somewhere between these two extreme values and is expressed either as a proper or decimal fraction. Examples of probability are:

that a length of copper wire has zero resistance at 100°C	0
that a fair, six-sided dice will stop with a 3 upwards	$\frac{1}{6}$ or 0.1667
that a fair coin will land with a head upwards	$\frac{1}{2}$ or 0.5
that a length of copper wire has some resistance at 100°C	1

If p is the probability of an event happening and q is the probability of the same event not happening, then the total probability is $p + q$ and is equal to unity, since it is an absolute certainty that the event either does or does not occur, i.e. $\mathbf{p + q = 1}$

Expectation

The **expectation**, E, of an event happening is defined in general terms as the product of the probability p of an event happening and the number of attempts made, n, i.e. $\boldsymbol{E = pn}$

Thus, since the probability of obtaining a 3 upwards when rolling a fair dice is $\frac{1}{6}$, the expectation of getting a 3 upwards on four throws of the dice is $\frac{1}{6} \times 4$, i.e. $\frac{2}{3}$

Thus expectation is the average occurrence of an event.

Dependent event

A **dependent event** is one in which the probability of an event happening affects the probability of another ever happening. Let 5 transistors be taken at random from a batch of 100 transistors for test purposes, and the probability of there being a defective transistor, p_1, be determined. At some later time, let another 5 transistors be taken at random from the 95 remaining transistors in the batch and the probability of there being a defective transistor, p_2, be determined. The value of p_2 is different from p_1 since batch size has effectively altered from 100 to 95, i.e. probability p_2 is dependent on probability p_1. Since transistors are drawn, and then another 5 transistors drawn without replacing the first 5, the second random selection is said to be **without replacement**.

Independent event

An independent event is one in which the probability of an event happening does not affect the probability of another event happening. If 5 transistors are taken at random from a batch of transistors and the probability of a defective transistor p_1 is determined and the process is repeated after the original 5 have been replaced in the batch to give p_2, then p_1 is equal to p_2. Since the 5 transistors are replaced between draws, the second selection is said to be **with replacement**.

Laws of probability

The addition law of probability

The addition law of probability is recognised by the word **'or'** joining the probabilities. If p_A is the probability of event A happening and p_B is the probability of event B happening, the probability of **event A or event B** happening is given by $p_A + p_B$. Similarly, the probability of events **A or B or C or** $\ldots N$ happening is given by

$$p_A + p_B + p_C + \cdots + p_N$$

The multiplication law of probability

The multiplication law of probability is recognised by the word **'and'** joining the probabilities. If p_A is the probability of event A happening and p_B is the probability of event B happening, the probability of **event A and event B** happening is given by $p_A \times p_B$. Similarly, the probability of events **A and B and C and** $\ldots N$ happening is given by

$$p_A \times p_B \times p_C \times \cdots \times p_N$$

For example, to determine the probability of selecting at random the winning horse in a race in which 10 horses are running:
Since only one of the ten horses can win, the probability of selecting at random the winning horse is $\dfrac{\text{number of winners}}{\text{number of horses}}$, i.e. $\mathbf{\dfrac{1}{10}}$ or **0.10**
To determine the probability of selecting at random the winning horses in both the first and second races if there are 10 horses in each race:
The probability of selecting the winning horse in the first race is $\dfrac{1}{10}$.
The probability of selecting the winning horse in the second race is $\dfrac{1}{10}$.
The probability of selecting the winning horses in the first **and** second race is given by the multiplication law of probability, i.e.

$$\textbf{probability} = \frac{1}{10} \times \frac{1}{10} = \mathbf{\frac{1}{100}} \text{ or } \mathbf{0.01}$$

In another example, the probability of a component failing in one year due to excessive temperature is $\dfrac{1}{20}$, due to excessive vibration is $\dfrac{1}{25}$ and due to excessive humidity is $\dfrac{1}{50}$.
Let p_A be the probability of failure due to excessive temperature, then

$p_A = \dfrac{1}{20}$ and $\overline{p_A} = \dfrac{19}{20}$ (where $\overline{p_A}$ is the probability of not failing)

Let p_B be the probability of failure due to excessive vibration, then

$$p_B = \frac{1}{25} \text{ and } \overline{p_B} = \frac{24}{25}$$

Let p_C be the probability of failure due to excessive humidity, then

$$p_C = \frac{1}{50} \text{ and } \overline{p_C} = \frac{49}{50}$$

The probability of a component failing due to excessive temperature **and** excessive vibration is given by:

$$p_A \times p_B = \frac{1}{20} \times \frac{1}{25} = \frac{1}{500} \text{ or } \mathbf{0.002}$$

The probability of a component failing due to excessive vibration **or** excessive humidity is:

$$p_B + p_C = \frac{1}{25} + \frac{1}{50} = \frac{3}{50} \text{ or } \mathbf{0.06}$$

The probability that a component will not fail due excessive temperature **and** will not fail due to excess humidity is:

$$\overline{p_A} \times \overline{p_C} = \frac{19}{20} \times \frac{49}{50} = \frac{931}{1000} \text{ or } \mathbf{0.931}$$

In another example, a batch of 40 components contains 5 which are defective. If a component is drawn at random from the batch and tested and then a second component is drawn at random, the probability of having one defective component, both with and without replacement, is determined as follows:
The probability of having one defective component can be achieved in two ways. If p is the probability of drawing a defective component and q is the probability of drawing a satisfactory component, then the probability of having one defective component is given by drawing a satisfactory component and then a defective component **or** by drawing a defective component and then a satisfactory one, i.e. by $q \times p + p \times q$

With replacement:

$$p = \frac{5}{40} = \frac{1}{8} \text{ and } q = \frac{35}{40} = \frac{7}{8}$$

Hence, probability of having one defective component is:

$$\frac{1}{8} \times \frac{7}{8} + \frac{7}{8} \times \frac{1}{8}, \text{ i.e. } \frac{7}{64} + \frac{7}{64} = \frac{7}{32} \text{ or } \mathbf{0.2188}$$

Without replacement:

$p_1 = \frac{1}{8}$ and $q_1 = \frac{7}{8}$ on the first of the two draws. The batch number is now 39 for the second draw, thus, $p_2 = \dfrac{5}{39}$ and $q_2 = \dfrac{35}{39}$

$$p_1 q_2 + q_1 p_2 = \frac{1}{8} \times \frac{35}{39} + \frac{7}{8} \times \frac{5}{39} = \frac{35 + 35}{312} = \frac{70}{312} \text{ or } \mathbf{0.2244}$$

79 The Binomial and Poisson Distributions

The binomial distribution

The binomial distribution deals with two numbers only, these being the probability that an event will happen, p, and the probability that an event will not happen, q. Thus, when a coin is tossed, if p is the probability of the coin landing with a head upwards, q is the probability of the coin landing with a tail upwards. $p + q$ must always be equal to unity. A binomial distribution can be used for finding, say, the probability of getting three heads in seven tosses of the coin, or in industry for determining defect rates as a result of sampling. One way of defining a binomial distribution is as follows:

'if p is the probability that an event will happen and q is the probability that the event will not happen, then the probabilities that the event will happen 0, 1, 2, 3,..., n times in n trials are given by the successive terms of the expansion of $(q+p)^n$, taken from left to right'.

The binomial expansion of $(q+p)^n$ is:

$$q^n + nq^{n-1}p + \frac{n(n-1)}{2!}q^{n-2}p^2 + \frac{n(n-1)(n-2)}{3!}q^{n-3}p^3 + \cdots$$

from Chapter 16.

For example, let a dice be rolled 9 times.
Let p be the probability of having a 4 upwards. Then $p = 1/6$, since dice have six sides.
Let q be the probability of not having a 4 upwards. Then $q = 5/6$. The probabilities of having a 4 upwards 0, 1, 2, .. n times are given by the successive terms of the expansion of $(q+p)^n$, taken from left to right.
From the binomial expansion:

$$(q+q)^9 = q^9 + 9q^8p + 36q^7p^2 + 84q^6p^3 + ..$$

The probability of having a 4 upwards no times is

$$q^9 = (5/6)^9 = \mathbf{0.1938}$$

The probability of having a 4 upwards once is

$$9q^8p = 9(5/6)^8(1/6) = \mathbf{0.3489}$$

The probability of having a 4 upwards twice is

$$36q^7p^2 = 36(5/6)^7(1/6)^2 = \mathbf{0.2791}$$

The probability of having a 4 upwards 3 times is

$$84q^6p^3 = 84(5/6)^6(1/6)^3 = \mathbf{0.1302}$$

The probability of having a 4 upwards less than 4 times is the sum of the probabilities of having a 4 upwards 0, 1, 2, and 3 times, i.e.

$$0.1938 + 0.3489 + 0.2791 + 0.1302 = \mathbf{0.9520}$$

Industrial inspection

In industrial inspection, p is often taken as the probability that a component is defective and q is the probability that the component is satisfactory. In this case, a binomial distribution may be defined as:

'the probabilities that 0, 1, 2, 3,..., n components are defective in a sample of n components, drawn at random from a large batch of components, are given by the successive terms of the expansion of $(q+p)^n$, taken from left to right'.

For example, a package contains 50 similar components and inspection shows that four have been damaged during transit. Let six components be drawn at random from the contents of the package.

The probability of a component being damaged, p, is 4 in 50, i.e. 0.08 per unit. Thus, the probability of a component not being damaged, q, is $1 - 0.08$, i.e. 0.92 The probability of there being 0, 1, 2,..., 6 damaged components is given by the successive terms of $(q + p)^6$, taken from left to right.

$$(q + p)^6 = q^6 + 6q^5 p + 15q^4 p^2 + 20q^3 p^3 + \cdots$$

The probability of one damaged component is $6q^5 p = 6 \times 0\ 92^5 \times 0.08 =$ **0.3164**. The probability of less than three damaged components is given by the sum of the probabilities of 0, 1 and 2 damaged components.

$$q^6 + 6q^5 p + 15q^4 p^2 = 0.92^6 + 6 \times 0.92^5 \times 0.08 + 15 \times 0.92^4 \times 0.08^2$$

$$= 0.6064 + 0.3164 + 0.0688 = \mathbf{0.9916}$$

The Poisson distribution

When the number of trials, n, in a binomial distribution becomes large (usually taken as larger than 10), the calculations associated with determining the values of the terms becomes laborious. If n is large and p is small, and the product $n\,p$ is less than 5, a very good approximation to a binomial distribution is given by the corresponding Poisson distribution, in which calculations are usually simpler.

The Poisson approximation to a binomial distribution may be defined as follows:

'the probabilities that an event will happen 0, 1, 2, 3, . ., n times in n trials are given by the successive terms of the expression

$$e^{-\lambda}\left(1 + \lambda + \frac{\lambda^2}{2!} + \frac{\lambda^3}{3!} + \ldots\right) \text{ taken from left to right'}$$

The symbol λ is the expectation of an event happening and is equal to $n\,p$

For example, let 3% of the gearwheels produced by a company be defective, and let a sample of 80 gearwheels be taken.

The sample number, n, is large, the probability of a defective gearwheel, p, is small and the product $n\,p$ is 80×0.03, i.e. 2.4, which is less than 5. Hence a Poisson approximation to a binomial distribution may be used. The expectation of a defective gearwheel, $\lambda = n\,p = 2.4$

The probabilities of 0, 1, 2,. . . defective gearwheels are given by the successive terms of the expression $e^{-\lambda}\left(1 + \lambda + \frac{\lambda^2}{2!} + \frac{\lambda^3}{3!} + \cdots\right)$ taken from left to right, i.e. by $e^{-\lambda}$, $\lambda e^{-\lambda}$, $\frac{\lambda^2 e^{-\lambda}}{2!}$, . . Thus:

the probability of no defective gearwheels is $e^{-\lambda} = e^{-2.4} = \mathbf{0.0907}$

the probability of 1 defective gearwheel is $\lambda e^{-\lambda} = 2.4e^{-2.4} = \mathbf{0.2177}$

the probability of 2 defective gearwheels is $\frac{\lambda^2 e^{-\lambda}}{2!} = \frac{2.4^2 e^{-2.4}}{2 \times 1} = \mathbf{0.2613}$

The probability of having more than 2 defective gearwheels is 1−(the sum of the probabilities of having 0, 1, and 2 defective gearwheels), i.e.

$$1 - (0.0907 + 0.2177 + 0.2613), \text{ that is, } \mathbf{0.4303}$$

The principal use of a Poisson distribution is to determine the theoretical probabilities when p, the probability of an event happening, is known, but q, the probability of the event not happening is unknown. For example, the average number of goals scored per match by a football team can be calculated, but it is not possible to quantify the number of goals which were not scored. In this type of problem, a Poisson distribution may be defined as follows:

'the probabilities of an event occurring 0, 1, 2, 3.... times are given by the successive terms of the expression $e^{-\lambda}\left(1 + \lambda + \frac{\lambda^2}{2!} + \frac{\lambda^3}{3!} + \cdots\right)$, *taken from left to right'*

The symbol λ is the value of the average occurrence of the event.
For example, a production department has 35 similar milling machines. The number of breakdowns on each machine averages 0.06 per week.
Since the average occurrence of a breakdown is known but the number of times when a machine did not break down is unknown, a Poisson distribution must be used. The expectation of a breakdown for 35 machines is 35×0.06, i.e. 2.1 breakdowns per week. The probabilities of a breakdown occurring 0,1, 2,... times are given by the successive terms of the expression $e^{-\lambda}\left(1 + \lambda + \frac{\lambda^2}{2!} + \frac{\lambda^3}{3!} + \ldots\right)$, taken from left to right. Hence:

$$\text{the probability of no breakdowns } e^{-\lambda} = e^{-2.1} = \mathbf{0.1225}$$

$$\text{the probability of 1 breakdown is } \lambda e^{-\lambda} = 2.1e^{-2.1} = \mathbf{0.2572}$$

$$\text{the probability of 2 breakdowns is } \frac{\lambda^2 e^{-\lambda}}{2!} = \frac{2.1^2 e^{-2.1}}{2 \times 1} = \mathbf{0.2700}$$

The probability of less than 3 breakdowns per week is the sum of the probabilities of 0, 1 and 2 breakdowns per week,

i.e. $0.1225 + 0.2572 + 0.2700$, i.e. **0.6497**

---

80 The Normal Distribution

Introduction to the normal distribution

When data is obtained, it can frequently be considered to be a sample (i.e. a few members) drawn at random from a large population (i.e. a set having many members). If the sample number is large, it is theoretically possible to choose class intervals which are very small, but which still have a number of members falling within each class. A frequency polygon of this data then has a large number of small line segments and approximates to a continuous curve. Such a curve is called a **frequency or a distribution curve**.

An extremely important symmetrical distribution curve is called the **normal curve** and is as shown in Figure 80.1. This curve can be described by a mathematical equation and is the basis of much of the work done in more advanced statistics. Many natural occurrences such as the heights or weights of a group of people, the sizes of components produced by a particular machine and the life length of certain components approximate to a normal distribution.

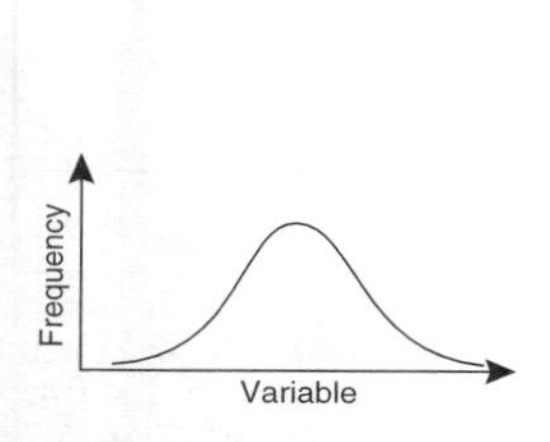

Figure 80.1

Figure 80.2

Normal distribution curves can differ from one another in the following four ways:

(a) by having different mean values
(b) by having different values of standard deviations
(c) the variables having different values and different units and
(d) by having different areas between the curve and the horizontal axis.

A normal distribution curve is **standardised** as follows:

(a) The mean value of the unstandardized curve is made the origin, thus making the mean value, $\bar{x}$, zero.
(b) The horizontal axis is scaled in standard deviations. This is done by letting $z = \dfrac{x - \bar{x}}{\sigma}$, where z is called the **normal standard variate**, x is the value of the variable, $\bar{x}$ is the mean value of the distribution and σ is the standard deviation of the distribution.
(c) The area between the normal curve and the horizontal axis is made equal to unity.

When a normal distribution curve has been standardised, the normal curve is called a **standardised normal curve** or a **normal probability curve**, and any normally distributed data may be represented by the **same** normal probability curve.

The area under part of a normal probability curve is directly proportional to probability and the value of the shaded area shown in Figure 80.2 can be determined by evaluating:

$$\int \frac{1}{\sqrt{(2\pi)}} e^{(z^2/2)}\, dz, \quad \text{where } z = \frac{x - \bar{x}}{\sigma}$$

To save repeatedly determining the values of this function, tables of partial areas under the standardised normal curve are available in many mathematical formulae books, and such a table is shown in Table 80.1.

For example, let the mean height of 500 people be 170 cm and the standard deviation be 9 cm. Assuming the heights are normally distributed, the number

Table 80.1 Partial areas under the standardised normal curve

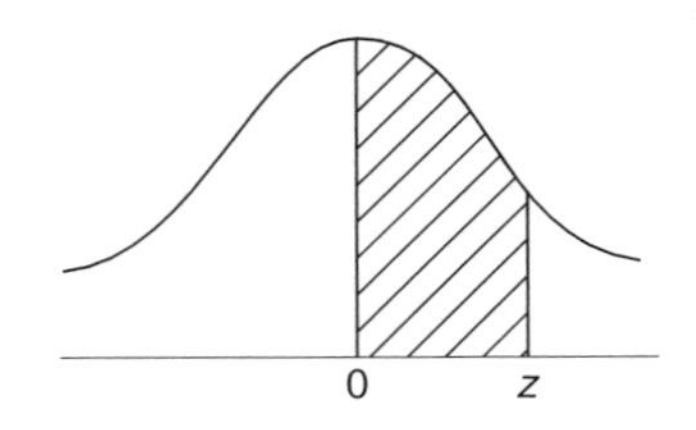

$z = \frac{x - \bar{x}}{\sigma}$	0	1	2	3	4	5	6	7	8	9
0.0	0.0000	0.0040	0.0080	0.0120	0.0159	0.0199	0.0239	0.0279	0.0319	0.0359
0.1	0.0398	0.0438	0.0478	0.0517	0.0557	0.0596	0.0636	0.0678	0.0714	0.0753
0.2	0.0793	0.0832	0.0871	0.0910	0.0948	0.0987	0.1026	0.1064	0.1103	0.1141
0.3	0.1179	0.1217	0.1255	0.1293	0.1331	0.1388	0.1406	0.1443	0.1480	0.1517
0.4	0.1554	0.1591	0.1628	0.1664	0.1700	0.1736	0.1772	0.1808	0.1844	0.1879
0.5	0.1915	0.1950	0.1985	0.2019	0.2054	0.2086	0.2123	0.2157	0.2190	0.2224
0.6	0.2257	0.2291	0.2324	0.2357	0.2389	0.2422	0.2454	0.2486	0.2517	0.2549
0.7	0.2580	0.2611	0.2642	0.2673	0.2704	0.2734	0.2760	0.2794	0.2823	0.2852
0.8	0.2881	0.2910	0.2939	0.2967	0.2995	0.3023	0.3051	0.3078	0.3106	0.3133
0.9	0.3159	0.3186	0.3212	0.3238	0.3264	0.3289	0.3315	0.3340	0.3365	0.3389
1.0	0.3413	0.3438	0.3451	0.3485	0.3508	0.3531	0.3554	0.3577	0.3599	0.3621
1.1	0.3643	0.3665	0.3686	0.3708	0.3729	0.3749	0.3770	0.3790	0.3810	0.3830
1.2	0.3849	0.3869	0.3888	0.3907	0.3925	0.3944	0.3962	0.3980	0.3997	0.4015
1.3	0.4032	0.4049	0.4066	0.4082	0.4099	0.4115	0.4131	0.4147	0.4162	0.4177
1.4	0.4192	0.4207	0.4222	0.4236	0.4251	0.4265	0.4279	0.4292	0.4306	0.4319
1.5	0.4332	0.4345	0.4357	0.4370	0.4382	0.4394	0.4406	0.4418	0.4430	0.4441
1.6	0.4452	0.4463	0.4474	0.4484	0.4495	0.4505	0.4515	0.4525	0.4535	0.4545
1.7	0.4554	0.4564	0.4573	0.4582	0.4591	0.4599	0.4608	0.4616	0.4625	0.4633
1.8	0.4641	0.4649	0.4656	0.4664	0.4671	0.4678	0.4686	0.4693	0.4699	0.4706
1.9	0.4713	0.4719	0.4726	0.4732	0.4738	0.4744	0.4750	0.4756	0.4762	0.4767
2.0	0.4772	0.4778	0.4783	0.4785	0.4793	0.4798	0.4803	0.4808	0.4812	0.4817
2.1	0.4821	0.4826	0.4830	0.4834	0.4838	0.4842	0.4846	0.4850	0.4854	0.4857
2.2	0.4861	0.4864	0.4868	0.4871	0.4875	0.4878	0.4881	0.4884	0.4882	0.4890
2.3	0.4893	0.4896	0.4898	0.4901	0.4904	0.4906	0.4909	0.4911	0.4913	0.4916
2.4	0.4918	0.4920	0.4922	0.4925	0.4927	0.4929	0.4931	0.4932	0.4934	0.4936
2.5	0.4938	0.4940	0.4941	0.4943	0.4945	0.4946	0.4948	0.4949	0.4951	0.4952
2.6	0.4953	0.4955	0.4956	0.4957	0.4959	0.4960	0.4961	0.4962	0.4963	0.4964
2.7	0.4965	0.4966	0.4967	0.4968	0.4969	0.4970	0.4971	0.4972	0.4973	0.4974
2.8	0.4974	0.4975	0.4076	0.4977	0.4977	0.4978	0.4979	0.4980	0.4980	0.4981
2.9	0.4981	0.4982	0.4982	0.4983	0.4984	0.4984	0.4985	0.4985	0.4986	0.4986
3.0	0.4987	0.4987	0.4987	0.4988	0.4988	0.4989	0.4989	0.4989	0.4990	0.4990
3.1	0.4990	0.4991	0.4991	0.4991	0.4992	0.4992	0.4992	0.4992	0.4993	0.4993
3.2	0.4993	0.4993	0.4994	0.4994	0.4994	0.4994	0.4994	0.4995	0.4995	0.4995
3.3	0.4995	0.4995	0.4995	0.4996	0.4996	0.4996	0.4996	0.4996	0.4996	0.4997
3.4	0.4997	0.4997	0.4997	0.4997	0.4997	0.4997	0.4997	0.4997	0.4997	0.4998
3.5	0.4998	0.4998	0.4998	0.4998	0.4998	0.4998	0.4998	0.4998	0.4998	0.4998
3.6	0.4998	0.4998	0.4999	0.4999	0.4999	0.4999	0.4999	0.4999	0.4999	0.4999
3.7	0.4999	0.4999	0.4999	0.4999	0.4999	0.4999	0.4999	0.4999	0.4999	0.4999
3.8	0.4999	0.4999	0.4999	0.4999	0.4999	0.4999	0.4999	0.4999	0.4999	0.4999
3.9	0.5000	0.5000	0.5000	0.5000	0.5000	0.5000	0.5000	0.5000	0.5000	0.5000

of people likely to have heights between 150 cm and 195 cm is determined as follows:

The mean value, $\bar{x}$, is 170 cm and corresponds to a normal standard variate value, z, of zero on the standardised normal curve. A height of 150 cm has a z-value given by $z = \frac{x - \bar{x}}{\sigma}$ standard deviations, i.e. $\frac{150 - 170}{9}$ or -2.22 standard deviations. Using a table of partial areas beneath the standardised normal curve (see Table 80.1), a z-value of -2.22 corresponds to an area of 0.4868 between the mean value and the ordinate $z = -2.22$. The negative z-value shows that it lies to the left of the $z = 0$ ordinate.

This area is shown shaded in Figure 80.3(a). Similarly, 195 cm has a z-value of $\frac{195 - 170}{9}$ that is 2.78 standard deviations. From Table 80.1, this value of z corresponds to an area of 0.4973, the positive value of z showing that it lies to the right of the $z = 0$ ordinate. This area is shown shaded in Figure 80.3(b). The total area shaded in Figures 80.3(a) and (b) is shown in Figure 80.3(c) and is $0.4868 + 0.4973$, i.e. 0.9841 of the total area beneath the curve.

However, the area is directly proportional to probability. Thus, the probability that a person will have a height of between 150 and 195 cm is 0.9841. For a group of 500 people, 500×0.9841, i.e. **492 people are likely to have heights in this range**. The value of 500×0.9841 is 492.05, but since answers based on a normal probability distribution can only be approximate, results are usually given correct to the nearest whole number.

Similarly, the number of people likely to have heights of less than 165 cm is determined as follows:

A height of 165 cm corresponds to $\frac{165 - 170}{9}$, i.e. -0.56 standard deviations.

The area between $z = 0$ and $z = -0.56$ (from Table 80.1) is 0.2123, shown shaded in Figure 80.4(a). The total area under the standardised normal curve

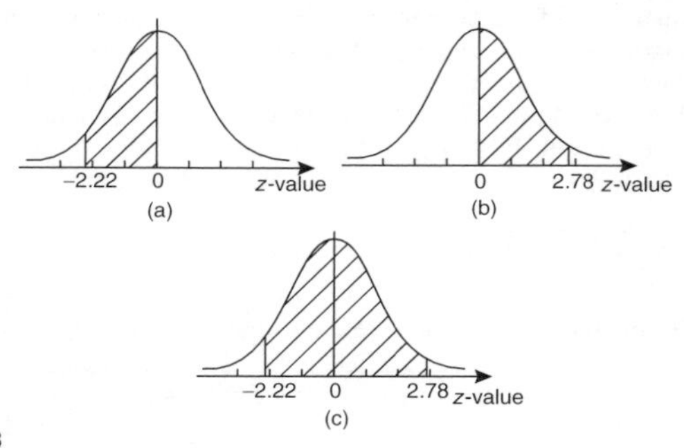

Figure 80.3

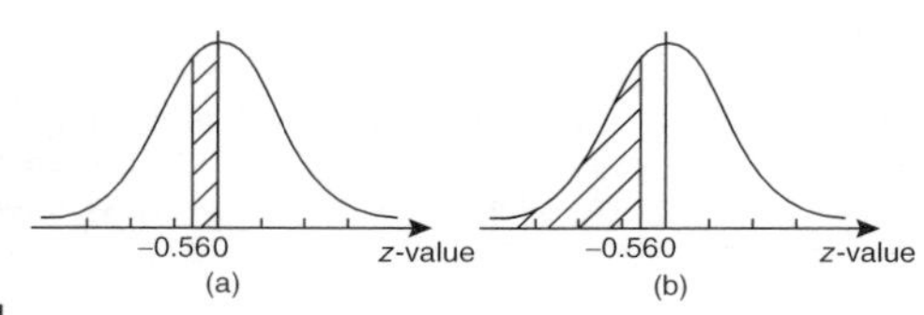

Figure 80.4

is unity and since the curve is symmetrical, it follows that the total area to the left of the $z = 0$ ordinate is 0.5000. Thus the area to the left of the $z = -0.56$ ordinate ('left' means 'less than', 'right' means 'more than') is $0.5000 - 0.2123$, i.e. 0.2877 of the total area, which is shown shaded in Figure 80.4(b). The area is directly proportional to probability and since the total area beneath the standardised normal curve is unity, the probability of a person's height being less than 165 cm is 0.2877. For a group of 500 people, 500×0.2877, i.e. **144 people are likely to have heights of less than 165 cm**.

Testing for a normal distribution

It should never be assumed that because data is continuous it automatically follows that it is normally distributed. One way of checking that data is normally distributed is by using **normal probability paper**, often just called **probability paper**. This is special graph paper which has linear markings on one axis and percentage probability values from 0.01 to 99.99 on the other axis (see Figure 80.5). The divisions on the probability axis are such that a straight line graph results for normally distributed data when percentage cumulative frequency values are plotted against upper class boundary values. If the points do not lie in a reasonably straight line, then the data is not normally distributed.
The mean value and standard deviation of normally distributed data may be determined using normal probability paper. For normally distributed data, the area beneath the standardised normal curve and a z-value of unity (i.e. one standard deviation) may be obtained from Table 80.1. For one standard deviation, this area is 0.3413, i.e. 34.13%. An area of ± 1 standard deviation is symmetrically placed on either side of the $z = 0$ value, i.e. is symmetrically placed on either side of the 50 per cent cumulative frequency value. Thus an area corresponding to ± 1 standard deviation extends from percentage cumulative frequency values of $(50 + 34.13)\%$ to $(50 - 34.13)\%$, i.e. from 84.13% to 15.87%. For most purposes, these values are taken as 84% and 16%. Thus, when using normal probability paper, the standard deviation of the distribution is given by:

$$\frac{\begin{pmatrix}\text{variable value for 84\%} \\ \text{cumulative frequency}\end{pmatrix} - \begin{pmatrix}\text{variable value for 16\%} \\ \text{cumalative frequency}\end{pmatrix}}{2}$$

For example, the data given below refers to the masses of 50 copper ingots.

Class mid-point value (kg)	29.5	30.5	31.5	32.5	33.5	34.5	35.5	36.5	37.5	38.5
Frequency	2	4	6	8	9	8	6	4	2	1

To test the normality of a distribution, the upper class boundary/percentage cumulative frequency values are plotted on normal probability paper. The upper class boundary values are: 30, 31, 32, ..., 38, 39. The corresponding cumulative frequency values (for 'less than' the upper class boundary values) are: 2, $(4 + 2) = 6$, $(6 + 4 + 2) = 12$, 20, 29, 37, 43, 47, 49 and 50.

The corresponding percentage cumulative frequency values are $\frac{2}{50} \times 100 = 4$, $\frac{6}{50} \times 100 = 12$, 24, 40, 58, 74, 86, 94, 98 and 100%

The co-ordinates of upper class boundary/percentage cumulative frequency values are plotted as shown in Figure 80.5. When plotting these values, it will always be found that the co-ordinate for the 100% cumulative frequency value cannot be plotted, since the maximum value on the probability scale is 99.99. **Since the points plotted in Figure 80.5 lie very nearly in a straight line, the data is approximately normally distributed**.

The mean value and standard deviation can be determined from Figure 80.5. Since a normal curve is symmetrical, the mean value is the value of the variable corresponding to a 50% cumulative frequency value, shown as point P on the graph. This shows that **the mean value is 33.6 kg**. The standard deviation is determined using the 84% and 16% cumulative frequency values, shown as Q and R in Figure 80.5. The variable values for Q and R are 35.7 and 31.4 respectively; thus two standard deviations correspond to 35.7 − 31.4, i.e. 4.3, showing that the standard deviation of the distribution is approximately $\frac{4.3}{2}$ i.e. **2.15 standard deviations**.

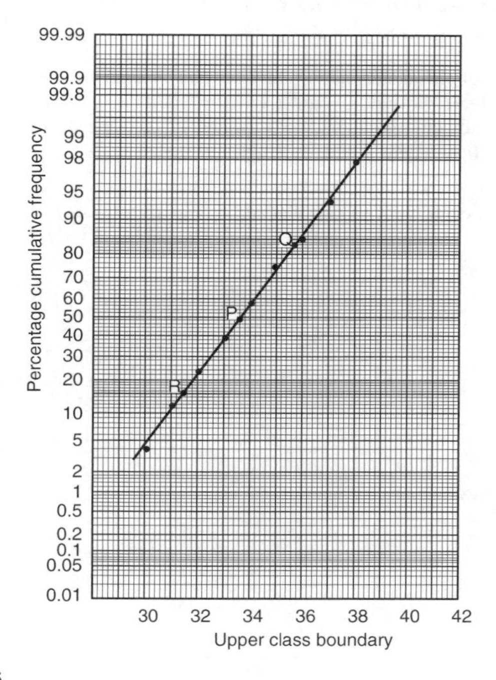

Figure 80.5

81 Linear Correlation

Introduction to linear correlation

Correlation is a measure of the amount of association existing between two variables. For linear correlation, if points are plotted on a graph and all the points lie on a straight line, then **perfect linear correlation** is said to exist. When a straight line having a positive gradient can reasonably be drawn through points on a graph **positive or direct linear correlation** exists, as shown in Figure 81.1(a). Similarly, when a straight line having a negative gradient can reasonably be drawn through points on a graph, **negative or inverse linear correlation** exists, as shown in Figure 81.1(b). When there is no apparent relationship between co-ordinate values plotted on a graph then **no correlation** exists between the points, as shown in Figure 81.1(c). In statistics, when two variables are being investigated, the location of the co-ordinates on a rectangular co-ordinate system is called a **scatter diagram** — as shown in Figure 81.1.

The product-moment formula for determining the linear correlation coefficient

The amount of linear correlation between two variables is expressed by a **coefficient of correlation**, given the symbol r. This is defined in terms of the deviations of the co-ordinates of two variables from their mean values and is given by the **product-moment formula** which states:

$$\textbf{coefficient of correlation, } r = \frac{\Sigma xy}{\sqrt{\{(\Sigma x^2)(\Sigma y^2)\}}} \qquad (1)$$

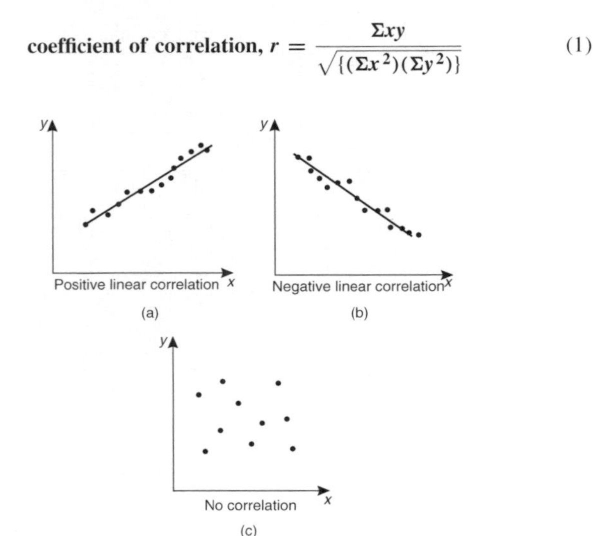

Figure 81.1

where the x-values are the values of the deviations of co-ordinates X from $\overline{X}$, their mean value and the y-values are the values of the deviations of co-ordinates Y from $\overline{Y}$, their mean value, i.e. $x = (X - \overline{X})$ and $y = (Y - \overline{Y})$. The results of this determination give values of r lying between $+1$ and -1, where $+1$ indicates perfect direct correlation, -1 indicates perfect inverse correlation and 0 indicates that no correlation exists. Between these values, the smaller the value of r, the less is the amount of correlation which exists. Generally, values of r in the ranges 0.7 to 1 and -0.7 to -1 show that a fair amount of correlation exists.

For example, in an experiment to determine the relationship between force on a wire and the resulting extension, the following data is obtained:

Force (N)	10	20	30	40	50	60	70
Extension (mm)	0.22	0.40	0.61	0.85	1.20	1.45	1.70

The linear coefficient of correlation for this data is obtained as follows:
Let X be the variable force values and Y be the dependent variable extension values. The coefficient of correlation is given by:

$$r = \frac{\Sigma xy}{\sqrt{\{(\Sigma x^2)(\Sigma y^2)\}}}$$

where $x = (X - \overline{X})$ and $y = (Y - \overline{Y})$, $\overline{X}$ and $\overline{Y}$ being the mean values of the X and Y values respectively. Using a tabular method to determine the quantities of this formula gives:

X	Y	$x = (X - \overline{X})$	$y = (Y - \overline{Y})$	xy	x^2	y^2
10	0.22	−30	−0.699	20.97	900	0.489
20	0.40	−20	−0.519	10.38	400	0.269
30	0.61	−10	−0.309	3.09	100	0.095
40	0.85	0	−0.069	0	0	0.005
50	1.20	10	0.281	2.81	100	0.079
60	1.45	20	0.531	10.62	400	0.282
70	1.70	30	0.781	23.43	900	0.610
$\Sigma X = 280$ $\overline{X} = \frac{280}{7}$ $= 40$	$\Sigma Y = 6.43$ $\overline{Y} = \frac{6.43}{7}$ $\overline{Y} = 0.919$			$\Sigma xy =$ 71.30	$\Sigma x^2 =$ 2800	$\Sigma y^2 =$ 1.829

$$\text{Thus } r = \frac{71.3}{\sqrt{[2800 \times 1.829]}} = \mathbf{0.996}$$

This shows that a **very good direct correlation exists** between the values of force and extension.

The significance of a coefficient of correlation

When the value of the coefficient of correlation has been obtained from the product moment formula, some care is needed before coming to conclusions based on this result. Checks should be made to ascertain the following two points:

(a) that a 'cause and effect' relationship exists between the variables; it is relatively easy, mathematically, to show that some correlation exists between, say, the number of ice creams sold in a given period of time and the number of chimneys swept in the same period of time, although there is no relationship between these variables;
(b) that a linear relationship exists between the variables; the product-moment formula given above is based on linear correlation. Perfect non-linear correlation may exist (for example, the co-ordinates exactly following the curve $y = x^3$), but this gives a low value of coefficient of correlation since the value of r is determined using the product-moment formula, based on a linear relationship.

---

82 Linear Regression

Introduction to linear regression

Regression analysis, usually termed **regression**, is used to draw the line of 'best fit' through co-ordinates on a graph. The techniques used enable a mathematical equation of the straight line form $y = mx + c$ to be deduced for a given set of co-ordinate values, the line being such that the sum of the deviations of the co-ordinate values from the line is a minimum, i.e. it is the line of 'best fit'. When a regression analysis is made, it is possible to obtain two lines of best fit, depending on which variable is selected as the dependent variable and which variable is the independent variable. For example, in a resistive electrical circuit, the current flowing is directly proportional to the voltage applied to the circuit. There are two ways of obtaining experimental values relating the current and voltage. Either, certain voltages are applied to the circuit and the current values are measured, in which case the voltage is the independent variable and the current is the dependent variable; or, the voltage can be adjusted until a desired value of current is flowing and the value of voltage is measured, in which case the current is the independent value and the voltage is the dependent value.

The least-squares regression lines

For a given set of co-ordinate values, (X_1, Y_1), (X_2, Y_2),..., (X_N, Y_N) let the X values be the independent variables and the Y-values be the dependent values. Also let D_1,..., D_N be the vertical distances between the line shown as PQ in Figure 82.1 and the points representing the co-ordinate values. The least-squares regression line, i.e. the line of best fit, is the line which makes the value of $D_1^2 + D_2^2 + \cdots + D_N^2$ a minimum value.
The equation of the least-squares regression line is usually written as $Y = a_0 + a_1X$, where a_0 is the Y-axis intercept value and a_1 is the gradient of the line (analogous to c and m in the equation $y = mx + c$). The values of a_0 and a_1 to make the sum of the 'deviations squared' a minimum can be obtained

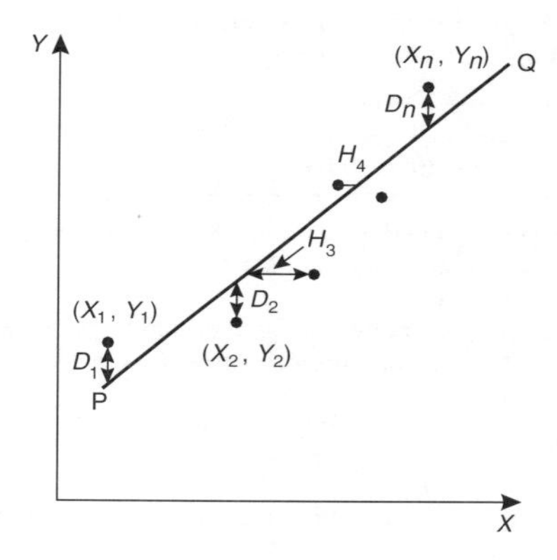

Figure 82.1

from the two equations:

$$\sum Y = a_0 N + a_1 \sum X \tag{1}$$

$$\sum (XY) = a_0 \sum X + a_1 \sum X^2 \tag{2}$$

where X and Y are the co-ordinate values, N is the number of co-ordinates and a_0 and a_1 are called the **regression coefficients** of Y on X. Equations (1) and (2) are called the **normal equations** of the regression line of Y on X. The regression line of Y on X is used to estimate values of Y for given values of X.

If the Y-values (vertical-axis) are selected as the independent variables, the horizontal distances between the line shown as PQ in Figure 82.1 and the co-ordinate values (H_3, H_4, etc.) are taken as the deviations. The equation of the regression line is of the form: $X = b_0 + b_1 Y$ and the normal equations become:

$$\sum X = b_0 N + b_1 \sum Y \tag{3}$$

$$\sum (XY) = b_0 \sum Y + b_1 \sum Y^2 \tag{4}$$

where X and Y are the co-ordinate values, b_0 and b_1 are the regression coefficients of X on Y and N is the number of co-ordinates. These normal equations are of the regression line of X on Y, which is slightly different to the regression line of Y on X. The regression line of X on Y is used to estimate values of X for given values of Y. The regression line of Y on X is used to determine any value of Y corresponding to a given value of X. If the value of Y lies within the range of Y-values of the extreme co-ordinates, the process of finding the corresponding value of X is called **linear interpolation**. If it lies outside of the range of Y-values of the extreme co-ordinates then the process is called **linear extrapolation** and the assumption must be made that the line of best fit extends outside of the range of the co-ordinate values given. By using the regression line of X on Y, values of X corresponding to given values of Y may be found by either interpolation or extrapolation.

For example, the experimental values relating centripetal force and radius, for a mass travelling at constant velocity in a circle, are as shown:

Force (N)	5	10	15	20	25	30	35	40
Radius (cm)	55	30	16	12	11	9	7	5

Let the radius be the independent variable X, and the force be the dependent variable Y. (This decision is usually based on a 'cause' corresponding to X and an 'effect' corresponding to Y).
The equation of the regression line of force on radius is of the form $Y = a_0 + a_1X$ and the constants a_0 and a_1 are determined from the normal equations:

$$\sum Y = a_0N + a_1 \sum X \quad \text{and} \quad \sum XY = a_0 \sum X + a_1 \sum X^2$$

(from equations (1) and (2))
Using a tabular approach to determine the values of the summations gives:

Radius, X	Force, Y	X^2	XY	Y^2
55	5	3025	275	25
30	10	900	300	100
16	15	256	240	225
12	20	144	240	400
11	25	121	275	625
9	30	81	270	900
7	35	49	245	1225
5	40	25	200	1600
$\sum X = 145$	$\sum Y = 180$	$\sum X^2 = 4601$	$\sum XY = 2045$	$\sum Y^2 = 5100$

Thus $180 = 8a_0 + 145a_1$ and $2045 = 145a_0 + 4601a_1$
Solving these simultaneous equations gives $a_0 = 33.7$ and $a_1 = -0.617$, correct to 3 significant figures. Thus the equation of the regression line of force on radius is: $\boldsymbol{Y = 33.7 - 0.617\,X}$
Thus the force, Y, at a radius of, say, 40 cm, is

$$Y = 33.7 - 0.617(40) = 9.02,$$

i.e. **the force at a radius of 40 cm is 9.02 N**
The equation of the regression line of radius on force is of the form $X = b_0 + b_1Y$ and the constants b_0 and b_1 are determined from the normal equations:

$$\sum X = b_0N + b_1 \sum Y \quad \text{and} \quad \sum XY = b_0 \sum Y + b_1 \sum Y^2$$

(from equations (3) and (4))
The values of the summations have been obtained above giving:

$$145 = 8b_0 + 180b_1 \quad \text{and} \quad 2045 = 180b_0 + 5100b_1$$

Solving these simultaneous equations gives $b_0 = 44.2$ and $b_1 = -1.16$, correct to 3 significant figures. Thus the equation of the regression line of radius on force is: $\boldsymbol{X = 44.2 - 1.16\,Y}$

Thus, the radius, X, when the force is, say, 32 newtons is

$$X = 44.2 - 1.16(32) = 7.08,$$

i.e. **the radius when the force is 32 N is 7.08 cm**.

83 Sampling and Estimation Theories

Introduction

The concepts of elementary sampling theory and estimation theories introduced in this chapter will be provide the basis for a more detailed study of inspection, control and quality control techniques used in industry. Such theories can be quite complicated; in this chapter a full treatment of the theories and the derivation of formulae have been omitted for clarity-basic concepts only have been developed.

Sampling distributions

In statistics, it is not always possible to take into account all the members of a set and in these circumstances, a **sample**, or many samples, are drawn from a population. Usually when the word sample is used, it means that a **random sample** is taken. If each member of a population has the same chance of being selected, then a sample taken from that population is called random. A sample that is not random is said to be **biased** and this usually occurs when some influence affects the selection.

When it is necessary to make predictions about a population based on random sampling, often many samples of, say, N members are taken, before the predictions are made. If the mean value and standard deviation of each of the samples is calculated, it is found that the results vary from sample to sample, even though the samples are all taken from the same population. In the theories introduced in the following sections, it is important to know whether the differences in the values obtained are due to chance or whether the differences obtained are related in some way. If M samples of N members are drawn at random from a population, the mean values for the M samples together form a set of data. Similarly, the standard deviations of the M samples collectively form a set of data. Sets of data based on many samples drawn from a population are called **sampling distributions**. They are often used to describe the chance fluctuations of mean values and standard deviations based on random sampling.

The sampling distribution of the means

Suppose that it is required to obtain a sample of two items from a set containing five items. If the set is the five letters A, B, C, D and E, then the different samples which are possible are:

$$AB, AC, AD, AE, BC, BD, BE, CD, CE \text{ and } DE,$$

that is, ten different samples. The number of possible different samples in this case is given by $\frac{5 \times 4}{2 \times 1}$ i.e. 10. Similarly, the number of different ways in which a sample of three items can be drawn from a set having ten members can be shown to be $\frac{10 \times 9 \times 8}{3 \times 2 \times 1}$ i.e. 120. It follows that when a small sample is drawn from a large population, there are very many different combinations of members possible. With so many different samples possible, quite a large variation can occur in the mean values of various samples taken from the same population.

Usually, the greater the number of members in a sample, the closer will be the mean value of the sample to that of the population. Consider the set of numbers 3, 4, 5, 6 and 7. For a sample of 2 members, the lowest value of the mean is $\frac{3+4}{2}$, i.e. 3.5; the highest is $\frac{6+7}{2}$, i.e. 6.5, giving a range of mean values of $6.5 - 3.5 = 3$. For a sample of 3 members, the range is, $\frac{3+4+5}{3}$ to $\frac{5+6+7}{3}$ that is, 2. As the number in the sample increases, the range decreases until, in the limit, if the sample contains all the members of the set, the range of mean values is zero. When many samples are drawn from a population and a sample distribution of the mean values of the samples is formed, the range of the mean values is small provided the number in the sample is large. Because the range is small it follows that the standard deviation of all the mean values will also be small, since it depends on the distance of the mean values from the distribution mean. The relationship between the standard deviation of the mean values of a sampling distribution and the number in each sample can be expressed as follows:

Theorem 1 *'If all possible samples of size N are drawn from a finite population, N_p, without replacement, and the standard deviation of the mean values of the sampling distribution of means is determined, then:*

$$\sigma_{\overline{x}} = \frac{\sigma}{\sqrt{N}}\sqrt{\left(\frac{N_p - N}{N_p - 1}\right)}$$

where $\sigma_{\overline{x}}$ is the standard deviation of the sampling distribution of means and σ is the standard deviation of the population'

The standard deviation of a sampling distribution of mean values is called the **standard error of the means**, thus

$$\textbf{standard error of the means, } \sigma_{\overline{x}} = \frac{\sigma}{\sqrt{N}}\sqrt{\left(\frac{N_p - N}{N_p - 1}\right)} \qquad (1)$$

Equation (1) is used for a finite population of size N_p and/or for sampling without replacement. The word 'error' in the 'standard error of the means' does not mean that a mistake has been made but rather that there is a degree of uncertainty in predicting the mean value of a population based on the mean values of the samples. The formula for the standard error of the means is true for all values of the number in the sample, N. When N_p is very large compared with N or when the population is infinite (this can be considered to be the case when sampling is done with replacement), the correction factor

$\sqrt{\left(\frac{N_p - N}{N_p - 1}\right)}$ approaches unity and equation (1) becomes:

$$\sigma_{\overline{x}} = \frac{\sigma}{\sqrt{N}} \tag{2}$$

Equation (2) is used for an infinite population and/or for sampling with replacement.

Theorem 2 *'If all possible samples of size N are drawn from a population of size N_p and the mean value of the sampling distribution of means $\mu_{\overline{x}}$ is determined then*

$$\mu_{\overline{x}} = \mu \tag{3}$$

where μ is the mean value of the population'

In practice, all possible samples of size N are not drawn from the population. However, if the sample size is large (usually taken as 30 or more), then the relationship between the mean of the sampling distribution of means and the mean of the population is very near to that shown in equation (3). Similarly, the relationship between the standard error of the means and the standard deviation of the population is very near to that shown in equation (2).

Another important property of a sampling distribution is that when the sample size, N, is large, **the sampling distribution of means approximates to a normal distribution**, of mean value $\mu_{\overline{x}}$ and standard deviation $\sigma_{\overline{x}}$. This is true for all normally distributed populations and also for populations that are not normally distributed provided the population size is at least twice as large as the sample size. This property of normality of a sampling distribution is based on a special case of the 'central limit theorem', an important theorem relating to sampling theory. Because the sampling distribution of means and standard deviations is normally distributed, the table of the partial areas under the standardised normal curve (shown in Table 80.1 on page 394) can be used to determine the probabilities of a particular sample lying between, say, ± 1 standard deviation, and so on.

For example, the heights of 3000 people are normally distributed with a mean of 175 cm and a standard deviation of 8 cm. Random samples are taken of 40 people. The standard deviation and the mean of the sampling distribution of means if sampling is done (a) with replacement, and (b) without replacement, may be predicted as follows:

For the population: number of members, $N_p = 3000$;

standard deviation, $\sigma = 8$ cm;

mean, $\mu = 175$ cm

For the samples: number in each sample, $N = 40$

(a) When sampling is done **with replacement**, the total number of possible samples (two or more can be the same) is infinite. Hence, from equation (2) the **standard error of the mean (i.e. the standard deviation of the sampling distribution of means)**

$$\sigma_{\overline{x}} = \frac{\sigma}{\sqrt{N}} = \frac{8}{\sqrt{40}} = \mathbf{1.265\ cm}$$

From equation (3), **the mean of the sampling distribution**

$$\mu_{\overline{x}} = \mu = \mathbf{175\ cm}$$

(b) When sampling is done **without replacement**, the total number of possible samples is finite and hence equation (1) applies. Thus **the standard error of the means**

$$\sigma_{\bar{x}} = \frac{\sigma}{\sqrt{N}}\sqrt{\left(\frac{N_p - N}{N_p - 1}\right)} = \frac{8}{\sqrt{40}}\sqrt{\left(\frac{3000 - 40}{3000 - 1}\right)}$$

$$= (1.265)(0.9935) = \mathbf{1.257\ cm}$$

As stated, following equation (3), provided the sample size is large, the mean of the sampling distribution of means is the same for both finite and infinite populations. Hence, from equation (3), $\mu_{\bar{x}} = \mathbf{175\ cm}$

The estimation of population parameters based on a large sample size

When a population is large, it is not practical to determine its mean and standard deviation by using the basic formulae for these parameters. In fact, when a population is infinite, it is impossible to determine these values. For large and infinite populations the values of the mean and standard deviation may be estimated by using the data obtained from samples drawn from the population.

Point and interval estimates

An estimate of a population parameter, such as mean or standard deviation, based on a single number is called a **point estimate**. An estimate of a population parameter given by two numbers between which the parameter may be considered to lie is called an **interval estimate**. Thus if an estimate is made of the length of an object and the result is quoted as 150 cm, this is a point estimate. If the result is quoted as 150 ± 10 cm, this is an interval estimate and indicates that the length lies between 140 and 160 cm. Generally, a point estimate does not indicate how close the value is to the true value of the quantity and should be accompanied by additional information on which its merits may be judged. A statement of the error or the precision of an estimate is often called its **reliability**. In statistics, when estimates are made of population parameters based on samples, usually interval estimates are used. The word estimate does not suggest that we adopt the approach 'let's guess that the mean value is about..', but rather that a value is carefully selected and the degree of confidence which can be placed in the estimate is given in addition.

Confidence intervals

It is stated earlier that when samples are taken from a population, the mean values of these samples are approximately normally distributed, that is, the mean values forming the sampling distribution of means is approximately normally distributed. It is also true that if the standard deviation of each of the samples is found, then the standard deviations of all the samples are approximately normally distributed, that is, the standard deviations of the sampling distribution of standard deviations are approximately normally distributed. Parameters such as the mean or the standard deviation of a sampling distribution are called **sampling statistics**, S. Let μ_S be the mean value of a sampling statistic of the sampling distribution, that is, the mean value of the means of the samples or the mean value of the standard deviations of the samples. Also let σ_S be the standard deviation of a sampling statistic of the sampling distribution, that is, the standard deviation of the means of the samples or the standard deviation of the standard deviations of the samples. Because the sampling distribution of the means and of the standard deviations are normally distributed, it is possible to predict the probability of the sampling statistic lying in the intervals:

mean $\pm$ 1 standard deviation,

mean $\pm$ 2 standard deviations,

or mean $\pm$ 3 standard deviations,

by using tables of the partial areas under the standardised normal curve given in Table 80.1 on page 394. From this table, the area corresponding to a z-value of $+1$ standard deviation is 0.3413, thus the area corresponding to ± 1 standard deviation is 2×0.3413, that is, 0.6826. Thus the percentage probability of a sampling statistic lying between the mean ± 1 standard deviation is 68.26%. Similarly, the probability of a sampling statistic lying between the mean ± 2 standard deviations is 95.44% and of lying between the mean ± 3 standard deviations is 99.74%.

The values 68.26%, 95.44% and 99.74% are called the **confidence levels** for estimating a sampling statistic. A confidence level of 68.26% is associated with two distinct values, these being, $S -$ (1 standard deviation), i.e. $S - \sigma_S$ and $S +$ (1 standard deviation), i.e. $S + \sigma_S$.These two values are called the **confidence limits** of the estimate and the distance between the confidence limits is called the **confidence interval**. A confidence interval indicates the expectation or confidence of finding an estimate of the population statistic in that interval, based on a sampling statistic. The list in Table 83.1 is based on values given in Table 80.1, and gives some of the confidence levels used in practice and their associated z-values; (some of the values given are based on interpolation). When the table is used in this context, z-values are usually indicated by 'z_C' and are called the **confidence coefficients**.

Any other values of confidence levels and their associated confidence coefficients can be obtained using Table 80.1.

For example, to determine the confidence coefficient corresponding to a confidence level of 98.5%:

98.5% is equivalent to a per unit value of 0.9850. This indicates that the area under the standardised normal curve between $-z_C$ and $+z_C$, i.e. corresponding to $2z_C$, is 0.9850 of the total area. Hence the area between the mean value and z_C is $\frac{0.9850}{2}$ i.e. 0.4925 of the total area. The z-value

Table 83.1

Confidence level, %	99	98	96	95	90	80	50
Confidence coefficient, z_C	2.58	2.33	2.05	1.96	1.645	1.28	0.6745

corresponding to a partial area of 0.4925 is 2.43 standard deviations from Table 80.1. Thus, **the confidence coefficient corresponding to a confidence limit of 98.5% is 2.43**

(a) Estimating the mean of a population when the standard deviation of the population is known

When a sample is drawn from a large population whose standard deviation is known, the mean value of the sample, $\overline{x}$, can be determined. This mean value can be used to make an estimate of the mean value of the population, μ. When this is done, the estimated mean value of the population is given as lying between two values, that is, lying in the confidence interval between the confidence limits. If a high level of confidence is required in the estimated value of μ, then the range of the confidence interval will be large. For example, if the required confidence level is 96%, then from Table 83.1 the confidence interval is from $-z_C$ to $+z_C$, that is, $2 \times 2.05 = 4.10$ standard deviations wide. Conversely, a low level of confidence has a narrow confidence interval and a confidence level of, say, 50%, has a confidence interval of 2×0.6745, that is 1.3490 standard deviations. The 68.26% confidence level for an estimate of the population mean is given by estimating that the population mean, μ, is equal to the same mean, $\overline{x}$, and then stating the confidence interval of the estimate. Since the 68.26% confidence level is associated with '± 1 standard deviation of the means of the sampling distribution', then the 68.26% confidence level for the estimate of the population mean is given by: $\overline{x} \pm \sigma_{\overline{x}}$

In general, any particular confidence level can be obtained in the estimate, by using $\overline{x} \pm z_C \sigma_{\overline{x}}$, where z_C is the confidence coefficient corresponding to the particular confidence level required. Thus for a 96% confidence level, the confidence limits of the population mean are given by $\overline{x} \pm 2.05\sigma_{\overline{x}}$

Since only one sample has been drawn, the standard error of the means, $\sigma_{\overline{x}}$,is not known. However, it is shown earlier that

$$\sigma_{\overline{x}} = \frac{\sigma}{\sqrt{N}} \sqrt{\left(\frac{N_p - N}{N_p - 1}\right)}$$

Thus, **the confidence limits of the mean of the population are**:

$$\overline{x} \pm \frac{z_C \sigma}{\sqrt{N}} \sqrt{\left(\frac{N_p - N}{N_p - 1}\right)} \qquad (4)$$

for a finite population of size N_p

The **confidence limits for the mean of the population are**:

$$\overline{x} \pm \frac{z_C \sigma}{\sqrt{N}} \qquad (5)$$

for an infinite population .

Thus for a sample of size N and mean $\overline{x}$, drawn from an infinite population having a standard deviation of σ, the mean value of the population is estimated to be, for example, $\overline{x} \pm \dfrac{2.33\sigma}{\sqrt{N}}$ for a confidence level of 98%. This indicates that the mean value of the population lies between $\overline{x} - \dfrac{2.33\sigma}{\sqrt{N}}$ and $\overline{x} + \dfrac{2.33\sigma}{\sqrt{N}}$, with 98% confidence in this prediction.

For example, it is found that the standard deviation of the diameters of rivets produces by a certain machine over a long period of time is 0.018 cm. The diameters of a random sample of 100 rivets produced by this machine in a day have a mean value of 0.476 cm. If the machine produces 2500 rivets a day, (a) the 90% confidence limits, and (b) the 97% confidence limits for an estimate of the mean diameter of all the rivets produced by the machine in a day, is determined as follows:

For the population: standard deviation, $\sigma = 0.018$ cm

number in the population, $N_p = 2500$

For the sample: number in the sample, $N = 100$

mean, $\overline{x} = 0.476$ cm

There is a finite population and the standard deviation of the population is known, hence expression (4) is used.

(a) For a 90% confidence level, the value of z_C, the confidence coefficient, is 1.645 from Table 83.1. Hence, the estimate of the confidence limits of the population mean, μ, is:

$$0.476 \pm \left(\frac{(1.645)(0.018)}{\sqrt{100}}\right)\sqrt{\left(\frac{2500-100}{2500-1}\right)}$$

i.e. $0.476 \pm (0.00296)(0.9800) = 0.476 \pm 0.0029$ cm

Thus, **the 90% confidence limits are 0.473 cm and 0.479 cm**

This indicates that if the mean diameter of a sample of 100 rivets is 0.476 cm, then it is predicted that the mean diameter of all the rivets will be between 0.473 cm and 0.479 cm and this prediction is made with confidence that it will be correct nine times out of ten.

(b) For a 97% confidence level, the value of z_C has to be determined from a table of partial areas under the standardised normal curve given in Table 80.1, as it is not one of the values given in Table 83.1. The total area between ordinates drawn at $-z_C$ and $+z_C$ has to be 0.9700. Because the standardised normal curve is symmetrical, the area between $z_C = 0$ and z_C is $\dfrac{0.9700}{2}$, i.e. 0.4850. From Table 80.1 an area of 0.4850 corresponds to a z_C value of 2.17.

Hence, the estimated value of the confidence limits of the population mean is between

$$\overline{x} \pm \frac{z_C\sigma}{\sqrt{N}}\sqrt{\left(\frac{N_p - N}{N_p - 1}\right)} = 0.476 \pm \left(\frac{(2.17)(0.018)}{\sqrt{100}}\right)\sqrt{\left(\frac{2500-100}{2500-1}\right)}$$

$$= 0.476 \pm (0.0039)(0.9800) = 0.476 \pm 0.0038$$

Thus, **the 97% confidence limits are 0.472 cm and 0.480 cm**
It can be seen that the higher value of confidence level required in part (b) results in a larger confidence interval.

(b) Estimating the mean and standard deviation of a population from sample data

The standard deviation of a large population is not known and, in this case, several samples are drawn from the population. The mean of the sampling distribution of means, $\mu_{\overline{x}}$ and the standard deviation of the sampling distribution of means (i.e. the standard error of the means), $\sigma_{\overline{x}}$, may be determined. The confidence limits of the mean value of the population, μ, are given by:

$$\mu_{\overline{x}} \pm z_C \sigma_{\overline{x}} \qquad (6)$$

where z_C is the confidence coefficient corresponding to the confidence level required.
To make an estimate of the standard deviation, σ, of a normally distributed population:

(i) a sampling distribution of the standard deviations of the samples is formed, and
(ii) the standard deviation of the sampling distribution is determined by using the basic standard deviation formula.

This standard deviation is called the standard error of the standard deviations and is usually signified by σ_S. If s is the standard deviation of a sample, then the confidence limits of the standard deviation of the population are given by:

$$s \pm z_C \sigma_S \qquad (7)$$

where z_C is the confidence coefficient corresponding to the required confidence level.
For example, several samples of 50 fuses selected at random from a large batch are tested when operating at a 10% overload current and the mean time of the sampling distribution before the fuses failed is 16.50 minutes. The standard error of the means is 1.4 minutes. The estimated mean time to failure of the batch of fuses for a confidence level of 90% is determined as follows:

For the sampling distribution: the mean, $\mu_{\overline{x}} = 16.50$,

the standard error of the means, $\sigma_{\overline{x}} = 1.4$

The estimated mean of the population is based on sampling distribution data only and so expression (6) is used.
For an 90% confidence level, $z_C = 1.645$ (from Table 83.1), thus
$\mu_{\overline{x}} \pm z_C\sigma_{\overline{x}} = 16.50 \pm (1.645)(1.4) = 16.50 \pm 2.30$ minutes.
Thus, **the 90% confidence level of the mean time to failure is from 14.20 minutes to 18.80 minutes**.

Estimating the mean of a population based on a small sample size

The methods used earlier to estimate the population mean and standard deviation rely on a relatively large sample size, usually taken as 30 or more. This is because when the sample size is large the sampling distribution of a parameter is approximately normally distributed. When the sample size is small, usually taken as less than 30, the earlier techniques used for estimating the population parameters become more and more inaccurate as the sample size becomes smaller, since the sampling distribution no longer approximates to a normal distribution. Investigations were carried out into the effect of small sample sizes on the estimation theory by W. S. Gosset in the early twentieth century and, as a result of his work, tables are available which enable a realistic estimate to be made, when sample sizes are small. In these tables, the t-value is determined from the relationship $t = \frac{(\bar{x} - \mu)}{s}\sqrt{(N-1)}$ where $\bar{x}$ is the mean value of a sample, μ is the mean value of the population from which the sample is drawn, s is the standard deviation of the sample and N is the number of independent observations in the sample. He published his findings under the pen name of 'Student', and these tables are often referred to as the '**Student's t distribution**'

The confidence limits of the mean value of a population based on a small sample drawn at random from the population are given by

$$\bar{x} \pm \frac{t_C\, s}{\sqrt{(N-1)}} \qquad (8)$$

In this estimate, t_C is called the confidence coefficient for small samples, analogous to z_C for large samples, s is the standard deviation of the sample, $\bar{x}$ is the mean value of the sample and N is the number of members in the sample. Table 83.2 is called 'percentile values for Student's t distribution'. The columns are headed t_p where p is equal to 0.995, 0.99, 0.975, ..., 0.55. For a confidence level of, say, 95%, the column headed $t_{0.95}$ is selected and so on. The rows are headed with the Greek letter 'nu', ν, and are numbered from 1 to 30 in steps of 1, together with the numbers 40, 60, 120 and ∞. These numbers represent a quantity called the **degrees of freedom**, which is defined as follows:

'the sample number, N, minus the number of population parameters which must be estimated for the sample'.

When determining the t-value, given by $t = \frac{(\bar{x} - \mu)}{s}\sqrt{(N-1)}$, it is necessary to know the sample parameters $\bar{x}$ and s and the population parameter μ. $\bar{x}$ and s can be calculated for the sample, but usually an estimate has to be made of the population mean μ, based on the sample mean value. The number of degrees of freedom, ν, is given by the number of independent observations in the sample, N, minus the number of population parameters which have to be estimated, k, i.e. $\nu = N - k$. For the equation $t = \frac{(\bar{x} - \mu)}{s}\sqrt{(N-1)}$, only μ has to be estimated, hence $k = 1$, and $\nu = N - 1$

When determining the mean of a population based on a small sample size, only one population parameter is to be estimated, and hence ν can always be taken as $(N - 1)$.

For example, a sample of 12 measurements of the diameter of a bar are made and the mean of the sample is 1.850 cm. The standard deviation of the samples is 0.16 mm. The (a) the 90% confidence limits and (b) the 70% confidence limits for an estimate of the actual diameter of the bar, is determined as follows:

Table 83.2 Percentile values (t_p) for Student's t distribution with ν degrees of freedom (shaded area $= p$)

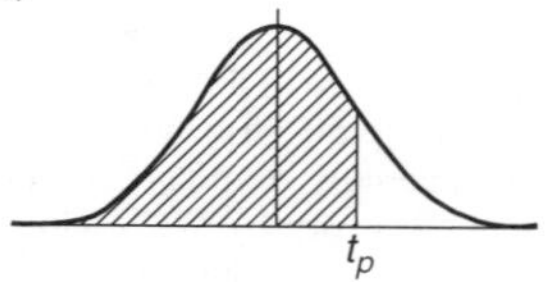

ν	$t_{0.995}$	$t_{0.99}$	$t_{0.975}$	$t_{0.95}$	$t_{0.90}$	$t_{0.80}$	$t_{0.75}$	$t_{0.70}$	$t_{0.60}$	$t_{0.55}$
1	63.66	31.82	12.71	6.31	3.08	1.376	1.000	0.727	0.325	0.158
2	9.92	6.96	4.30	2.92	1.89	1.061	0.816	0.617	0.289	0.142
3	5.84	4.54	3.18	2.35	1.64	0.978	0.765	0.584	0.277	0.137
4	4.60	3.75	2.78	2.13	1.53	0.941	0.741	0.569	0.271	0.134
5	4.03	3.36	2.57	2.02	1.48	0.920	0.727	0.559	0.267	0.132
6	3.71	3.14	2.45	1.94	1.44	0.906	0.718	0.553	0.265	0.131
7	3.50	3.00	2.36	1.90	1.42	0.896	0.711	0.549	0.263	0.130
8	3.36	2.90	2.31	1.86	1.40	0.889	0.706	0.546	0.262	0.130
9	3.25	2.82	2.26	1.83	1.38	0.883	0.703	0.543	0.261	0.129
10	3.17	2.76	2.23	1.81	1.37	0.879	0.700	0.542	0.260	0.129
11	3.11	2.72	2.20	1.80	1.36	0.876	0.697	0.540	0.260	0.129
12	3.06	2.68	2.18	1.78	1.36	0.873	0.695	0.539	0.259	0.128
13	3.01	2.65	2.16	1.77	1.35	0.870	0.694	0.538	0.259	0.128
14	2.98	2.62	2.14	1.76	1.34	0.868	0.692	0.537	0.258	0.128
15	2.95	2.60	2.13	1.75	1.34	0.866	0.691	0.536	0.258	0.128
16	2.92	2.58	2.12	1.75	1.34	0.865	0.690	0.535	0.258	0.128
17	2.90	2.57	2.11	1.74	1.33	0.863	0.689	0.534	0.257	0.128
18	2.88	2.55	2.10	1.73	1.33	0.862	0.688	0.534	0.257	0.127
19	2.86	2.54	2.09	1.73	1.33	0.861	0.688	0.533	0.257	0.127
20	2.84	2.53	2.09	1.72	1.32	0.860	0.687	0.533	0.257	0.127
21	2.83	2.52	2.08	1.72	1.32	0.859	0.686	0.532	0.257	0.127
22	2.82	2.51	2.07	1.72	1.32	0.858	0.686	0.532	0.256	0.127
23	2.81	2.50	2.07	1.71	1.32	0.858	0.685	0.532	0.256	0.127
24	2.80	2.49	2.06	1.71	1.32	0.857	0.685	0.531	0.256	0.127
25	2.79	2.48	2.06	1.71	1.32	0.856	0.684	0.531	0.256	0.127
26	2.78	2.48	2.06	1.71	1.32	0.856	0.684	0.531	0.256	0.127
27	2.77	2.47	2.05	1.70	1.31	0.855	0.684	0.531	0.256	0.127
28	2.76	2.47	2.05	1.70	1.31	0.855	0.683	0.530	0.256	0.127
29	2.76	2.46	2.04	1.70	1.31	0.854	0.683	0.530	0.256	0.127
30	2.75	2.46	2.04	1.70	1.31	0.854	0.683	0.530	0.256	0.127
40	2.70	2.42	2.02	1.68	1.30	0.851	0.681	0.529	0.255	0.126
60	2.66	2.39	2.00	1.67	1.30	0.848	0.679	0.527	0.254	0.126
120	2.62	2.36	1.98	1.66	1.29	0.845	0.677	0.526	0.254	0.126
∞	2.58	2.33	1.96	1.645	1.28	0.842	0.674	0.524	0.253	0.126

For the sample: the sample size, $N = 12$; mean, $\bar{x} = 1.850$ cm;

standard deviation $s = 0.16$ mm $= 0.016$ cm

Since the sample number is less than 30, the small sample estimate as given in expression (8) must be used. The number of degrees of freedom, i.e. sample size minus the number of estimations of population parameters to be made, is 12 − 1, i.e. 11

(a) The percentile value corresponding to a confidence coefficient value of $t_{0.90}$ and a degree of freedom value of $\nu = 11$ can be found by using Table 83.2, and is 1.36, i.e. $t_C = 1.36$. The estimated value of the mean of the population is given by

$$\bar{x} \pm \frac{t_C s}{\sqrt{(N-1)}} = 1.850 \pm \frac{(1.36)(0.016)}{\sqrt{11}} = 1.850 \pm 0.0066 \text{ cm}$$

Thus, **the 90% confidence limits are 1.843 cm and 1.857 cm**
This indicates that the actual diameter is likely to lie between 1.843 cm and 1.857 cm and that this prediction stands a 90% chance of being correct.

(b) The percentile value corresponding to $t_{0.70}$ and to $\nu = 11$ is obtained from Table 83.2, and is 0.540, i.e. $t_C = 0.540$.
The estimated value of the 70% confidence limits is given by:

$$\bar{x} \pm \frac{t_C s}{\sqrt{(N-1)}} = 1.850 \pm \frac{(0.540)(0.016)}{\sqrt{11}} = 1.850 \pm 0.0026 \text{ cm}$$

Thus, **the 70% confidence limits are 1.847 cm and 1.853 cm**, i.e. the actual diameter of the bar is between 1.847 cm and 1.853 cm and this result has a 70% probability of being correct.

Laplace Transforms

84 Introduction to Laplace Transforms

Introduction

The solution of most electrical circuit problems can be reduced ultimately to the solution of differential equations. The use of **Laplace transforms** provides an alternative method to those discussed in Chapters 70 to 74 for solving linear differential equations.

Definition of a Laplace transform

The Laplace transform of the function $f(t)$ is defined by the integral $\int_0^\infty \mathrm{e}^{-st} f(t)\,\mathrm{d}t$, where s is a parameter assumed to be a real number.

Common notations used for the Laplace transform

There are various commonly used notations for the Laplace transform of $f(t)$ and these include:

(i) $\mathcal{L}\{f(t)\}$ or $L\{f(t)\}$
(ii) $\mathcal{L}(f)$ or Lf
(iii) $\overline{f}(s)$ or $f(s)$

Also, the letter p is sometimes used instead of s as the parameter. The notation adopted in this book will be $f(t)$ for the original function and $\mathcal{L}f(t)$ for its Laplace transform

Hence, from above: $\boxed{\mathcal{L}\{f(t)\} = \int_0^\infty e^{-st} f(t)\mathrm{d}t}$ (1)

Linearity property of the Laplace transform

From equation (1),

$$\mathcal{L}\{kf(t)\} = \int_0^\infty \mathrm{e}^{-st} kf(t)\mathrm{d}t = k\int_0^\infty \mathrm{e}^{-st} f(t)\mathrm{d}t$$

i.e. $\mathcal{L}\{kf(t)\} = k\mathcal{L}\{f(t)\}$ (2)

where k is any constant

Similarly, $\mathcal{L}\{af(t) + bg(t)\} = \int_0^\infty \mathrm{e}^{-st}(af(t) + bg(t))\mathrm{d}t$

$$= a\int_0^\infty \mathrm{e}^{-st} f(t)\,\mathrm{d}t + b\int_0^\infty \mathrm{e}^{-st} g(t)\,\mathrm{d}t$$

i.e. $\mathcal{L}\{af(t) + bg(t)\} = a\mathcal{L}\{f(t)\} + b\mathcal{L}\{g(t)\}$, (3)
where a and b are any real constants

The Laplace transform is termed a **linear operator** because of the properties shown in equations (2) and (3).

Laplace transforms of elementary functions

Using the definition of the Laplace transform in equation (1) a number of elementary functions may be transformed, as summarised in Table 84.1.
For example,

$$\mathcal{L}\left\{1+2t-\frac{1}{3}t^4\right\} = \mathcal{L}\{1\}+2\mathcal{L}\{t\}-\frac{1}{3}\mathcal{L}\{t^4\} \text{ from equations (2) and (3)}$$

$$= \frac{1}{s}+2\left(\frac{1}{s^2}\right)-\frac{1}{3}\left(\frac{4!}{s^{4+1}}\right)$$

from (i), (vi) and (viii) of Table 84.1

$$= \frac{1}{s}+\frac{2}{s^2}-\frac{1}{3}\left(\frac{4.3.2.1}{s^5}\right) = \mathbf{\frac{1}{s}+\frac{2}{s^2}-\frac{8}{s^5}}$$

In another example,

$$\mathcal{L}\{5e^{2t}-3e^{-t}\} = 5\mathcal{L}(e^{2t})-3\mathcal{L}\{e^{-t}\}, \text{ from equations (2) and (3)}$$

$$= 5\left(\frac{1}{s-2}\right)-3\left(\frac{1}{s--1}\right) \text{ from (iii) of Table 84.1}$$

$$= \frac{5}{s-2}-\frac{3}{s+1} = \frac{5(s+1)-3(s-2)}{(s-2)(s+1)} = \mathbf{\frac{2s+11}{s^2-s-2}}$$

Table 84.1 Elementary standard Laplace transforms

	Function $f(t)$	*Laplace transforms* $\mathcal{L}\{f(t)\} = \int_0^\infty e^{-st}f(t)\,dt$
(i)	1	$\frac{1}{s}$
(ii)	k	$\frac{k}{s}$
(iii)	e^{at}	$\frac{1}{s-a}$
(iv)	$\sin at$	$\frac{a}{s^2+a^2}$
(v)	$\cos at$	$\frac{s}{s^2+a^2}$
(vi)	t	$\frac{1}{s^2}$
(vii)	t^2	$\frac{2!}{s^3}$
(viii)	$t^n\ (n=1,2,3,\ldots)$	$\frac{n!}{s^{n+1}}$
(ix)	$\cosh at$	$\frac{s}{s^2-a^2}$
(x)	$\sinh at$	$\frac{a}{s^2-a^2}$

In another example,

$$\mathcal{L}\{6\sin 3t - 4\cos 5t\} = 6\mathcal{L}\{\sin 3t\} - 4\mathcal{L}\{\cos 5t\}$$

$$= 6\left(\frac{3}{s^2+3^2}\right) - 4\left(\frac{s}{s^2+5^2}\right)$$

from (iv) and (v) of Table 84.1

$$= \mathbf{\frac{18}{s^2+9} - \frac{4s}{s^2+25}}$$

In another example,

$$\mathcal{L}\{2\cosh 2\theta - \sinh 3\theta\} = 2\mathcal{L}\{\cosh 2\theta\} - \{\sinh 3\theta\}$$

$$= 2\left(\frac{s}{s^2-2^2}\right) - \left(\frac{3}{s^2-3^2}\right)$$

from (ix) and (x) of Table 84.1

$$= \mathbf{\frac{2s}{s^2-4} - \frac{3}{s^2-9}}$$

---

85 Properties of Laplace Transforms

The Laplace transform of $e^{at}f(t)$

From Chapter 84, the definition of the Laplace transform of $f(t)$ is:

$$\mathcal{L}\{f(t)\} = \int_0^\infty e^{-st} f(t)\,dt \qquad (1)$$

Thus $$\mathcal{L}\{e^{at} f(t)\} = \int_0^\infty e^{-st}(e^{at} f(t))\,dt = \int_0^\infty e^{-(s-a)} f(t)\,dt \qquad (2)$$

(where a is a real constant)

Hence the substitution of $(s-a)$ for s in the transform shown in equation (1) corresponds to the multiplication of the original function $f(t)$ by e^{at}. This is known as a **shift theorem**.

Laplace transforms of the form $e^{at}f(t)$

A summary of Laplace transforms of the form e^{at}f(t) is shown in Table 85.1

For example, from (i) of Table 85.1,

$$\mathcal{L}\{2t^4e^{3t}\} = 2\mathcal{L}\{t^4e^{3t}\} = 2\left(\frac{4!}{(s-3)^{4+1}}\right) = \frac{2(4)(3)(2)}{(s-3)^5} = \mathbf{\frac{48}{(s-3)^5}}$$

Table 85.1 Laplace transforms of the form $e^{at}f(t)$

Function $e^{at}f(t)$ *(a is a real constant)*	*Laplace transform* $\mathscr{L}\{e^{at}f(t)\}$
(i) $e^{at}t^n$	$\dfrac{n!}{(s-a)^{n+1}}$
(ii) $e^{at}\sin\omega t$	$\dfrac{\omega}{(s-a)^2+\omega^2}$
(iii) $e^{at}\cos\omega t$	$\dfrac{s-a}{(s-a)^2+\omega^2}$
(iv) $e^{at}\sinh\omega t$	$\dfrac{\omega}{(s-a)^2-\omega^2}$
(v) $e^{at}\cosh\omega t$	$\dfrac{s-a}{(s-a)^2-\omega^2}$

In another example, from (iii) of Table 85.1,

$$\mathscr{L}\{4e^{3t}\cos 5t\} = 4\mathscr{L}\{e^{3t}\cos 5t\} = 4\left(\frac{s-3}{(s-3)^2+5^2}\right)$$

$$= \frac{4(s-3)}{s^2-6s+9+25} = \mathbf{\frac{4(s-3)}{s^2-6s+34}}$$

The Laplace transforms of derivatives

(a) First derivative

$$\left.\begin{aligned} &\mathscr{L}\{f'(t)\} = s\mathscr{L}\{f(t)\} - f(0) \\ \text{or}\quad &\mathscr{L}\left\{\frac{dy}{dx}\right\} = sL\{y\} - y(0) \end{aligned}\right\} \qquad (3)$$

where $y(0)$ is the value of y at $x = 0$

(b) Second derivative

$$\left.\begin{aligned} &\mathscr{L}\{f''(t)\} = s^2\mathscr{L}\{f(t)\} - sf(0) - f'(0) \\ \text{or}\quad &\mathscr{L}\left\{\frac{d^2y}{dx^2}\right\} = s^2\mathscr{L}\{y\} - sy(0) - y'(0) \end{aligned}\right\} \qquad (4)$$

where $y'(0)$ is the value of $\dfrac{dy}{dx}$ at $x = 0$.

Equations (3) and (4) are important and are used in the solution of differential equations (see Chapter 87) and simultaneous differential equations (Chapter 88).

The initial and final value theorems

There are several Laplace transform theorems used to simplify and interpret the solution of certain problems. Two such theorems are the initial value theorem and the final value theorem.

(a) The initial value theorem states:

$$\boxed{\lim_{t\to 0}[f(t)] = \lim_{s\to\infty}[s\mathcal{L}\{f(t)\}]}$$

For example, to verify the initial value theorem for the voltage function $(5 + 2\cos 3t)$ volts:

Let $\quad f(t) = 5 + 2\cos 3t$

$$\mathcal{L}\{f(t)\} = \mathcal{L}\{5 + 2\cos 3t\} = \frac{5}{s} + \frac{2s}{s^2+9}$$

from (ii) and (v) of Table 84.1, page 415.

By the initial value theorem,

$$\lim_{t\to 0}[f(t)) = \lim_{s\to\infty}[s\mathcal{L}\{f(t)\}]$$

i.e. $$\lim_{t\to 0}[5 + 2\cos 3t] = \lim_{s\to\infty}\left[s\left(\frac{5}{s} + \frac{2s}{s^2+9}\right)\right] = \lim_{s\to\infty}\left[5 + \frac{2s^2}{s^2+9}\right]$$

i.e. $$5 + 2(1) = 5 + \frac{2\infty^2}{\infty^2+9} = 5 + 2$$

i.e. $7 = 7$, which verifies the theorem in this case.
The initial value of the voltage is thus **7 V.**

(b) The final value theorem states:

$$\boxed{\lim_{t\to\infty}[f(t)] = \lim_{s\to 0}[s\mathcal{L}\{f(t)\}]}$$

For example, to verify the final value theorem for the function $(2 + 3e^{-2t}\sin 4t)$ cm, which represents the displacement of a particle:
Let $f(t) = 2 + 3e^{-2t}\sin 4t$

$$\mathcal{L}\{f(t)\} = \mathcal{L}\{2 + 3e^{-2t}\sin 4t\} = \frac{2}{s} + 3\left(\frac{4}{(s - -2)^2 + 4^2}\right)$$

$$= \frac{2}{s} + \frac{12}{(s+2)^2+16}$$

from (ii) of Table 84.1, page 415 and (ii) of Table 85.1 on page 417.
By the final value theorem,

$$\lim_{t\to\infty}[f(t)] = \lim_{s\to 0}[s\mathcal{L}\{f(t)\}]$$

i.e. $$\lim_{t\to\infty}[2 + 3e^{-2t}\sin 4t] = \lim_{s\to 0}\left[s\left(\frac{2}{s} + \frac{12}{(s+2)^2+16}\right)\right]$$

$$= \lim_{s\to 0}\left[2 + \frac{12s}{(s+2)^2+16}\right]$$

i.e. $$2 + 0 = 2 + 0$$

i.e. **2 = 2**, which verifies the theorem in this case.

The final value of the displacement is thus 2 cm.

The initial and final value theorems are used in pulse circuit applications where the response of the circuit for small periods of time, or the behaviour immediately after the switch is closed, are of interest. The final value theorem is particularly useful in investigating the stability of systems (such as in automatic aircraft-landing systems) and is concerned with the steady state response for large values of time t, i.e. after all transient effects have died away.

86 Inverse Laplace Transforms

Definition of the inverse Laplace transform

If the Laplace transform of a function $f(t)$ is $F(s)$, i.e. $\mathscr{L}\{f(t)\} = F(s)$, then $f(t)$ is called the **inverse Laplace transform** of $F(s)$ and is written as

$$\mathscr{L} f(t) = \mathscr{L}^{-1}\{F(s)\}$$

For example, since $\mathscr{L}\{1\} = \frac{1}{s}$ then $\mathscr{L}^{-1}\left\{\frac{1}{s}\right\} = \mathbf{1}$

In another example,

since $$\mathscr{L}\{\sin at\} = \frac{a}{s^2+a^2}$$

then $$\mathscr{L}^{-1}\left\{\frac{a}{s^2+a^2}\right\} = \mathbf{\sin at},$$ and so on.

Inverse Laplace transforms of simple functions

Tables of Laplace transforms, such as the tables in Chapters 84 and 85 (see pages 415 and 417) may be used to find inverse Laplace transforms.

For example, from (iv) of Table 84.1, $\mathscr{L}^{-1}\left\{\frac{a}{s^2+a^2}\right\} = \sin at$,

Hence $$\mathscr{L}^{-1}\left\{\frac{1}{s^2+9}\right\} = \mathscr{L}^{-1}\left\{\frac{1}{s^2+3^2}\right\} = \frac{1}{3}\mathscr{L}^{-1}\left\{\frac{3}{s^2+3^2}\right\} = \mathbf{\frac{1}{3}\sin 3t}$$

In another example, $$\mathscr{L}^{-1}\left\{\frac{5}{3s-1}\right\} = \mathscr{L}^{-1}\left\{\frac{5}{3\left(s-\frac{1}{3}\right)}\right\}$$

$$= \frac{5}{3}\mathscr{L}^{-1}\left\{\frac{1}{\left(s-\frac{1}{3}\right)}\right\} = \mathbf{\frac{5}{3}e^{(1/3)t}}$$

from (iii) of Table 84.1

In another example, to determine $\mathcal{L}^{-1}\left\{\frac{3}{s^4}\right\}$:
From (viii) of Table 84.1, if s is to have a power of 4 then $n = 3$.

Thus $\mathcal{L}^{-1}\left\{\frac{3!}{s^4}\right\} = t^3$ i.e. $\mathcal{L}^{-1}\left\{\frac{6}{s^4}\right\} = t^3$

Hence $\mathcal{L}^{-1}\left\{\frac{3}{s^4}\right\} = \frac{1}{2}\mathcal{L}^{-1}\left\{\frac{6}{s^4}\right\} = \mathbf{\frac{1}{2}t^3}$

In another example, $\mathcal{L}^{-1}\left\{\frac{7s}{s^2+4}\right\} = 7\mathcal{L}^{-1}\left\{\frac{s}{s^2+2^2}\right\}$

$= \mathbf{7\cos 2t}$, from (v) of Table 84.1

In another example,

$$\mathcal{L}^{-1}\left\{\frac{3}{s^2-4s+13}\right\} = \mathcal{L}^{-1}\left\{\frac{3}{(s-2)^2+3^2}\right\}$$

$$= \mathbf{e^{2t}\sin 3t}, \text{ from (ii) of Table 85.1}$$

In another example,

$$\mathcal{L}^{-1}\left\{\frac{4s-3}{s^2-4s-5}\right\} = \mathcal{L}^{-1}\left\{\frac{4s-3}{(s-2)^2-3^2}\right\} = \mathcal{L}^{-1}\left\{\frac{4(s-2)+5}{(s-2)^2-3^2}\right\}$$

$$= \mathcal{L}^{-1}\left\{\frac{4(s-2)}{(s-2)^2-3^2}\right\} + \mathcal{L}^{-1}\left\{\frac{5}{(s-2)^2-3^2}\right\},$$

$$= 4e^{2t}\cosh 3t + \mathcal{L}^{-1}\left\{\frac{\frac{5}{3}(3)}{(s-2)^2-3^2}\right\},$$

from (v) of Table 85.1

$$= \mathbf{4e^{2t}\cosh 3t + \frac{5}{3}e^{2t}\sinh 3t},$$

from (iv) of Table 85.1.

Inverse Laplace transforms using partial fractions

Sometimes the function whose inverse is required is not recognizable as a standard type, such as those listed in Tables 84.1 and 85.1. In such cases it may be possible, by using partial fractions, to resolve the function into simpler fractions that may be inverted on sight. For example, the function,

$$F(s) = \frac{2s-3}{s(s-3)}$$

cannot be inverted on sight from Table 84.1. However, by using partial fractions, $\frac{2s-3}{s(s-3)} \equiv \frac{1}{s} + \frac{1}{s-3}$ which may be inverted as $1 + e^{3t}$ from (i) and (iii) of Table 84.1.
Partial fractions are discussed in Chapter 14, and a summary of the forms of partial fractions is given in Table 14.1 on page 61.

For example, to determine $\mathcal{L}^{-1}\left\{\dfrac{4s-5}{s^2-s-2}\right\}$:

$$\frac{4s-5}{s^2-s-2} \equiv \frac{4s-5}{(s-2)(s+1)} \equiv \frac{A}{(s-2)} + \frac{B}{(s+1)} \equiv \frac{A(s+1)+B(s-2)}{(s-2)(s+1)}$$

Hence $\quad 4s-5 \equiv A(s+1)+B(s-2)$

When $s = 2$, $\qquad 3 = 3A$, from which, $A = 1$

When $s = -1$, $\quad -9 = -3B$, from which, $B = 3$

$$\text{Hence} \quad \mathcal{L}^{-1}\left\{\frac{4s-5}{s^2-s-2}\right\} \equiv \mathcal{L}^{-1}\left\{\frac{1}{s-2}+\frac{3}{s+1}\right\}$$

$$= \mathcal{L}^{-1}\left\{\frac{1}{s-2}\right\} + \mathcal{L}^{-1}\left\{\frac{3}{s+1}\right\}$$

$$= \boldsymbol{e^{2t} + 3e^{-t}}, \text{ from (iii) of Table 84.1}$$

In another example, to determine $\mathcal{L}^{-1}\left\{\dfrac{5s^2+8s-1}{(s+3)(s^2+1)}\right\}$:

$$\frac{5s^2+8s-1}{(s+3)(s^2+1)} \equiv \frac{A}{s+3} + \frac{Bs+C}{(s^2+1)} \equiv \frac{A(s^2+1)+(Bs+C)(s+3)}{(s+3)(s^2+1)}$$

Hence $5s^2+8s-1 \equiv A(s^2+1)+(Bs+C)(s+3)$
When $s = -3$, $\quad 20 = 10\,A$, from which, $A = 2$
Equating s^2 terms gives: $\quad 5 = A + B$, from which, $B = 3$, since $A = 2$
Equating s terms gives: $\quad 8 = 3B + C$, from which, $C = -1$, since $B = 3$

$$\text{Hence} \quad \mathcal{L}^{-1}\left\{\frac{5s^2+8s-1}{(s+3)(s^2+1)}\right\} \equiv \mathcal{L}^{-1}\left\{\frac{2}{s+3}+\frac{3s-1}{s^2+1}\right\}$$

$$\equiv \mathcal{L}^{-1}\left\{\frac{2}{s+3}\right\} + \mathcal{L}^{-1}\left\{\frac{3s}{s^2+1}\right\}$$

$$- \mathcal{L}^{-1}\left\{\frac{1}{s^2+1}\right\}$$

$$= \boldsymbol{2e^{-3t} + 3\cos t - \sin t},$$

from (iii), (v) and (iv) of Table 84.1

---

87 The Solution of Differential Equations Using Laplace Transforms

Introduction

An alternative method of solving differential equations to that used in Chapters 70 to 74 is possible by using Laplace transforms.

Procedure to solve differential equations by using Laplace transforms

(i) Take the Laplace transform of both sides of the differential equation by applying the formulae for the Laplace transforms of derivatives (i.e. equations (3) and (4) of Chapter 85) and, where necessary, using a list of standard Laplace transforms, such as Tables 84.1 and 85.1 on pages 415 and 417.
(ii) Put in the given initial conditions, i.e. $y(0)$ and $y'(0)$
(iii) Rearrange the equation to make $\mathcal{L}\{y\}$ the subject.
(iv) Determine y by using, where necessary, partial fractions, and taking the inverse of each term by using Tables 84.1 and 85.1 on pages 415 and 417.

For example, to solve the differential equation $2\dfrac{d^2y}{dx^2} + 5\dfrac{dy}{dx} - 3y = 0$, given that when $x = 0$, $y = 4$ and $\dfrac{dy}{dx} = 9$:

Using the above procedure:

(i) $2\mathcal{L}\left\{\dfrac{d^2y}{dx^2}\right\} + 5\mathcal{L}\left\{\dfrac{dy}{dx}\right\} - 3\mathcal{L}\{y\} = \mathcal{L}\{0\}$

$2[s^2\mathcal{L}\{y\} - sy(0) - y'(0)] + 5[s\mathcal{L}\{y\} - y(0)] - 3\mathcal{L}\{y\} = 0,$

from equations (3) and (4) of Chapter 85

(ii) $y(0) = 4$ and $y'(0) = 9$

Thus $\quad 2[s^2\mathcal{L}\{y\} - 4s - 9] + 5[s\mathcal{L}\{y\} - 4] - 3\mathcal{L}\{y\} = 0$

i.e. $\quad 2s^2\mathcal{L}\{y\} - 8s - 18 + 5s\mathcal{L}\{y\} - 20 - 3\mathcal{L}\{y\} = 0$

(iii) Rearranging gives: $(2s^2 + 5s - 3)\mathcal{L}\{y\} = 8s + 38$

i.e. $\mathcal{L}\{y\} = \dfrac{8s + 38}{2s^2 + 5s - 3}$

(iv) $y = \mathcal{L}^{-1}\left\{\dfrac{8s + 38}{2s^2 + 5s - 3}\right\}$

$$\frac{8s + 38}{2s^2 + 5s - 3} \equiv \frac{8s + 38}{(2s - 1)(s + 3)} \equiv \frac{A}{2s - 1} + \frac{B}{s + 3}$$

$$\equiv \frac{A(s + 3) + B(2s - 1)}{(2s - 1)(s + 3)}$$

Hence $\quad 8s + 38 = A(s + 3) + B(2s - 1)$

When $s = \frac{1}{2}$, $\quad 42 = 3\frac{1}{2}A$, from which, $A = 12$

When $s = -3$, $\quad 14 = -7B$, from which, $B = -2$

Hence $\quad y = \mathcal{L}^{-1}\left\{\dfrac{8s + 38}{2s^2 + 5s - 3}\right\} = \mathcal{L}^{-1}\left\{\dfrac{12}{2s - 1} - \dfrac{2}{s + 3}\right\}$

$$= \mathscr{L}^{-1}\left\{\frac{12}{2\left(s-\frac{1}{2}\right)}\right\} - {}^{-1}\left\{\frac{2}{s+3}\right\}$$

Hence $\boldsymbol{y = 6\,e^{\frac{1}{2}x} - 2\,e^{-3x}}$, from (iii) of Table 84.1

In another example, to solve $\dfrac{d^2y}{dx^2} - 3\dfrac{dy}{dx} = 9$, given that when $x = 0$, $y = 0$ and $\dfrac{dy}{dx} = 0$:

(i) $\mathscr{L}\left\{\dfrac{d^2y}{dx^2}\right\} - 3\mathscr{L}\left\{\dfrac{dy}{dx}\right\} = \mathscr{L}\{9\}$

Hence $[s^2\mathscr{L}\{y\} - sy(0) - y'(0)] - 3[s\mathscr{L}(y - y(0)] = \dfrac{9}{s}$

(ii) $y(0) = 0$ and $y'(0) = 0$

Hence $s^2\{y\} - 3s\{y\} = \dfrac{9}{s}$

(iii) Rearranging gives: $(s^2 - 3s)\mathscr{L}\{y\} = \dfrac{9}{s}$

i.e. $\mathscr{L}\{y\} = \dfrac{9}{s(s^2-3s)} = \dfrac{9}{s^2(s-3)}$

(iv) $y = \mathscr{L}^{-1}\left\{\dfrac{9}{s^2(s-3)}\right\}$

$$\frac{9}{s^2(s-3)} \equiv \frac{A}{s} + \frac{B}{s^2} + \frac{C}{s-3} \equiv \frac{A(s)(s-3) + B(s-3) + Cs^2}{s^2(s-3)}$$

Hence $9 \equiv A(s)(s-3) + B(s-3) + Cs^2$

When $s = 0$, $9 = -3B$, from which , $B = -3$

When $s = 3$, $9 = 9C$, from which, $C = 1$

Equating s^2 terms gives: $0 = A + C$, from which, $A = -1$, since $C = 1$

Hence $\mathscr{L}^{-1}\left\{\dfrac{9}{s^2(s-3)}\right\} = \mathscr{L}^{-1}\left\{-\dfrac{1}{s} - \dfrac{3}{s^2} + \dfrac{1}{s-3}\right\}$

$= -1 - 3x + e^{3x}$,

from (i), (vi) and (iii) of Table 84.1

i.e $\boldsymbol{y = e^{3x} - 3x - 1}$

88 The Solution of Simultaneous Differential Equations Using Laplace Transforms

Introduction

It is sometimes necessary to solve simultaneous differential equations. An example occurs when two electrical circuits are coupled magnetically where the equations relating the two currents i_1 and i_2 are typically:

$$L_1 \frac{\mathrm{d}i_1}{\mathrm{d}t} + M \frac{\mathrm{d}i_2}{\mathrm{d}t} + R_1 i_1 = E_1$$

$$L_2 \frac{\mathrm{d}i_2}{\mathrm{d}t} + M \frac{\mathrm{d}i_1}{\mathrm{d}t} + R_2 i_2 = 0$$

where L represents inductance, R resistance, M mutual inductance and E_1 the p.d. applied to one of the circuits.

Procedure to solve simultaneous differential equations using Laplace transforms

(i) Take the Laplace transform of both sides of each simultaneous equation by applying the formulae for the Laplace transforms of derivatives (i.e. equations (3) and (4) of Chapter 85, page 417) and using a list of standard Laplace transforms, as in Table 84.1, page 415 and Table 85.1, page 417.

(ii) Put in the initial conditions, i.e. $x(0)$, $y(0)$, $x'(0)$, $y'(0)$

(iii) Solve the simultaneous equations for $\mathcal{L}\{y\}$ and $\mathcal{L}\{x\}$ by the normal algebraic method.

(iv) Determine y and x by using, where necessary, partial fractions, and taking the inverse of each term.

For example, to solve the following pair of simultaneous differential equations

$$\frac{\mathrm{d}y}{\mathrm{d}t} + x = 1$$

$$\frac{\mathrm{d}x}{\mathrm{d}t} - y + 4\mathrm{e}^t = 0$$

given that at $t = 0$, $x = 0$ and $y = 0$, using the above procedure:

(i) $$\mathcal{L}\left\{\frac{\mathrm{d}y}{\mathrm{d}t}\right\} + \mathcal{L}\{x\} = \mathcal{L}\{1\} \qquad (1)$$

$$\mathcal{L}\left\{\frac{\mathrm{d}x}{\mathrm{d}t}\right\} - \mathcal{L}\{y\} + 4\mathcal{L}\{\mathrm{e}^t\} = 0 \qquad (2)$$

Equation (1) becomes:

$$[s\mathcal{L}\{y\} - y(0)] + \mathcal{L}\{x\} = \frac{1}{s} \qquad (1')$$

from equation (3), page 417 and Table 84.1, page 415
Equation (2) becomes:

$$[s\mathcal{L}\{x\} - x(0)] - \mathcal{L}\{y\} = -\frac{4}{s-1} \qquad (2')$$

(ii) $x(0) = 0$ and $y(0) = 0$ hence

Equation (1′) becomes: $$s\mathcal{L}\{y\} + \mathcal{L}\{x\} = \frac{1}{s} \qquad (1'')$$

and equation (2′) becomes: $$s\mathcal{L}\{x\} - \mathcal{L}\{y\} = -\frac{4}{s-1}$$

or $$-\mathcal{L}\{y\} + s\mathcal{L}\{x\} = -\frac{4}{s-1} \qquad (2'')$$

(iii) $1 \times$ equation (1) and $s \times$ equation (2″) gives:

$$s\mathcal{L}\{y\} + \mathcal{L}\{x\} = \frac{1}{s} \qquad (3)$$

$$-s\mathcal{L}\{y\} + s^2\mathcal{L}\{x\} = -\frac{4s}{s-1} \qquad (4)$$

Adding equations (3) and (4) gives:

$$(s^2+1)\mathcal{L}\{x\} = \frac{1}{s} - \frac{4s}{s-1} = \frac{(s-1) - s(4s)}{s(s-1)} = \frac{-4s^2+s-1}{s(s-1)}$$

from which, $$\mathcal{L}\{x\} = \frac{-4s^2+s-1}{s(s-1)(s^2+1)}$$

Using partial fractions

$$\frac{-4s^2+s-1}{s(s-1)(s^2+1)} \equiv \frac{A}{s} + \frac{B}{(s-1)} + \frac{Cs+D}{(s^2+1)}$$

$$= \frac{A(s-1)(s^2+1) + Bs(s^2+1) + (Cs+D)s(s-1)}{s(s-1)(s^2+1)}$$

Hence $-4s^2 + s - 1 = A(s-1)(s^2+1) + Bs(s^2+1) + (Cs+D)s(s-1)$

When $s = 0$, $-1 = -A$ hence $\boldsymbol{A = 1}$
When $s = 1$, $-4 = 2B$ hence $\boldsymbol{B = -2}$
Equating s^3 coefficients:

$$0 = A + B + C \text{ hence } \boldsymbol{C = 1} \text{ (since } A = 1 \text{ and } B = -2)$$

Equating s^2 coefficients:

$$-4 = -A + D - C \text{ hence } \boldsymbol{D = -2} \text{ (since } A = 1 \text{ and } C = 1)$$

Thus $$\mathcal{L}\{x\} = \frac{-4s^2+s-1}{s(s-1)(s^2+1)} = \frac{1}{s} - \frac{2}{(s-1)} + \frac{s-2}{(s^2+1)}$$

(iv) Hence $x = \mathcal{L}^{-1}\left\{\frac{1}{s} - \frac{2}{(s-1)} + \frac{s-2}{(s^2+1)}\right\}$

$$= \mathcal{L}^{-1}\left\{\frac{1}{s} - \frac{2}{(s-1)} + \frac{s}{(s^2+1)} - \frac{2}{(s^2+1)}\right\}$$

i.e. $\boldsymbol{x = 1 - 2e^t + \cos t - 2\sin t}$ from Table 84.1, page 415

The second equation given in the question is $\frac{dx}{dt} - y + 4e^t = 0$ from which,

$$y = \frac{dx}{dt} + 4e^t = \frac{d}{dt}(1 - 2e^t + \cos t - 2\sin t) + 4e^t$$

$$= -2e^t - \sin t - 2\cos t + 4e^t$$

i.e. $$\boldsymbol{y = 2e^t - \sin t - 2\cos t}$$

[Alternatively, to determine y, return to equations $(1'')$ and $(2'')$]

Fourier Series

89 Fourier Series for Periodic Functions of Period 2π

Introduction

Fourier series provides a method of analysing periodic functions into their constituent components. Alternating currents and voltages, displacement, velocity and acceleration of slider-crank mechanisms and acoustic waves are typical practical examples in engineering and science where periodic functions are involved and often requiring analysis.

Periodic functions

A function $f(x)$ is said to be **periodic** if $f(x+T) = f(x)$ for all values of x, where T is some positive number. T is the interval between two successive repetitions and is called the **period** of the functions $f(x)$. For example, $y = \sin x$ is periodic in x with period 2π since $\sin x = \sin(x+2\pi) = \sin(x+4\pi)$, and so on. In general, if $y = \sin \omega t$ then the period of the waveform is $2\pi/\omega$. The function shown in Figure 89.1 is also periodic of period 2π and is defined by:

$$f(x) = \begin{cases} -1, & \text{when } -\pi < x < 0 \\ 1, & \text{when } 0 < x < \pi \end{cases}$$

If a graph of a function has no sudden jumps or breaks it is called a **continuous function**, examples being the graphs of sine and cosine functions. However, other graphs make finite jumps at a point or points in the interval. The square wave shown in Figure 89.1 has finite discontinuities at $x = \pi$, 2π, 3π, and so on. A great advantage of Fourier series over other series is that it can be applied to functions that are discontinuous as well as those which are continuous.

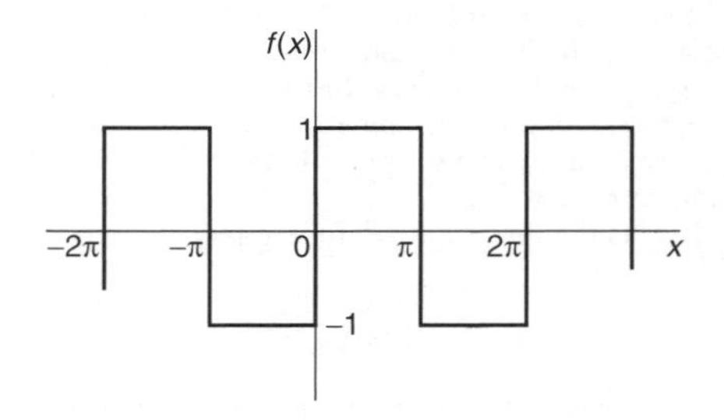

Figure 89.1

Fourier series

The basis of a Fourier series is that all functions of practical significance which are defined in the interval $-\pi \le x \le \pi$ can be expressed in terms of a convergent trigonometric series of the form:

$$f(x) = a_0 + a_1 \cos x + a_2 \cos 2x + a_3 \cos 3x + \ldots$$
$$+ b_1 \sin x + b_2 \sin 2x + b_3 \sin 3x + \ldots$$

when $a_0, a_1, a_2, \ldots b_1, b_2, \ldots$ are real constants,

$$\text{i.e.} \quad \boldsymbol{f(x) = a_0 + \sum_{n=1}^{\infty} (a_n \cos nx + b_n \sin nx)} \qquad (1)$$

where for the range $-\pi$ to π:

$$a_0 = \frac{1}{2\pi} \int_{-\pi}^{\pi} f(x)\,dx$$
$$a_n = \frac{1}{\pi} \int_{-\pi}^{\pi} f(x) \cos nx\,dx \quad (n = 1, 2, 3, \ldots)$$
$$\text{and} \quad b_n = \frac{1}{\pi} \int_{-\pi}^{\pi} f(x) \sin nx\,dx \quad (n = 1, 2, 3, \ldots)$$

a_0, a_n and b_n are called the **Fourier coefficients** of the series and if these can be determined, the series of equation (1) is called the **Fourier series** corresponding to $f(x)$.
An alternative way of writing the series is by using the $a \cos x + b \sin x = c \sin(x + \alpha)$ relationship introduced in chapter 31,

$$\text{i.e.} \quad f(x) = a_0 + c_1 \sin(x + \alpha_1) + c_2 \sin(2x + \alpha_2) + \ldots + c_n \sin(nx + \alpha_n),$$

where a_0 is a constant, $c_1 = \sqrt{a_1^2 + b_1^2}, \ldots c_n = \sqrt{a_n^2 + b_n^2}$ are the amplitudes of the various components, and phase angle $\alpha_n = \tan^{-1} \dfrac{a_n}{b_n}$.
For the series of equation (1): the term $(a_1 \cos x + b_1 \sin x)$ or $c_1 \sin(x + \alpha_1)$ is called the **first harmonic** or the **fundamental**, the term $(a_2 \cos 2x + b_2 \sin 2x)$ or $c_2 \sin(2x + \alpha_2)$ is called the **second harmonic**, and so on.
For an exact representation of a complex wave, an infinite number of terms are, in general, required. In many practical cases, however, it is sufficient to take the first few terms only.
For example, to obtain a Fourier series for the periodic function $f(x)$ defined as:

$$f(x) = \begin{cases} -k, & \text{when } -\pi < x < 0 \\ +k, & \text{when } 0 < x < \pi \end{cases}$$

(The function is periodic outside of this range with period 2π):
The square wave function defined is shown in Figure 89.2. Since $f(x)$ is given by two different expressions in the two halves of the range the integration is

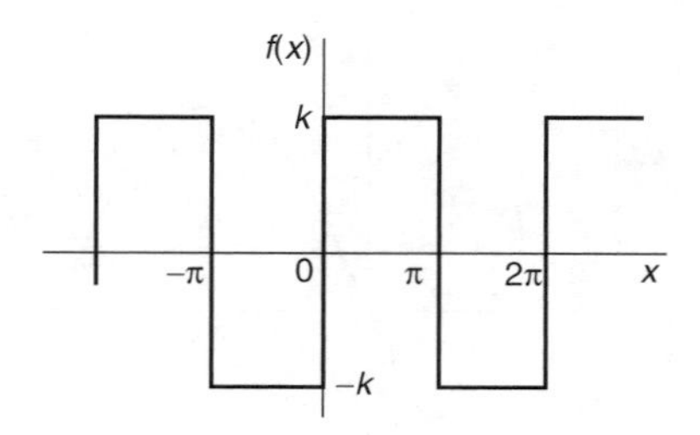

Figure 89.2

performed in two parts, one from $-\pi$ to 0 and the other from 0 to π.

From above: $$a_0 = \frac{1}{2\pi}\int_{-\pi}^{\pi} f(x)\,dx = \frac{1}{2\pi}\left[\int_{-\pi}^{0} -k\,dx + \int_{0}^{\pi} k\,dx\right]$$

$$= \frac{1}{2\pi}\{[-kx]_{-\pi}^{0} + [kx]_{0}^{\pi}\} = 0$$

[a_0 is in fact the **mean value** of the waveform over a complete period of 2π and this could have been deduced on sight from Figure 89.2]

$$a_n = \frac{1}{\pi}\int_{-\pi}^{\pi} f(x)\cos nx\,dx = \frac{1}{\pi}\left\{\int_{-\pi}^{0} -k\cos nx\,dx + \int_{0}^{\pi} k\cos nx\,dx\right\}$$

$$= \frac{1}{\pi}\left\{\left[\frac{-k\sin nx}{n}\right]_{-\pi}^{0} + \left[\frac{k\sin nx}{n}\right]_{0}^{\pi}\right\} = 0$$

Hence $a_1, a_2, a_3, \ldots$ are all zero (since $\sin 0 = \sin(-n\pi) = \sin n\pi = 0$), and therefore no cosine terms will appear in the Fourier series.

$$b_n = \frac{1}{\pi}\int_{-\pi}^{\pi} f(x)\sin nx\,dx = \frac{1}{\pi}\left\{\int_{-\pi}^{0} -k\sin nx\,dx + \int_{0}^{\pi} k\sin nx\,dx\right\}$$

$$= \frac{1}{\pi}\left\{\left[\frac{k\cos nx}{n}\right]_{-\pi}^{0} + \left[\frac{-k\cos nx}{n}\right]_{0}^{\pi}\right\}$$

When n is odd: $$b_n = \frac{k}{\pi}\left\{\left[\left(\frac{1}{n}\right) - \left(-\frac{1}{n}\right)\right] + \left[-\left(-\frac{1}{n}\right) - \left(-\frac{1}{n}\right)\right]\right\}$$

$$= \frac{k}{\pi}\left\{\frac{2}{n} + \frac{2}{n}\right\} = \frac{4k}{n\pi}$$

Hence $$b_1 = \frac{4k}{\pi},\ b_3 = \frac{4k}{3\pi},\ b_5 = \frac{4k}{5\pi},\quad \text{and so on}$$

When n is even: $$b_n = \frac{k}{\pi}\left\{\left[\frac{1}{n} - \frac{1}{n}\right] + \left[-\frac{1}{n} - \left(-\frac{1}{n}\right)\right]\right\} = 0$$

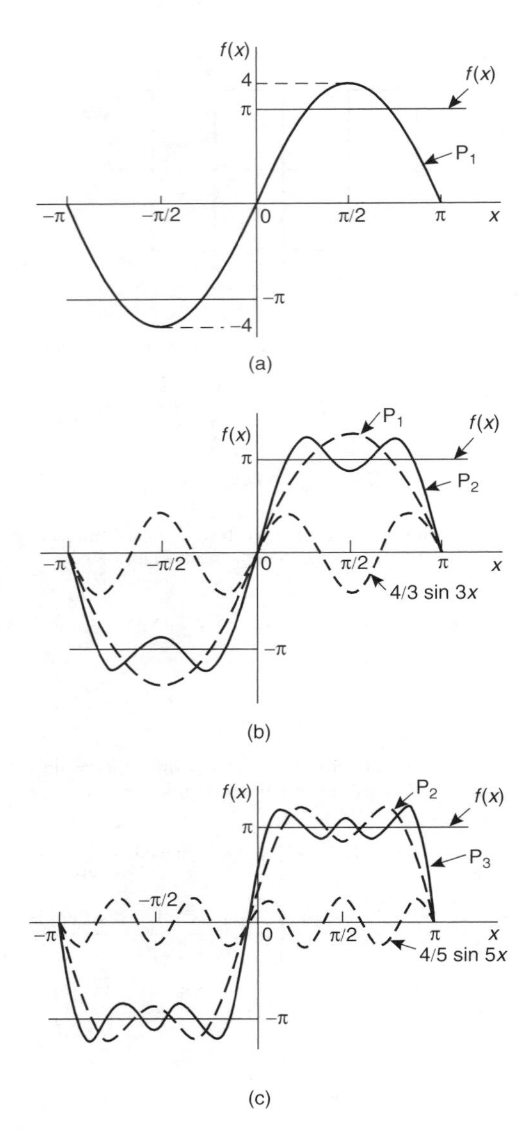
f(x)
4
π
f(x)
P1
−π
−π/2
0
π/2
π
x
−π
−4
(a)
P1
f(x)
f(x)
π
P2
−π
−π/2
0
π/2
π
x
4/3 sin 3x
−π
(b)
f(x)
P2
f(x)
π
P3
−π/2
−π
0
π/2
π
x
4/5 sin 5x
−π
(c)

Figure 89.3

Hence, from equation (1), the Fourier series for the function shown in Figure 89.2 is given by:

$$f(x) = a_0 + \sum_{n=1}^{\infty}(a_n \cos nx + b_n \sin nx) = 0 + \sum_{n=1}^{\infty}(0 + b_n \sin nx)$$

i.e. $$f(x) = \frac{4k}{\pi}\sin x + \frac{4k}{3\pi}\sin 3x + \frac{4k}{5\pi}\sin 5x + ..$$

i.e. $$\boldsymbol{f(x) = \frac{4k}{\pi}\left(\sin x + \frac{1}{3}\sin 3x + \frac{1}{5}\sin 5x + \cdots\right)}$$

If $k = \pi$ in the above Fourier then: $f(x) = 4(\sin x + \frac{1}{3}\sin 3x + \frac{1}{5}\sin 5x + \ldots)$
$4\sin x$ is termed the first partial sum of the Fourier series of $f(x)$, $\left(4\sin x + \frac{4}{3}\sin 3x\right)$ is termed the second partial sum of the Fourier series, and $\left(4\sin x + \frac{4}{3}\sin 3x + \frac{4}{5}\sin 5x\right)$ is termed the third partial sum, and so on.

Let $P_1 = 4\sin x$, $P_2 = \left(4\sin x + \frac{4}{3}\sin 3x\right)$

and $P_3 = \left(4\sin x + \frac{4}{3}\sin 3x + \frac{4}{5}\sin 5x\right)$.

Graphs of P_1, P_2 and P_3, obtained by drawing up tables of values, and adding waveforms, are shown in Figures 89.3(a) to (c) and they show that the series is convergent, i.e. continually approximating towards a definite limit as more and more partial sums are taken, and in the limit will have the sum $f(x) = \pi$. Even with just three partial sums, the waveform is starting to approach the rectangular wave the Fourier series is representing.

---

90 Fourier Series for a Non-periodic Function Over Range 2π

Expansion of non-periodic functions

If a function $f(x)$ is not periodic then it cannot be expanded in a Fourier series for **all** values of x. However, it is possible to determine a Fourier series to represent the function over any range of width 2π.
Given a non-periodic function, a new function may be constructed by taking the values of $f(x)$ in the given range and then repeating them outside of the given range at intervals of 2π. Since this new function is, by construction, periodic with period 2π, it may then be expanded in a Fourier series for all values of x. For example, the function $f(x) = x$ is not a periodic function. However, if a Fourier series for $f(x) = x$ is required then the function is constructed outside of this range so that it is periodic with period 2π as shown by the broken lines in Figure 90.1.
For non-periodic functions, such as $f(x) = x$, the sum of the Fourier series is equal to $f(x)$ at all points in the given range but it is not equal to $f(x)$ at points outside of the range.

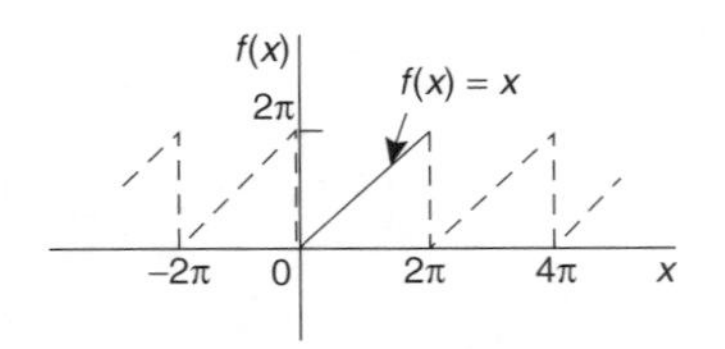

Figure 90.1

For determining a Fourier series of a non-periodic function over a range 2π, exactly the same formulae for the Fourier coefficients are used as in Chapter 89.

For example, to determine the Fourier series to represent the function $f(x) = 2x$ in the range $-\pi$ to $+\pi$:

The function $f(x) = 2x$ is not periodic. The function is shown in the range $-\pi$ to π in Figure 90.2 and is then constructed outside of that range so that it is periodic of period 2π (see broken lines) with the resulting saw-tooth waveform.

For a Fourier series: $f(x) = a_0 + \sum_{n=1}^{\infty}(a_n \cos nx + b_n \sin nx)$

From Chapter 89,

$$a_0 = \frac{1}{2\pi}\int_{-\pi}^{\pi} f(x)\,\mathrm{d}x = \frac{1}{2\pi}\int_{-\pi}^{\pi} 2x\,\mathrm{d}x = \frac{2}{2\pi}\left[\frac{x^2}{2}\right]_{-\pi}^{\pi} = 0$$

$$\begin{aligned}
a_n &= \frac{1}{\pi}\int_{-\pi}^{\pi} f(x)\cos nx\,\mathrm{d}x = \frac{1}{\pi}\int_{-\pi}^{\pi} 2x\cos nx\,\mathrm{d}x \\
&= \frac{2}{\pi}\left[\frac{x\sin nx}{n} - \int \frac{\sin nx}{n}\,\mathrm{d}x\right]_{-\pi}^{\pi} \quad \text{by parts (see Chapter 62)} \\
&= \frac{2}{\pi}\left[\frac{x\sin nx}{n} + \frac{\cos nx}{n^2}\right]_{-\pi}^{\pi} \\
&= \frac{2}{\pi}\left[\left(0 + \frac{\cos n\pi}{n^2}\right) - \left(0 + \frac{\cos n(-\pi)}{n^2}\right)\right] = 0
\end{aligned}$$

$$b_n = \frac{1}{\pi}\int_{-\pi}^{\pi} f(x)\sin nx\,\mathrm{d}x = \frac{1}{\pi}\int_{-\pi}^{\pi} 2x\sin nx\,\mathrm{d}x$$

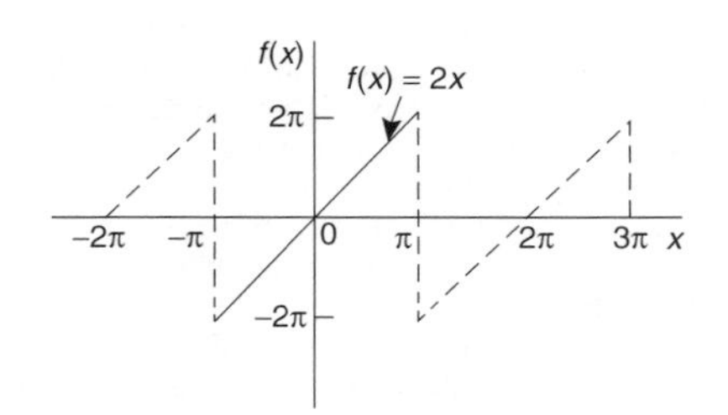

Figure 90.2

$$= \frac{2}{\pi}\left[\frac{-x\cos nx}{n} - \int\left(\frac{-\cos nx}{n}\right)\mathrm{d}x\right]_{-\pi}^{\pi} \quad \text{by parts}$$

$$= \frac{2}{\pi}\left[\frac{-x\cos nx}{n} + \frac{\sin nx}{n^2}\right]_{-\pi}^{\pi}$$

$$= \frac{2}{\pi}\left[\left(\frac{-\pi\cos n\pi}{n} + \frac{\sin n\pi}{n^2}\right) - \left(\frac{-(-\pi)\cos n(-\pi)}{n} + \frac{\sin n(-\pi)}{n^2}\right)\right]$$

$$= \frac{2}{\pi}\left[\frac{-\pi\cos n\pi}{n} - \frac{\pi\cos(-n\pi)}{n}\right] = \frac{-4}{n}\cos n\pi$$

When n is odd, $b_n = \frac{4}{n}$. Thus $b_1 = 4$, $b_3 = \frac{4}{3}$, $b_5 = \frac{4}{5}$, and so on.
When n is even, $b_n = \frac{-4}{n}$. Thus $b_2 = -\frac{4}{2}$, $b_4 = -\frac{4}{4}$, $b_6 = -\frac{4}{6}$, and so on.

Thus $$f(x) = 2x = 4\sin x - \tfrac{4}{2}\sin 2x + \tfrac{4}{3}\sin 3x - \tfrac{4}{4}\sin 4x + \tfrac{4}{5}\sin 5x - \tfrac{4}{6}\sin 6x + \ldots$$

i.e. $$\mathbf{2x = 4\left(\sin x - \tfrac{1}{2}\sin 2x + \tfrac{1}{3}\sin 3x - \tfrac{1}{4}\sin 4x + \tfrac{1}{5}\sin 5x - \tfrac{1}{6}\sin 6x + \ldots\right)}$$

for values of $f(x)$ between $-\pi$ and π. For values of $f(x)$ outside the range $-\pi$ to $+\pi$ the sum of the series is not equal to $f(x)$.

---

91 Even and Odd Functions and Half-range Fourier Series

Even and odd functions

A function $y = f(x)$ is said to be **even** if $f(-x) = f(x)$ for all values of x. Graphs of even functions are always **symmetrical about the *y*-axis** (i.e. is a mirror image). Two examples of even functions are $y = x^2$ and $y = \cos x$ as shown in Figure 37.12, page 193.
A function $y = f(x)$ is said to be **odd** if $f(-x) = -f(x)$ for all values of x. Graphs of odd functions are always **symmetrical about the origin**. Two examples of odd functions are $y = x^3$ and $y = \sin x$ as shown in Figure 37.13, page 193.
Many functions are neither even nor odd.

Fourier cosine series

The Fourier series of an **even periodic** function $f(x)$ having period 2π contains **cosine terms only** (i.e. contains no sine terms) and may contain a constant term. Hence

$$\boldsymbol{f(x) = a_0 + \sum_{n=1}^{\infty} a_n \cos nx}$$

where $\boldsymbol{a_0} = \frac{1}{2\pi}\int_{-\pi}^{\pi} f(x)\mathrm{d}x = \boldsymbol{\frac{1}{\pi}\int_0^{\pi} f(x)\mathrm{d}x}$ (due to symmetry)

and

$$\boldsymbol{a_n} = \frac{1}{\pi}\int_{-\pi}^{\pi} f(x)\cos nx\,\mathrm{d}x = \boldsymbol{\frac{2}{\pi}\int_0^{\pi} f(x)\cos nx\,\mathrm{d}x}$$

For example, to determine the Fourier series for the periodic function defined by:

$$f(x) = \begin{cases} -2, & \text{when } -\pi < x < -\frac{\pi}{2} \\ 2, & \text{when } -\frac{\pi}{2} < x < \frac{\pi}{2} \\ -2, & \text{when } \frac{\pi}{2} < x < \pi \end{cases} \quad \text{and has a period of } 2\pi$$

The square wave shown in Figure 91.1 is an even function since it is symmetrical about the $f(x)$ axis.

Hence from above, the Fourier series is given by:

$f(x) = a_0 + \sum_{n=1}^{\infty} a_n \cos nx$ (i.e. the series contains no sine terms).

$$a_0 = \frac{1}{\pi}\int_0^{\pi} f(x)\,\mathrm{d}x = \frac{1}{\pi}\left\{\int_0^{\pi/2} 2\,\mathrm{d}x + \int_{\pi/2}^{\pi} -2\,\mathrm{d}x\right\}$$

$$= \frac{1}{\pi}\left\{[2x]_0^{\pi/2} + [-2x]_{\pi/2}^{\pi}\right\} = \frac{1}{\pi}[(\pi) + [(-2\pi) - (-\pi)]] = 0$$

$$a_n = \frac{2}{\pi}\int_0^{\pi} f(x)\cos nx\,\mathrm{d}x = \frac{2}{\pi}\left\{\int_0^{\pi/2} 2\cos nx\,\mathrm{d}x + \int_{\pi/2}^{\pi} -2\cos nx\,\mathrm{d}x\right\}$$

$$= \frac{4}{\pi}\left\{\left[\frac{\sin nx}{n}\right]_0^{\pi/2} + \left[\frac{-\sin nx}{n}\right]_{\pi/2}^{\pi}\right\}$$

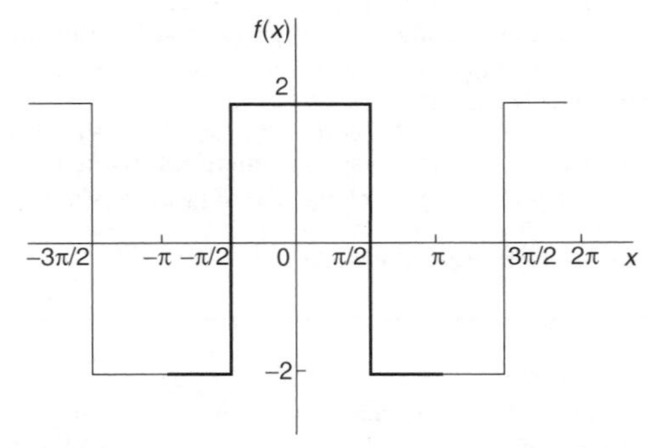

Figure 91.1

$$= \frac{4}{\pi}\left\{\left(\frac{\sin(\pi/2)n}{n} - 0\right) + \left(0 - \frac{-\sin(\pi/2)n}{n}\right)\right\}$$

$$= \frac{4}{\pi}\left(\frac{2\sin(\pi/2)n}{n}\right) = \frac{8}{\pi n}\left(\sin\frac{n\pi}{2}\right)$$

When n is even, $a_n = 0$. When n is odd,

$$a_n = \frac{8}{\pi n} \text{ for } n = 1, 5, 9, \ldots \text{ and } a_n = \frac{-8}{\pi n} \text{ for } n = 3, 7, 11, \ldots$$

Hence $a_1 = \frac{8}{\pi}$, $\quad a_3 = \frac{-8}{3\pi}$, $\quad a_5 = \frac{8}{5\pi}$, and so on

Hence the Fourier series for the waveform of Figure 91.1 is given by:

$$\mathbf{f(x) = \frac{8}{\pi}\left(\cos x - \frac{1}{3}\cos 3x + \frac{1}{5}\cos 5x - \frac{1}{7}\cos 7x + \ldots\right)}$$

Fourier sine series

The Fourier series of an **odd** periodic function $f(x)$ having period 2π contains sine terms only (i.e. contains no constant term and no cosine terms).

Hence $\quad \mathbf{f(x) = \sum_{n=1}^{\infty} b_n \sin nx}$

where $\qquad \mathbf{b_n} = \frac{1}{\pi}\int_{-\pi}^{\pi} f(x)\sin nx\,dx = \mathbf{\frac{2}{\pi}\int_0^{\pi} f(x)\sin nx\,dx}$

For example, to obtain the Fourier series for the square wave shown in Figure 91.2:
The square wave is an odd function since it is symmetrical about the origin.

Hence, from above, the Fourier series is given by: $f(x) = \sum_{n=1}^{\infty} b_n \sin nx$

The function is defined by: $f(x) = \begin{cases} -2, & \text{when } -\pi < x < 0 \\ 2, & \text{when } 0 < x < \pi \end{cases}$

$$b_n = \frac{2}{\pi}\int_0^{\pi} f(x)\sin nx\,dx = \frac{2}{\pi}\int_0^{\pi} 2\sin nx\,dx = \frac{4}{\pi}\left[\frac{-\cos nx}{n}\right]_0^{\pi}$$

$$= \frac{4}{\pi}\left[\left(\frac{-\cos n\pi}{n}\right) - \left(-\frac{1}{n}\right)\right] = \frac{4}{\pi n}(1 - \cos n\pi)$$

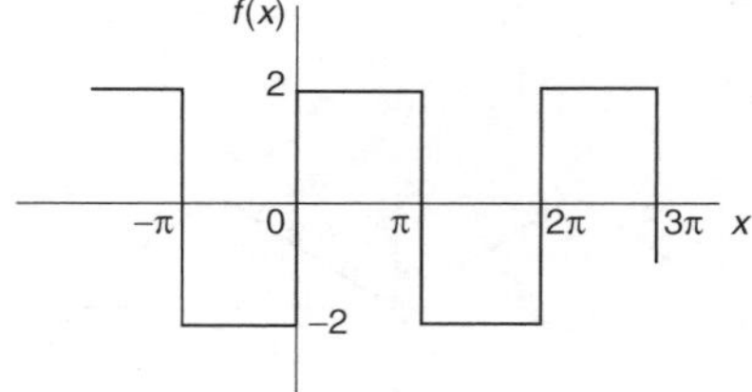

Figure 91.2

When n is even, $b_n = 0$. When n is odd, $b_n = \frac{4}{\pi n}(1 - (-1)) = \frac{8}{\pi n}$

Hence $b_1 = \frac{8}{\pi}$, $b_3 = \frac{8}{3\pi}$, $b_5 = \frac{8}{5\pi}$, and so on

Hence the Fourier series is:

$$\mathbf{f(x) = \frac{8}{\pi}\left(\sin x + \frac{1}{3}\sin 3x + \frac{1}{5}\sin 5x + \frac{1}{7}\sin 7x + \ldots\right)}$$

Half range Fourier series

When a function is defined over the range say 0 to π instead of from 0 to 2π it may be expanded in a series of sine terms only or of cosine terms only. The series produced is called a **half-range Fourier series**.

(a) If a **half-range cosine series** is required for the function $f(x) = x$ in the range 0 to π then an **even** periodic function is required. In Figure 91.3, $f(x) = x$ is shown plotted from $x = 0$ to $x = \pi$. Since an even function is symmetrical about the $f(x)$ axis the line AB is constructed as shown. If the triangular waveform produced is assumed to be periodic of period 2π outside of this range then the waveform is as shown in Figure 91.3. When a half range cosine series is required then the Fourier coefficients a_0 and a_n are calculated as earlier, i.e.

$$\mathbf{f(x) = a_0 + \sum_{n=1}^{\infty} a_n \cos nx}$$

where $\quad \mathbf{a_0 = \frac{1}{\pi}\int_0^{\pi} f(x)\,dx}$ and $\mathbf{a_n = \frac{2}{\pi}\int_0^{\pi} f(x)\cos nx\,dx}$

For example, to determine the half-range Fourier cosine series to represent the function $f(x) = x$ in the range $0 \le x \le \pi$:

When $f(x) = x$, $\quad a_0 = \frac{1}{\pi}\int_0^{\pi} f(x)\,dx = \frac{1}{\pi}\int_0^{\pi} x\,dx = \frac{1}{\pi}\left[\frac{x^2}{2}\right]_0^{\pi} = \frac{\pi}{2}$

$$a_n = \frac{2}{\pi}\int_0^{\pi} f(x)\cos nx\,dx = \frac{2}{\pi}\int_0^{\pi} x\cos nx\,dx$$

$$= \frac{2}{\pi}\left[\frac{x\sin nx}{n} + \frac{\cos nx}{n^2}\right]_0^{\pi} \text{ by parts}$$

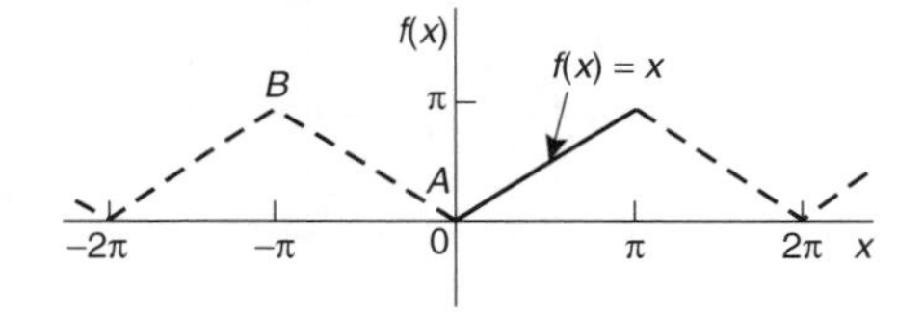

Figure 91.3

$$= \frac{2}{\pi}\left[\left(\frac{\pi \sin n\pi}{n} + \frac{\cos n\pi}{n^2}\right) - \left(0 + \frac{\cos 0}{n^2}\right)\right]$$

$$= \frac{2}{\pi}\left(0 + \frac{\cos n\pi}{n^2} - \frac{\cos 0}{n^2}\right) = \frac{2}{\pi n^2}(\cos n\pi - 1)$$

When n is even, $a_n = 0$

When n is odd, $a_n = \frac{2}{\pi n^2}(-1 - 1) = \frac{-4}{\pi n^2}$

Hence $a_1 = \frac{-4}{\pi}$, $a_3 = \frac{-4}{\pi 3^2}$, $a_5 = \frac{-4}{\pi 5^2}$, and so on. Hence the half range Fourier cosine series is given by:

$$\mathbf{f(x) = x = \frac{\pi}{2} - \frac{4}{\pi}\left(\cos x + \frac{1}{3^2}\cos 3x + \frac{1}{5^2}\cos 5x + \ldots\right)}$$

(b) If a **half-range sine series** is required for the function $f(x) = x$ in the range 0 to π then an odd periodic function is required. In Figure 91.4, $f(x) = x$ is shown plotted from $x = 0$ to $x = \pi$. Since an odd function is symmetrical about the origin the line CD is constructed as shown. If the sawtooth waveform produced is assumed to be periodic of period 2π outside of this range, then the waveform is as shown in Figure 91.4. When a half-range sine series is required then the Fourier coefficient b_n is calculated as earlier, i.e.

$$\mathbf{f(x) = \sum_{n=1}^{\infty} b_n \sin nx} \quad \text{where} \quad \mathbf{b_n = \frac{2}{\pi}\int_0^{\pi} f(x)\sin nx\, dx}$$

For example, to determine the half-range Fourier sine series to represent the function $f(x) = x$ in the range $0 \le x \le \pi$:

When $f(x) = x$, $b_n = \frac{2}{\pi}\int_0^{\pi} f(x)\sin nx\, dx = \frac{2}{\pi}\int_0^{\pi} x \sin nx\, dx$

$$= \frac{2}{\pi}\left[\frac{-x\cos nx}{n} + \frac{\sin nx}{n^2}\right]_0^{\pi} \text{ by parts}$$

$$= \frac{2}{\pi}\left[\left(\frac{-\pi\cos n\pi}{n} + \frac{\sin n\pi}{n^2}\right) - (0 + 0)\right] = -\frac{2}{n}\cos n\pi$$

When n is odd, $b_n = \frac{2}{n}$. Hence $b_1 = \frac{2}{1}$, $b_3 = \frac{2}{3}$, $b_5 = \frac{2}{5}$ and so on.

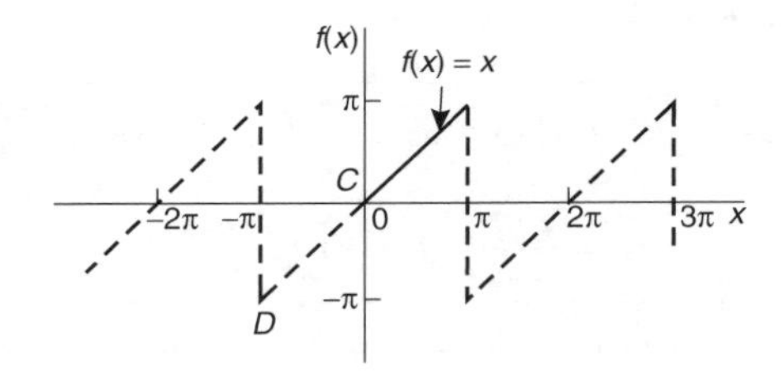

Figure 91.4

When n is even, $b_n = -\frac{2}{n}$. Hence $b_2 = -\frac{2}{2}$, $b_4 = -\frac{2}{4}$, $b_6 = -\frac{2}{6}$ and so on.

Hence the half-range Fourier sine series is given by:

$$\mathbf{f(x) = x = 2\left(\sin x - \tfrac{1}{2}\sin 2x + \tfrac{1}{3}\sin 3x - \tfrac{1}{4}\sin 4x + \tfrac{1}{5}\sin 5x - \cdots\right)}$$

---

92 Fourier Series Over Any Range

Expansion of a periodic function of period L

A periodic function $f(x)$ of period L repeats itself when x increases by L, i.e. $f(x + L) = f(x)$. The change from functions dealt with previously having period 2π to functions having period L is not difficult since it may be achieved by a change of variable.

To find a Fourier series for a function $f(x)$ in the range $-\frac{L}{2} \le x \le \frac{L}{2}$ a new variable u is introduced such that $f(x)$, as a function of u, has period 2π. If $u = \frac{2\pi x}{L}$ then, when $x = -\frac{L}{2}$, $u = -\pi$ and when $x = \frac{L}{2}$, $u = +\pi$. Also, let $f(x) = f\left(\frac{Lu}{2\pi}\right) = F(u)$. The Fourier series for $F(u)$ is given by:

$$F(u) = a_0 + \sum_{n=1}^{\infty}(a_n \cos nu + b_n \sin nu), \text{ where } a_0 = \frac{1}{2\pi}\int_{-\pi}^{\pi} F(u)\,\mathrm{d}u,$$

$$a_n = \frac{1}{\pi}\int_{-\pi}^{\pi} F(u)\cos nu\,\mathrm{d}u \quad \text{and} \quad b_n = \frac{1}{\pi}\int_{-\pi}^{\pi} F(u)\sin nu\,\mathrm{d}u$$

It is however more usual to change the above formulae to terms of x. Since $u = \frac{2\pi x}{L}$, then $\mathrm{d}u = \frac{2\pi}{L}\,\mathrm{d}x$, and the limits of integration are $-\frac{L}{2}$ to $+\frac{L}{2}$ instead of from $-\pi$ to $+\pi$. Hence the Fourier series expressed in terms of x is given by:

$$\boxed{\mathbf{f(x) = a_0 + \sum_{n=1}^{\infty}\left[a_n \cos\left(\frac{2\pi nx}{L}\right) + b_n \sin\left(\frac{2\pi nx}{L}\right)\right]}}$$

where, in the range $-\frac{L}{2}$ to $+\frac{L}{2}$:

$$\boxed{\mathbf{a_0 = \frac{1}{L}\int_{-L/2}^{L/2} f(x)\,dx, \quad a_n = \frac{2}{L}\int_{-L/2}^{L/2} f(x)\cos\left(\frac{2\pi nx}{L}\right)dx}}$$

and

$$\boxed{\mathbf{b_n = \frac{2}{L}\int_{-L/2}^{L/2} f(x)\sin\left(\frac{2\pi nx}{L}\right)dx}}$$

(The limits of integration may be replaced by any interval of length L, such as from 0 to L).

For example, if the voltage from a square wave generator is of the form:

$v(t) = \begin{cases} 0, & -4 < t < 0 \\ 10, & 0 < t < 4 \end{cases}$ and has a period of 8 ms, then the Fourier series is obtained as follows:

The square wave is shown in Figure 92.1. The Fourier series is of the form:

$$v(t) = a_0 + \sum_{n=1}^{\infty} \left[a_n \cos\left(\frac{2\pi n t}{L}\right) + b_n \sin\left(\frac{2\pi n t}{L}\right) \right]$$

$$a_0 = \frac{1}{L}\int_{-L/2}^{L/2} v(t)\,\mathrm{d}t = \frac{1}{8}\int_{-4}^{4} v(t)\,\mathrm{d}t$$

$$= \frac{1}{8}\left\{\int_{-4}^{0} 0\,\mathrm{d}t + \int_{0}^{4} 10\,\mathrm{d}t\right\} = \frac{1}{8}[10t]_0^4 = 5$$

$$a_n = \frac{2}{L}\int_{-L/2}^{L/2} v(t)\cos\left(\frac{2\pi n t}{L}\right)\mathrm{d}t = \frac{2}{8}\int_{-4}^{4} v(t)\cos\left(\frac{2\pi n t}{8}\right)\mathrm{d}t$$

$$= \frac{1}{4}\left\{\int_{-4}^{0} 0\cos\left(\frac{\pi n t}{4}\right)\mathrm{d}t + \int_{0}^{4} 10\cos\left(\frac{\pi n t}{4}\right)\mathrm{d}t\right\}$$

$$= \frac{1}{4}\left[\frac{10\sin\left(\frac{\pi n t}{4}\right)}{\left(\frac{\pi n}{4}\right)}\right]_0^4$$

$$= \frac{10}{\pi n}[\sin \pi n - \sin 0] = 0 \text{ for } n = 1, 2, 3, \ldots$$

$$b_n = \frac{2}{L}\int_{-L/2}^{L/2} v(t)\sin\left(\frac{2\pi n t}{L}\right)\mathrm{d}t = \frac{2}{8}\int_{-4}^{4} v(t)\sin\left(\frac{2\pi n t}{8}\right)\mathrm{d}t$$

$$= \frac{1}{4}\left\{\int_{-4}^{0} 0\sin\left(\frac{\pi n t}{4}\right)\mathrm{d}t + \int_{0}^{4} 10\sin\left(\frac{\pi n t}{4}\right)\mathrm{d}t\right\}$$

$$= \frac{1}{4}\left[\frac{-10\cos\left(\frac{\pi n t}{4}\right)}{\left(\frac{\pi n}{4}\right)}\right]_0^4 = \frac{-10}{\pi n}[\cos \pi n - \cos 0]$$

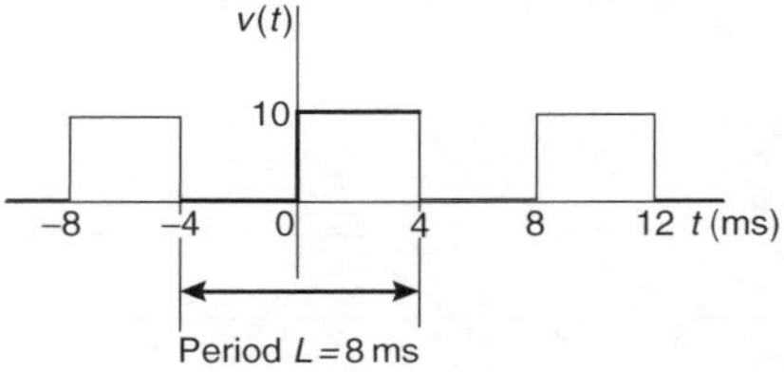

Figure 92.1

When n is even, $b_n = 0$

When n is odd, $b_1 = \frac{-10}{\pi}(-1-1) = \frac{20}{\pi}, \quad b_3 = \frac{-10}{3\pi}(-1-1) = \frac{20}{3\pi},$

$b_5 = \frac{20}{5\pi}$, and so on

Thus the Fourier series for the function $v(t)$ is given by:

$$\boldsymbol{v(t) = 5 + \frac{20}{\pi}\left[\sin\left(\frac{\pi t}{4}\right) + \frac{1}{3}\sin\left(\frac{3\pi t}{4}\right) + \frac{1}{5}\sin\left(\frac{5\pi t}{4}\right) + \cdots\right]}$$

Half-range Fourier series for functions defined over range L

By making the substitution $u = \frac{\pi x}{L}$, the range $x = 0$ to $x = L$ corresponds to the range $u = 0$ to $u = \pi$. Hence a function may be expanded in a series of either cosine terms or sine terms only, i.e. a **half-range Fourier series**.
A **half-range cosine series** in the range 0 to L can be expanded as:

$$\boldsymbol{f(x) = a_0 + \sum_{n=1}^{\infty} a_n \cos\left(\frac{n\pi x}{L}\right)}$$

where $$\boldsymbol{a_0 = \frac{1}{L}\int_0^L f(x)\,dx \quad \text{and} \quad a_n = \frac{2}{L}\int_0^L f(x)\cos\left(\frac{n\pi x}{L}\right)dx}$$

For example, the half-range Fourier cosine series for the function $f(x) = x$ in the range $0 \le x \le 2$ is obtained as follows:
A half-range Fourier cosine series indicates an even function. Thus the graph of $f(x) = x$ in the range 0 to 2 is shown in Figure 92.2 and is extended outside of this range so as to be symmetrical about the $f(x)$ axis as shown by the broken lines.

For a half-range cosine series: $f(x) = a_0 + \sum_{n=1}^{\infty} a_n \cos\left(\frac{n\pi x}{L}\right)$

$$a_0 = \frac{1}{L}\int_0^L f(x)\,dx = \frac{1}{2}\int_0^2 x\,dx = \frac{1}{2}\left[\frac{x^2}{2}\right]_0^2 = 1$$

$$a_n = \frac{2}{L}\int_0^L f(x)\cos\left(\frac{n\pi x}{L}\right)dx = \frac{2}{2}\int_0^2 x\cos\left(\frac{n\pi x}{2}\right)dx$$

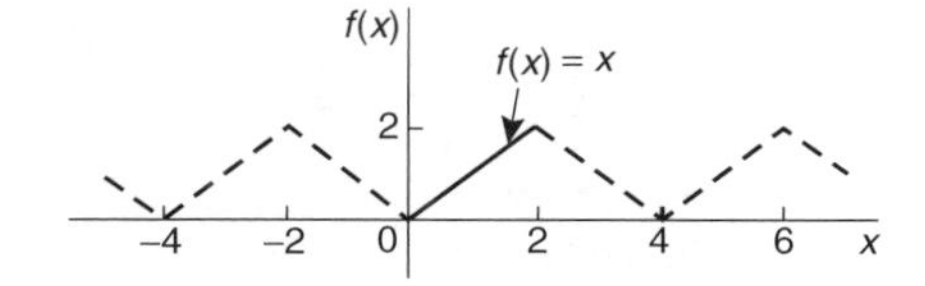

Figure 92.2

$$= \left[\frac{x\sin\left(\frac{n\pi x}{2}\right)}{\left(\frac{n\pi}{2}\right)} + \frac{\cos\left(\frac{n\pi x}{2}\right)}{\left(\frac{n\pi}{2}\right)^2}\right]_0^2$$

$$= \left[\left(\frac{2\sin n\pi}{\left(\frac{n\pi}{2}\right)} + \frac{\cos n\pi}{\left(\frac{n\pi}{2}\right)^2}\right) - \left(0 + \frac{\cos 0}{\left(\frac{n\pi}{2}\right)^2}\right)\right]$$

$$= \left[\frac{\cos n\pi}{\left(\frac{n\pi}{2}\right)^2} - \frac{1}{\left(\frac{n\pi}{2}\right)^2}\right] = \left(\frac{2}{\pi n}\right)^2 (\cos n\pi - 1)$$

When n is even, $a_n = 0$

$a_1 = \frac{-8}{\pi^2}$, $a_3 = \frac{-8}{\pi^2 3^2}$, $a_5 = \frac{-8}{\pi^2 5^2}$, and so on.

Hence the half-range Fourier cosine series for $f(x)$ in the range 0 to 2 is given by:

$$\mathbf{f(x) = 1 - \frac{8}{\pi^2}\left[\cos\left(\frac{\pi x}{2}\right) + \frac{1}{3^2}\cos\left(\frac{3\pi x}{2}\right) + \frac{1}{5^2}\cos\left(\frac{5\pi x}{2}\right) + \cdots\right]}$$

A **half-range sine series** in the range 0 to L can be expanded as:

$$f(x) = \sum_{n=1}^{\infty} b_n \sin\left(\frac{n\pi x}{L}\right) \text{ where } b_n = \frac{2}{L}\int_0^L f(x)\sin\left(\frac{n\pi x}{L}\right) dx$$

---

93 A Numerical Method of Harmonic Analysis

Introduction

Many practical waveforms can be represented by simple mathematical expressions, and, by using Fourier series, the magnitude of their harmonic components determined, as shown in Chapters 89 to 92. For waveforms not in this category, analysis may be achieved by numerical methods.

Harmonic analysis is the process of resolving a periodic, non-sinusoidal quantity into a series of sinusoidal components of ascending order of frequency.

Harmonic analysis on data given in tabular or graphical form

The Fourier coefficients a_0, a_n and b_n used in Chapters 89 to 92 all require functions to be integrated, i.e.

$$a_0 = \frac{1}{2\pi}\int_{-\pi}^{\pi} f(x)\,dx = \frac{1}{2\pi}\int_0^{2\pi} f(x)\,dx$$

$= \text{mean value of } f(x) \text{ in the range } -\pi \text{ to } \pi \text{ or } 0 \text{ to } 2\pi$

$$a_n = \frac{1}{\pi}\int_{-\pi}^{\pi} f(x)\cos nx\,dx = \frac{1}{\pi}\int_{0}^{2\pi} f(x)\cos nx\,dx$$

= twice the mean value of $f(x)\cos nx$ in the range 0 to 2π

$$b_n = \frac{1}{\pi}\int_{-\pi}^{\pi} f(x)\sin nx\,dx = \frac{1}{\pi}\int_{0}^{2\pi} f(x)\sin nx\,dx$$

= twice the mean value of $f(x)\sin nx$ in the range 0 to 2π

However, irregular waveforms are not usually defined by mathematical expressions and thus the Fourier coefficients cannot be determined by using calculus. In these cases, approximate methods, such as the **trapezoidal rule**, can be used to evaluate the Fourier coefficients.

Most practical waveforms to be analysed are periodic. Let the period of a waveform be 2π and be divided into p equal parts as shown in Figure 93.1. The width of each interval is thus $\frac{2\pi}{p}$. Let the ordinates be labelled $y_0, y_1, y_2, \ldots y_p$ (note that $y_0 = y_p$). The trapezoidal rule states:

$$\text{Area} = \begin{pmatrix}\text{width of}\\ \text{interval}\end{pmatrix}\left[\frac{1}{2}\begin{pmatrix}\text{first + last}\\ \text{ordinate}\end{pmatrix} + \begin{matrix}\text{sum of remaining}\\ \text{ordinates}\end{matrix}\right]$$

$$\approx \frac{2\pi}{p}\left[\frac{1}{2}(y_0 + y_p) + y_1 + y_2 + y_3 + \ldots\right]$$

Since $y_0 = y_p$, then $\frac{1}{2}(y_0 + y_p) = y_0 = y_p$. Hence area $\approx \frac{2\pi}{p}\sum_{k=1}^{p} y_k$

$$\text{Mean value} = \frac{\text{area}}{\text{length of base}} \approx \frac{1}{2\pi}\left(\frac{2\pi}{p}\right)\sum_{k=1}^{p} y_k \approx \frac{1}{p}\sum_{k=1}^{p} y_k$$

However, a_0 = mean value of $f(x)$ in the range 0 to 2π

Thus $\boldsymbol{a_0 \approx \frac{1}{p}\sum_{k=1}^{p} y_k}$ (1)

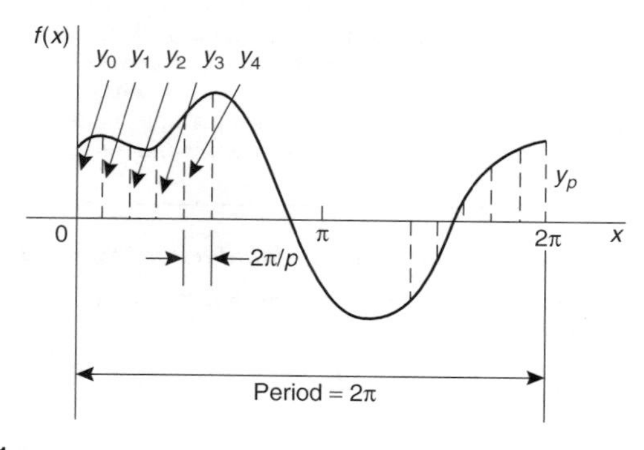

Figure 93.1

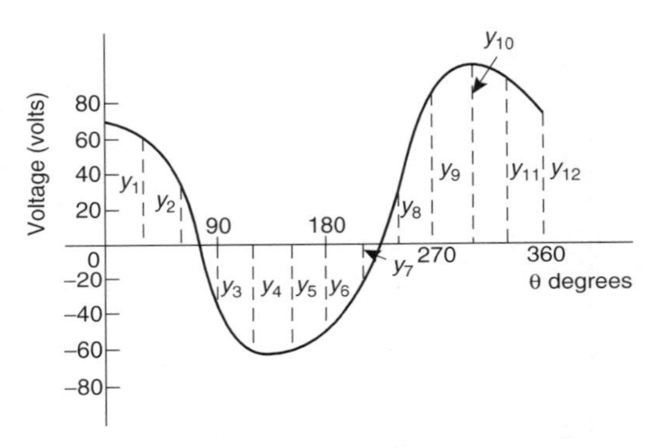

Figure 93.2

Similarly, a_n = twice the mean value of $f(x)\cos nx$ in the range 0 to 2π,

thus $$\boldsymbol{a_n \approx \frac{2}{p}\sum_{k=1}^{p} y_k \cos nx_k} \qquad (2)$$

and b_n = twice the mean value of $f(x)\sin nx$ in the range 0 to 2π,

thus $$\boldsymbol{b_n \approx \frac{2}{p}\sum_{k=1}^{p} y_k \sin nx_k} \qquad (3)$$

For example, a graph of voltage V against angle θ is shown in Figure 93.2. The values of the ordinates $y_1, y_2, y_3, \ldots$. are 62, 35, −38, −64, −63, −52, −28, 24, 80, 96, 90 and 70, the 12 equal intervals each being of width 30°. (If a larger number of intervals are used, results having a greater accuracy are achieved).
The voltage may be analysed into its first three constituent components as follows:
The data is tabulated in the proforma shown in Table 93.1.

From equation (1), $a_0 \approx \frac{1}{p}\sum_{k=1}^{p} y_k = \frac{1}{12}(212) = 17.67$ (since $p = 12$)

From equation (2), $a_n \approx \frac{2}{p}\sum_{k=1}^{p} y_k \cos nx_k$ hence $a_1 \approx \frac{2}{12}(417.94) = 69.66$,

$$a_2 \approx \frac{2}{12}(-39) = -6.50 \text{ and } a_3 \approx \frac{2}{12}(-49) = -8.17$$

From equation (3), $b_n \approx \frac{2}{p}\sum_{k=1}^{p} y_k \sin nx_k$ hence $b_1 \approx \frac{2}{12}(-278.53) = -46.42$,

$$b_2 \approx \frac{2}{12}(29.43) = 4.91 \text{ and } b_3 \approx \frac{2}{12}(55) = 9.17$$

Table 93.1

Ordinates	θ	V	$\cos\theta$	$V\cos\theta$	$\sin\theta$	$V\sin\theta$	$\cos 2\theta$	$V\cos 2\theta$	$\sin 2\theta$	$V\sin 2\theta$	$\cos 3\theta$	$V\cos 3\theta$	$\sin 3\theta$	$V\sin 3\theta$
y_1	30	62	0.866	53.69	0.5	31	0.5	31	0.866	53.69	0	0	1	62
y_2	60	35	0.5	17.5	0.866	30.31	−0.5	−17.5	0.866	30.31	−1	−35	0	0
y_3	90	−38	0	0	1	−38	−1	38	0	0	0	0	−1	38
y_4	120	−64	−0.5	32	0.866	−55.42	−0.5	32	−0.866	55.42	1	−64	0	0
y_5	150	−63	−0.866	54.56	0.5	−31.5	0.5	−31.5	−0.866	54.56	0	0	1	−63
y_6	180	−52	−1	52	0	0	1	−52	0	0	−1	52	0	0
y_7	210	−28	−0.866	24.25	−0.5	14	0.5	−14	0.866	−24.25	0	0	−1	28
y_8	240	24	−0.5	−12	−0.866	−20.78	−0.5	−12	0.866	20.78	1	24	0	0
y_9	270	80	0	0	−1	−80	−1	−80	0	0	0	0	1	80
y_{10}	300	96	0.5	48	−0.866	−83.14	−0.5	−48	−0.866	−83.14	−1	−96	0	0
y_{11}	330	90	0.866	77.94	−0.5	−45	0.5	45	−0.866	−77.94	0	0	−1	−90
y_{12}	360	70	1	70	0	0	1	70	0	0	1	70	0	0
		$\sum_{k=1}^{12} y_k = 212$		$\sum_{k=1}^{12} y_k\cos\theta_k = 417.94$		$\sum_{k=1}^{12} y_k\sin\theta_k = -278.53$		$\sum_{k=1}^{12} y_k\cos 2\theta_k = -39$		$\sum_{k=1}^{12} y_k\sin 2\theta_k = 29.43$		$\sum_{k=1}^{12} y_k\cos 3\theta_k = -49$		$\sum_{k=1}^{12} y_k\sin 3\theta_k = 55$

Substituting these values into the Fourier series:

$$f(x) = a_0 + \sum_{n=1}^{\infty}(a_n \cos nx + b_n \sin nx)$$

gives: $\quad \boldsymbol{v = 17.67 + 69.66\cos\theta - 6.50\cos 2\theta - 8.17\cos 3\theta + \ldots}$

$$\boldsymbol{-46.42\sin\theta + 4.91\sin 2\theta + 9.17\sin 3\theta + \ldots} \quad (4)$$

Note that in equation (4), $(-46.42\sin\theta + 69.66\cos\theta)$ comprises the fundamental, $(4.91\sin 2\theta - 6.50\cos 2\theta)$ comprises the second harmonic and

$(9.17\sin 3\theta - 8.17\cos 3\theta)$ comprises the third harmonic.

It is shown in Chapter 31 that: $a\sin\omega t + b\cos\omega t = R\sin(\omega t + \alpha)$ where $a = R\cos\alpha$, $b = R\sin\alpha$, $R = \sqrt{a^2 + b^2}$ and $\alpha = \tan^{-1}\frac{b}{a}$

Hence equation (4) is equivalent to:

$$\boldsymbol{v = 17.67 + 83.71\sin(\theta + 2.16) + 8.15\sin(2\theta - 0.92)}$$

$$\boldsymbol{+12.28\sin(3\theta - 0.73)\ \text{volts}}$$

which is the form normally used with complex waveforms.

Complex waveform considerations

It is sometimes possible to predict the harmonic content of a waveform on inspection of particular waveform characteristics.

(i) If a periodic waveform is such that the area above the horizontal axis is equal to the area below then the mean value is zero. Hence $a_0 = 0$ (see Figure 93.3(a)).

(ii) An **even function** is symmetrical about the vertical axis and contains **no sine terms** (see Figure 93.3(b)).

(iii) An **odd function** is symmetrical about the origin and contains **no cosine terms** (see Figure 93.3(c)).

(iv) $f(x) = f(x + \pi)$ represents a waveform which repeats after half a cycle and **only even harmonics** are present (see Figure 93.3(d)).

(v) $f(x) = -f(x + \pi)$ represents a waveform for which the positive and negative cycles are identical in shape and **only odd harmonics** are present (see Figure 93.3(e)).

For example, an alternating current i amperes is shown in Figure 93.4. The waveform is analysed into its constituent harmonics as far as and including the fifth harmonic, taking 30° intervals, as follows:

With reference to Figure 93.4, the following characteristics are noted:

(i) The mean value is zero since the area above the θ axis is equal to the area below it. Thus the constant term, or d.c. component, $a_0 = 0$

(ii) Since the waveform is symmetrical about the origin the function i is odd, which means that there are no cosine terms present in the Fourier series.

(iii) The waveform is of the form $f(\theta) = -f(\theta + \pi)$ which means that only odd harmonics are present.

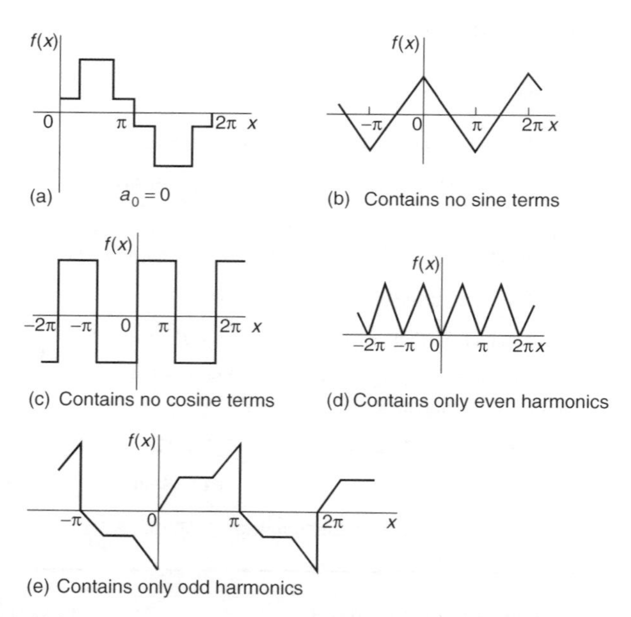

Figure 93.3

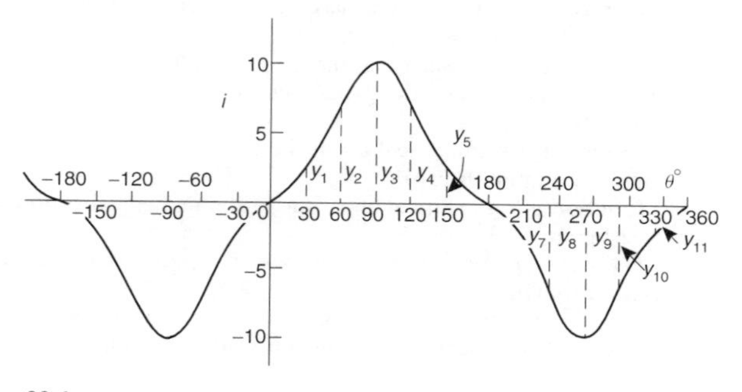

Figure 93.4

Investigating waveform characteristics has thus saved unnecessary calculations and in this case the Fourier series has only odd sine terms present, i.e.

$$i = b_1 \sin\theta + b_3 \sin 3\theta + b_5 \sin 5\theta + ..$$

A proforma, similar to Table 93.1, but without the 'cosine terms' columns and without the 'even sine terms' columns in shown in Table 93.2 up to, and including, the fifth harmonic, from which the Fourier coefficients b_1,

Table 93.2

Ordinate	θ	i	$\sin\theta$	$i\sin\theta$	$\sin 3\theta$	$i\sin 3\theta$	$\sin 5\theta$	$i\sin 5\theta$
Y_1	30	2	0.5	1	1	2	0.5	1
Y_2	60	7	0.866	6.06	0	0	−0.866	−6.06
Y_3	90	10	1	10	−1	−10	1	10
Y_4	120	7	0.866	6.06	0	0	−0.866	−6.06
Y_5	150	2	0.5	1	1	2	0.5	1
Y_6	180	0	0	0	0	0	0	0
Y_7	210	−2	−0.5	1	−1	2	−0.5	1
Y_8	240	−7	−0.866	6.06	0	0	0.866	−6.06
Y_9	270	−10	−1	10	1	−10	−1	10
Y_{10}	300	−7	−0.866	6.06	0	0	0.866	−6.06
Y_{11}	330	−2	−0.5	1	−1	2	−0.5	1
Y_{12}	360	0	0	0	0	0	0	0
			$\sum_{k=1}^{12} y_k \sin\theta_k = 48.24$		$\sum_{k=1}^{12} y_k \sin 3\theta_k = -12$		$\sum_{k=1}^{12} y_k \sin 5\theta_k = -0.24$	

b_3 and b_5 can be determined. Twelve co-ordinates are chosen and labelled $y_1, y_2, y_3, \ldots y_{12}$ as shown in Figure 93.4.

From equation (3), $b_n = \dfrac{2}{p}\sum_{k=1}^{p} i_k \sin n\theta_k$, where $p = 12$

Hence $\quad b_1 \approx \dfrac{2}{12}(48.24) = 8.04, \quad b_3 \approx \dfrac{2}{12}(-12) = -2.00,$

and $\quad b_5 \approx \dfrac{2}{12}(-0.24) = -0.04$

Thus the Fourier series for current i is given by:

$$\boldsymbol{i = 8.04\sin\theta - 2.00\sin 3\theta - 0.04\sin 5\theta}$$

Index

Abscissa, 155
Acute angle, 109
  angled triangle, 111
Adjoint of matrix, 235
Algebra, 16
  Boolean, 244
Algebraic method of successive
    approximations, 77
Alternate angles, 110
Amplitude, 136
And-function, 244
And-gate, 255
Angle between two vectors, 212
Angles, of any magnitude, 129
  types and properties of, 109
Angles of depression, 119
  elevation, 119
Angular measurement, 109
  velocity, 139
Approximations, 14
Arc of circle, 92, 93
Area between curves, 334
  under a curve, 330
Areas of irregular figures, 102
  plane figures, 87
  sector, 93
  similar shapes, 91
  triangles, 126
Argand diagram, 220
Argument, 222
Arithmetic, 1
Arithmetic mean, 381
  progression, 65
Array, 231
Astroid, 283
Asymptotes, 195
Auxiliary equation, 361
Average value, 380
  integration, 336
  of waveform, 106

Base, 10
Binary numbers, 80
Binomial distribution, 389
  series, 67, 68
  practical problems, 70
Bisection method, 75
BODMAS, 3, 19
Boolean algebra, 244
  laws and rules of, 248
Boundary conditions, 354
Boyle's law, 20
Brackets, 19

Calculator, 14, 50, 52, 120
Calculus, 264
Cancelling, 4, 5
Cardioid, 185, 283
Cartesian axes, 155
  complex numbers, 219
Cartesian co-ordinates, 122
Catenary, 57
Centroids, 340
Chain rule, 274
Charles' law, 20
Chord, 92
Circle, 88, 91, 186
  equation of, 94
  properties of, 92
Circumference, 92
Class intervals, 378
  limits, 378
Coefficient of correlation, 398
  proportionality, 20
  significance of, 399
Cofactor, 234
Combinational logic networks, 257
Combination of waveforms, 207
Common difference, 65
  logarithms, 46
  ratio, 65
Complementary angles, 110
  function, 364
Completing the square, 36, 44
Complex numbers, 220
  addition and subtraction, 220
  applications of, 224
  conjugate, 221
  De Moivre theorem, 226
  equations, 221
  exponential form, 228
  multiplication and division, 221, 223
  polar form, 221
  powers of, 226
  roots of, 227
  waveform considerations, 445
Compound angle formulae, 148
  angles, 148
Computer numbering systems, 80
Cone, 96
  frustum of, 99
Confidence coefficients, 407
  intervals, 406
  limits, 408

Congruent triangles, 112
Continued fractions, 24
Continuous data, 373
  function, 191, 427
Contour map, 303
Convergence, 50, 71
Convergents, 24
Conversion of a sin $\omega t + b \cos \omega t$ into $R\sin(\omega t + \alpha)$, 149
Conversion tables, 15
Co-ordinates, 155
Correlation, 398
Corresponding angles, 110
Cosecant, 116
Cosh, 55
Coshec, 55
Cosine, 116
  rule, 125
  wave, 130, 134
  wave production, 133
Cotangent, 116
Coth, 55
Cramer's rule, 241
Cross products, 215
Cubic equation, 177, 186
Cuboid, 96
Cumulative frequency distribution, 378, 379
Curve sketching, guide to, 198
Cycles of logarithmic graph paper, 166
Cycloid, 283
Cylinder, 96

Deciles, 386
Decimal places, 7
  system, 80
Decimals, 7, 80
Definite integrals, 307
Degrees, 109
  of freedom, 411
De Moivre's theorem, 226
De Morgan's laws, 250
Denary system, 80
Denominator, 4
Dependent event, 387
Depression, angle of, 119
Derivative, 266
  of Laplace transforms, 417
Determinants, 233, 234
  solving simultaneous equations, 238
Determination of law, 158, 160
  involving logarithms, 162
Diameter, 92
Differential calculus, 264
  coefficient, 266
Differential equations, 353
  $a\frac{d^2y}{dx^2} + b\frac{dy}{dx} + cy = 0$, 360
  $a\frac{d^2y}{dx^2} + b\frac{dy}{dx} + cy = f(x)$, 363
  $\frac{dy}{dx} = f(x)$, 354
  $\frac{dy}{dx} = f(y)$, 355
  $\frac{dy}{dx} = f(x) \cdot f(y)$, 355
  $\frac{dy}{dx} + Py = Q$, 358
  $P\frac{dy}{dx} = Q$, 357
  using Laplace transforms, 421
Differentiation, 264, 266
  applications of, 276
  from first principles, 266
  function of a function, 274
  hyperbolic functions, 275
  implicit functions, 286
  inverse hyperbolic, 290, 293
    trigonometric, 290
  logarithmic, 288
  methods of, 271
  of ax^x, 267
  of e^{ax} and ln ax, 270
  of parametric equations, 283
  of products, 272
  of quotients, 273
  of sine and cosine functions, 268
  partial, 294
  successive, 274
Digits, 7
Direction cosines, 214
Direct proportion, 6, 20
Discontinuous functions, 191
Discrete data, 373
Dividend, 21
Divisor, 21
D-operator, 360
Dot product, 212
Double angles, 152

Element of matrix, 231
Elevation, angle of, 119
Ellipse, 186, 283
Equation, 25
  complex, 221
  hyperbolic, 58
  indicial, 48
  quadratic, 35
  of a circle, 94
  simple, 25
  simultaneous, 29
  trigonometric, 142
Errors, 14
Expectation, 387
Explicit function, 286

Exponent, 12
Exponential functions, 49, 187
  form of complex number, 228
Equilateral triangle, 111
Extrapolation, 159
Euler-Cauchey method, 370
Euler's method, 367
Evaluation of formulae, 16
Even function, 56, 192
  Fourier series, 433

Factorisation, 19, 35, 44
Factors, 2
Factor theorem, 21
Family of curves, 353
Final value theorem, 418
First moment of area, 341
Formulae, 16
  transposition of, 32
Fourier coefficients, 428
  cosine series, 433
  sine series, 435
Fourier series, 427
  even and odd functions, 433
  half range, 436, 440
  non-periodic, period 2π, 431
  over any range, 438
  periodic, period 2π, 427
Fractions, 4
  continued, 24
  partial, 61
Frequency, 139
  distribution, 377, 379
  polygon, 378, 379
Frustum, 99
Functional notation, 264, 266
Function of a function, 274
Functions and their curves, 185
  of two variables, 299
Fundamental, 428

Gaussian elimination method, 242
General solution of differential
    equation, 354
Geometric progression, 65
Geometry, 109
Gradient of graphs, 155
  of a curve, 264
Graphs, 155
Graphical solution of equations, 170
  cubic equations, 177
  linear and quadratic equations
    simultaneously, 176
  quadratic equations, 170
  simultaneous equations, 170
Graphs of exponential functions, 51
  hyperbolic functions, 57
  logarithmic functions, 48
  polar curves, 178
  straight lines, 155
  trigonometric functions, 129
  $y = ab^x$, 163, 168
  $y = ae^{kx}$, 163, 168
  $y = ax^n$, 162, 166
Graphs with logarithmic scales, 166
Grouped data, presentation of, 377

Half range Fourier series, 436, 440
Harmonic analysis, 441
  numerical method, 441
Harmonics, 428
Hexadecimal numbers, 83
Hexagon, 86, 90
H.C.F., 2
Heptagon, 86
Histogram, 378, 379, 382
Homogeneous first order differential
    equations, 357
Hooke's law, 20
Horizontal bar charts, 373
Hyperbola, 186, 283
  rectangular, 186, 283
Hyperbolic equations, 58
  functions, 55, 145
    differentiation of, 275
    graphs of, 57
    inverse, 290
  identities, 58, 146
  logarithms, 46, 52
  substitutions, integration, 310
Hypotenuse, 115

Identity, 25
  hyperbolic, 58
  trigonometric, 141
Imaginary number, 219
Implicit differentiation, 286
  function, 286
Improper fraction, 4
Independent event, 387
Indices, 10, 18
Indicial equations, 48
Industrial inspection, 390
Inequalities, 40
Initial value theorem, 418
Integers, 1
Integral calculus, 264
Integrating factor, 358
Integration, 305
  algebraic substitutions, 308
  areas, 330
  by partial fractions, 314
  by parts, 318
  centroids, 340

definite, 307
introduction to, 305
mean and r.m.s. values, 336
numerical, 326
reduction formulae, 320
second moments of area, 346
standard, 306
$\tan\frac{\theta}{2}$ substitution, 316
using trigonometric and hyperbolic substitutions, 310
volumes, 338
Interior angles, 110, 111
Interpolation, 158
Interval estimates, 406
Inverse functions, 193, 290
hyperbolic functions, 290
Laplace transforms, 419
using partial fractions, 420
of matrix, 233, 235
proportion, 6, 20
trigonometric functions, 194, 290
Invert-gate, 256
Irregular areas and volumes, 102
Isosceles triangle, 111
Iterative methods, 74

Karnaugh maps, 251

Lagging angles, 136
Lamina, 340
Laplace transforms, 414
definition of, 414
derivatives of, 417
inverse, 419
notations used, 414
of elementary functions, 415, 416
to solve differential equations, 421
to solve simultaneous differential equations, 424
Laws of Boolean algebra, 248
growth and decay, 53
indices, 10, 18
logarithms, 46, 288
precedence, 3, 19
probability, 387
L.C.M., 2
Leading angles, 136
Least-squares regression line, 400
Leibniz notation, 266
L'Hopital's rule, 74
Limiting value, 74, 265
Linear and quadratic equations simultaneously, 39
Linear correlation, 398
extrapolation, 159, 401
first order differential equation, 358
interpolation, 158, 401
regression, 400
Logarithmic differentiation, 288
forms of inverse hyperbolic functions, 292
function, 46, 185
graphs, 48, 186
scale, 166
Logarithms, 46
laws of, 46, 288
Logic circuits, 255
Log-linear graph paper, 168
Log-log graph paper, 166
Long division, 2

Maclaurin's theorem, 71
numerical integration, 73
Mantissa, 12
Matrices, 231
addition and subtraction of, 231
for solving simultaneous equations, 235
multiplication of, 232
Matrix notation, 231
Maximum and minimum values, 171, 277, 299
applications of, 279
Mean value, 381, 408, 410
by integration, 336
of waveforms, 106
Measures of central tendency, 380
Median, 381
Mensuration, 86
Mid-ordinate rule, 103, 327
Minor of matrix, 234
Mixed numbers, 4
Mode, 381
Modulus, 41, 212, 222
Moment of force, 217
Multiple, 2

Nand-gate, 256
Napierian logarithms, 44, 52
Natural logarithms, 46, 52
Newton–Raphson method, 79
Nor-gate, 256
Normal curve, 392
distribution, 392
testing for, 396
equations, 401
probability paper, 396
standard variate, 393
Normals, 281
Norm of vector, 212
Nose-to-tail method, 200
Not-function, 244
Not-gate, 256
Number sequences, 64

Numerator, 4
Numerical integration, 73, 326
 method of harmonic analysis, 441
Numerical method for first order differential equations, 367

Obtuse angle, 109
 angled triangle, 111
Octagon, 86, 90
Octal numbers, 81
Odd functions, 56, 193
 Fourier series, 433
Ogive, 378, 379
Ohm's law, 20
Order of precedence, 3, 19
Ordinate, 155
Or-function, 244
Or-gate, 256
Osborne's rule, 58, 147

Pappus' theorem, 344
Parabola, 171, 283
Parallel axis theorem, 347
 lines, 110
Parallelogram, 86, 88
 method, 200
Parametric differentiation, 283
 equations, 283
Partial areas under normal curve, 394
 differentiation, 294
 fractions, 61
  integration of, 314
  for Laplace transforms, 420
Particular integral, 364
 solution of differential equation, 354
Pascal's triangle, 67
Pentagon, 86
Percentage component bar charts, 375
 relative frequency, 373
Percentages, 9
Percentiles, 386
Perfect square, 36
Period, 135, 191, 427
Periodic functions, 135, 191, 427
 plotting of, 207
 time, 139
Perpendicular axis theorem, 348
Phasor, 138
Pictograms, 373
Pie diagram, 376
Planimeter, 103
Point estimate, 406
Points of inflexion, 277
Poisson distribution, 391
Polar co-ordinates, 122
 curves, 178, 187
 form of complex number, 221
Polygon, 86
Polynomial division, 20
Population, 373
Power, 10
Power series for e^x, 50
Practical problems, binomial theorem, 70
 involving straight line graphs, 158
 maximum/minimum values, 279
 quadratic equations, 38
 simple equations, 27
 simultaneous equations, 31
 triangles, 127
Principal value, 222
Prismoidal rule, 105
Probability, 386
 laws of, 387
 paper, 396
Product-moment formula, 398
Product rule, differentiation, 272
Proper fraction, 4
Pyramid, 96
 frustum of, 99
Pythagoras' theorem, 115

Quadratic equations, 35
 formula, 38
 inequalities, 44
 graphs, 171, 186
 practical problems, 38
Quadrilaterals, 86
Quartiles, 385
Quotient rule, differentiation, 273
Quotients, 24, 42

Radian, 93, 109
Radix, 80
Radius, 92
 of gyration, 346
Rates of change, 276, 298
Ratio and proportion, 6
Reciprocal, 10
 of matrix, 233, 235
 ratios, 117
Rectangle, 86, 88
Rectangular axes, 155
 co-ordinates, 125
 hyperbola, 186, 283
 prism, 96
Reduction formulae, 320
 of non-linear laws to linear form, 160
Reflex angle, 110
Regions, 45
Regression, 400
 coefficients, 401
Relationship between trigonometric and hyperbolic functions, 145
Relative frequency, 373
 velocity, 205

Reliability, 406
Remainder theorem, 23
Resolution of phasors, by calculation, 209
  vectors, 202
Rhombus, 87
Right angle, 109
  angled triangle, 111, 118
R.m.s. values, 336

Saddle points, 299, 301
Sample, 403
Sampling distributions, 403
  statistics, 407
  theories, 403
Scalar multiplication, matrices, 232
  product, 212
    practical applications, 214
  quantity, 199
Scalene triangle, 111
Scatter diagram, 398
Secant, 116
Sech, 55
Second moment of area, 346
Sector of a circle, 88, 92
Segment of circle, 92
Semicircle, 88, 92
Semi-interquartile range, 386
Sequences, 64
Series for cosh x and sinh x, 60
  e^x, 50
Set, 373
Shift theorem, Laplace transforms, 416
Short division, 2
Significant figures, 7
Similar triangles, 113
Simple equations, 25
  practical problems, 27
Simpson's rule, 103, 328
Simultaneous differential equations, by
    Laplace transforms, 424
  equations, 29
  by Cramer's rule, 241
    determinants, 238
    Gaussian elimination, 242
    matrices, 235
  practical problems, 31
Sine, 116
  rule, 125
  wave, 106, 130, 134
    production of, 133
Sinh, 55
Sinusoidal form $A\sin(\omega t \pm \alpha)$, 138
Slope of straight line, 156
Small changes, 282, 298
Space diagram, 206
Sphere, 96
  frustum of, 100
Square, 86, 88
  root, 10
Standard derivatives, 271
  deviation, 383, 410
  error of the means, 404
  form, 12
  integrals, 305
Stationary points, 278
Statistics, 373
Straight line graphs, 155
  practical problems, 158
Student's t distribution, 411, 412
Successive differentiation, 274
Sum to infinity, 66
Supplementary angles, 110
Surd, 118
Surface areas of solids, 95
  frustum of, 99
Switching circuits, 244

Tally diagram, 377, 378
Tangent, 92, 116
  wave, 130
Tangents to curves, 281
Tanh, 55
Tan $\frac{\theta}{2}$ substitution, 316
Taylor's series, 368
Theorem of Pappus, 344
  Pythagoras, 115
Total differential, 297
Transformations, 187
Transpose matrix, 235
Transposition of formulae, 32
Transversal, 110
Trapezium, 87, 88
Trapezoidal rule, 103, 326, 442
Triangle, 86, 88
Triangles, area of, 126
  congruent, 112
  construction of, 114
  properties of, 111
  similar, 113
  solution of right angled, 118
Trigonometric equations, 142
  functions, 145, 186
  identities, 141
  ratios, 116
    evaluation of, 120
    fractional and surd forms, 117
  substitutions, 310
  waveforms, 129
Trigonometry, 115
  practical situations, 127
Truth table, 244
Turning point, 171, 277
Two-state device, 244

Ungrouped data, presentation of, 373
Unit matrix, 233
  triad, 211
Universal logic gates, 260

Vector addition, 199
  products, 215
    practical applications, 217
  subtraction, 203
Vectors, 199
  resolution of, 202
Velocity and acceleration, 276
Vertical bar chart, 373
Vertically opposite angles, 110
Volumes of frusta, 99
  irregular solids, 105
  similar shapes, 102
  solids, 95
  solids of revolution, 338

Wallis's formula, 325
Waveforms, combination of, 207
Work done, 214

Zone of a sphere, 100